AF685523

SPEECH CODING ALGORITHMS

SPEECH CODING ALGORITHMS

Foundation and Evolution of Standardized Coders

WAI C. CHU
Mobile Media Laboratory
DoCoMo USA Labs
San Jose, California

WILEY-INTERSCIENCE
A JOHN WILEY & SONS, INC., PUBLICATION

Copyright © 2003 by John Wiley & Sons, Inc. All rights reserved.

Published by John Wiley & Sons, Inc., Hoboken, New Jersey.
Published simultaneously in Canada.

No part of this publication may be reproduced, stored in a retrieval system, or transmitted in any form or by any means, electronic, mechanical, photocopying, recording, scanning, or otherwise, except as permitted under Section 107 or 108 of the 1976 United States Copyright Act, without either the prior written permission of the Publisher, or authorization through payment of the appropriate per-copy fee to the Copyright Clearance Center, Inc., 222 Rosewood Drive, Danvers, MA 01923, 978-750-8400, fax 978-750-4470, or on the web at www.copyright.com. Requests to the Publisher for permission should be addressed to the Permissions Department, John Wiley & Sons, Inc., 111 River Street, Hoboken, NJ 07030, (201) 748-6011, fax (201) 748-6008, e-mail: permreq@wiley.com.

Limit of Liability/Disclaimer of Warranty: While the publisher and author have used their best efforts in preparing this book, they make no representations or warranties with respect to the accuracy or completeness of the contents of this book and specifically disclaim any implied warranties of merchantability or fitness for a particular purpose. No warranty may be created or extended by sales representatives or written sales materials. The advice and strategies contained herein may not be suitable for your situation. You should consult with a professional where appropriate. Neither the publisher nor author shall be liable for any loss of profit or any other commercial damages, including but not limited to special, incidental, consequential, or other damages.

For general information on our other products and services please contact our Customer Care Department within the U.S. at 877-762-2974, outside the U.S. at 317-572-3993 or fax 317-572-4002.

Wiley also publishes its books in a variety of electronic formats. Some content that appears in print, however, may not be available in electronic format.

Library of Congress Cataloging-in-Publication Data:

Chu, Wai C. —
Speech coding algorithms: Foundation and evolution of standardized coders

ISBN 0-471-37312-5

Printed in the United States of America

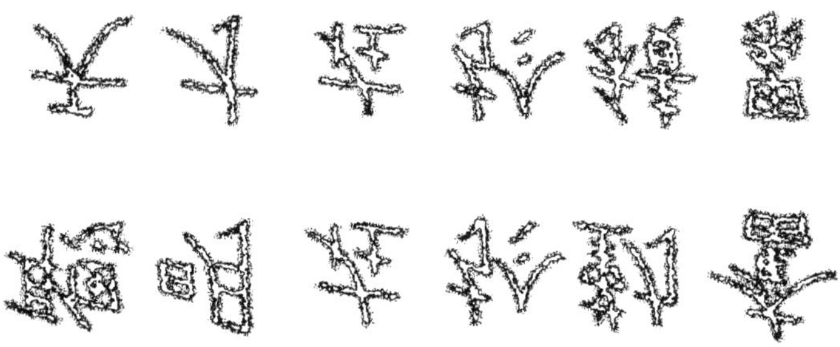

Intelligence is the fruit of industriousness
Accretion of knowledge creates genii

A Chinese proverb

CONTENTS

PREFACE

My first contact with speech coding was in 1993 when I was a Field Application Engineer at Texas Instruments, Inc. Soon after joining the company I was assigned to design a demo prototype for the digital telephone answering device project. Initially I was in charge of hardware including circuit design and printed circuit board layout. The core of the board consisted of a microcontroller sending commands to a mixed signal processor, where all the signal processing tasks—including speech coding—were performed. In those days a major concern was the excessive cost associated with random-access memory (RAM), and compressing the digital speech before storing was almost a mandatory requirement, as this greatly improved cost-effectiveness.

Soon after the hardware was finished, the focus switched to software (or firmware) design, mainly dealing with the control of various on-board peripheral devices. My true interest, however, was the program code inside the mixed signal processor, which was developed by a separate team of "advanced" engineers. I was told that voice signals were compressed using a code-excited linear prediction (CELP) algorithm. Also, it was possible to play back fixed announcement messages—such as numbers and days of the week—with the messages stored in the linear prediction coding (LPC) format. I had no idea what these algorithms were, nor how they worked to compress speech. However, I was eager to learn the details, and decided to go back to school and pursue a PhD with concentration in speech coding.

This book is the result of my personal experience as a researcher and practitioner in the field of speech coding. Four years ago I decided to put in extra hours, usually late nights and early mornings as well as weekends, to organize the literature in speech coding and develop it into a logical presentation in terms of content and terminology. Speech coding has evolved into a highly matured branch of signal processing, with deployment in a plethora of products, such as, cellular phones, answering machines, communication devices, and more recently, voice over internet protocol (VoIP). It is obvious that a thorough textbook is necessary for students, professors, and engineering professionals to handle the subject appropriately. My sincere hope is that the availability of a book that collects many of the techniques used in speech coding and presents them in an accessible fashion will create excitement and enthusiasm, ensuring continuous rapid advances in the field.

Philosophy and Approach

Speech Coding Algorithms reflects the core subject of the book, since most coding techniques are implemented as algorithms, or computational procedures performed by a processor. However, this is by no means an exhaustive documentation of all methods developed in this field; it is rather the study of the most successful techniques, defined as those incorporated in a standard. By doing so we concentrate our effort on understanding the most influential ideas, which is a rather efficient manner to navigate this vast territory of knowledge.

In my own personal learning curve, I found that there is a different and refreshing lesson to be found in every standard. To understand a new standard it is often necessary to look back into the developed techniques adopted by past standards or studies. Attempting to learn by reading the official documentation describing the standard is very often a frustrating experience, since the assumption made in preparing those materials is that the audience consists of experts in the subject, and hence the logical order and justification of a given approach is routinely omitted. Therefore the origin and the reason behind a certain practice cannot be fully understood. This might not be a problem if one's objective is to implement the algorithm without comprehending it. However, for those researchers eager to delve deeply into its roots, alternative reference sources must be explored, which can be a strenuous and prolonged process. In this book I have summarized the knowledge acquired over an extended period of time, with the intention of filling the void between principles and implementations.

In writing this book, a balance is sought between theory and practice, and between intuition and rigor. Theoretical ideas are included only if they are used to solve practical problems, and thorough proofs are provided. Speech coding is related to human perception, and therefore a degree of fuzziness exists, in the sense that no absolute right or wrong can be established for certain situations; in other words, no mathematical proofs are obtainable. In these cases, solutions are often found and justified on an intuitive basis. For the most part, the book is meant to be pragmatic, since the discussed techniques are widely used in industry.

Prerequisites

The minimum background required to understand the book is explained, with reference to popular textbooks where the relevant subjects can be found.

- Advanced calculus, including complex variables [Churchill and Brown, 1990].
- Discrete-time signals and systems, Fourier transforms, z-transforms, filtering, and convolution [Oppenheim and Schafer, 1989; Stearns and Hush, 1990].
- Random variables and stochastic processes, expectation, probability, and wide-sense stationarity [Papoulis, 1991; Peebles 1993].
- Linear algebra, including linear equations, matrices, and vectors [Strang, 1988].
- Experience with high-level programming using a language such as C.

The above list is covered in most undergraduate Electrical Engineering curricula; with this background, the book is self-contained.

Organization

The text is divided into 19 chapters. Chapter 1 provides an overview of the subjects covered, with references to various aspects of speech coding, standards, algorithms, and comments on notation and terminology. Chapter 2 is a review of some signal processing techniques, some are very general, but others are less known outside speech coding literature. Chapter 3 contains some foundation for stochastic processes and models, which are important for an understanding of the theoretical aspects. Chapter 4 is about linear prediction, the integral part of almost all modern speech coders. Chapter 5 reviews the various aspects of scalar quantization, which are utilized routinely by many speech coding algorithms. One of the earliest digital coding techniques is pulse code modulation (PCM); it and its variants are the topic of Chapter 6. Chapter 7 deals with vector quantization, which has become more and more important for the achievement of high efficiency in coding systems. Linear prediction coefficients (LPC) are normally quantized for transmission as part of the compressed bit-stream; Chapter 8 covers the various methods for scalar quantization of these coefficients. One of the landmarks in low bit-rate speech coding is the linear prediction coding (LPC) algorithm, discussed in Chapter 9. Chapter 10 is devoted to regular pulse excitation coders, with a thorough description of the GSM 6.10 standard. Principles of code-excited linear prediction (CELP) are given in Chapter 11, covering the various aspects of analysis-by-synthesis, signal calculation, postfilter design, and efficiency. Chapters 12 and 13 present the structure of two standardized CELP coders: FS1016 and IS54, respectively; these are both milestones in speech coding development. Chapter 14 is dedicated to the G.728 low-delay CELP standard, with thorough explanations of strategies for delay reduction and detailed structures of the coder. Vector quantization of LPC is included in Chapter 15, representing a huge advance with respect to scalar quantization techniques covered in Chapter 8, and methods used by various standardized coders are analyzed. The highly influential algebraic CELP (ACELP) algorithm is covered in Chapter 16, where several ACELP-based standards are described, with focus on the G.729 standard. The mixed excitation linear prediction (MELP) algorithm is discussed in Chapter 17, and is shown to be an improvement upon the LPC coder, covered in Chapter 9. Chapter 18 is devoted to the IS96 variable bit-rate CELP algorithm, which is a source-controlled multimode coder with the operating mode selected by the input characteristics of the speech signal. Finally, Chapter 19 is concerned with various methods to assess the quality of speech signals, especially those processed by a speech coding algorithm.

The following table summarizes the chapters and their prerequisites.

Chapter	Title	Prerequisites
1	Introduction	
2	Signal Processing Techniques	1
3	Stochastic Processes and Models	
4	Linear Prediction	1, 2, 3
5	Scalar Quantization	
6	Pulse Code Modulation and its Variants	4, 5
7	Vector Quantization	5
8	Scalar Quantization of Linear Prediction Coefficients	4, 5
9	Linear Prediction Coding	4, 8
10	Regular-Pulse Excitation Coders	4, 8
11	Code-Excited Linear Prediction	2, 4
12	The Federal Standard Version of CELP	2, 8, 11
13	Vector Sum Excited Linear Prediction	8, 12
14	Low-Delay CELP	4, 11
15	Vector Quantization of Linear Prediction Coefficients	7 , 8
16	Algebraic CELP	7, 12, 15
17	Mixed Excitation Linear Prediction	9, 15
18	Source-Controlled Variable Bit-Rate CELP	11
19	Speech Quality Assessment	1

Acknowledgments

Throughout my professional career, I have had the opportunity to work with and learn from a number of people whom I should like to publicly acknowledge. My former advisor Dr. Nirmal K. Bose at the Pennsylvania State University had provided me with invaluable instruction, trust, and friendship during my graduate studies; his methodical style, hard-working spirit, and commitment toward education have served as a role model to follow. I am grateful to my former supervisor Dr. Tandhoni S. Rao at Texas Instruments Inc., who had guided me through projects involving adaptive filters, speech coding, and programming of digital signal processors.

I would like to dedicate this book to my parents who have always encouraged my academic interests and provided the moral support throughout my life and career. I am deeply indebted to my cousin Chi-Ming Chu and wife Kam-Chi Chu for their help and support during my graduate studies at Stevens Tech; their industriousness and candid spirit have given me a great deal of positive influence.

I am particularly indebted to my wife Laura for her love and patience, and for thoroughly reviewing and proofreading the first version of the manuscript.

I am grateful to the Wiley team for their professionalism and help during the production of this book; special thanks to George Telecki (Executive Editor) and Rosalyn Farkas (Associate Editor). I am also most grateful to Dr. Andreas Spanias and Dr. Allen Levesque for their encouraging comments and constructive critiques—both early reviewers of the manuscript. I also wish to thank my former colleague at Texas Instruments Inc., Wai-Ming Lai for her help in examining some chapters of the text.

Last but not least, this book is dedicated to Universidad Simón Bolivar, the school where I received most of my early engineering education. *Universidad Simón Bolivar me ha dado generosamente el vigor, la fortaleza, y la sabiduria necesaria para conquistar obstáculos y dominar dificultades tanto en la ingenieria como en la vida. Espero dar con este libro a los aspirantes en esta rama de la ingenieria lo mismo que me ha dado la respectuosa universidad.*

Feedback

A book of this length is certain to contain errors and omissions. While attempts were made to provide a highly understandable and correct content, there are doubtless many places where improvements are possible. Feedback is welcome to the author via email at wcc2@ieee.org. Please note that a personal reply to all messages might not be possible.

Wai C. Chu

ACRONYMS

2-D	Two-dimensional
3GPP	Third generation partnership project
AbS	Analysis-by-synthesis
ACELP	Algebraic code-excited linear prediction
ACR	Absolute category rating
ADPCM	Adaptive differential pulse code modulation
AES	Audio Engineering Society
ANSI	American National Standards Institute
APCM	Adaptive pulse code modulation
AR	Autoregressive
ARMA	Autoregressive moving average
CCITT	International Telegraph and Telephone Consultative Committee (replaced by ITU-T)
CCR	Comparison category rating
CDMA	Code division multiple access
CELP	Code-excited linear prediction
CEPT	Conference of European Posts and Telephones
CS-ACELP	Conjugate structure algebraic code-excited linear prediction
DC	Direct current
DCR	Degradation category rating
DFT	Discrete Fourier transform
DMOS	Degradation mean opinion score
DoD	U.S. Department of Defense
DPCM	Differential pulse code modulation
DSP	Digital signal processing/processor

DTAD	Digital telephone answering device
DTFT	Discrete time Fourier transform
DTMF	Dual-tone multifrequency
EFR	Enhanced full rate
ETSI	European Telecommunications Standards Institute
FFT	Fast Fourier transform
FIR	Finite impulse response
FM	Frequency modulation
FS	Federal Standard
GLA	Generalized Lloyd algorithm
GSM	Groupe Speciale Mobile
ICASSP	International Conference on Acoustics, Speech, and Signal Processing
IDFT	Inverse discrete Fourier transform
IEC	International Electrotechnical Commission
IEEE	Institute of Electrical and Electronics Engineers
IIR	Infinite impulse response
ISO	International Organization for Standardization
ITU	International Telecommunications Union
ITU–R	ITU–Radiocommunication Sector
ITU–T	ITU–Telecommunications Standardization Sector
LAR	Log area ratio
LD-CELP	Low-delay code-excited linear prediction
LMS	Least mean square
LP	Linear prediction
LPC	Linear prediction coding/coefficient
LSF	Line spectral frequency
LSP	Line spectral pair
LTI	Linear time-invariant
MA	Moving average
MIPS	Millions of instructions per second
MNB	Measuring normalizing block
MOS	Mean opinion score
MP–MLQ	Multipulse–maximum likelihood quantization
MPEG	Moving Picture Expert Group
MSE	Mean square error
MSVQ	Multistage vector quantization
NCS	National Communications System
PAQM	Perceptual audio quality measure
PC	Personal computer
PCM	Pulse code modulation
PDF	Probability density function
PESQ	Perceptual evaluation of speech quality
PG	Prediction gain
PMF	Probability mass function

PSD	Power spectral density
PSQM	Perceptual speech quality measure
PVQ	Predictive vector quantization
QCELP	Qualcomm code-excited linear prediction
RAM	Random access memory
RC	Reflection coefficient
RCR	Research and Development Center for Radio Systems of Japan
RMS	Root mean square
ROM	Read only memory
RPE–LTP	Regular pulse excited–long-term prediction
RV	Random variable
SD	Spectral distortion
SNR	Signal to noise ratio
SPG	Segmental prediction gain
SSE	Sum of squared error
SSNR	Segmental signal to noise ratio
TDMA	Time division multiple access
TI	Texas Instruments
TIA	Telecommunications Industry Association
TTS	Text to speech
UMTS	Universal Mobile Telecommunications System
VBR	Variable bit-rate
VoIP	Voice over internet protocol
VQ	Vector quantization
VSELP	Vector sum excited linear prediction
WSS	Wide sense stationary

NOTATION

$\lfloor \cdot \rfloor$	Floor operation: returns the highest integer just below the operand
$\lceil \cdot \rceil$	Ceiling operation: returns the lowest integer just above the operand
$(\cdot)^*$	Complex conjugate
$(\cdot)^{\mathrm{T}}$	Transpose
$(\cdot)^{\mathrm{B}}$	Backward arrangement (vector)
$\leftarrow$	Assignment
$\equiv$	Equivalent by definition
$\emptyset$	Empty set
$\cup$	Union
$\cap$	Interception
$\prod$	Product
$\sum$	Sum
$*$	Convolution
$\oplus$	Exclusive or
$\boldsymbol{R}$	Bold italic type implies a field or set
$\mathbf{a}, \mathbf{R}$	Bold type implies a matrix or vector
$\mathbf{0}$	Zero vector
$\mathbf{I}$	Identity matrix
$A\{\cdot\}$	Time average
$E\{\cdot\}$	Expectation
$\mathrm{Re}\{\cdot\}$	Real part
$\mathrm{Im}\{\cdot\}$	Imaginary part
$\arg(\cdot)$	Argument (phase)
$\cos(\cdot)$	Cosine

$\mathrm{ctg}(\cdot)$	Cotangent
$\exp(\cdot)$	Exponential
$\mathrm{grd}(\cdot)$	Group delay
$\lg(\cdot)$	Base 2 logarithm
$\ln(\cdot)$	Natural logarithm
$\log(\cdot)$	Base 10 logarithm
$\mathrm{mod}(\cdot)$	Modulo operation
$\mathrm{sgn}(\cdot)$	Sign function, returns ± 1 depending on the operand
$\sin(\cdot)$	Sine

$\mathrm{sinc}(x) = \sin(\pi x)/(\pi x)$

bps	Bits per second
dB	Decibel
Hz	Hertz
kbps	Kilo-bit per second
kHz	Kilo-Hertz

CHAPTER 1

INTRODUCTION

In general, *speech coding* is a procedure to represent a digitized speech signal using as few bits as possible, maintaining at the same time a reasonable level of speech quality. A not so popular name having the same meaning is *speech compression*. Speech coding has matured to the point where it now constitutes an important application area of signal processing. Due to the increasing demand for speech communication, speech coding technology has received augmenting levels of interest from the research, standardization, and business communities. Advances in microelectronics and the vast availability of low-cost programmable processors and dedicated chips have enabled rapid technology transfer from research to product development; this encourages the research community to investigate alternative schemes for speech coding, with the objectives of overcoming deficiencies and limitations. The standardization community pursues the establishment of standard speech coding methods for various applications that will be widely accepted and implemented by the industry. The business communities capitalize on the ever-increasing demand and opportunities in the consumer, corporate, and network environments for speech processing products.

Speech coding is performed using numerous steps or operations specified as an algorithm. An algorithm is any well-defined computational procedure that takes some value, or set of values, as input and produces some value, or set of values, as output. An algorithm is thus a sequence of computational steps that transform the input into the output. Many signal processing problems—including speech coding—can be formulated as a well-specified computational problem; hence, a particular coding scheme can be defined as an algorithm. In general, an algorithm is specified with a set of instructions, providing the computational steps needed to perform a task. With these instructions, a computer or processor can execute them so as to complete the coding task. The instructions can also be translated to the structure of a digital circuit, carrying out the computation directly at the hardware level.

The purpose of this book is to explain the theoretical issues and implementational techniques related to the fascinating field of speech coding. The topics of discussion are focused on some of the well-established and widely used speech coding standards. By studying the most successful standards and understanding their principles, performance, and limitations, it is possible to apply a particular technique to a given situation according to the underlying constraints—with the ultimate goal being the development of next-generation algorithms, with improvements in all aspects.

This chapter is organized as follows: an overview of speech coding is provided first, with the structure, properties, and applications of speech coders explained; the different classes of speech coders are described next, followed by speech production and modeling, covering properties of speech signals and a very simple coding system. High-level explanation of the human auditory system is given, where the system properties are used to develop efficient coding schemes. Activities of standard bodies and many standardized coders are discussed in the next section, followed by issues related to analysis and implementation of algorithms. A brief summary is given at the end of the chapter.

1.1 OVERVIEW OF SPEECH CODING

This section describes the structure, properties, and applications of speech coding technology.

Structure of a Speech Coding System

Figure 1.1 shows the block diagram of a speech coding system. The continuous-time analog speech signal from a given source is digitized by a standard connection

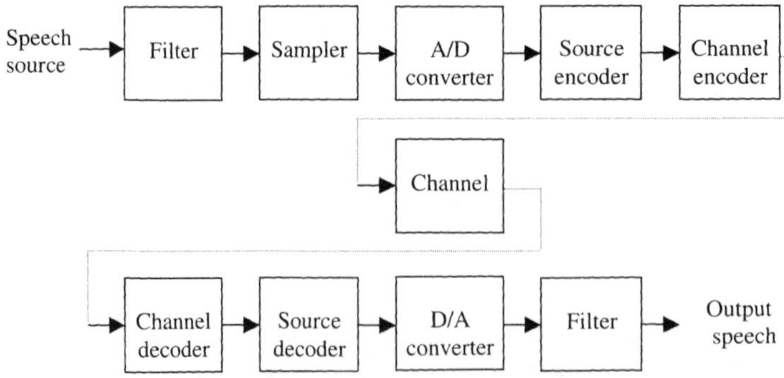

Figure 1.1 Block diagram of a speech coding system.

of filter (eliminates aliasing), sampler (discrete-time conversion), and analog-to-digital converter (uniform quantization is assumed). The output is a discrete-time speech signal whose sample values are also discretized. This signal is referred to as the *digital speech.*

Traditionally, most speech coding systems were designed to support telecommunication applications, with the frequency contents limited between 300 and 3400 Hz. According to the Nyquist theorem, the sampling frequency must be at least twice the bandwidth of the continuous-time signal in order to avoid aliasing. A value of 8 kHz is commonly selected as the standard sampling frequency for speech signals. To convert the analog samples to a digital format using uniform quantization and maintaining toll quality [Jayant and Noll, 1984]—the digital speech will be roughly indistinguishable from the bandlimited input—more than 8 bits/sample is necessary. The use of 16 bits/sample provides a quality that is considered high. Throughout this book, the following parameters are assumed for the digital speech signal:

$$\textbf{Sampling frequency} = \textbf{8 kHz},$$
$$\textbf{Number of bits per sample} = \textbf{16}.$$

This gives rise to

$$\textbf{Bit-rate} = \textbf{8 kHz} \cdot \textbf{16 bits} = \textbf{128 kbps}.$$

The above bit-rate, also known as input bit-rate, is what the source encoder attempts to reduce (Figure 1.1). The output of the source encoder represents the encoded digital speech and in general has substantially lower bit-rate than the input. The linear prediction coding algorithm (Chapter 9), for instance, has an output rate of 2.4 kbps, a reduction of more than 53 times with respect to the input.

The encoded digital speech data is further processed by the channel encoder, providing error protection to the bit-stream before transmission to the communication channel, where various noise and interference can sabotage the reliability of the transmitted data. Even though in Figure 1.1 the source encoder and channel encoder are separated, it is also possible to jointly implement them so that source and channel encoding are done in a single step.

The channel decoder processes the error-protected data to recover the encoded data, which is then passed to the source decoder to generate the output digital speech signal, having the original rate. This output digital speech signal is converted to continuous-time analog form through standard procedures: digital-to-analog conversion followed by antialiasing filtering.

In this book, the emphasis is on design of the source encoder and source decoder. For simplicity, they are referred to as the encoder and decoder, respectively (Figure 1.2). The input speech (a discrete-time signal having a bit-rate of 128 kbps) enters the encoder to produce the encoded bit-stream, or compressed speech data. Bit-rate of the bit-stream is normally much lower than that of the input

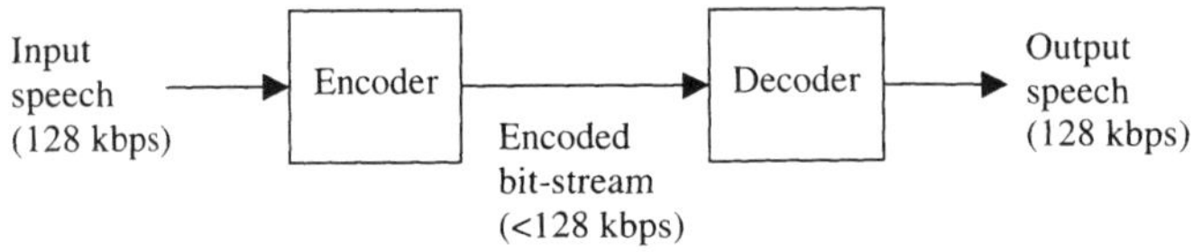

Figure 1.2 Block diagram of a speech coder.

speech. The decoder takes the encoded bit-stream as its input to produce the output speech signal, which is a discrete-time signal having the same rate as the input speech. As we will see later in this book, many diverse approaches can be used to design the encoder/decoder pair. Different methods provide differing speech quality and bit-rate, as well as implementational complexity.

The encoder/decoder structure represented in Figure 1.2 is known as a *speech coder*, where the input speech is encoded to produce a low-rate bit-stream. This bit-stream is input to the decoder, which constructs an approximation of the original signal.

Desirable Properties of a Speech Coder

The main goal of speech coding is either to maximize the perceived quality at a particular bit-rate, or to minimize the bit-rate for a particular perceptual quality. The appropriate bit-rate at which speech should be transmitted or stored depends on the cost of transmission or storage, the cost of coding (compressing) the digital speech signal, and the speech quality requirements. In almost all speech coders, the reconstructed signal differs from the original one. The bit-rate is reduced by representing the speech signal (or parameters of a speech production model) with reduced precision and by removing inherent redundancy from the signal, resulting therefore in a *lossy* coding scheme. Desirable properties of a speech coder include:

- *Low Bit-Rate.* The lower the bit-rate of the encoded bit-stream, the less bandwidth is required for transmission, leading to a more efficient system. This requirement is in constant conflict with other good properties of the system, such as speech quality. In practice, a trade-off is found to satisfy the necessity of a given application.
- *High Speech Quality.* The decoded speech should have a quality acceptable for the target application. There are many dimensions in quality perception, including intelligibility, naturalness, pleasantness, and speaker recognizability. See Chapter 19 for a thorough discussion on speech quality and techniques to assess it.
- *Robustness Across Different Speakers / Languages.* The underlying technique of the speech coder should be general enough to model different speakers (adult male, adult female, and children) and different languages adequately. Note that this is not a trivial task, since each voice signal has its unique characteristics.

- *Robustness in the Presence of Channel Errors.* This is crucial for digital communication systems where channel errors will have a negative impact on speech quality.
- *Good Performance on Nonspeech Signals (i.e., telephone signaling).* In a typical telecommunication system, other signals might be present besides speech. Signaling tones such as dual-tone multifrequency (DTMF) in keypad dialing and music are often encountered. Even though low bit-rate speech coders might not be able to reproduce all signals faithfully, it should not generate annoying artifacts when facing these alternate signals.
- *Low Memory Size and Low Computational Complexity.* In order for the speech coder to be practicable, costs associated with its implementation must be low; these include the amount of memory needed to support its operation, as well as computational demand. Speech coding researchers spend a great deal of effort to find out the most efficient realizations.
- *Low Coding Delay.* In the process of speech encoding and decoding, delay is inevitably introduced, which is the time shift between the input speech of the encoder with respect to the output speech of the decoder. An excessive delay creates problems with real-time two-way conversations, where the parties tend to "talk over" each other. Thorough discussion on coding delay is given next.

About Coding Delay

Consider the delay measured using the topology shown in Figure 1.3. The delay obtained in this way is known as *coding delay*, or *one-way coding delay* [Chen, 1995], which is given by the elapsed time from the instant a speech sample arrives at the encoder input to the instant when the same speech sample appears at the decoder output. The definition does not consider exterior factors, such as communication distance or equipment, which are not controllable by the algorithm designer. Based on the definition, the coding delay can be decomposed into four major components (see Figure 1.4):

1. *Encoder Buffering Delay.* Many speech encoders require the collection of a certain number of samples before processing. For instance, typical linear prediction (LP)-based coders need to gather one frame of samples ranging from 160 to 240 samples, or 20 to 30 ms, before proceeding with the actual encoding process.

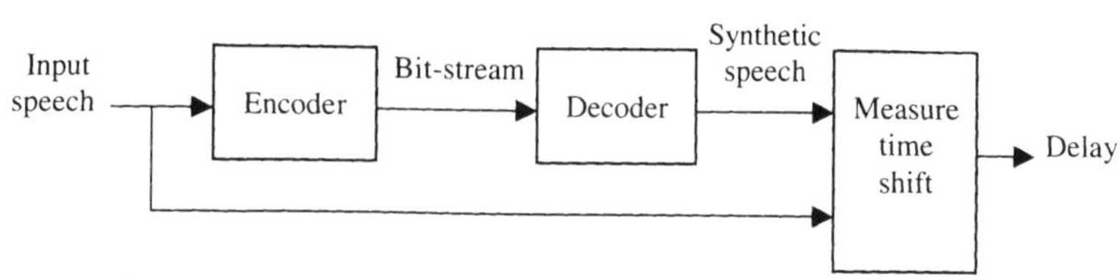

Figure 1.3 System for delay measurement.

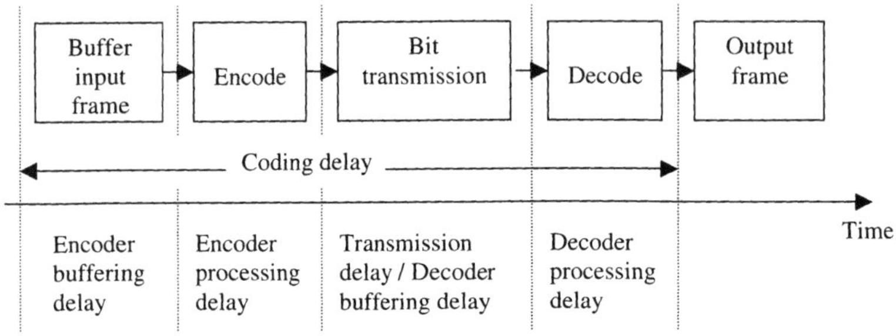

Figure 1.4 Illustration of the components of coding delay.

2. *Encoder Processing Delay.* The encoder consumes a certain amount of time to process the buffered data and construct the bit-stream. This delay can be shortened by increasing the computational power of the underlying platform and by utilizing efficient algorithms. The processing delay must be shorter than the buffering delay, otherwise the encoder will not be able to handle data from the next frame.

3. *Transmission Delay.* Once the encoder finishes processing one frame of input samples, the resultant bits representing the compressed bit-stream are transmitted to the decoder. Many transmission modes are possible and the choice depends on the particular system requirements. For illustration purposes, we will consider only two transmission modes: *constant* and *burst*. Figure 1.5 depicts the situations for these modes.

In constant mode the bits are transmitted synchronously at a fixed rate, which is given by the number of bits corresponding to one frame divided by the length of the frame. Under this mode, transmission delay is equal to encoder buffering delay: bits associated with the frame are fully transmitted at the instant when bits of the next frame are available. This mode of operation is dominant for most classical digital communication systems, such as wired telephone networks.

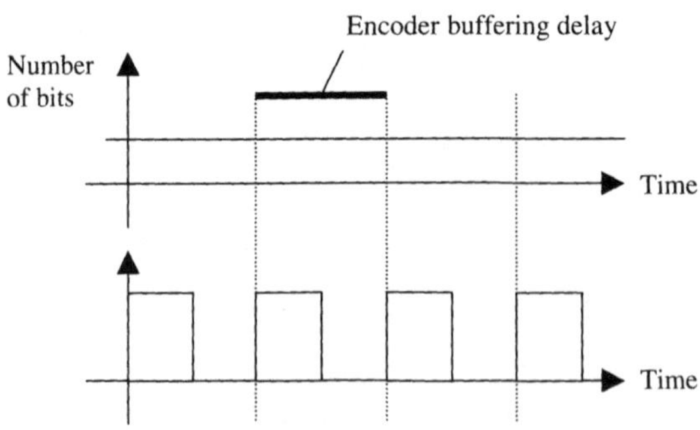

Figure 1.5 Plots of bit-stream transmission pattern for constant mode (*top*) and burst mode (*bottom*).

In burst mode all bits associated with a particular frame are completely sent within an interval that is shorter than the encoder buffering delay. In the extreme case, all bits are released right after they become available, leading to a negligibly small transmission delay. This mode is inherent to packetized network and the internet, where data are grouped and sent as packets.

Transmission delay is also known as *decoder buffering delay*, since it is the amount of time that the decoder must wait in order to collect all bits related to a particular frame so as to start the decoding process.

4. *Decoder Processing Delay*. This is the time required to decode in order to produce one frame of synthetic speech. As for the case of the encoder processing delay, its upper limit is given by the encoder buffering delay, since a whole frame of synthetic speech data must be completed within this time frame in order to be ready for the next frame.

As stated earlier, one of the good attributes of a speech coder is measured by its coding delay, given by the sum of the four described components. As an algorithm designer, the task is to reduce the four delay components to a minimum. In general, the encoder buffering delay has the greatest impact: it determines the upper limit for the rest of the delay components. A long encoding buffer enables a more thorough evaluation of the signal properties, leading to higher coding efficiency and hence lower bit-rate. This is the reason why most low bit-rate coders often have high delay. Thus, coding delay in most cases is a trade-off with respect to the achievable bit-rate.

In the ideal case where infinite computational power is available, the processing delays (encoder and decoder) can be made negligible with respect to the encoder buffering delay. Under this assumption, the coding delay is equal to two times the encoder buffering delay if the system is transmitting in constant mode. For burst mode, the shortest possible coding delay is equal to the encoder buffering delay, where it is assumed that all output bits from the encoder are sent instantaneously to the decoder. These values are idealistic in the sense that it is achievable only if the processing delay is zero or the computational power is infinite: the underlying platform can find the results instantly once the required amount of data is collected. These ideal values are frequently used for benchmarking purposes, since they represent the lower bound of the coding delay. In the simplest form of delay comparison among coders, only the encoder buffering delay is cited. In practice, a reasonable estimate of the coding delay is to take 2.5 to 3 and 1.5 to 2.5 times the frame interval (encoder buffering delay) for constant mode transmission and burst mode transmission, respectively.

Applications of Speech Coders

Speech coding has an important role in modern voice-enabled technology, particularly for digital speech communication, where quality and complexity have a direct impact on the marketability and cost of the underlying products or services. There are many speech coding standards designed to suit the need of a given application. Some examples are as follows:

- *FS1015 LPC* (*Chapter 9*). This coder was created in 1984 to provide secure communication in military applications. On a battlefield, the messages must be sent in such a way that the enemy cannot understand them. By deploying a secret coding scheme, the transmitted messages are safeguarded.
- *TIA IS54 VSELP* (*Chapter 13*). This coder was standardized in 1989 for time division multiple access (TDMA) digital cellular telephony in North America.
- *ETSI AMR ACELP* (*Chapter 16*). This coder was standardized in 1999 as part of the Universal Mobile Telecommunications System (UMTS) linked to the 3rd Generation Partnership Project (3GPP).

More recently, with the explosive growth of the internet, the potential market of voice over internet protocol (voice over IP, or VoIP) has lured many companies to develop products and services around the concept. In sharp contrast with conventional telephony, the internet carries voice traffic as data packets over a packet-switched data network instead of a synchronous stream of binary data. To residential customers, a major benefit of internet telephony is lower bills for long-distance voice calls. To corporations, VoIP allows integration of data and voice into a single network, which is translated into substantial cost saving and administration efficiency. According to one study [Thomsen and Jani, 2000], VoIP traffic grew by almost 900% from 1998 to 1999 and is projected to grow another 5000% by 2004. Speech coding will play a central role in this revolution.

Another smaller-scale area of application includes voice storage or digital recording, with some outstanding representatives being the digital telephone answering device (DTAD) and solid-state recorders. For these products to be competitive in the marketplace, their costs must be driven to a minimum. By compressing the digital speech signal before storage, longer-duration voice messages can be recorded for a given amount of memory chips, leading to improved cost effectiveness.

Techniques developed for speech coding have also been applied to other application areas such as speech synthesis, audio coding, speech recognition, and speaker recognition. Due to the weighty position that speech coding occupies in modern technology, it will remain in the center of attention for years to come.

1.2 CLASSIFICATION OF SPEECH CODERS

The task of classifying modern speech coders is not simple and is often confusing, due to the lack of clear separation between various approaches. This section presents some existent classification criteria. Readers must bear in mind that this is a constantly evolving area and new classes of coders will be created as alternative techniques are introduced.

TABLE 1.1 Classification of Speech Coders According to Bit-Rate

Category	Bit-Rate Range
High bit-rate	>15 kbps
Medium bit-rate	5 to 15 kbps
Low bit-rate	2 to 5 kbps
Very low bit-rate	<2 kbps

Classification by Bit-Rate

All speech coders are designed to reduce the reference bit-rate of 128 kbps toward lower values. Depending on the bit-rate of the encoded bit-stream, it is common to classify the speech coders according to Table 1.1. As we will see later in this chapter and throughout the book, different coding techniques lead to different bit-rates. A given method works fine at a certain bit-rate range, but the quality of the decoded speech will drop drastically if it is decreased below a certain threshold. The minimum bit-rate that speech coders will achieve is limited by the information content of the speech signal. Judging from the recoverable message rate from a linguistic perspective for typical speech signals, it is reasonable to say that the minimum lies somewhere around 100 bps. Current coders can produce good quality at 2 kbps and above, suggesting that there is plenty of room for future improvement.

Classification by Coding Techniques

Waveform Coders

An attempt is made to preserve the original shape of the signal waveform, and hence the resultant coders can generally be applied to any signal source. These coders are better suited for high bit-rate coding, since performance drops sharply with decreasing bit-rate. In practice, these coders work best at a bit-rate of 32 kbps and higher.

Signal-to-noise ratio (SNR, Chapter 19) can be utilized to measure the quality of waveform coders. Some examples of this class include various kinds of pulse code modulation (PCM, Chapter 6) and adaptive differential PCM (ADPCM).

Parametric Coders

Within the framework of parametric coders, the speech signal is assumed to be generated from a *model*, which is controlled by some *parameters*. During encoding, parameters of the model are estimated from the input speech signal, with the parameters transmitted as the encoded bit-stream. This type of coder makes no attempt to preserve the original shape of the waveform, and hence SNR is a useless quality measure. Perceptual quality of the decoded speech is directly related to the accuracy and sophistication of the underlying model. Due to this limitation, the coder is signal specific, having poor performance for nonspeech signals.

There are several proposed models in the literature. The most successful, however, is based on *linear prediction*. In this approach, the human speech production mechanism is summarized using a time-varying filter (Section 1.3), with the coefficients of the filter found using the linear prediction analysis procedure (Chapter 4). This is the only type of parametric coder considered in this book.

This class of coders works well for low bit-rate. Increasing the bit-rate normally does not translate into better quality, since it is restricted by the chosen model. Typical bit-rate is in the range of 2 to 5 kbps. Example coders of this class include linear prediction coding (LPC, Chapter 9) and mixed excitation linear prediction (MELP, Chapter 17).

Hybrid Coders

As its name implies, a hybrid coder combines the strength of a waveform coder with that of a parametric coder. Like a parametric coder, it relies on a speech production model; during encoding, parameters of the model are located. Additional parameters of the model are optimized in such a way that the decoded speech is as close as possible to the original waveform, with the closeness often measured by a perceptually weighted error signal. As in waveform coders, an attempt is made to match the original signal with the decoded signal in the time domain.

This class dominates the medium bit-rate coders, with the code-excited linear prediction (CELP, Chapter 11) algorithm and its variants the most outstanding representatives. From a technical perspective, the difference between a hybrid coder and a parametric coder is that the former attempts to quantize or represent the excitation signal to the speech production model, which is transmitted as part of the encoded bit-stream. The latter, however, achieves low bit-rate by discarding all detail information of the excitation signal; only coarse parameters are extracted.

A hybrid coder tends to behave like a waveform coder for high bit-rate, and like a parametric coder at low bit-rate, with fair to good quality for medium bit-rate.

Single-Mode and Multimode Coders

Single-mode coders are those that apply a specific, fixed encoding mechanism at all times, leading to a constant bit-rate for the encoded bit-stream. Examples of such coders are pulse code modulation (PCM, Chapter 6) and regular-pulse-excited long-term prediction (RPE-LTP, Chapter 10).

Multimode coders were invented to take advantage of the dynamic nature of the speech signal, and to adapt to the time-varying network conditions. In this configuration, one of several distinct coding modes is selected, with the selection done by *source control*, when it is based on the local statistics of the input speech, or *network control*, when the switching obeys some external commands in response to network needs or channel conditions.

Figure 1.6 shows the block diagram of a multimode coder with source control. In this system several coding modes are selected according to the properties of the signal at a given interval of time. In an open-loop system, the modes are selected

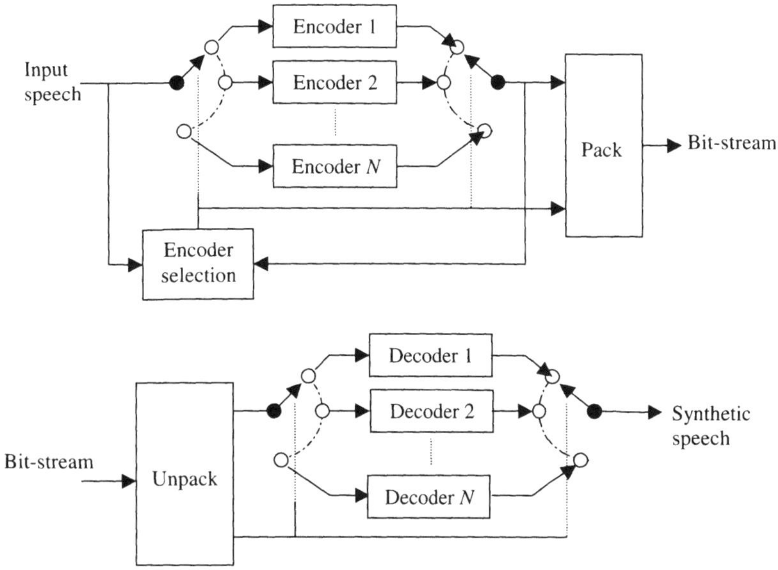

Figure 1.6 Encoder (*top*) and decoder (*bottom*) of a source-controlled multimode coder.

by solely analyzing the input signal. While in a closed-loop approach, encoded outcomes of each mode are taken into account in the final decision. The mode selection information is transmitted as part of the bit-stream, which is used by the decoder to select the proper mode.

Most multimode coders have variable bit-rate, where each mode has a particular, fixed value. Keeping the bit-rate varied allows more flexibility, leading to improved efficiency and a significant reduction in average bit-rate. Examples of multimode coders include the TIA IS96 variable bit-rate CELP coder (Chapter 18), which is source controlled in nature; and the ETSI AMR ACELP coder (Chapter 16), which is a network-controlled version.

1.3 SPEECH PRODUCTION AND MODELING

In this section, the origin and types of speech signals are explained, followed by the modeling of the speech production mechanism. Principles of parametric speech coding are illustrated using a simple example, with the general structure of speech coders described at the end.

Origin of Speech Signals

The speech waveform is a sound pressure wave originating from controlled movements of anatomical structures making up the human speech production

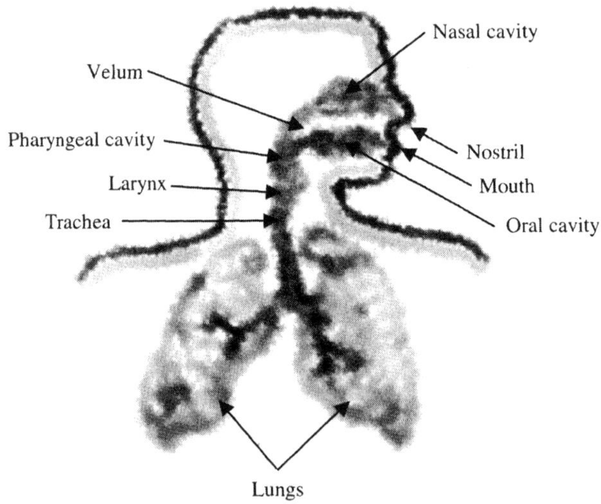

Figure 1.7 Diagram of the human speech production system.

system. A simplified structural view is shown in Figure 1.7. Speech is basically generated as an acoustic wave that is radiated from the nostrils and the mouth when air is expelled from the lungs with the resulting flow of air perturbed by the constrictions inside the body. It is useful to interpret speech production in terms of acoustic filtering. The three main cavities of the speech production system are nasal, oral, and pharyngeal forming the main acoustic filter. The filter is excited by the air from the lungs and is loaded at its main output by a radiation impedance associated with the lips.

The *vocal tract* refers to the pharyngeal and oral cavities grouped together. The nasal tract begins at the velum and ends at the nostrils of the nose. When the velum is lowered, the nasal tract is acoustically coupled to the vocal tract to produce the nasal sounds of speech.

The form and shape of the vocal and nasal tracts change continuously with time, creating an acoustic filter with time-varying frequency response. As air from the lungs travels through the tracts, the frequency spectrum is shaped by the frequency selectivity of these tracts. The resonance frequencies of the vocal tract tube are called *formant frequencies* or simply *formants*, which depend on the shape and dimensions of the vocal tract.

Inside the larynx is one of the most important components of the speech production system—the vocal cords. The location of the cords is at the height of the "Adam's apple"—the protrusion in the front of the neck for most adult males. Vocal cords are a pair of elastic bands of muscle and mucous membrane that open and close rapidly during speech production. The speed by which the cords open and close is unique for each individual and define the feature and personality of the particular voice.

Classification of Speech Signals

Roughly speaking, a speech signal can be classified as *voiced* or *unvoiced*. Voiced sounds are generated when the vocal cords vibrate in such a way that the flow of air from the lungs is interrupted periodically, creating a sequence of pulses to excite the vocal tract. With the vocal cords stationary, the turbulence created by the flow of air passing through a constriction of the vocal tract generates unvoiced sounds. In time domain, voiced sound is characterized by strong periodicity present in the signal, with the fundamental frequency referred to as the *pitch frequency*, or simply *pitch*. For men, pitch ranges from 50 to 250 Hz, while for women the range usually falls somewhere in the interval of 120 to 500 Hz. Unvoiced sounds, on the other hand, do not display any type of periodicity and are essentially random in nature.

To experiment with voice and unvoiced sounds and the involvement of the vocal cords, try placing your fingers on the front of your neck while you speak. Consider the "fa" sound as in "father." First, attempt to pronounce for a few seconds the "f" sound alone, which is a *consonant* in American English. Next, pronounce "a" for a few seconds, which is a *vowel*. How do your fingers feel for the two cases? In the first case you shouldn't feel any vibration in the front of your neck; while in the second case some pulsation is detected. Speak louder if you have problems feeling it. The oscillation is associated with the activities of the vocal cords and is present for the pronunciation of vowels.

Figure 1.8 shows an example speech waveform uttered by a male subject, where both voiced and unvoiced signals are present. It is possible to appreciate from this example the nonstationarity nature of speech signals, where statistics of the signal change constantly with time. We see that for the voiced frame, there is clear periodicity in time domain, where the signal repeats itself in a quasiperiodic pattern; and also in frequency domain, where a harmonic structure is observed. Note that the spectrum indicates dominant low-frequency contents, due mainly to the relatively low value of the pitch frequency. For the unvoiced frame, however, the signal is essentially random. From the spectrum we can see that there is a significant amount of high-frequency components, corresponding to rapidly changing signals.

It is necessary to indicate that the voiced / unvoiced classification might not be absolutely clear for all frames, since during transitions (voiced to unvoiced or vice versa) there will be randomness and quasiperiodicity that is difficult to judge as strictly voiced or strictly unvoiced.

For most speech coders, the signal is processed on a frame-by-frame basis, where a frame consists of a finite number of samples. The length of the frame is selected in such a way that the statistics of the signal remain almost constant within the interval. This length is typically between 20 and 30 ms, or 160 and 240 samples for 8-kHz sampling.

Modeling the Speech Production System

In general terms, a model is a simplified representation of the real world. It is designed to help us better understand the world in which we live and, ultimately,

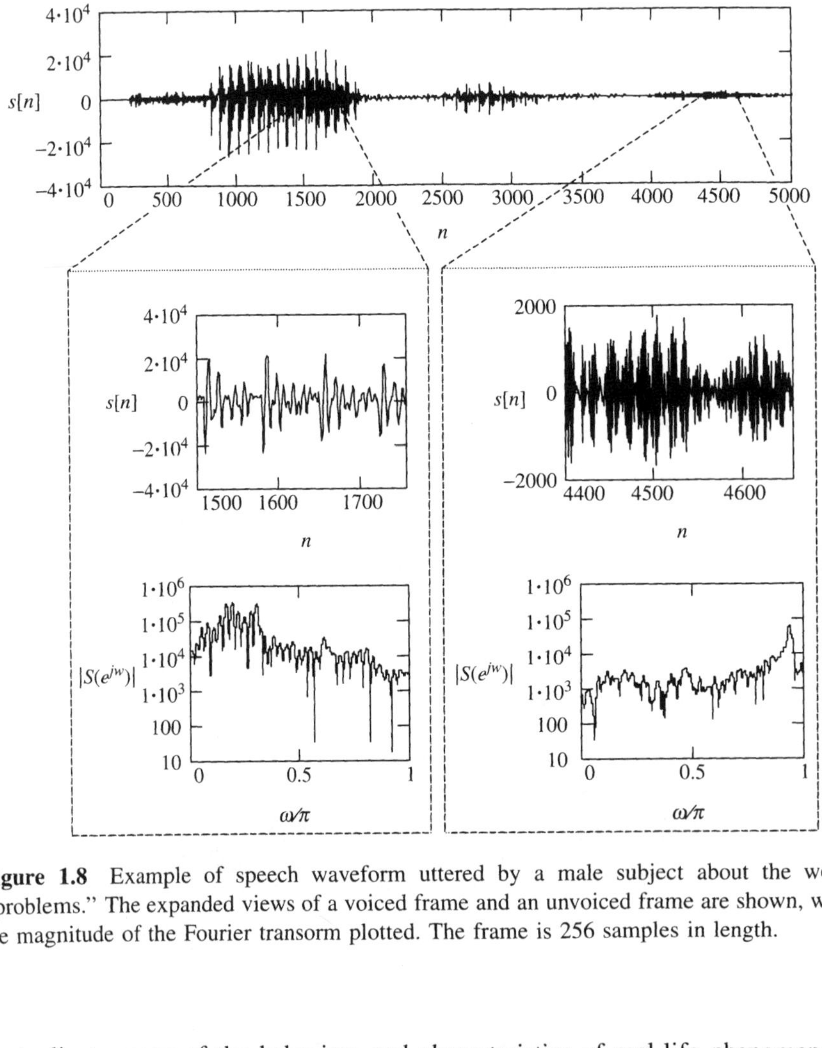

Figure 1.8 Example of speech waveform uttered by a male subject about the word "problems." The expanded views of a voiced frame and an unvoiced frame are shown, with the magnitude of the Fourier transorm plotted. The frame is 256 samples in length.

to duplicate many of the behaviors and characteristics of real-life phenomenon. However, it is incorrect to assume that the model and the real world that it represents are identical in every way. In order for the model to be successful, it must be able to replicate partially or completely the behaviors of the particular object or fact that it intends to capture or simulate. The model may be a physical one (i.e., a model airplane) or it may be a mathematical one, such as a formula.

The human speech production system can be modeled using a rather simple structure: the lungs—generating the air or energy to excite the vocal tract—are represented by a white noise source. The acoustic path inside the body with all its components is associated with a time-varying filter. The concept is illustrated in Figure 1.9. This simple model is indeed the core structure of many speech coding algorithms, as can be seen later in this book. By using a *system identification*

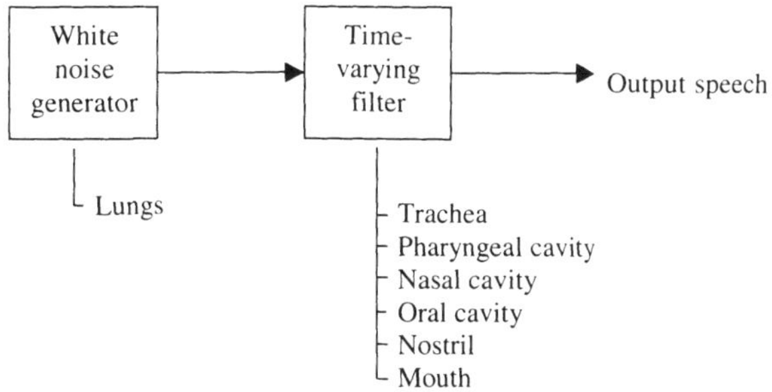

Figure 1.9 Correspondence between the human speech production system with a simplified system based on time-varying filter.

technique called *linear prediction* (Chapter 4), it is possible to estimate the parameters of the time-varying filter from the observed signal.

The assumption of the model is that the energy distribution of the speech signal in frequency domain is totally due to the time-varying filter, with the lungs producing an excitation signal having a flat-spectrum white noise. This model is rather efficient and many analytical tools have already been developed around the concept. The idea is the well-known *autoregressive model*, reviewed in Chapter 3.

A Glimpse of Parametric Speech Coding

Consider the speech frame corresponding to an unvoiced segment with 256 samples of Figure 1.8. Applying the samples of the frame to a linear prediction analysis procedure (Chapter 4), the coefficients of an associated filter are found. This filter has system function

$$H(z) = \frac{1}{1 + \sum_{i=1}^{10} a_i z^{-i}}$$

with the coefficients denoted by a_i, $i = 1$ to 10.

White noise samples are created using a unit variance Gaussian random number generator; when passing these samples (with appropriate scaling) to the filter, the output signal is obtained. Figure 1.10 compares the original speech frame, with two realizations of filtered white noise. As we can see, there is no time-domain correspondence between the three cases. However, when these three signal frames are played back to a human listener (converted to sound waves), the perception is almost the same!

How could this be? After all, they look so different in the time domain. The secret lies in the fact that they all have a similar magnitude spectrum, as plotted in Figure 1.11. As we can see, the frequency contents are similar, and since the human auditory system is not very sensitive toward phase differences, all three

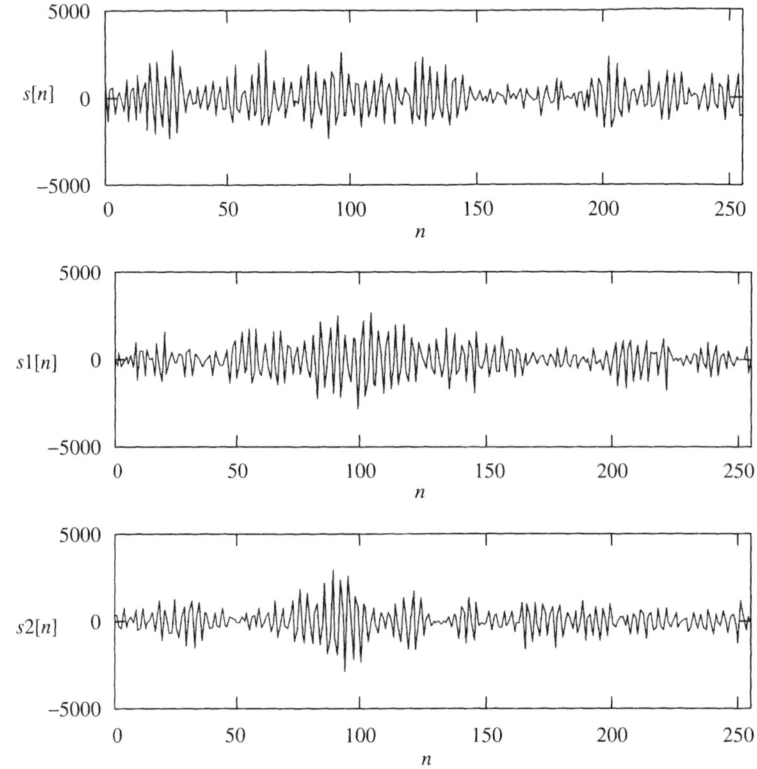

Figure 1.10 Comparison between an original unvoiced frame (*top*) and two synthesized frames.

frames sound almost identical (more on this in the next section). The original frequency spectrum is captured by the filter, with all its coefficients. Thus, the flat-spectrum white noise is shaped by the filter so as to produce signals having a spectrum similar to the original speech. Hence, linear prediction analysis is also known as a *spectrum estimation* technique.

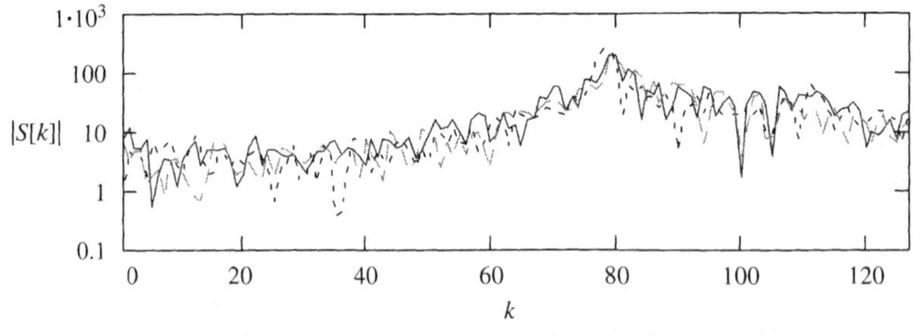

Figure 1.11 Comparison between the magnitude of the DFT for the three signal frames of Figure 1.10.

How can we use this trick for speech coding? As we know, the objective is to represent the speech frame with a lower number of bits. The original number of bits for the speech frame is

$$\text{Original number of bits} = 256 \text{ samples} \cdot 16 \text{ bits/sample} = 4096 \text{ bits}.$$

As indicated previously, by finding the coefficients of the filter using linear prediction analysis, it is possible to generate signal frames having similar frequency contents as the original, with almost identical sounds. Therefore, the frame can be represented alternatively using ten filter coefficients, plus a scale factor. The scale factor is found from the power level of the original frame. As we will see later in the book, the set of coefficients can be represented with less than 40 bits, while 5 bits are good enough for the scale factor. This leads to

$$\text{Alternative number of bits} = 40 \text{ bits} + 5 \text{ bits} = 45 \text{ bits}.$$

Therefore, we have achieved an order of magnitude saving in terms of the number of required bits by using this alternative representation, fulfilling in the process our objective of bit reduction. This simple speech coding procedure is summarized below.

- *Encoding*

  Derive the filter coefficients from the speech frame.

  Derive the scale factor from the speech frame.

  Transmit filter coefficients and scale factor to the decoder.

- *Decoding*

  Generate white noise sequence.

  Multiply the white noise samples by the scale factor.

  Construct the filter using the coefficients from the encoder and filter the scaled white noise sequence. Output speech is the output of the filter.

By repeating the above procedures for every speech frame, a time-varying filter is created, since its coefficients are changed from frame to frame. Note that this overly simplistic scheme is for illustration only: much more elaboration is necessary to make the method useful in practice. However, the core ideas for many speech coders are not far from this uncomplicated example, as we will see in later chapters.

General Structure of a Speech Coder

Figure 1.12 shows the generic block diagrams of a speech encoder and decoder. For the encoder, the input speech is processed and analyzed so as to extract a number of parameters representing the frame under consideration. These parameters are encoded or quantized with the binary indices sent as the compressed bit-stream

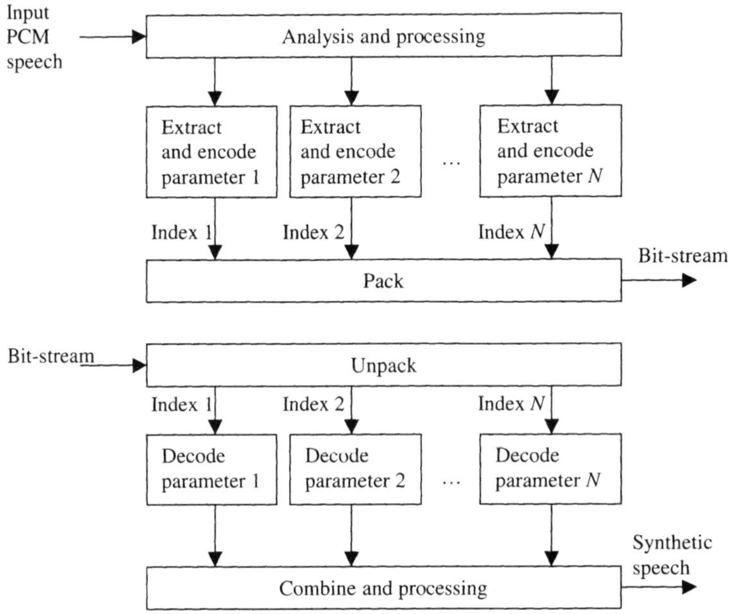

Figure 1.12 General structure of a speech coder. *Top*: Encoder. *Bottom*: Decoder.

(see Chapter 5 for concepts of quantization). As we can see, the indices are *packed* together to form the bit-stream; that is, they are placed according to certain predetermined order and transmitted to the decoder.

The speech decoder *unpacks* the bit-stream, where the recovered binary indices are directed to the corresponding parameter decoder so as to obtain the quantized parameters. These decoded parameters are combined and processed to generate the synthetic speech.

Similar block diagrams as in Figure 1.12 will be encountered many times in later chapters. It is the responsibility of the algorithm designer to decide the functionality and features of the various processing, analysis, and quantization blocks. Their choices will determine the performance and characteristic of the speech coder.

1.4 SOME PROPERTIES OF THE HUMAN AUDITORY SYSTEM

The way that the human auditory system works plays an important role in speech coding systems design. By understanding how sounds are perceived, resources in the coding system can be allocated in the most efficient manner, leading to improved cost effectiveness. In subsequent chapters we will see that many speech coding standards are tailored to take advantage of the properties of the human auditory system. This section provides an overview of the subject, summarizing several topics including the structure of the human auditory system, absolute threshold, masking, and phase perception.

Structure of the Human Auditory System

A simplified diagram of the human auditory system appears in Figure 1.13. The pinna (or informally the ear) is the surface surrounding the canal in which sound is funneled. Sound waves are guided by the canal toward the eardrum—a membrane that acts as an acoustic-to-mechanic transducer. The sound waves are then translated into mechanical vibrations that are passed to the cochlea through a series of bones known as the ossicles. Presence of the ossicles improves sound propagation by reducing the amount of reflection and is accomplished by the principle of impedance matching.

The cochlea is a rigid snail-shaped organ filled with fluid. Mechanical oscillations impinging on the ossicles cause an internal membrane, known as the basilar membrane, to vibrate at various frequencies. The basilar membrane is characterized by a set of frequency responses at different points along the membrane; and a simple modeling technique is to use a bank of filters to describe its behavior. Motion along the basilar membrane is sensed by the inner hair cells and causes neural activities that are transmitted to the brain through the auditory nerve.

The different points along the basilar membrane react differently depending on the frequencies of the incoming sound waves. Thus, hair cells located at different positions along the membrane are excited by sounds of different frequencies. The neurons that contact the hair cells and transmit the excitation to higher auditory centers maintain the frequency specificity. Due to this arrangement, the human auditory system behaves very much like a frequency analyzer; and system characterization is simpler if done in the frequency domain.

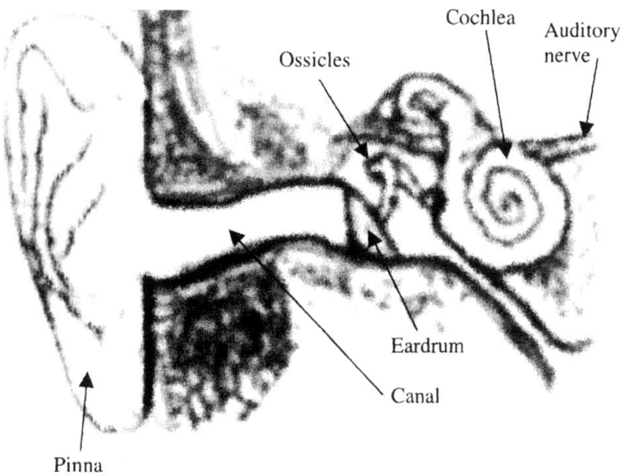

Figure 1.13 Diagram of the human auditory system.

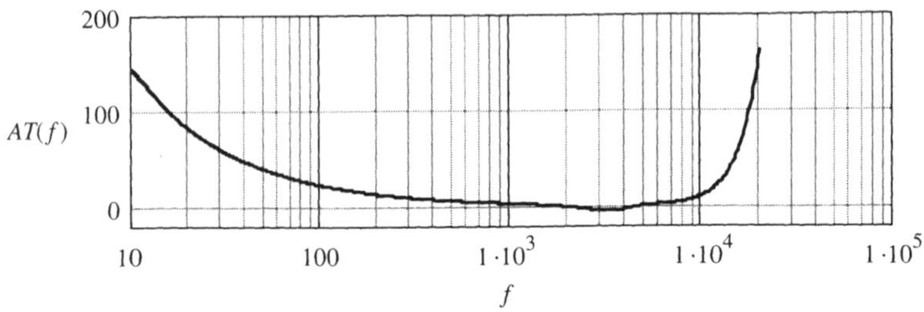

Figure 1.14 A typical absolute threshold curve.

Absolute Threshold

The absolute threshold of a sound is the minimum detectable level of that sound in the absence of any other external sounds. That is, it characterizes the amount of energy needed in a pure tone such that it can be detected by a listener in a noiseless environment. Figure 1.14 shows a typical absolute threshold curve, where the horizontal axis is frequency measured in hertz (Hz); while the vertical axis is the absolute threshold in decibels (dB), related to a reference intensity of 10^{-12} watts per square meter—a standard quantity for sound intensity measurement.

Note that the absolute threshold curve, as shown in Figure 1.14, reflects only the average behavior; the actual shape varies from person to person and is measured by presenting a tone of a certain frequency to a subject, with the intensity being tuned until the subject no longer perceive its presence. By repeating the measurements using a large number of frequency values, the absolute threshold curve results.

As we can see, human beings tend to be more sensitive toward frequencies in the range of 1 to 4 kHz, while thresholds increase rapidly at very high and very low frequencies. It is commonly accepted that below 20 Hz and above 20 kHz, the auditory system is essentially dysfunctional. These characteristics are due to the structures of the human auditory system: acoustic selectivity of the pinna and canal, mechanical properties of the eardrum and ossicles, elasticity of the basilar membrane, and so on.

We can take advantage of the absolute threshold curve in speech coder design. Some approaches are the following:

- Any signal with an intensity below the absolute threshold need not be considered, since it does not have any impact on the final quality of the coder.
- More resources should be allocated for the representation of signals within the most sensitive frequency range, roughly from 1 to 4 kHz, since distortions in this range are more noticeable.

Masking

Masking refers to the phenomenon where one sound is rendered inaudible because of the presence of other sounds. The presence of a single tone, for instance, can

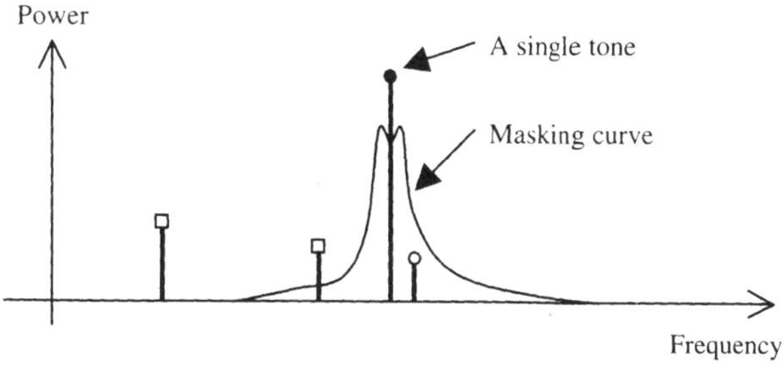

Figure 1.15 Example of the masking curve associated with a single tone. Based on the masking curve, examples of audible (□) and inaudible (○) tones are shown, which depend on whether the power is above or below the masking curve, respectively.

mask the neighboring signals—with the masking capability inversely proportional to the absolute difference in frequency. Figure 1.15 shows an example where a single tone is present; the tone generates a masking curve that causes any signal with power below it to become imperceptible. In general, masking capability increases with the intensity of the reference signal, or the single tone in this case.

The features of the masking curve depend on each individual and can be measured in practice by putting a subject in a laboratory environment and asking for his/her perception of a certain sound tuned to some amplitude and frequency values in the presence of a reference tone.

Masking can be explored for speech coding developments. For instance, analyzing the spectral contents of a signal, it is possible to locate the frequency regions that are most susceptible to distortion. An example is shown in Figure 1.16. In this case a typical spectrum is shown, which consists of a series of high- and low-power regions, referred to as peaks and valleys, respectively. An associated masking curve exists that follows the ups and downs of the original spectrum. Signals with power below the masking curve are inaudible; thus, in general, peaks can tolerate more distortion or noise than valleys.

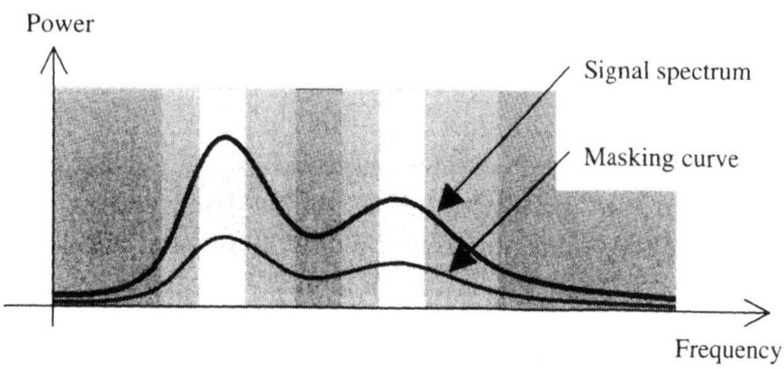

Figure 1.16 Example of a signal spectrum and the associated masking curve. Dark areas correspond to regions with relatively little tolerance to distortion, while clear areas correspond to regions with relatively high tolerance to distortion.

A well-designed coding scheme should ensure that the valleys are well preserved or relatively free of distortions; while the peaks can tolerate a higher amount of noise. By following this principle, effectiveness of the coding algorithm is improved, leading to enhanced output quality.

As we will see in Chapter 11, coders obeying the principle of code-excited linear prediction (CELP) rely on the perceptual weighting filter to weight the error spectrum during encoding; frequency response of the filter is time-varying and depends on the original spectrum of the input signal. The mechanism is highly efficient and is widely applied in practice.

Phase Perception

Modern speech coding technologies rely heavily on the application of perceptual characteristics of the human auditory system in various aspects of a quantizer's design and general architecture. In most cases, however, the focus on perception is largely confined to the magnitude information of the signal; the phase counterpart has mostly been neglected with the underlying assumption that human beings are phase deaf.

There is abundant evidence on phase deafness; for instance, a single tone and its time-shifted version essentially produce the same sensation; on the other hand, noise perception is chiefly determined by the magnitude spectrum. This latter example was already described in the last section for the design of a rudimentary coder and is the foundation of some early speech coders, such as the linear prediction coding (LPC) algorithm, studied in Chapter 9.

Even though phase has a minor role in perception, some level of phase preservation in the coding process is still desirable, since naturalness is normally increased. The code-excited linear prediction (CELP) algorithm, for instance, has a mechanism to retain phase information of the signal, covered in Chapter 11.

1.5 SPEECH CODING STANDARDS

This book focuses mainly on the study of the foundation and historical evolution of many standardized coders. As a matter of principle, a technique is included only if it is part of some standard. Standards exist because there are strong needs to have common means for communication: it is to everyone's best interest to be able to develop and utilize products and services based on the same reference.

By studying the supporting techniques of standardized coders, we are indeed concentrating our effort on understanding the most influential and successful ideas in this field of knowledge. Otherwise, we would have to spend an enormous amount of effort to deal with the endless papers, reports, and propositions in the literature; many of these might be immature, incomplete, or, in some instances, impractical. A standard, on the other hand, is developed by a team of experts over an extended period of time, with extensive testing and repeated evaluation to warrant that a set of requirements is met. Only organizations with vast resources can coordinate such endeavors. According to Cox [1995], the time required to complete a standard from beginning to end under the best of circumstances is around 4.5 years.

This does not mean that a standard is error-free or has no room for improvement. As a matter of fact, new standards often appear as improvement on the existing ones. In many instances, a standard represents the state-of-the-art at the time; in other terms, a reference for future improvement. The relentless research effort will continuously push existent technology toward unknown boundaries.

Standard Bodies

The standard bodies are organizations responsible for overseeing the development of standards for a particular application. Brief descriptions of some well-known standard bodies are given here.

- *International Telecommunications Union* (*ITU*). The Telecommunications Standardization Sector of the ITU (ITU-T) is responsible for creating speech coding standards for network telephony. This includes both wired and wireless networks.
- *Telecommunications Industry Association* (*TIA*). The TIA is in charge of promulgating speech coding standards for specific applications. It is part of the American National Standards Institute (ANSI). The TIA has successfully developed standards for North American digital cellular telephony, including time division multiple access (TDMA) and code division multiple access (CDMA) systems.
- *European Telecommunications Standards Institute* (*ETSI*). The ETSI has memberships from European countries and companies and is mainly an organization of equipment manufacturers. ETSI is organized by application; the most influential group in speech coding is the Groupe Speciale Mobile (GSM), which has several prominent standards under its belt.
- *United States Department of Defense* (*DoD*). The DoD is involved with the creation of speech coding standards, known as U.S. Federal standards, mainly for military applications.
- *Research and Development Center for Radio Systems of Japan* (*RCR*). Japan's digital cellular standards are created by the RCR.

The Standards Covered in this Book

As mentioned before, this book is dedicated to standardized coders. Table 1.2 contains the major standards developed up to 1999. The name of a standard begins with the acronym of the standard body responsible for development, followed by a label or number assigned to the coder (if available); at the end is the particular algorithm selected. The list in Table 1.2 is not meant to be exhaustive, and many other standards are available either for special purpose or private use by corporations.

TABLE 1.2 Summary of Major Speech Coding Standards

Year Finalized	Standard Name	Bit-Rate (kbps)	Applications
1972[a]	ITU-T G.711 PCM	64	General purpose
1984[b]	FS 1015 LPC	2.4	Secure communication
1987[b]	ETSI GSM 6.10 RPE-LTP	13	Digital mobile radio
1990[c]	ITU-T G.726 ADPCM	16, 24, 32, 40	General purpose
1990[b]	TIA IS54 VSELP	7.95	North American TDMA digital cellular telephony
1990[c]	ETSI GSM 6.20 VSELP	5.6	GSM cellular system
1990[c]	RCR STD-27B VSELP	6.7	Japanese cellular system
1991[b]	FS1016 CELP	4.8	Secure communication
1992[b]	ITU-T G.728 LD-CELP	16	General purpose
1993[b]	TIA IS96 VBR-CELP	8.5, 4, 2, 0.8	North American CDMA digital cellular telephony
1995[a]	ITU-T G.723.1 MP-MLQ / ACELP	5.3, 6.3	Multimedia communications, videophones
1995[b]	ITU-T G.729 CS-ACELP	8	General purpose
1996[a]	ETSI GSM EFR ACELP	12.2	General purpose
1996[a]	TIA IS641 ACELP	7.4	North American TDMA digital cellular telephony
1997[b]	FS MELP	2.4	Secure communication
1999[a]	ETSI AMR-ACELP	12.2, 10.2, 7.95, 7.40, 6.70, 5.90, 5.15, 4.75	General purpose telecommunication

[a]Coder is described only partially.
[b]Coder is fully explained.
[c]Coder is mentioned only briefly without detailed technical descriptions.

However, the major achievements in speech coding for the past thirty years are well represented by the coders on the list.

It is important to mention that the philosophy of this book is to explain the whys and hows of a specific algorithm; most importantly, to justify the selection of a particular technique for an application. Each standardized coder tends to have its own idiosyncrasies and minute operational tricks that might not be important for the understanding of the foundation of the algorithm and hence are often omitted. In order to develop a bit-stream compatible version of the coder, consultation with official documentation is mandatory. Even though many documents describing the standards are available to the general public, it doesn't mean that it is free for anyone to use. The standards are created through the efforts of individuals, and licensing royalties are the way that these individuals get compensated. On incorporating many of the techniques discussed in this book to commercial

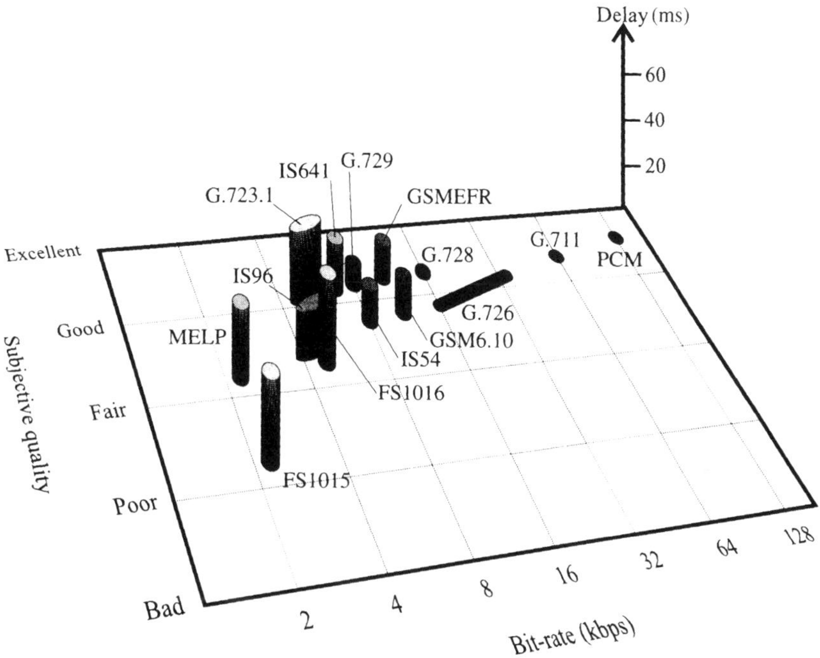

Figure 1.17 Performance comparison between some standardized coders.

products, readers must be aware of patent licenses and be ready to negotiate intellectual property rights agreements with the related organizations.

Figure 1.17 shows a quality/bit-rate/delay comparison plot for the various standards, with quality informally referring to how good the synthetic speech sounds as a result of the encoding/decoding process associated with a speech coder. The plot is for illustration purposes and does not mean to be absolutely precise, since quality measurements (Chapter 19) must be done under various conditions and, in many instances, it is difficult to establish a fair comparison. The data are compiled from various sources and give a rough idea of relative performance among the different coders. Delay is reflected by the height of a particular quality/bit-rate coordinate and refers to the encoder buffering delay.

Finally, the fact that certain proposed techniques have not become part of a standard does not mean that they are worthless. Sometimes there is a need for refinement; in other instances they are more suitable for special conditions. Since the standardization process is routinely linked to politics, power, and money, the adopted technology might not necessarily represent the best choice from a pure technical perspective. Serious researchers should be ready to learn from many sources and apply the acquired knowledge to optimize the solution for a particular problem.

1.6 ABOUT ALGORITHMS

A speech coder is generally specified as an algorithm, which is defined as a computational procedure that takes some input values to produce some output values. An algorithm can be implemented as software (i.e., a program to command a processor) or as hardware (direct execution through digital circuitry). With the widespread availability of low-cost high-performance digital signal processors (DSPs) and general-purpose microprocessors, many signal processing tasks—done in the old days using analog circuitry—are predominantly executed in the digital domain. Advantages of going digital are many: programmability, reliability, and the ability to handle very complex procedures, such as the operations involved in a speech coder, so complex that the analog world would have never dreamed of it. In this section the various aspects of algorithmic implementation are explained.

The Reference Code

It is the trend for most standard bodies to come up with a reference source code for their standards, where *code* refers to the algorithm or program written in text form. The source code is elaborated with some high-level programming language, with the C language being the most commonly used [Harbison and Steele, 1995]. In this reference code, the different components of the speech coding algorithm are implemented. Normally, there are two main functions: *encode* and *decode* taking care of the operations of the encoder and decoder, respectively.

The reference source code is very general and might not be optimized for speed or storage; therefore, it is an engineering task to adjust the code so as to suit a given platform. Since different processors have different strengths and weaknesses, the adjustment must be custom made; in many instances, this translates into assembly language programming. The task normally consists of changing certain parts of the algorithm so as to speed up the computational process or to reduce memory requirements.

Depending on the platform, the adjustment of the source code can be relatively easy or extremely hard; or it may even be unrealizable, if the available resources are not enough to cover the demand of the algorithm. A supercomputer, for instance, is a platform where there are abundant memory and computational power; minimum change is required to make an algorithm run under this environment. The personal computer (PC), on the other hand, has a moderate amount of memory and computational power; so adjustment is desirable to speed up the algorithm, but memory might not be such a big concern. A cellular handset is an example where memory and computational power are limited; the code must be adjusted carefully so that the algorithm runs within the restricted confinements.

To verify that a given implementation is accurate, standard bodies often provide a set of test vectors. That is, a given input test vector must produce a corresponding output vector. Any deviation will be considered a failure to conform to the specification.

About C and C++

The C programming language has become almost a mandatory medium in software development for many signal processing tasks. Its popularity is due to several factors: provision of a fairly complete set of facilities for dealing with a wide variety of applications—including low-level, high efficiency for implementation and portability across various computing platforms. Unfortunately, some of the advantages of C can also pose problems for programmers. For instance, the efficiency is largely due to the absence of confining rules that can lead to error-prone programming habits.

C++ is referred to as an *object-oriented* language and has closed many holes in the C language, providing better type checking and compile-time analysis. C++ programmers are forced to declare functions so that the compiler can check their use. On the other hand, systems designed using C++ are easier to express and understand, which is especially true for complex projects, where many programmers are involved.

At present the speech coding community relies heavily on the C programming language. Most standard bodies offer reference code written in C. The use of C++ is largely for maintainability and extensibility and often comes with some performance penalty, which is due to the many additional features of the language; the penalty involved might not be acceptable for resource-critical platforms. Recently, there are movements in the programming world regarding the creation of a new intermediate language between C and C++. It is essentially C++ but eliminates many of the cumbersome and often unnecessary features so as to boost efficiency but, at the same time, conserves the good programming guidelines of an object-oriented approach.

Fixed-Point and Floating-Point Implementation

One of the earliest decisions that must be made on the implementation of a signal processing system is whether the algorithms are going to be run on a fixed-point or floating-point platform. Fixed-point numbers refer to those having limited dynamic range; for instance, a 16-bit signed integer can represent a maximum of 65536 numbers within the interval of $[-32768, 32767]$.

Floating-point numbers, on the other hand, can represent extremely small numbers and extremely big numbers. The IEEE Standard for Binary Floating-Point Arithmetic (ISO/IEEE Std. 754-1985; see Harbison and Steele, [1995]), for instance, defines the following 32-bit floating-point number:

$$s \cdot 2^e \cdot \sum_{k=1}^{24} f_k \cdot 2^{-k},$$

where $s = \pm 1$ is the sign bit, e is the exponent in the range of $[-125, 128]$, and f_k, with $k = 1$ to 24, equal to 0 or 1 are the binary digits. The smallest positive number out of this scheme is 2^{-149} while the biggest is $3.403 \cdot 10^{38}$.

Ideally, all algorithms should be implemented with floating-point processors; in that way the rounding error after each operation will be negligibly small, and the hazard of numeric overflow is virtually eliminated. Unfortunately, floating-point processors are relatively expensive, due to the increased size of the processor's chip needed to support the more complex operations; also, power consumption is higher when compared to a fixed-point processor. For cost and power sensitive consumer appliances (i.e., the cellular handset), the fixed-point processor is almost the mandatory choice.

Software development for floating-point environment is straightforward; under most normal circumstances the numbers should remain within the wide dynamic range supported by the processor. This is not quite the case for a fixed-point environment: it is tricky and a considerable amount of time must be spent on finding out the range of each intermediate variable so as to ensure that all numbers remain within the limited range. Texas Instruments, Inc. [1990] offers some guidelines of fixed-point programming on a digital signal processor.

For early development and research, floating-point operations are normally assumed, so that effort can be concentrated on the algorithm, instead of being distracted by rounding errors and precision issues. After the operations of the algorithm are well tested in a floating-point environment, it will be translated to fixed-point (if that is the goal), which could be a time-consuming process; in some instances, part of the algorithm must be modified or adjusted in order to run properly.

Due to the fact that many speech coders are targeted to consumer products, the final cost becomes a primary concern. Thus, many standard bodies specify the reference code using fixed-point operations. In this way, the algorithm can run under a fixed-point environment in a straightforward manner.

Description of Algorithms

The notation used in this book for algorithmic descriptions are illustrated with a simple example. Given the array of samples $s[n]$, $n = 0$ to 200, consider the task of autocorrelation calculation:

$$R[l] = \sum_{n=l}^{200} s[n]s[n-l]$$

for $l = 20$ to 50, followed by peak search over the autocorrelation values within this interval and returning the peak value and its position as results.

The *block diagram* description of the aforementioned algorithm is given in Figure 1.18. As we can see, the input and output are clearly specified, with the

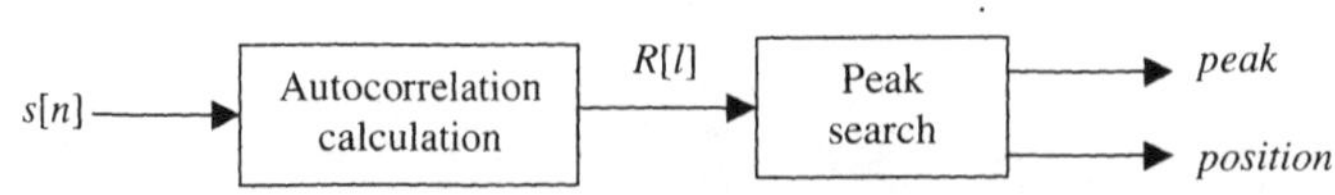

Figure 1.18 Example of algorithm and its block diagram.

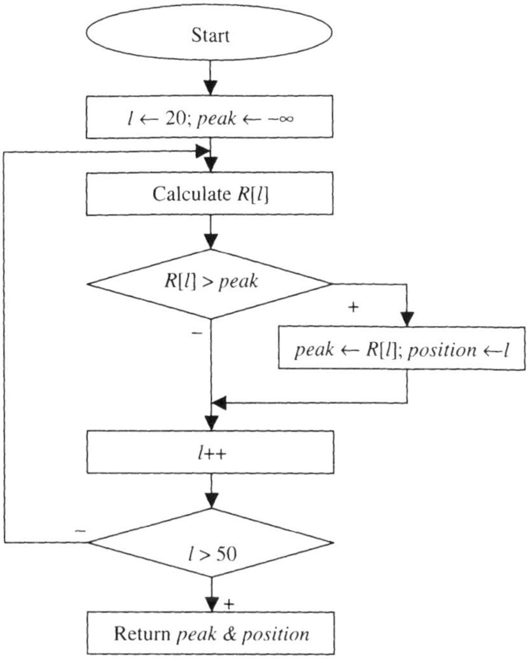

Figure 1.19 Example of algorithm and its flowchart.

involved operations grouped into blocks. This type of description provides high-level visualization of the operations and the relationships among the various components. Many implementational details, however, remain hidden.

A *flowchart* contains more details about how the algorithm is implemented. Figure 1.19 contains the flowchart of our example algorithm. It still preserves the block structure, but the meaning and ordering of each block are precise, allowing direct translation into program code. This type of description represents an intermediate level between a high-level block diagram and the actual program code.

Ultimately, the algorithm is translated into program code. In this book, we shall describe algorithms as programs written in a *pseudocode* that is very much like C. Anyone who has been exposed to programming in high-level language should have little trouble reading the code. For our example, the pseudocode description is as follows:

```
AUTO_PEAK(s)
peak ← −∞; position ← 20
for l ← 20 to 50
        R ← 0
        for n ← 1 to 200
                R ← s[n]*s[n−1] + R
```

```
        if R > peak
                peak ← R; position ← l
return peak, position
```

Note the use of assignment operator "←" instead of equal sign "=" like in many programming languages. The expression $a \leftarrow b$ means that the content of b is assigned to a; after the operation, b remains intact while a is modified (having same content as b). With the popularity of programming languages in recent years, the meaning of "=" has shifted from the traditional equal to that of assignment. To many mathematicians, it is unacceptable since confusion occurs with common practices in equation writing. To avoid problems with this regard, this book preserves the old meaning of "=", and assignments are explicitly indicated with "←".

Another note is the use of multicharacter variables. Typically, variables in the equations are expressed using single characters, which could be extracted from the English or Greek alphabets. With the increasing complexity of algorithms, the use of multicharacter variables is imperative to give a clear description of the problem at hand. Thus, we frequently see variable names such as *d1*, *d2*, *energy*, *peak*, *position*, and so on.

What separates pseudocode from "real" code is that, in the former, we employ whatever expressive method is most clear and concise to specify a given algorithm. Another difference is that for pseudocode, there is no concern with issues of software engineering, such as data abstraction, modularity, and error handling. Attention is directed only to the essence of the algorithm.

Conventions for pseudocode writing are the use of indentation to indicate block structure. The program constructs, such as **for**, **if**, and **return**, are represented using bold characters. Meanings of these statements are essentially the same as for standard C. Consult a C manual when in doubt. Also, the variables are local to the given procedure (such as n and l). We shall not use global variables without explicit indication.

Analysis of Algorithms

Analyzing an algorithm implies the prediction of resource requirements necessitated to run it. The resources are often measured in terms of memory and computation constituting the two fundamental cost components of digital hardware. These two components are explained next.

Memory Cost

Many types of memory devices are available for use in modern hardware. Most software developers think of memory as being either random-access (RAM) or read-only (ROM). But in fact there are subtypes of each and even hybrid memories; see Barr [1999] for detail descriptions. ROM is needed to hold the instructions corresponding to the program code and the supporting data; its contents are normally unchanged during execution and its size depends on the complexity of the algorithm. RAM is needed to store the input, output, and intermediate variables. Program code can usually be optimized so as to remove unnecessary operations, leading to reduced memory size. Also, an algorithm may be modified to use less memory, with possible speed penalty.

Computational Cost

Given a certain amount of input data, it is desirable to process them as quickly as possible so as to generate the corresponding output data. Depending on the selected technique, one algorithm can be more efficient than another. The running time is measured by the number of primitive operations required to complete the mission. For signal processing, it is common to count the number of sum (adding two numbers) and the number of product (multiplying two numbers) as measurements of the computational cost. These two quantities can be found for different algorithms and compared; the one offering the lowest counts is the most efficient.

Computational cost is often platform dependent; that is, counting the number of primitive operations alone might not make sense for a certain processor. For instance, the DSP families of Texas Instruments [1990, 1993] can often perform one addition and one multiplication in one step; thus, a more meaningful performance measure would be the maximum between the number of sums and the number of products. On the other hand, the Intel Pentium processor [Intel, 1997] can perform four operations in parallel; an algorithm running on this processor is normally modified to take advantage of the enhanced architecture, and an alternative performance measure is necessary for meaningful comparison. A commonly used reference measure between processors is *millions-of-instructions-per-second* (MIPS). The final performance, however, depends on other architectural features, as well as the specific algorithm; see Eyre [2001] for additional details.

1.7 SUMMARY AND REFERENCES

This introductory chapter provided an overview to the general aspects of speech coding, with guidelines to the rest of the materials covered in this book. The purpose, operation, and classification of speech coders are described; origin and modeling of speech signals are explained with revelation of the structure of a simple parametric coder. Structure of the human auditory system is analyzed, with the most important properties explained; these properties can be used to develop efficient coding schemes for speech. The mission of standard bodies and various aspects of algorithm design are described. Since speech coding is related to human perception, it is often not possible to outline an absolute design guideline. For instance, what is perceived as good by one person might not be so good for another person. In fact, experts disagree on methodologies and techniques applied to a given situation. Therefore, speech coding is a combination of art and science, in the sense that an engineering framework is applied but very often it is refined according to human perception, which cannot be absolutely justifiable from a mathematical perspective.

See Spanias [1994] and Kleijn and Paliwal [1995b] for alternative classification criteria, as well as a more diversified survey on existing speech coding technology. Das et al. [1995] provides detailed descriptions of multimode and variable bit-rate coders. See Rabiner and Schafer [1978] for discussions of acoustic modeling of speech production, as well as early modeling attempts using digital means. In Deller et al. [1993], similar acoustic modeling is described, together with an interesting historical recount on how a mechanical system was built for speech generation, using principles of acoustic filters in 1939. Many references are available for human auditory system and psychoacoustics; see Rabiner and Juang [1993] for an introduction and simple modeling; more extensive studies appear in Moore [1997] and Zwicker and Fastl [1999]; a more signal-oriented treatment of sound perception is found in Hartmann [1998].

Standardization procedures are discussed in Cox [1995]. Many issues related to algorithm analysis and design can be found in Cormen et al. [1990]. There are a plethora of books available for introductory C/C++ programming; see Harbison and Steele [1995] for reference in the features of C, and Eckel [2000] for C++ programming. An overview of the historical evolution of digital signal processors appears in Eyre and Bier [2000]. Appendix C contains some research directions in the speech coding arena.

CHAPTER 2

SIGNAL PROCESSING TECHNIQUES

The basic and commonly used signal processing techniques in speech coding are explained in this chapter, including pitch period estimation, all-pole/all-zero filters, and convolution. Some topics are very general while others are specific to speech processing.

Properties of speech signals constantly change with time. To process them effectively it is necessary to work on a frame-by-frame basis, where a frame consists of a certain number of samples. The actual duration of the frame is known as *length*. Typically, length is selected between 10 and 30 ms or 80 and 240 samples. Within this short interval, properties of the signal remain roughly constant. Thus, many signal processing techniques are adapted to this context when deployed to speech coding applications.

2.1 PITCH PERIOD ESTIMATION

One of the most important parameters in speech analysis, synthesis, and coding applications is the fundamental frequency, or pitch, of voiced speech. Pitch frequency is directly related to the speaker and sets the unique characteristic of a person. Voicing is generated when the airflow from the lungs is periodically interrupted by movements of the vocal cords. The time between successive vocal cord openings is called the fundamental period, or pitch period.

For men, the possible pitch frequency range is usually found somewhere between 50 and 250 Hz, while for women the range usually falls between 120 and 500 Hz. In terms of period, the range for a male is 4 to 20 ms, while for a female it is 2 to 8 ms.

Pitch period must be estimated at every frame. By comparing a frame with past samples, it is possible to identify the period in which the signal repeats itself, resulting in an estimate of the actual pitch period. Note that the estimation procedure makes sense only for voiced frames. Meaningless results are obtained for unvoiced frames due to their random nature.

Design of a pitch period estimation algorithm is a complex undertaking due to lack of perfect periodicity, interference with formants of the vocal tract, uncertainty of the starting instance of a voiced segment, and other real-world elements such as noise and echo. In practice, pitch period estimation is implemented as a trade-off between computational complexity and performance. Many techniques have been proposed for the estimation of pitch period and only a few are included here.

The Autocorrelation Method

Assume we want to perform the estimation on the signal $s[n]$, with n being the time index. We consider the frame that ends at time instant m, where the length of the frame is equal to N (i.e., from $n = m - N + 1$ to m). Then the autocorrelation value*

$$R[l, m] = \sum_{n=m-N+1}^{m} s[n]s[n-l] \qquad (2.1)$$

reflects the similarity between the frame $s[n]$, $n = m - N + 1$ to m, with respect to the time-shifted version $s[n - l]$, where l is a positive integer representing a time lag. The range of lag is selected so that it covers a wide range of pitch period values. For instance, for $l = 20$ to 147 (2.5 to 18.3 ms), the possible pitch frequency values range from 54.4 to 400 Hz at 8 kHz sampling rate. This range of l is applicable for most speakers and can be encoded using 7 bits, since there are $2^7 = 128$ values of pitch period.

By calculating the autocorrelation values for the entire range of lag, it is possible to find the value of lag associated with the highest autocorrelation representing the pitch period estimate, since, in theory, autocorrelation is maximized when the lag is equal to the pitch period. The method is summarized with the following pseudocode:

```
PITCH(m, N)
    peak ← 0
    for l ← 20 to 150
        autoc ← 0
        for n ← m − N + 1 to m
            autoc ← autoc + s[n]s[n − l]
```

*In Chapter 3, it is shown that the described quantity is actually an estimate of the true autocorrelation function $E\{s[n]s[n + l]\}$, without the scaling factor.

```
        if autoc > peak
            peak ← autoc
            lag ← l
return lag
```

It is important to mention that, in practice, the speech signal is often lowpass filtered before being used as input for pitch period estimation. Since the fundamental frequency associated with voicing is located in the low-frequency region (<500 Hz), lowpass filtering eliminates the interfering high-frequency components as well as out-of-band noise, leading to a more accurate estimate.

Example 2.1 The autocorrelation method is demonstrated here using the portion of voiced speech signal shown in Figure 2.1, which is clearly periodic. Computing the autocorrelation according to (2.1) for $l = 20$ to 150 gives the plot in Figure 2.2. As we can see, two strong peaks are obtained together with minor peaks. The lag corresponding to the highest peak is 71 and is the pitch period estimate for $m = 1500$ and $N = 180$. This estimate is close to the period of the signal in time domain.

Note that the next strong peak is located at a lag of 140, roughly doubling our pitch period estimate. This is expected since a periodic waveform with a period of T is also periodic with a period of $2T$, $3T$, . . . , and so on.

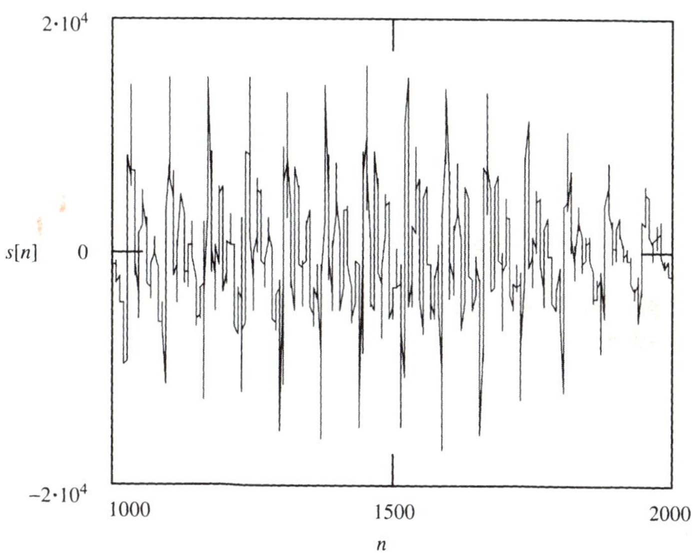

Figure 2.1 A voiced portion of a speech waveform used in pitch period estimation.

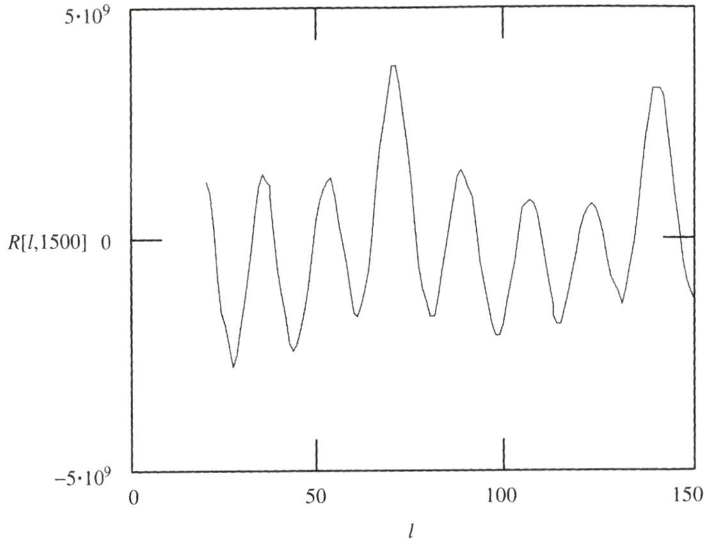

Figure 2.2 Autocorrelation values obtained from the waveform of Figure 2.1.

Magnitude Difference Function

One drawback of the autocorrelation method is the need for multiplication, which is relatively expensive for implementation, especially in those processors with limited functionality. To overcome this problem, the magnitude difference function is invented. This function is defined by

$$MDF[l, m] = \sum_{n=m-N+1}^{m} |s[n] - s[n-l]|. \tag{2.2}$$

For short segments of voiced speech it is reasonable to expect that $s[n] - s[n-l]$ is small for $l = 0, \pm T, \pm 2T, \ldots,$ with T being the signal's period. Thus, by computing the magnitude difference function for the lag range of interest, one can estimate the period by locating the lag value associated with the minimum magnitude difference. Note that no products are needed for the implementation of the present method. The following pseudocode summarizes the procedure:

```
PITCH_MD(m, N)
min ← ∞
for l ← 20 to 150
        mdf ← 0
        for n ← m − N + 1 to m
```

```
            mdf ← mdf + | s[n] − s[n − l] |
        if mdf < min
            min ← mdf
            lag ← l
return lag
```

Further computational saving is obtainable from the fact that the magnitude difference function is bounded. This fact is derived from (2.2) where $MDF[l, m] \geq 0$. From the same equation, each additional accumulation of term causes the result to be greater than or equal to the previous sum since each term is positive. Thus, it is not necessary to calculate the sum entirely; if the accumulated result at any instance during the iteration loop is greater than the minimum found so far, calculation stops and resumes with the next lag. The idea is implemented with the following:

```
PITCH_MD1(m, N)
 1. min ← ∞
 2. for l ← 20 to 150
 3.     mdf ← 0
 4.     for n ← m − N + 1 to m
 5.         mdf ← mdf + | s[n] − s[n − l] |
 6.         if mdf ≥ min break
 7.     if mdf < min
 8.         min ← mdf
 9.         lag ← l
10. return lag
```

In this new implementation, whenever the accumulated result (*mdf*, Line 5) exceeds the minimum found so far (*min*, Line 6), the loop is terminated and the algorithm moves on to verify the next lag l. On average, a substantial computational saving is achieved. Note that the approach is not applicable for autocorrelation computation, because it relies on the finding of a peak, which normally must evaluate the entire sum before any decision can be made. Low computational cost and lack of multiplication make the magnitude difference function attractive for practical applications.

Example 2.2 The same situation as in Example 2.1 is considered, where magnitude difference is computed for $l \in [20, 150]$. The plot is shown in Figure 2.3. Lowest *MDF* occurs at $l = 70$ with the next lowest *MDF* point located at $l = 139$. Compared with the results of Example 2.1, the present method yields a slightly lower estimate.

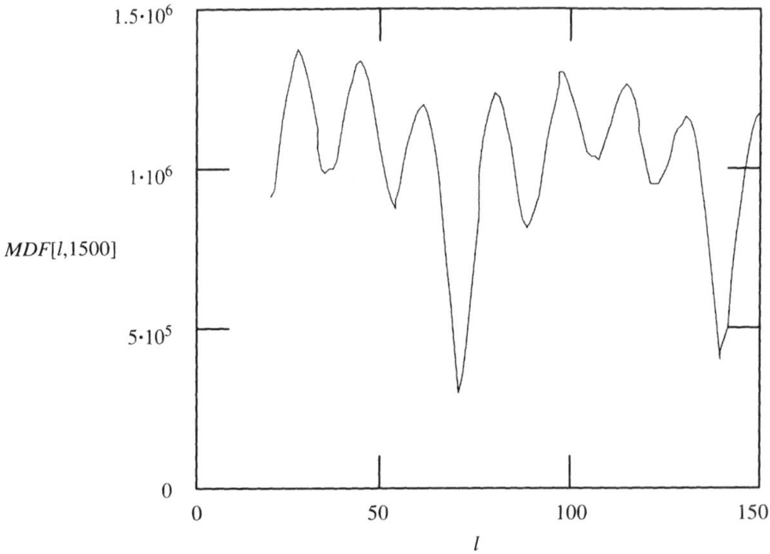

Figure 2.3 Magnitude difference values obtained from the waveform of Figure 2.1.

Fractional Pitch Period

The methods discussed earlier can only find integer-valued pitch periods. That is, the resultant period values are multiples of the sampling period $(8\,\text{kHz})^{-1} = 0.125$ ms. In many applications, higher resolution is necessary to achieve good performance. In fact, pitch period of the original continuous-time (before sampling) signal is a real number; thus, integer periods are only approximations introducing errors that might have negative impact on system performance.

Multirate signal processing techniques can be introduced to extend the resolution beyond the limits set by fixed sampling rate. Interpolation, for instance, is a widely used method, where the actual sampling rate is increased. Medan, Yair, and Chazan (1991) published an algorithm for pitch period determination, which is based on a simple linear interpolation technique. The method allows the finding of a real-valued pitch period and can be implemented efficiently in practice. This method is explained in detail as follows.

Optimal Integer-Valued Pitch Period

Consider a speech frame that ends at time instant $n = m$, with a length of N (Figure 2.4). The frame can be expressed by

$$s[n] = b \cdot s[n - N] + e[n], \quad m - N + 1 \leq n \leq m. \tag{2.3}$$

The above equation expresses $\{s[n], m - N + 1 \leq n \leq m\}$ as the sum between the product of a coefficient b with the frame $\{s[n - N], m - 2N + 1 \leq n \leq m - N\}$ and

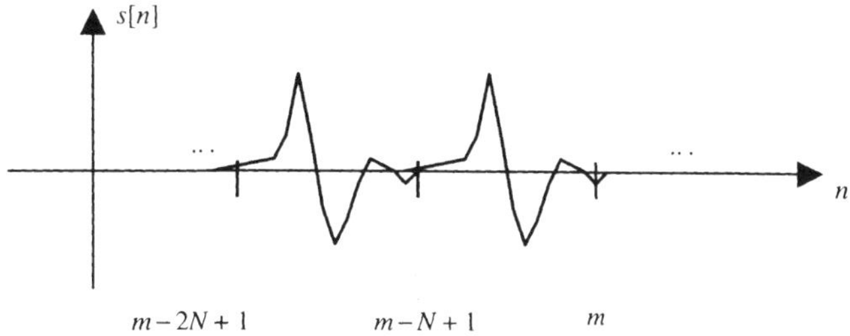

Figure 2.4 Signal frames in pitch period estimation.

the error signal* $e[n]$. Note from Figure 2.4 that two consecutive frames of length N are involved. The optimal pitch period at time instant $n = m$ can be defined as the particular value of N, denoted by N_o, that minimizes the normalized sum of squared error

$$J[m, N] = \frac{\sum_{n=m-N+1}^{m} (s[n] - bs[n-N])^2}{\sum_{n=m-N+1}^{m} s^2[n]}. \tag{2.4}$$

The normalization term (denominator) is required to compensate for the variable size of the speech segments involved and the uneven energy distribution over the pitch interval. Denoting N_o as our optimal pitch period, we have

$$N_o = \{N | J[m, N] \leq J[m, M], N_{\min} \leq N, M < N_{\max}\}, \tag{2.5}$$

where $N_{\min}$ and $N_{\max}$ are the minimum and maximum limits for the pitch period, respectively. These two are parameters of the algorithm that can be set according to the application. For instance, $N_{\min} = 20$ and $N_{\max} = 147$.

The optimal value of b can be found by differentiating J with respect to b and setting the result to zero. This gives

$$b = \frac{\sum_{n=m-N+1}^{m} s[n]s[n-N]}{\sum_{n=m-N+1}^{m} s^2[n-N]}. \tag{2.6}$$

* This indeed is the concept of linear prediction, where the signal of the current frame is predicted from the past, with the prediction calculated as the product of the past with a coefficient. Chapter 4 contains further details on the topic.

Substituting (2.6) in (2.4) and manipulating yields

$$J[m,N] = 1 - \frac{\left\{ \sum\limits_{n=m-N+1}^{m} s[n]s[n-N] \right\}^2}{\sum\limits_{n=m-N+1}^{m} s^2[n-N] \sum\limits_{n=m-N+1}^{m} s^2[n]}. \tag{2.7}$$

Optimal Fractional Pitch Period

Consider the continuous-valued pitch period T_o defined by

$$T_o = (N_o + \eta_o)T_s \tag{2.8}$$

where T_s is the sampling period $(8\,\text{kHz})^{-1} = 0.125$ ms and η_o is the fractional pitch period.

Here we assume $0 \leq \eta_o < 1$ so that

$$\frac{T_o}{T_s} - 1 < N_o \leq \frac{T_o}{T_s}, \tag{2.9}$$

or

$$N_o = \left\lfloor \frac{T_o}{T_s} \right\rfloor, \tag{2.10}$$

with $\lfloor \cdot \rfloor$ the floor function. Assume that the integer pitch period N_o is known. The problem of fractional pitch period estimation consists of the determination of $\eta = \eta_o$ so that

$$J[m, N_o + \eta] = 1 - \frac{\left\{ \sum\limits_{n=m-N_o+1}^{m} s[n]s[n-N_o-\eta] \right\}^2}{\sum\limits_{n=m-N_o+1}^{m} s^2[n-N_o-\eta] \sum\limits_{n=m-N_o+1}^{m} s^2[n]} \tag{2.11}$$

is minimized, with $0 \leq \eta < 1$. In (2.11) the discrete-time signal is being delayed by a real-valued amount and can be obtained with interpolation [Oppenheim and Schafer, 1989]. In this case, a simple linear combination is used to interpolate, which is given by

$$s[n+\eta] \equiv (1-\eta)s[n] + \eta s[n+1]. \tag{2.12}$$

Applying (2.12) to $s[n - N_o - \eta]$,

$$s[n - N_o - \eta] \equiv (1-\eta)s[n-N_o] + \eta s[n-N_o-1]. \tag{2.13}$$

Substituting (2.13) in (2.11) and expanding leads to

$$J[m, N_o + \eta] = 1 - \frac{\{(1-\eta)\alpha_1[m, N_o] + \eta\alpha_2[m, N_o]\}^2}{\alpha_3[m, N_o]\left\{(1-\eta)^2\alpha_4[m, N_o] + 2\eta(1-\eta)\alpha_6[m, N_o] + \eta^2\alpha_5[m, N_o]\right\}}, \tag{2.14}$$

where

$$\alpha_1[m, N_o] = \sum_{n=m-N_o+1}^{m} s[n]s[n - N_o], \tag{2.15}$$

$$\alpha_2[m, N_o] = \sum_{n=m-N_o+1}^{m} s[n]s[n - N_o - 1], \tag{2.16}$$

$$\alpha_3[m, N_o] = \sum_{n=m-N_o+1}^{m} s^2[n], \tag{2.17}$$

$$\alpha_4[m, N_o] = \sum_{n=m-N_o+1}^{m} s^2[n - N_o], \tag{2.18}$$

$$\alpha_5[m, N_o] = \sum_{n=m-N_o+1}^{m} s^2[n - N_o - 1], \tag{2.19}$$

$$\alpha_6[m, N_o] = \sum_{n=m-N_o+1}^{m} s[n - N_o]s[n - N_o - 1]. \tag{2.20}$$

It is left as an exercise to verify the validity of equations (2.14) to (2.20). The optimal fractional pitch period η_o is the one that minimizes (2.14). Differentiating (2.14) with respect to η and equating to zero, the optimal fractional pitch is found to be

$$\eta_o[m, N_o] = \frac{\alpha_2[m, N_o]\alpha_4[m, N_o] - \alpha_1[m, N_o]\alpha_6[m, N_o]}{\alpha_2[m, N_o](\alpha_4[m, N_o] - \alpha_6[m, N_o]) + \alpha_1[m, N_o](\alpha_5[m, N_o] - \alpha_6[m, N_o])}. \tag{2.21}$$

In some cases, η_o may fall outside the interval [0,1]. This happens when the integer pitch period N_o deviates by one sampling period from the value defined in (2.9). In such cases, the integer period is incremented by one when $\eta_o \geq 1$, and decremented by one when $\eta_o < 0$. Then η_o is recalculated from (2.21).

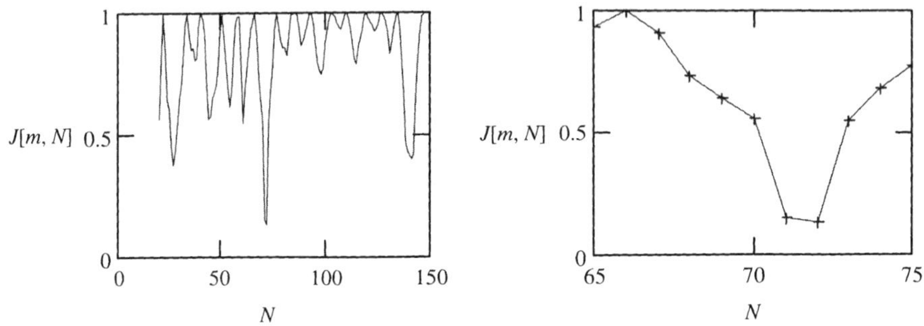

Figure 2.5 *Left*: Normalized sum of squared error obtained from the waveform of Figure 2.1. *Right*: Expanded view showing the minimum point at $N = 72$.

Summary of Algorithm

The Medan–Yair–Chazan method is summarized as follow:

Step 1. For $N \leftarrow N_{\min}$ to $N_{\max}$, compute the normalized sum of squared error from (2.7).

Step 2. Find the minimum value of $J[m, N]$; the corresponding value of N is the optimal integer pitch period N_o.

Step 3. Compute η_o from (2.21).

Step 4. If $\eta_o < 0$, $N_o \leftarrow N_o - 1$, go back to Step 3.

Step 5. If $\eta_o \geq 1$, $N_o \leftarrow N_o + 1$, go back to Step 3.

Minimization of $J[m, N]$ in Step 2 is equivalent to the maximization of the normalized autocorrelation

$$r[m, N] = \frac{\sum_{n=m-N+1}^{m} s[n]s[n-N]}{\sqrt{\sum_{n=m-N+1}^{m} s^2[n-N] \sum_{n=m-N+1}^{m} s^2[n]}} \tag{2.22}$$

since $J[m, N] \geq 0$. Thus, Step 2 can be replaced by the maximization of (2.22).

The normalized autocorrelation

$$r[m, N_o + \eta] = \frac{(1-\eta)\alpha_1[m, N_o] + \eta\alpha_2[m, N_o]}{\sqrt{\alpha_3[m, N_o]\left\{(1-\eta)^2\alpha_4[m, N_o] + 2\eta(1-\eta)\alpha_6[m, N_o] + \eta^2\alpha_5[m, N_o]\right\}}} \tag{2.23}$$

is often used to measure the amount of periodicity of the signal at the specified lag and is derived directly from (2.14).

Example 2.3 The same speech data as in Example 2.1 are utilized to illustrate fractional pitch period estimation. Figure 2.5 shows the normalized sum of squared error, where the global minimum is found at $N_o = 72$. From (2.21), the fractional pitch period is found to be equal to -0.224; since it is negative, N_o is decreased by one. Recalculating yields a fractional pitch period of 0.517. Thus, the final pitch period estimate is equal to 71.517. Note that this final result is consistent with the error plot in Figure 2.5, where the global minimum is more likely to be located between 71 and 72, than between 72 and 73.

Checking for Multiples of a Pitch Period

Consider an autocorrelation-based estimation procedure where a peak-picking strategy is applied. In this approach, autocorrelation of the signal is computed for a range of lag; the particular lag providing the highest autocorrelation is selected as the estimated pitch period.

This peak-picking approach might lead to erroneous outcome, whereas the result actually corresponds to multiples of the fundamental pitch period; this is mainly due to the fact that a periodic signal with period T is also periodic with periods $2T$, $3T$, ..., and so on, since the signal repeats itself for those time intervals. Thus, in the ideal case, the autocorrelation plot develops peaks at regular intervals separated by the period T. Figure 2.6 shows an example of an ideal autocorrelation plot, together with the plot of a real-world quasiperiodic signal, such as a voiced speech frame. In many practical situations, the fundamental period (shortest period

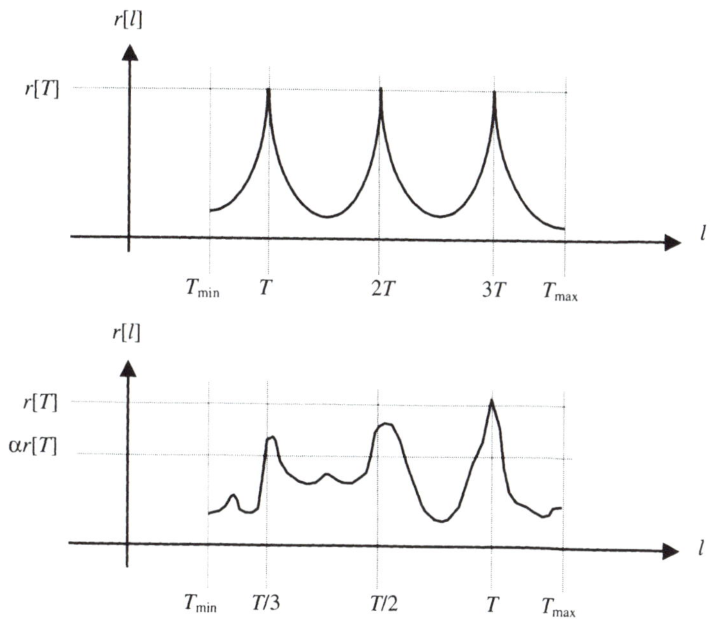

Figure 2.6 *Top*: Autocorrelation plot of an ideal periodic signal with period T. *Bottom*: Typical autocorrelation plot of a real-world quasiperiodic signal.

associated with strong autocorrelation) is occluded from the autocorrelation plot due to various conditions, such as:

- Period of the signal is not constant; that is, its value changes with time, such as most speech frames.
- Limited time resolution of a discrete-time system.
- Noise and distortion applied to the signal.

Following a peak-picking strategy without further analyzing the data can lead to disasters in speech coding applications, because the quality of the synthetic speech relies heavily on an accurate estimation. An abrupt change in the value of the pitch period for consecutive frames, for instance, introduces highly annoying artifacts. Thus, given a certain estimated period, it is desirable to verify whether the period itself is actually multiples of some fundamental period.

A simple procedure is presented here that allows multiplicity check. The main idea is to verify the autocorrelation values at lags of T/i, $i = 2, 3, 4, \ldots$, where T is the estimated pitch period. If $r[T/i] > \alpha r[T]$, where $r[\cdot]$ is the autocorrelation

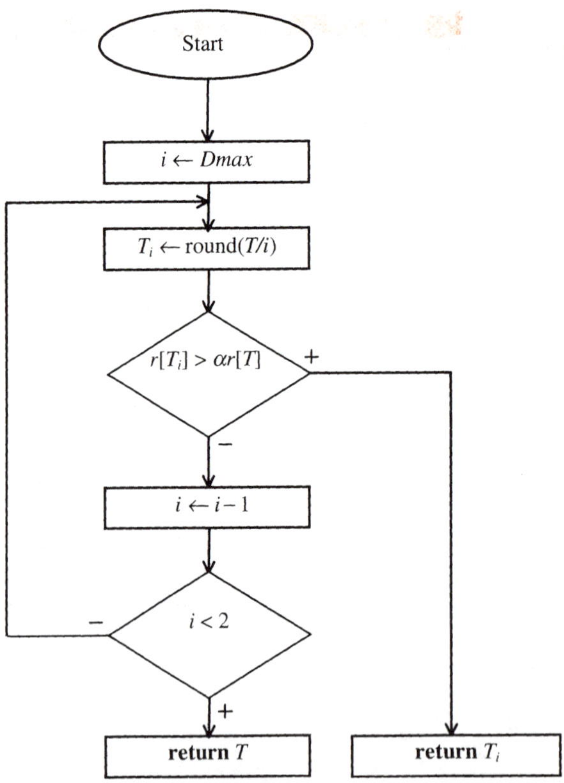

Figure 2.7 Flowchart of an algorithm for multiplicity check of pitch periods.

function and $\alpha < 1$ is a positive scaling constant, then the estimated pitch period becomes T/i. The purpose of α is to lower the peak autocorrelation value $r[T]$ so as to form a decision threshold; this is necessary since $r[T]$ is the peak within the search range. A value of α in the interval [0.5, 1] is a reasonable choice in practice. Figure 2.7 shows the flowchart of the proposed algorithm, where the inputs are the candidate pitch period T and the autocorrelation function $r[l]$. The algorithm starts by dividing the input period by a range of denominators, denoted by i; with i beginning at *Dmax*, which is a constant integer determining the minimum possible pitch period estimate. A value of *Dmax* within the interval of [5, 10] is appropriate for most practical purposes. Intermediate check points are found by dividing T by i and rounding the results. If the autocorrelation value at the check point is greater than that at T multiplied by the scaling factor α, $T_i = \text{round}(T/i)$ is returned as the fundamental period sought, where the round$(\cdot)$ operator rounds a number to the nearest integer. Otherwise, the denominator i is reduced by one and the operation repeated until $i < 2$. Even though the autocorrelation function is indicated here, other approaches such as the magnitude difference function can be included with little modification.

Note that once the algorithm finds a suitable period satisfying the threshold constraint, it will end and return the result; thus, it starts searching from the shortest lag, or highest denominator, since the purpose is to locate the fundamental pitch period, corresponding to the lowest value.

2.2 ALL-POLE AND ALL-ZERO FILTERS

The filters with system function

$$H(z) = \frac{1}{A(z)} = \frac{1}{1 + \sum_{i=1}^{M} a_i z^{-i}} \tag{2.24}$$

or

$$A(z) = 1 + \sum_{i=1}^{M} a_i z^{-i} \tag{2.25}$$

are of particular importance to speech coding. In (2.24) an all-pole filter is described, since only poles are present, while the all-zero filter has the system function given in (2.25). As we can see, $H(z)$ and $A(z)$ are the inverse of each other. The constant M is the order of the filter and the a_i are the filter's coefficients. These filters appear in all linear-prediction-based speech coders. As explained in Chapter 4, M is also known as the prediction order, while the a_i are referred to as the linear prediction coefficients.

Direct Form Realization

With $x[n]$ being the input to the filter and $y[n]$ the output, the time-domain difference equation corresponding to (2.24) is

$$y[n] = x[n] - \sum_{i=1}^{M} a_i y[n-i] \tag{2.26}$$

and for (2.25)

$$y[n] = x[n] + \sum_{i=1}^{M} a_i x[n-i]. \tag{2.27}$$

Figure 2.8 shows the signal flow graphs of the above difference equations. Filters implemented in this manner are called direct form. Note that the impulse response of an all-pole filter has an infinite number of samples with nontrivial values due to the fact that the scaled and delayed version of the output samples are added back to the input samples. This is referred to as an infinite-impulse-response (IIR) filter. For the all-zero filter, however, the impulse response only has $M+1$ nontrivial samples (the rest are zeros) and is known as a finite-impulse-response (FIR) filter.

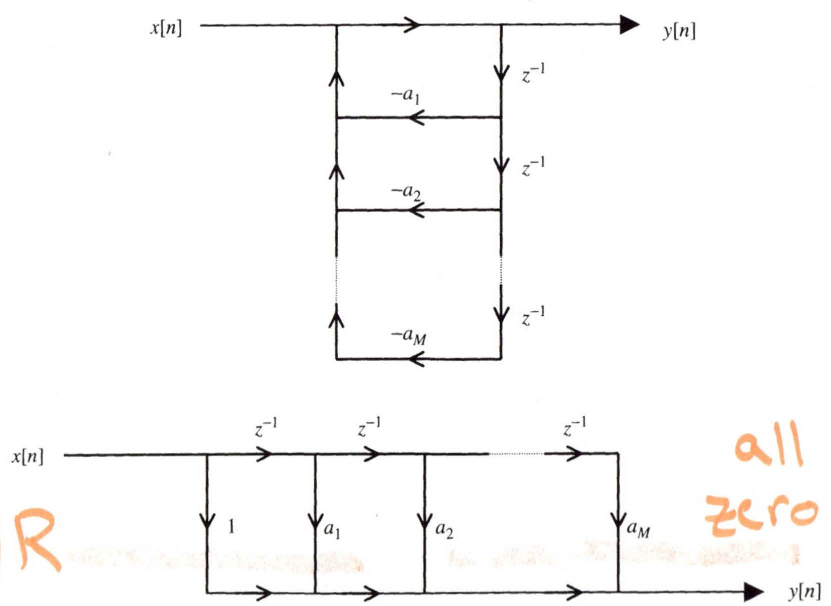

Figure 2.8 Signal flow graph for direct form implementation of an all-pole filter (*top*) and all-zero filter (*bottom*).

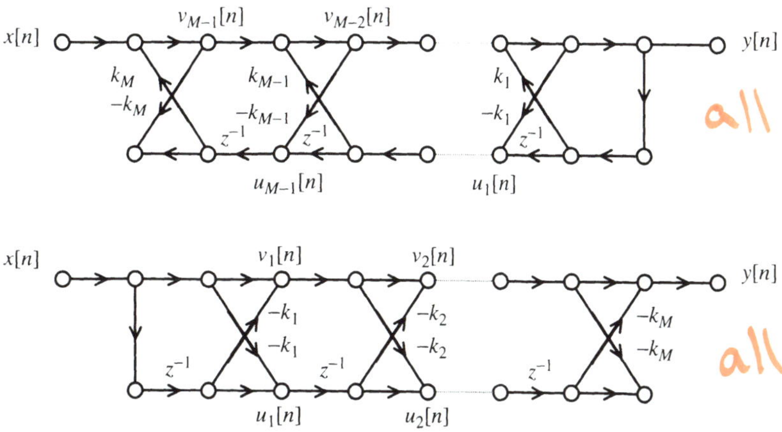

Figure 2.9 Signal flow graph for lattice implementation of an all-pole filter (*top*) and all-zero filter (*bottom*).

Lattice Realization

Figure 2.9 shows an alternative realization for the filters, called the lattice structure. The parameters $k_1, \ldots, k_M$ are known as the reflection coefficients. The reflection coefficients can be found from the direct form coefficients $(a_1, \ldots, a_M)$ through the computational loop specified below.

For $l = M, M-1, \ldots, 1$:

$$k_l = -a_l^{(l)}, \tag{2.28}$$

$$a_i^{(l-1)} = \frac{a_i^{(l)} + k_l a_{l-i}^{(l)}}{1 - k_l^2} \quad i = 1, 2, \ldots, l-1, \tag{2.29}$$

where $a_i = a_i^{(M)}$. The above relations are obtained directly by deriving the input–output difference equation of the lattice form and comparing to that of the direct form. Chapter 4 presents a derivation of (2.28) and (2.29). For the all-pole filter, the set of equations

$$\begin{aligned}
v_{M-1}[n] &= x[n] + k_M u_{M-1}[n-1], \\
v_{M-2}[n] &= v_{M-1}[n] + k_{M-1} u_{M-2}[n-1], \\
&\vdots \\
v_1[n] &= v_2[n] + k_2 u_1[n-1], \\
y[n] &= v_1[n] + k_1 y[n-1], \\
u_1[n] &= -k_1 y[n] + y[n-1], \\
u_2[n] &= -k_2 v_1[n] + u_1[n-1], \\
&\vdots \\
u_{M-1}[n] &= -k_{M-1} v_{M-2}[n] + u_{M-2}[n-1]
\end{aligned} \tag{2.30}$$

are solved successively to find out the output sequence $y[n]$. While for the all-zero filter,

$$\begin{aligned} v_1[n] &= x[n] - k_1 x[n-1], \\ u_1[n] &= -k_1 x[n] + x[n-1], \\ v_2[n] &= v_1[n] - k_2 u_1[n-1], \\ u_2[n] &= -k_2 v_1[n] + u_1[n-1], \\ &\vdots \\ y[n] &= v_{M-1}[n] - k_M u_{M-1}[n-1]. \end{aligned} \tag{2.31}$$

Example 2.4 Given the filter's coefficients, $a_1 = -0.9$, $a_2 = 0.64$, and $a_3 = -0.576$, the difference equations for direct form implementation are

$$y[n] = x[n] + 0.9y[n-1] - 0.64y[n-2] + 0.576y[n-3]$$

for the all-pole filter, and

$$y[n] = x[n] - 0.9x[n-1] + 0.64x[n-2] - 0.576x[n-3]$$

for the all-zero filter. Reflection coefficients are found to be:

$$\begin{aligned} k_3 &= -a_3 = 0.576, \\ a_1^{(2)} &= \frac{a_1^{(3)} + k_3 a_2^{(3)}}{1 - k_3^2} = \frac{a_1 + k_3 a_2}{1 - k_3^2} = -0.79518, \\ a_2^{(2)} &= \frac{a_2^{(3)} + k_3 a_1^{(3)}}{1 - k_3^2} = \frac{a_2 + k_3 a_1}{1 - k_3^2} = 0.181975, \\ k_2 &= -a_2^{(2)} = -0.181975, \\ a_1^{(1)} &= \frac{a_1^{(2)} + k_2 a_1^{(2)}}{1 - k_2^2} = -0.67276, \\ k_1 &= 0.67276. \end{aligned}$$

The difference equations for lattice form implementation are

$$\begin{aligned} v_2[n] &= x[n] + 0.576u_2[n-1], \\ v_1[n] &= v_2[n] - 0.182u_1[n-1], \\ y[n] &= v_1[n] + 0.6728y[n-1], \\ u_1[n] &= -0.6728y[n] + y[n-1], \\ u_2[n] &= 0.182v_1[n] + u_1[n-1], \end{aligned}$$

for the all-pole filter, and

$$\begin{aligned} v_1[n] &= x[n] - 0.6728x[n-1], \\ u_1[n] &= -0.6728x[n] + x[n-1], \\ v_2[n] &= v_1[n] + 0.182u_1[n-1], \\ u_2[n] &= 0.182v_1[n] + u_1[n-1], \\ y[n] &= v_2[n] - 0.576u_2[n-1], \end{aligned}$$

for the all-zero filter.

Comparison Between the Two Realizations

Direct form realization is often the preferred approach in practice due to its simplicity and lower computational requirement (Exercise 2.8). The lattice structure, however, does provide some advantage.

To appreciate the benefit offered by the lattice form realization, some background from Chapter 4 is needed. Readers are free to skip this paragraph and reread it later after familiarizing themselves with the material in Chapter 4. During linear prediction analysis, the method used to solve the normal equation could be the Levinson–Durbin algorithm—where the linear prediction coefficients (LPC or direct form coefficients) and reflection coefficients are both returned upon completion. Likewise, it could as well be the Leroux–Gueguen algorithm—where only the reflection coefficients are obtained. The lattice structure allows processing to be performed directly using the reflection coefficients, without converting them to LPCs; this is desirable for systems with limited numerical precision since precision loss during conversion may lead to filter instability. Also note that using the reflection coefficients allows a straightforward supervision of stability status, since the condition $|k_i| \leq 1$ can easily be monitored. This is less of a concern for systems with sufficient numerical precision, and direct form is customarily implemented due to diminished computational burden.

Calculation of Output Sequence on a Frame-by-Frame Basis

For practical implementation of speech coding, the signal is processed on a frame-by-frame basis. In filtering, for instance, the input signal is partitioned into frames having N samples according to

$$x_r[n] = \begin{cases} x[n + rN], & 0 \leq n \leq N-1, \\ 0, & \text{otherwise}, \end{cases} \tag{2.32}$$

where $x[n]$ is the input signal, defined for $-\infty < n < \infty$. Each frame is indexed by the variable r, with $r = 0, 1, 2, \ldots$, and the rth frame is denoted by $x_r[n]$. Figure 2.10 illustrates the notation; note that n actually "wrap-around" from frame to frame.

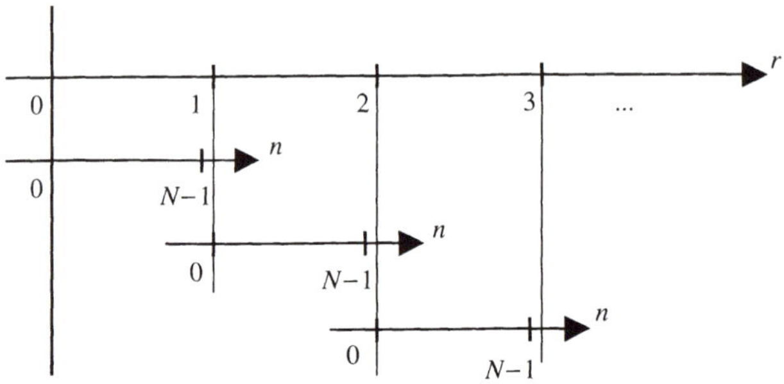

Figure 2.10 Illustration of time notation.

Thus, each frame consists of N samples, with $n = 0$ to $N - 1$ addressing samples inside the frame.

Two methods are presented next, allowing the computation of the filter output on a frame-by-frame basis. We will only consider direct form realization of all-pole filters; however, the techniques can be applied in a straightforward manner to other configurations and filters.

State-Save Method

This method saves the state of the current frame for use by the next frame. The state of the filter in this case refers to the values stored in the delay elements (the z^{-1} in the signal flow graphs, Figure 2.8). From (2.26), the procedure is executed frame-by-frame with

$$y_r[n] = y_{r-1}[n+N], \qquad -M \le n \le -1, \tag{2.33}$$

$$y_r[n] = x_r[n] - \sum_{i=1}^{M} a_i y_r[n-i], \quad 0 \le n \le N-1; \tag{2.34}$$

that is, M output values are saved and used to compute the next frame.

Zero-Input Zero-State Method

This method also saves the current state for future use; however, it separates the filter's output into two different responses and computes them separately. The *zero-input response* is the output of the filter due exclusively to the state or history of the filter with no input, or in other terms, zero input. The *zero-state response* is the output of the filter due to the input frame, with the assumption that the filter has zero initial state. The two responses are added together to form the overall response. This approach is possible because of the linearity of the filter and can easily be

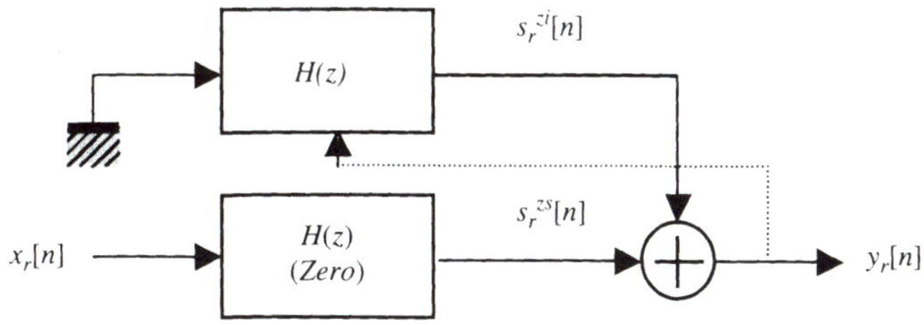

Figure 2.11 Illustration of the zero-input zero-state method.

shown to produce the same final result as the state-save method. Figure 2.11 illustrates the technique where two filters are involved. The first filter contains the initial state carried over from a prior frame and is indicated in the block diagram by the arrow entering the bottom of the filter, meaning that the overall response (from the prior frame) is used to initialize the state of the filter at $n = 0$. In addition, note that the filter input is "grounded," a symbol borrowed from electronic circuit diagrams to imply the fact that input to the filter is zero. The second filter is marked with "zero" and has an initial state of zero; that is, the values stored in the delay elements z^{-1} are all zeros at $n = 0$. From (2.26) the method is executed frame-by-frame with

- Zero-input response (s^{zi}):

$$s_r^{zi}[n] = y_{r-1}[n + N], \qquad -M \le n \le -1, \tag{2.35}$$

$$s_r^{zi}[n] = -\sum_{i=1}^{M} a_i s_r^{zi}[n - i], \quad 0 \le n \le N - 1. \tag{2.36}$$

- Zero-state response (s^{zs}):

$$s_r^{zs}[n] = 0, \qquad -M \le n \le -1, \tag{2.37}$$

$$s_r^{zs}[n] = x_r[n] - \sum_{i=1}^{M} a_i s_r^{zs}[n - i], \quad 0 \le n \le N - 1. \tag{2.38}$$

- Overall response (y):

$$y_r[n] = s_r^{zi}[n] + s_r^{zs}[n], \quad 0 \le n \le N - 1. \tag{2.39}$$

So why would we bother with the extra complications associated with this method (Excerise 2.9)? The answer becomes clear when we study the CELP coder in Chapter 11.

2.3 CONVOLUTION

Given the linear time-invariant (LTI) system with impulse response $h[n]$, and denoting the system's input as $x[n]$ and the output as $y[n]$, we have

$$y[n] = \sum_{k=-\infty}^{\infty} x[k]h[n-k] = \sum_{k=-\infty}^{\infty} x[n-k]h[k]. \tag{2.40}$$

The above equation is known as the *convolution sum* between $x[n]$ and $h[n]$ and is one of the fundamental relations in signal processing. In this section, we will explore the usage of the convolution sum for the calculation of the output sequence on a frame-by-frame basis, with emphasis on the all-pole filter.

Impulse Response of an All-Pole Filter

A straightforward way to find the impulse response sequence is by using the time-domain difference equation (2.26), when the input is a single impulse: $x[n] = \delta[n]$. That is,

$$h[n] = \delta[n] - \sum_{i=1}^{M} a_i h[n-i]. \tag{2.41}$$

Proceeding on a sample-by-sample basis, we have

$$\begin{aligned}
h[n] &= 0, \quad n < 0,\\
h[0] &= 1,\\
h[1] &= -a_1 h[0] = -a_1,\\
h[2] &= -a_1 h[1] - a_2,\\
&\vdots\\
h[M-1] &= -a_1 h[M-2] - \cdots - a_{M-2}h[1] - a_{M-1},\\
h[M] &= -\sum_{i=1}^{M} a_i h[M-i],\\
&\vdots\\
h[N-1] &= -\sum_{i=1}^{M} a_i h[N-1-i].
\end{aligned} \tag{2.42}$$

Thus, the impulse response sequence is determined by the filter coefficients.

All-Pole Filter: Calculation of Output Sequence on a Frame-by-Frame Basis Using the Zero-Input Zero-State Method

Due to the IIR nature of the all-pole filter, it is in general not possible to use a segment of the impulse response to compute the filter's output on a frame-by-frame

TABLE 2.1 Amount of Computation Spent by Each Individual Output Sample in the Convolution Sum

Sample	Sums	Number of Products
$y[0]$	0	1
$y[1]$	1	2
$\vdots$	$\vdots$	$\vdots$
$y[N-1]$	$N-1$	N

basis. However, it is possible to find the zero-state response using a segment of the impulse response to compute the convolution sum. Here, the zero-input zero-state method described in last section is modified to accommodate the convolution.

Given the impulse response sequence $h[n]$, $n = 0, \ldots, N-1$, the zero-state response (2.38) is found with

$$s_r^{zs}[n] = \sum_{k=0}^{n} x_r[k]h[n-k], \quad 0 \le n \le N-1, \tag{2.43}$$

which is the convolution sum adapted from (2.40). Equation (2.43) provides an alternative to compute the zero-state response of the frame. The total number of sums and products needed to find the N samples of the output sequence are summarized in Table 2.1. By adding the numbers in each column, we have

$$\text{Number of sums} = (N-1)N/2, \tag{2.44}$$

$$\text{Number of products} = N(N+1)/2; \tag{2.45}$$

representing the total computational costs involved with this approach. Table 2.1 is obtained by counting the number of sums and products in the following equations, obtained from (2.43):

$$\begin{aligned} s_r^{zs}[0] &= x_r[0]h[0], \\ s_r^{zs}[1] &= x_r[0]h[1] + x_r[1]h[0], \\ &\vdots \\ s_r^{zs}[N-1] &= x_r[0]h[N-1] + \cdots + x_r[N-1]h[0]. \end{aligned} \tag{2.46}$$

The above equations can be written in matrix form:

$$\mathbf{s}_r^{zs} = \mathbf{H} \cdot \mathbf{x}_r, \tag{2.47}$$

where

$$\mathbf{s}_r^{zs} = \begin{bmatrix} s_r^{zs}[0] & s_r^{zs}[1] & \cdots & s_r^{zs}[N-1] \end{bmatrix}^T \tag{2.48}$$

is the $N \times 1$ zero-state response vector,

$$\mathbf{x}_r = [x_r[0] \quad x_r[1] \quad \cdots \quad x_r[N-1]]^T \tag{2.49}$$

is the $N \times 1$ input vector, and

$$\mathbf{H} = \begin{bmatrix} h[0] & 0 & \cdots & 0 \\ h[1] & h[0] & \cdots & 0 \\ \vdots & \vdots & \ddots & \vdots \\ h[N-1] & h[N-2] & \cdots & h[0] \end{bmatrix} \tag{2.50}$$

is the $N \times N$ impulse response matrix.

Comparing the computational cost associated with the convolution sum ((2.44) and (2.45)) to that involved with a direct application of the time-domain difference equation ((2.38), Exercise 2.9)), one can see that the number of operations of the latter is less than the former. Thus, the practicality of the convolution sum approach is in doubt. Use of the convolution sum in the calculation of the zero-state response can reduce the computational cost significantly for the case of CELP coders, where the zero-state response must be computed in a repetitive manner (Chapter 11).

Recursive Convolution

Given the sequence $x[n]$, $n = 0, \ldots, S(L-1) + N - 1$, with S, L, and N positive integers, we define the following L sequences:

$$\begin{aligned} x^{(0)}[n] &= x[n + (L-1)S], \\ x^{(1)}[n] &= x[n + (L-2)S], \\ &\vdots \\ x^{(L-1)}[n] &= x[n]. \end{aligned} \tag{2.51}$$

For $n = 0$ to $N - 1$, in general, we write

$$x^{(l)}[n] = x[n + (L - l - 1)S]; \quad l = 0, 1, \ldots, L-1; \quad n = 0, 1, \ldots, n-1. \tag{2.52}$$

Figure 2.12 illustrates the relationship between $x[n]$ and $x^{(l)}[n]$. Note that

$$x^{(l+1)}[n] = x^{(l)}[n - S]; \quad S \le n \le N - 1. \tag{2.53}$$

That is, $x^{(l)}[n]$ is obtained by extracting N samples from the sequence $x[n]$ from different positions. Given a system with impulse response $h[n]$, it is desired to find

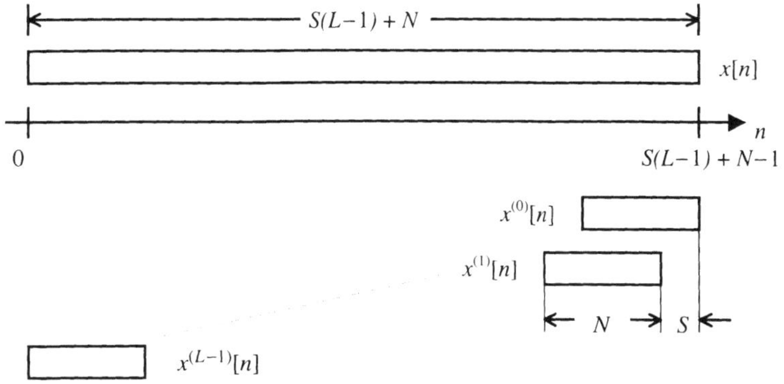

Figure 2.12 Obtaining the input sequences in recursive convolution.

the zero-state responses corresponding to the input sequences $x^{(l)}[n]$. Let's denote these responses as $y^{(l)}[n]$, in matrix form:

$$\mathbf{y}^{(l)} = \mathbf{H} \cdot \mathbf{x}^{(l)}. \tag{2.54}$$

The above equation suggests the computation of the system responses by applying an independent convolution operation to each input sequence. However, the total computation can be reduced, since the input sequences share common samples. To explore the situation, consider

$$y^{(l)}[n-S] = \sum_{k=0}^{n-S} x^{(l)}[n-S-k]h[k]; \tag{2.55}$$

then

$$\begin{aligned} &y^{(l+1)}[n] - y^{(l)}[n-S] \\ &\quad = \sum_{k=0}^{n-S} \left(x^{(l+1)}[n-k] - x^{(l)}[n-S-k]\right)h[k] + \sum_{k=n-S+1}^{n} x^{(l+1)}[n-k]h[k]. \end{aligned} \tag{2.56}$$

Using (2.53), we come to the conclusion that

$$y^{(l+1)}[n] = \begin{cases} \sum_{k=0}^{n} x^{(l+1)}[n-k]h[k], & 0 \le n \le S-1, \\ y^{(l)}[n-S] + \sum_{k=n-S+1}^{n} x^{(l+1)}[n-k]h[k], & S \le n \le N-1. \end{cases} \tag{2.57}$$

Therefore, if the lth response is available, only the first S samples of the $(l+1)$st response need to be computed via the usual convolution. The last $(N-S)$ samples can be found using a simpler, less complex operation. The equation is known as *recursive convolution*, since the convolution of one sequence can be found from the previous response in a recursive fashion.

Recursive convolution is widely used by CELP-type speech coders (Chapter 11), where the zero-state responses must be found from a codebook having overlapping codevectors. The computational procedure shown in (2.57) improves the efficiency dramatically, enabling the practical implementation of these types of coders. See Exercise 2.12 for computational cost and comparison with regular convolution.

The Case of Single-Shift

When $S = 1$, a rather simple expression arises. From (2.57)

$$y^{(l+1)}[n] = \begin{cases} x^{(l+1)}[0]h[0], & n = 0 \\ y^{(l)}[n-1] + x^{(l+1)}[0]h[n], & 1 \le n \le N-1. \end{cases} \tag{2.58}$$

With the definition that $y^{(l)}[-1] = 0$, we have

$$y^{(l+1)}[n] = y^{(l)}[n-1] + x^{(l+1)}[0]h[n] \tag{2.59}$$

for $0 \le n \le N-1$. From (2.52) we can derive an alternative expression

$$y^{(l+1)}[n] = y^{(l)}[n-1] + x[L-l-2]h[n] \tag{2.60}$$

or

$$y^{(l)}[n] = y^{(l-1)}[n-1] + x[L-l-1]h[n]. \tag{2.61}$$

In (2.59), (2.60), and (2.61) it is assumed that each $y^{(l)}[n]$, $l = 0$ to $L-1$, occupies separate arrays. That is, different memory space is needed for each sequence. In-place computation can be performed using the following pseudocode, with the assumption that $y^{(0)}[n]$ is available and is initially stored in the $y[n]$ array.

```
for l ← 1 to L − 1
        for n ← N − 1 downto 1
                y[n] ← y[n − 1] + x[L − l − 1]h[n]
        y[0] ← x[L − l − 1]h[0]
        // New sequence is available: do something.
```

Note how the sample of the new sequence replaces the old one by going backward in n. This method is applied if there is no need to store all L sequences in memory, leading to substantial memory cost reduction.

2.4 SUMMARY AND REFERENCES

This chapter presented several fundamental techniques that are widely used in signal processing. In the next chapters, we will see their application in actual implementations of speech coders. For a general reference on digital signal processing, see Oppenheim and Schafer [1989]. Additional algorithms for convolution can be found in Burrus and Parks [1985].

Pitch period estimation was an intense research topic back in the 1960s and 1970s. See Sondhi [1968] for evaluation of three pitch estimation methods; in Rabiner et al. [1976], a comparative study of seven estimation algorithms is described. Since the search procedure associated with pitch period estimation is quite computationally demanding, many implementations put priority on complexity reduction; alternative techniques are given in subsequent chapters.

EXERCISES

2.1 An effective and simple technique to improve the accuracy of the autocorrelation method in pitch period estimation is *center clipping*. In this method, a clipping function is applied to the speech signal prior to autocorrelation calculation. One such function is

$$f(x) = \begin{cases} x + c, & x < -c \\ 0, & -c \le x \le c \\ x - c, & x > c \end{cases}$$

where c is a positive constant known as the clipping limit. Typically, the clipping limits are set to $\pm 30\%$ of the absolute maximum of the waveform. One problem associated with the autocorrelation method is that the first formant frequency, which is often near or even below the fundamental pitch frequency, can interfere with its detection. If the first formant is particularly strong, a competing periodicity will be present in the autocorrelation values.

Clipping reduces the interference due to formant frequencies since the magnitude spectrum is flattened. By clipping the signal, low-amplitude samples are eliminated, leaving only the high-amplitude peaks of the waveform where most information related to pitch harmonics is located. The resultant magnitude spectrum is less affected by the formant frequencies of the vocal cavity, and the harmonic peaks will have more uniform amplitude. Since the spectrum is flattened, the contribution of formant frequency components to the periodicity present in the autocorrelation function is reduced, making pitch period estimation more accurate.

Using a portion of voiced speech waveform, implement the autocorrelation method utilizing the clipped speech signal as input. Plot the resultant autocorrelation curve and compare to the case of no clipping.

2.2 In speech coding applications, it is common to use only the low-frequency portion of the signal for pitch period estimation. Thus, the input signal is first lowpass filtered, with a typical bandwidth between 500 and 800 Hz. Implement this technique by first designing a lowpass filter. Plot the autocorrelation curve and compare to the case where the lowpass filter is absent.

2.3 Decimation is the process of lowpass filtering a signal followed by downsampling [Oppenheim and Schafer, 1989]. Consider the pitch period estimation algorithm where the signal is first decimated by a factor of 2; a first pitch period estimation is found in the decimated domain. Then a refined result is obtained in the original domain by searching the neighborhood near the first estimate. Specify the algorithm by drawing the block diagram and writing down all relevant equations. What is the advantage of this approach?

2.4 Autocorrelation can also be computed by

$$R[l, m] = \sum_{n=m-N+1}^{m} s[n]s[n+l].$$

In practice, both approaches yield similar results. Explain possible advantages or disadvantages between the two methods.

2.5 The normalized autocorrelation function, defined by

$$r[l, m] = \frac{\sum_{n=m-N+1}^{m} s[n]s[n-l]}{\sqrt{\sum_{n=m-N+1}^{m} s^2[n] \sum_{n=m-N+1}^{m} s^2[n-l]}},$$

is often employed for pitch period estimation. Due to the addition of the normalizing term (denominator), the resultant correlation values are compensated for changing signal amplitudes, leading to more precise estimations.

Using a portion of a voiced speech waveform, apply the normalized autocorrelation method and compare with the original approach.

2.6 Given the samples $x[0], x[1], \ldots, x[N-1]$ and assuming that they are sorted in ascending order of magnitude

$$x[0] \leq x[1] \leq \cdots \leq x[N-1],$$

then the *sample median* $\tilde{x}$ of these numbers is defined by

$$\tilde{x} = \begin{cases} x[(N+1)/2], & N \text{ odd}, \\ (x[N/2] + x[N/2+1])/2, & N \text{ even}. \end{cases}$$

Median filtering can be applied as an alternative to eliminate multiples of a pitch period. For instance, suppose the sequence of pitch period under consideration is 50, 51, 100, 52, 49, and the value being processed is 100; the median filter chooses the sample median of the five numbers and returns 51 as a result. After median filtering, the new sequence becomes 50, 51, 51, 52, 49. The technique is in fact a low-complexity alternative for the removal of multiplicity in a sequence of pitch period.

(a) Obtain a sequence of pitch period values by analyzing a speech signal with the autocorrelation method. Apply the median filter and compare the input-output values; change the number of samples under consideration by the median filter and record its effects.

(b) Discuss the advantages/disadvantages of the method when compared to the approach discussed in the present chapter.

2.7 Given the second order filter with $a_1 = -0.9$, $a_2 = 0.6$,

(a) Find the reflection coefficients.

(b) Find the difference equations corresponding to direct form and lattice form realizations for both the all-pole and all-zero configurations.

(c) Via a substitution/elimination process, manipulate the lattice equations into one single equation relating the output to the input. Show at the end that direct form and lattice form produce the exact same output.

2.8 Given the Mth order all-pole filter, find out the computational complexity associated with direct form realization and lattice realization. The answer should be expressed as the number of additions and multiplications per output sample. Which realization is more efficient? Repeat for an all-zero filter.

2.9 Find out the number of additions and products required per frame for the state-save method and the zero-input zero-state method. Express the answers in terms of the filter's order (M) and the frame length (N), with the assumption that $N > M$. *Partial answer: The latter requires twice the amount of computation as the former.*

2.10 On a sample-by-sample basis, find the impulse response of an all-zero filter using the time-domain difference equation. Express the answer in terms of the filter's coefficients. How many nontrivial samples are there?

2.11 Find out the computational cost involved with the calculation of N samples of the impulse response $h[n]$ of an all-pole filter ($n = 0$ to $N - 1$) using the time-domain difference equation, with $N > M$. For computational cost, fill out Table 2.2 and add the numbers in each column to yield the total number of sums and products. Show that

$$\text{Number of sums} = \frac{(M-2)(M-1)}{2} + (N-M)(N-1),$$

$$\text{Number of products} = \frac{(M-2)(M-1)}{2} + (N-M)M.$$

TABLE 2.2 Computational Cost for Exercise 2.11

n	Number of Sums	Number of Products
0	0	0
1	0	0
2	1	1
3		
$\vdots$		
$M-1$		
$M \leq n \leq N-1$		

2.12 **(a)** Show that for a direct convolution sum, computational costs involved with an N-sample sequence are

$$\text{Number of sums} = (N-1)N/2,$$
$$\text{Number of products} = N(N+1)/2.$$

(b) In the application of recursive convolution, show that the computational costs are given below.

$$\text{Number of sums} = S(2N-S-1)/2,$$
$$\text{Number of products} = S(2N-S+1)/2.$$

What happen when $S = N$? For $N = 40$, compare the computational cost of the two schemes when $S = 1$ and $S = 2$.

2.13 Within the context of recursive convolution, consider the alternative definition for input sequences:

$$x^{(l)}[n] = x[n + lS]; \quad l = 0, 1, \ldots, L-1; \quad n = 0, 1, \ldots, n-1.$$

Derive the equation for recursive convolution based on this definition.

CHAPTER 3

STOCHASTIC PROCESSES AND MODELS

This chapter is devoted to the study of stochastic processes or random signals and their analysis through statistical signal processing. Speech is very often modeled as random with certain properties that can be captured using a simple model. By estimating the parameters of the underlying model, information related to the signal can be represented using alternative media.

Power spectral density plays an important role in speech processing due to the fact that the human auditory system relies heavily on the power distribution in the frequency domain. Many diverse methods have been developed in the past to estimate the power spectral density from signal samples, a vast field known as *spectrum estimation*. Here the discussion is limited to the practical methods and procedures. A relatively simple method known as the periodogram is first presented followed by the autoregressive model. With the model, the signal spectrum is assumed to take on a specific functional form, controlled by a few parameters. The spectral estimation problem is then one of estimating the unknown parameters of the model rather than estimating the spectrum itself. Substituting the parameters into the model leads to the actual signal spectrum. The approach is known as the *parametric method* of spectral estimation.

The autocorrelation function is often estimated first from the signal sequence while dealing with parametric spectral estimation based on the autoregressive model. In fact, it is shown that the autocorrelation function and the power spectral density form a Fourier transform pair. Different methods used to estimate the autocorrelation function are presented with an analysis and discussion of advantages and disadvantages of each method. Other signal models are described at the end of the chapter, which are not very commonly applied to speech coding applications due to the complexity involved.

In dealing with these topics, many authors consider the general case of complex-valued signals. Since the underlying signal is real for most speech processing applications, we will deal exclusively with real-valued signals for simplicity.

3.1 POWER SPECTRAL DENSITY

The power spectral density (PSD), also referred to as the power spectrum, is a description of the second-order statistics of a stochastic process in the frequency domain. Here the definition of PSD is provided, and its relationship with the autocorrelation function is found; it is shown that the autocorrelation function is the time-domain counterpart of the power spectral density.

Average Power of a Deterministic Signal

The average power of the deterministic signal $x[n]$ is given by

$$P = \lim_{N\to\infty} \frac{1}{2N+1} \sum_{n=-N}^{N} |x[n]|^2. \tag{3.1}$$

By defining

$$x_N[n] = \begin{cases} x[n], & |n| \le N, \\ 0, & \text{otherwise}, \end{cases} \tag{3.2}$$

we have

$$P = \lim_{N\to\infty} \frac{1}{2N+1} \sum_{n=-\infty}^{\infty} |x_N[n]|^2 = \frac{1}{2\pi}\int_{-\pi}^{\pi} \lim_{N\to\infty}\left(\frac{|X_N(e^{j\omega})|^2}{2N+1}\right) d\omega, \tag{3.3}$$

where

$$x_N[n] \overset{F}{\longleftrightarrow} X_N(e^{j\omega}). \tag{3.4}$$

That is, they form a Fourier transform pair:

$$X_N(e^{j\omega}) = \sum_{n=-\infty}^{\infty} x_N[n] e^{-j\omega n}, \tag{3.5}$$

$$x_N[n] = \frac{1}{2\pi}\int_{-\pi}^{\pi} X_N(e^{j\omega}) e^{j\omega n}\, d\omega. \tag{3.6}$$

The Parseval theorem is used to derive (3.3); see Exercise 3.1 for a proof.

Average Power of a Stochastic Process

For the stochastic process* $x[n]$, (3.3) only represents the power in one sample realization. By taking the expected value, we obtain the average power P for the random signal as follows:

$$P = \frac{1}{2\pi} \int_{-\pi}^{\pi} \lim_{N\to\infty} \left(\frac{E\{|X_N(e^{j\omega})|^2\}}{2N+1} \right) d\omega \tag{3.7}$$

with $E\{\cdot\}$ the expectation operator.

Definition of Power Spectral Density

The power spectral density function $S(e^{j\omega})$ of a stochastic process is defined in general by

$$P = \frac{1}{2\pi} \int_{-\pi}^{\pi} S(e^{j\omega})\, d\omega, \tag{3.8}$$

where P is the average power of the stochastic process $x[n]$. Comparing (3.7) to (3.8) we arrive at the following relation for the PSD:

$$S(e^{j\omega}) = \lim_{N\to\infty} \left(\frac{E\{|X_N(e^{j\omega})|^2\}}{2N+1} \right). \tag{3.9}$$

Average Power as Time Average of the Second Moment

Applying the expectation operator to (3.1) leads to

$$P = \lim_{N\to\infty} \frac{1}{2N+1} \sum_{n=-N}^{N} E\{x^2[n]\} = A\{E\{x^2[n]\}\}, \tag{3.10}$$

where $x[n]$ is assumed to be real. Therefore, the average power of the stochastic process $x[n]$ is given by the time average of its second moment. For a wide-sense stationary (WSS) process, $E\{x^2[n]\}$ is constant with n and so is the average power P. The time average operator $A\{\cdot\}$ is defined by

$$A\{\cdot\} = \lim_{N\to\infty} \frac{1}{2N+1} \sum_{n=-N}^{N} (\cdot). \tag{3.11}$$

* The simplified notation for stochastic process is used here, where $x[n]$ for n fixed is a random variable. A stricter notation would be $x[n, \zeta]$, where ζ is a variable representing the outcome of an experiment. In this latter notation, if n and ζ are fixed, $x[n, \zeta]$ is a number. See Papoulis [1991] for details.

We are especially interested in the WSS process since the resultant mathematics are simple and tractable. For speech coding, the WSS assumption can be applied to short intervals.

Theorem 3.1. Given a stochastic process $x[n]$ with autocorrelation function

$$R[n_1, n_2] = E\{x[n_1]x[n_2]\}, \tag{3.12}$$

then

$$S(e^{j\omega}) = \sum_{l=-\infty}^{\infty} A\{R[n, n+l]\}e^{-j\omega l}, \tag{3.13}$$

and

$$A\{R[n, n+l]\} = \frac{1}{2\pi}\int_{-\pi}^{\pi} S(e^{j\omega})e^{j\omega l}\,d\omega. \tag{3.14}$$

Equations (3.13) and (3.14) show that $S(e^{j\omega})$ and $A\{R[n, n+l]\}$ form a Fourier transform pair, denoted by

$$A\{R[n, n+l]\} \stackrel{F}{\longleftrightarrow} S(e^{j\omega}). \tag{3.15}$$

Proof. From (3.4),

$$X_N(e^{j\omega}) = \sum_{n=-\infty}^{\infty} x_N[n]e^{-j\omega n} = \sum_{n=-N}^{N} x[n]e^{-j\omega n}. \tag{3.16}$$

Substituting (3.16) in (3.9) gives

$$\begin{aligned} S(e^{j\omega}) &= \lim_{N\to\infty}\frac{1}{2N+1}E\left\{\sum_{n_1=-N}^{N} x[n_1]e^{j\omega n_1}\sum_{n_2=-N}^{N} x[n_2]e^{-j\omega n_2}\right\} \\ &= \lim_{N\to\infty}\frac{1}{2N+1}\sum_{n_1=-N}^{N}\sum_{n_2=-N}^{N} E\{x[n_1]x[n_2]\}e^{-j\omega(n_2-n_1)}. \end{aligned} \tag{3.17}$$

The expectation within the summation of the above equation is identified as the autocorrelation function of $x[n]$, (3.12). Thus,

$$S(e^{j\omega}) = \lim_{N\to\infty}\frac{1}{2N+1}\sum_{n_1=-N}^{N}\sum_{n_2=-N}^{N} R[n_1, n_2]e^{-j\omega(n_2-n_1)}. \tag{3.18}$$

Now consider the change of variables with $n = n_1$ and $l = n_2 - n_1 = n_2 - n$. Equation (3.18) becomes

$$S(e^{j\omega}) = \sum_{l=-\infty}^{\infty} \left\{ \lim_{N\to\infty} \frac{1}{2N+1} \sum_{n=-N}^{N} R[n, n+l] \right\} e^{-j\omega l}. \tag{3.19}$$

The quantity within braces is recognized as the time average of the autocorrelation function. Thus, the theorem is proved.

Theorem 3.2. Given a WSS stochastic process with autocorrelation function $R[l] = E\{x[n]x[n+l]\}$, then

$$S(e^{j\omega}) = \sum_{l=-\infty}^{\infty} R[l]e^{-j\omega l} \tag{3.20}$$

and

$$R[l] = \frac{1}{2\pi}\int_{-\pi}^{\pi} S(e^{j\omega})e^{j\omega l}\,d\omega; \tag{3.21}$$

that is,

$$R[l] \overset{F}{\longleftrightarrow} S(e^{j\omega}). \tag{3.22}$$

See Exercise 3.2 for a proof of this result.

Example 3.1: White Noise White noise is a stochastic process characterized by a constant PSD, given by

$$S(e^{j\omega}) = \sigma^2, \tag{3.23}$$

where σ^2 represents the variance of the signal. The autocorrelation function is therefore

$$R[l] = \sigma^2\delta[l]. \tag{3.24}$$

That is, the autocorrelation function is a delta function. If follows from (3.24) that a white noise signal must have zero mean (Exercise 3.3) and two samples from different time instances are uncorrelated (Exercise 3.4).

White noise is generated in practice using a random number generator and, in most cases, is either uniformly distributed or normally distributed (Gaussian). Gaussian distribution is preferred in certain applications due to its analytic elegance: linear combination of any number of independent normal random variables with zero mean always leads to a normal random variable. Also, the central limit theorem states that under certain general conditions, the sum of independent random variables (any distribution) results in a normally distributed random variable

[Papoulis, 1991]. In order to approach the behavior of theoretical white noise, the random number generator must possess certain "good" qualities, namely, the sequence of numbers generated must be statistically independent from each other. Many tests exist that provide measurement of the property of random number generators [Banks and Carson, 1984].

Theorem 3.3. Correlation Relations Between the Input and Output of a Linear Time-Invariant (LTI) System. We are given an LTI system with impulse response $h[n]$. The system input is the WSS process $x[n]$ with output $y[n]$. Then

$$R_{yx}[l] = h[l] * R_x[l], \tag{3.25}$$

$$R_{xy}[l] = h[-l] * R_x[l], \tag{3.26}$$

$$R_y[l] = h[l] * R_{xy}[l], \tag{3.27}$$

$$R_y[l] = h[l] * h[-l] * R_x[l], \tag{3.28}$$

where $R_{xy}[n_1, n_0] = E\{x[n_1]\, y[n_0]\}$ is the cross-correlation between $x[n]$ and $y[n]$. When $x[n]$ and $y[n]$ are jointly WSS—as in the present case—the cross-correlation depends only on $l = n_1 - n_0$; hence, $R_{xy}[l] = E\{x[n]y[n-l]\}$.

The last equation indicates that the autocorrelation function of the output process is a twofold convolution of the input autocorrelation function with the system's impulse response.

Proof. From the convolution sum,

$$y[n] = \sum_{k=-\infty}^{\infty} h[k]x[n-k]. \tag{3.29}$$

Evaluating the above relation at $n = n_1$ and multiplying both sides by $x[n_0]$ gives

$$y[n_1]x[n_0] = \sum_{k=-\infty}^{\infty} h[k]x[n_1 - k]x[n_0]. \tag{3.30}$$

Taking the expectation,

$$R_{yx}[n_1, n_0] = \sum_{k=-\infty}^{\infty} h[k]R_x[n_1 - n_0 - k]. \tag{3.31}$$

Since $x[n]$ is a WSS process, R_{yx} is a function only of the difference $n_1 - n_0$, implying that $x[n]$ and $y[n]$ are jointly stationary. Substituting the variable $l = n_1 - n_0$ produces

$$R_{yx}[l] = \sum_{k=-\infty}^{\infty} h[k]R_x[l-k], \tag{3.32}$$

which is (3.25).

Equation (3.26) is a direct consequence of (3.25) and the symmetric properties of cross-correlation and autocorrelation. By using a procedure parallel to the one used to find (3.25) (multiplying both sides of (3.29) by $y[n_0]$ instead of $x[n_0]$), (3.27) is derived. Finally, (3.28) is found by substituting (3.26) in (3.27).

Theorem 3.4: Power Spectral Density of the Output of an LTI System. We are given an LTI system with transfer function $H(e^{j\omega})$. The system input is the WSS process $x[n]$ with PSD $S_x(e^{j\omega})$, and the output process is $y[n]$. Then

$$S_y(e^{j\omega}) = |H(e^{j\omega})|^2 S_x(e^{j\omega}) \tag{3.33}$$

is the PSD of the output process $y[n]$. This result is obtained by applying the Fourier transform to (3.28). The function $|H(e^{j\omega})|^2$ is sometimes referred to as the power transfer function.

3.2 PERIODOGRAM

Consider an N-point sequence $x[n], n = 0, \ldots, N-1$. The periodogram $I_N(e^{j\omega})$ is defined to be

$$I_N(e^{j\omega}) = \frac{1}{N}|X_N(e^{j\omega})|^2, \tag{3.34}$$

where

$$X_N(e^{j\omega}) = \sum_{n=0}^{N-1} x[n]e^{-jn\omega} \tag{3.35}$$

is the Fourier transform of the finite-length sequence $x[n]$.

When the finite-length sequence is selected through a window sequence $w[n]$, that is,

$$X_N(e^{j\omega}) = \sum_{n=0}^{N-1} w[n]x[n]e^{-jn\omega}, \tag{3.36}$$

the resultant frequency function, as defined in (3.34), is known as the modified periodogram, or simply periodogram.

Theorem 3.5. We are given the N-point real sequence $x[n], n = 0, \ldots, N-1$. Then

$$I_N(e^{j\omega}) = \sum_{l=-(N-1)}^{N-1} R[l]e^{-j\omega l} \tag{3.37}$$

with

$$R[l] = \frac{1}{N} \sum_{m=0}^{N-1} w[m+l]w[m]x[m+l]x[m] \tag{3.38}$$

being the autocorrelation function of the sequence $w[n]x[n]$. Thus, the periodogram is related to the autocorrelation function through the Fourier transform equation (3.37).

Proof. From (3.34),

$$\begin{aligned} I_N(e^{j\omega}) &= \frac{1}{N} X_N(e^{j\omega}) X_N^*(e^{j\omega}) \\ &= \frac{1}{N} \sum_{n=0}^{N-1} \sum_{m=0}^{N-1} w[n]w[m]x[n]x[m]e^{-j\omega(n-m)} \\ &= \frac{1}{N} \sum_{m=0}^{N-1} \sum_{l=-m}^{N-m-1} w[m]w[m+l]x[m]x[m+l]e^{-j\omega l}. \end{aligned} \tag{3.39}$$

Note that $w[n]$ is zero outside the interval $n \in [0, N-1]$; hence,

$$I_N(e^{j\omega}) = \frac{1}{N} \sum_{l=-(N-1)}^{N-1} \sum_{m=0}^{N-1} w[m]w[m+l]x[m]x[m+l]e^{-j\omega l}, \tag{3.40}$$

which completes the proof.

Comparing (3.37) with (3.20) one can reach the conclusion that the periodogram is similar to the power spectral density. Indeed, the periodogram is an estimate of the PSD using a finite number of samples from the signal source, with the estimate being an approximate calculation of the true function. Due to its simplicity, the periodogram is often used in practice to study the signal source of interest in the frequency domain. References are given at the end of the chapter where more extensive discussion regarding the statistical properties of periodogram can be found.

The choice of the window depends on frequency resolution and spectral leakage. The ideal window spectrum is an impulse, which would require a window sequence of infinite length. Many options are available for finite-length window, with the Hamming window being one of the most widely used. In practice, the sample mean of the finite-length sequence is often subtracted before computing the periodogram. This avoids leakage due to the zero-frequency component that interferes with the low-frequency zone.

3.3 AUTOREGRESSIVE MODEL

A model is used for any hypothesis that may be applied to explain or describe the hidden laws that are supposed to govern or constrain the generation of some data of interest. One common method for modeling random signals is to represent them as the output of an all-pole linear filter driven by white noise. Since the power spectrum of the filter output is given by the constant noise spectrum multiplied by the squared magnitude of the filter (see (3.33)), random signals with desired spectral characteristics can be produced by choosing a filter with an appropriate denominator polynomial.

The sequence values $x[n], x[n-1], \ldots, x[n-M]$ represent the realization of an autoregressive (AR) process of order M if it satisfies the difference equation

$$x[n] + a_1 x[n-1] + \cdots + a_M x[n-M] = v[n], \tag{3.41}$$

where the constants $a_1, a_2, \ldots, a_M$ are known as the AR parameters and $v[n]$ represents a white noise process; the above equation can be written as

$$x[n] = -a_1 x[n-1] - a_2 x[n-2] - \cdots - a_M x[n-M] + v[n]. \tag{3.42}$$

Therefore, the present value of the process, $x[n]$, is equal to a linear combination of past values of the process, $x[n-1], \ldots, x[n-M]$, plus an error term $v[n]$. The process $x[n]$ is said to be regressed on $x[n-1], x[n-2], \ldots, x[n-M]$; in particular, $x[n]$ is regressed on previous values of itself, hence the name "autoregressive."

System Function of the AR Process Analyzer

Taking the z-transform of (3.41) and manipulating yields

$$H_A(z) = \frac{V(z)}{X(z)} = \sum_{i=0}^{M} a_i z^{-i}, \tag{3.43}$$

where $H_A(z)$ denotes the system function of the AR analyzer, which is a filter that takes $x[n]$ as input and produces $v[n]$ at its output. The parameter a_0 is equal to one in the above equation. Thus, the AR analyzer transforms an AR process at its input to white noise at its output. Figure 3.1 shows the direct form realization [Oppenheim and Schafer, 1989] of the AR analyzer. Note that the AR process analyzer is an all-zero filter and hence of FIR nature.

System Function of the AR Process Synthesizer

With the white noise $v[n]$ acting as input, we can use the system function given by

$$H_S(z) = \frac{X(z)}{V(z)} = \frac{1}{H_A(z)} = \frac{1}{\sum_{i=0}^{M} a_i z^{-i}} \tag{3.44}$$

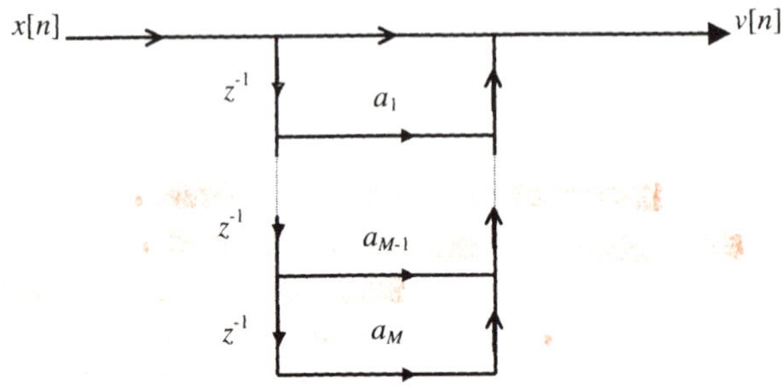

Figure 3.1 Direct form realization of the AR process analyzer filter.

to synthesize the AR process $x[n]$. Direct form realization is shown in Figure 3.2. Note that the AR process synthesizer is an all-pole filter whose impulse response length is infinite (IIR). The synthesizer takes white noise as input and produces an AR signal at its output. From (3.44) we see that the system function of the analyzer is the inverse of the system function for the synthesizer; we can also write

$$H_S(z) = \frac{1}{(1 - p_1 z^{-1})(1 - p_2 z^{-1}) \cdots (1 - p_M z^{-1})}, \tag{3.45}$$

where $p_1, p_2, \ldots, p_M$ are poles of $H_S(z)$ and are roots of the characteristic equation

$$1 + a_1 z^{-1} + a_2 z^{-2} + \cdots + a_M z^{-M} = 0. \tag{3.46}$$

Thus, an AR process is synthesized by filtering white noise using an all-pole filter.

PSD of an AR Process

As explained earlier, an AR process is the output of the LTI system characterized by $H_S(z)$, when the input is a white noise process. From (3.23) it follows that the input

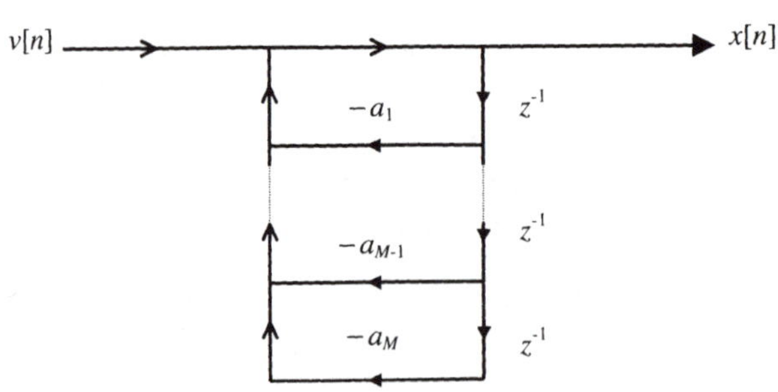

Figure 3.2 Direct form realization of the AR process synthesizer filter.

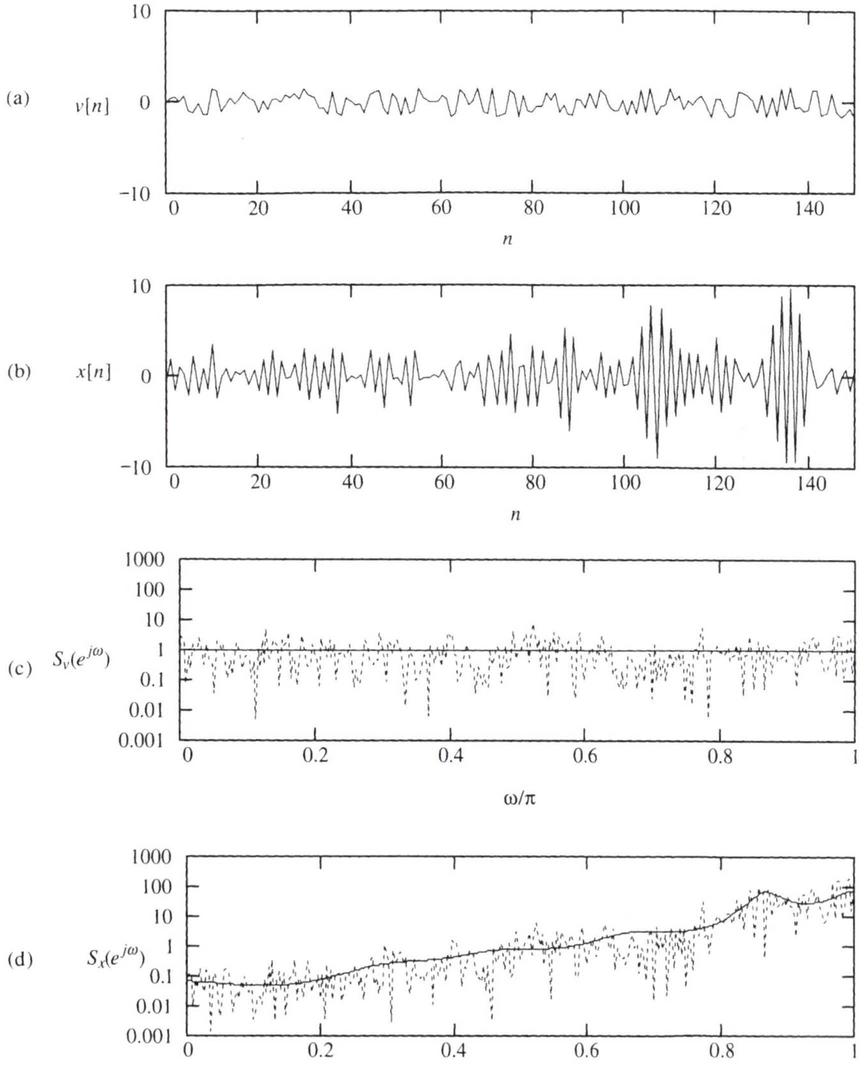

Figure 3.3 Signal plots for (a) white noise and (b) AR signal (150 samples). PSD (solid) and periodogram (dots, calculated with 400 samples) plots for (c) white noise and (d) AR signal.

PSD is constant and equal to σ_v^2, the variance of the input signal $v[n]$. From (3.33) the PSD of the output AR process $x[n]$ is

$$S_x(e^{j\omega}) = |H_S(e^{j\omega})|^2 \sigma_v^2. \tag{3.47}$$

That is, the PSD of an AR process is given by the product between the magnitude squared of the transfer function of the synthesizer and the variance of the input white noise.

Example 3.2 White noise is generated using a random number generator with uniform distribution and unit variance. This signal is then filtered by an AR synthesizer with

$a_1 = 1.534$	$a_2 = 1$	$a_3 = 0.587$	$a_4 = 0.347$	$a_5 = 0.08$
$a_6 = -0.061$	$a_7 = -0.172$	$a_8 = -0.156$	$a_9 = -0.157$	$a_{10} = -0.141$

Segments of the signals are plotted in Figure 3.3. Note that for white noise, correlation is almost nonexistent between adjacent samples; that is, signal values are independent from each other. For the AR signal, however, a strong correlation exists between adjacent samples, where the value of the signal at a given time instant tends to follow the close-by samples.

From the same figure, the theoretical PSDs of the two signals are plotted together with the periodogram using $N = 400$ samples and a rectangular window. The periodogram values, even though noise-like, fluctuate around the theoretical functions, confirming the fact that it is indeed an estimate of the PSD. Note that, for white noise, the theoretical PSD is constant and equal to one (3.23) while for the AR signal, it is given by (3.47). The periodogram is significantly different from the theoretical PSD because a single ensemble realization of the random process is considered in the experiment. By evaluating a large number of realizations, the average result will converge toward the theoretical functions.

Other observations can be drawn from the signal plots. First, the AR signal is "colored," meaning that its PSD is not flat, as opposed to that of the white noise, with the shape or contour of the spectrum being determined by the synthesizer. If the AR signal is filtered by the corresponding analyzer, white noise can be obtained at its output. Thus, the analyzer filter is often referred as the "whitener," which decorrelates an input signal by flattening its power spectrum.

Normal Equation

Since $v[n]$ represents a white noise sample at time instant n, it is not correlated with $x[n-l]$ for $l \geq 1$. That is,

$$E\{v[n]x[n-l]\} = 0; \quad l = 1, 2, \ldots, M. \tag{3.48}$$

Multiplying both sides of (3.41) by $v[n]$ and taking expectation yields

$$E\{v[n]x[n]\} = \sigma_v^2. \tag{3.49}$$

That is, the cross-correlation between $x[n]$ and $v[n]$ is given by the variance of $v[n]$. Multiplying both sides of (3.41) by $x[n-l]$, $l = 0, 1, \ldots, M$, and taking expectation yields the system of equations

$$\begin{aligned} R_x[0] + a_1 R_x[1] + \cdots + a_M R_x[M] &= \sigma_v^2, \\ R_x[1] + a_1 R_x[0] + \cdots + a_M R_x[M-1] &= 0, \\ &\vdots \\ R_x[M] + a_1 R_x[M-1] + \cdots + a_M R_x[0] &= 0. \end{aligned} \tag{3.50}$$

Or in matrix form,

$$\begin{pmatrix} R_x[0] & R_x[1] & \cdots & R_x[M] \\ R_x[1] & R_x[0] & \cdots & R_x[M-1] \\ \vdots & \vdots & \ddots & \vdots \\ R_x[M] & R_x[M-1] & \cdots & R_x[0] \end{pmatrix} \begin{pmatrix} 1 \\ a_1 \\ \vdots \\ a_M \end{pmatrix} = \begin{pmatrix} \sigma_v^2 \\ 0 \\ \vdots \\ 0 \end{pmatrix}. \tag{3.51}$$

The above equation is known as the normal equation* for WSS AR processes. Given the autocorrelation sequence $R_x[0], R_x[1], \ldots, R_x[M]$, (3.51) can be solved to yield the model parameters a_i. See Exercises 3.7 and 3.8 for alternative forms of the equation.

3.4 AUTOCORRELATION ESTIMATION

In the previous sections we mentioned that the periodogram is an estimate of the PSD. It was also shown that the autocorrelation function and PSD form a Fourier transform pair. Based on this fact, the autocorrelation function can be estimated first from the signal; the Fourier transform is then computed for the purpose of spectrum estimation.

As we will see in later chapters, the autocorrelation function plays an important role in linear prediction analysis: a procedure used to calculate the autoregressive parameters, or linear prediction coefficients of the signal model. Thus, it is important to study the different estimation methods available for autocorrelation.

Since speech is nonstationary, the autocorrelation values must be estimated and changed for every short interval of time; that is, their values are recalculated in each signal frame. Fundamentally, two types of procedure exist: *nonrecursive* and *recursive*. The difference between the two types of estimation methods is analogous to digital filters: FIR and IIR. In a nonrecursive approach the window used for extraction has finite length, while an infinite-length window is used for recursive methods. The use of either of these techniques depends on the particular application.

*The normal equation is also known as the Yule–Walker equation or Wiener–Hopf equation in the literature.

The autocorrelation function of a real discrete-time signal $x[n]$ at lag l is defined by*

$$R_x[l] = A\{x[n]x[n+l]\} = \lim_{N\to\infty} \frac{1}{2N+1} \sum_{n=-N}^{N} x[n]x[n+l]. \tag{3.52}$$

Implementations of various estimators are explained next.

Nonrecursive Estimation Methods

Nonrecursive methods are based on a well-defined window sequence $w[n]$ to extract the signal frame of interest for further processing, with the Hamming window being one of the most widely used. The causal Hamming window is defined with

$$w[n] = \begin{cases} 0.54 - 0.46\cos\left(\frac{2\pi n}{N-1}\right), & 0 \le n \le N-1, \\ 0, & \text{otherwise}, \end{cases} \tag{3.53}$$

with N being the window length (number of nonzero samples). Figure 3.4 shows a plot of the window sequence. In practice, the values of the window sequence are often stored in memory.

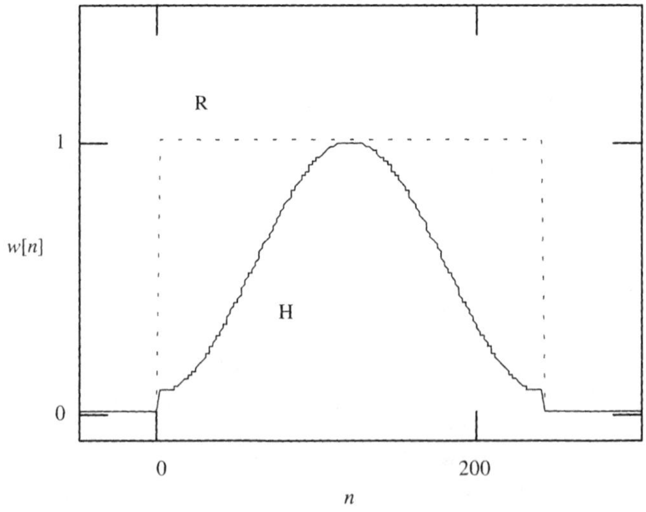

Figure 3.4 Plots of the rectangular and Hamming window, with a length of $N = 240$.

*In the application of the theory of stochastic processes, we will assume that the signals under consideration are "ergodic in correlation," meaning that the time average $A\{x[n]x[n+l]\}$ is equal to the expectation $E\{x[n]x[n+l]\}$.

Assume we want to perform the estimation on the frame that ends at time instant m, where the length of the frame is equal to N (i.e., from $n = m - N + 1$ to m). One approach is

$$R[l, m] = \frac{1}{N} \sum_{n=-\infty}^{\infty} x[n]w[m-n]x[n+l]w[m-n-l]. \tag{3.54}$$

Figure 3.5 illustrates the situations for $l > 0$ and $l < 0$. Note that the window sequence $w[n]$ is causal. Taking into account the limits of the summation, the above equation can be manipulated to yield

$$R[l, m] = \frac{1}{N} \sum_{n=m-N+1+|l|}^{m} x[n]w[m-n]x[n-|l|]w[m-n+|l|]. \tag{3.55}$$

The estimator represented by (3.54) and (3.55) has the following properties:

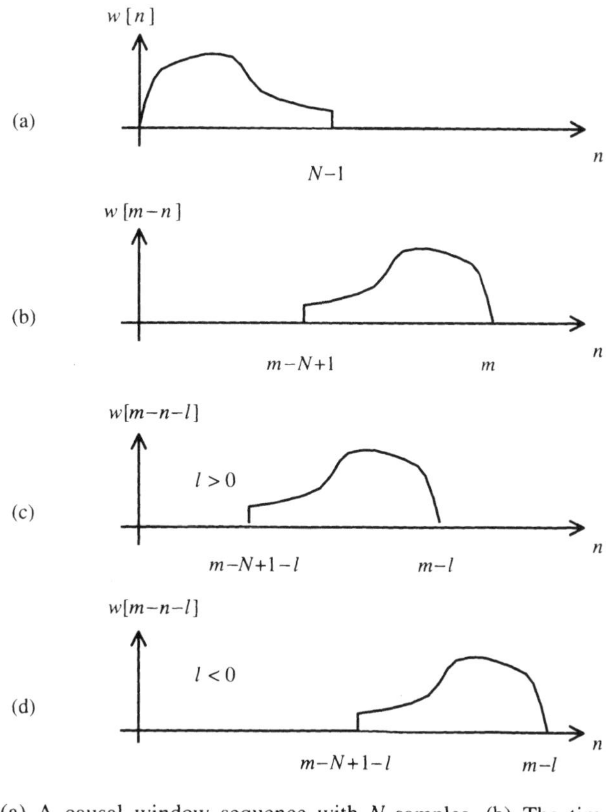

Figure 3.5 (a) A causal window sequence with N samples. (b) The time-reversed and -shifted window $w[m-n]$. (c) $w[m-n-l]$ for $l > 0$. (d) $w[m-n-l]$ for $l < 0$.

- $R[l, m]$ is a *biased* estimator of $R_x[l]$. If $x[n]$ is a realization of a WSS, ergodic, random process, then by taking the expected value of (3.55) we find (assuming a rectangular window)

$$E\{R[l, m]\} = \frac{1}{N} \sum_{n=m-N+1+|l|}^{m} E\{x[n]x[n-|l|]\} = \frac{N-|l|}{N} R_x[l]. \tag{3.56}$$

  In statistical terms, the estimator is biased when $E\{R[l, m]\} \neq R_x[l]$ with the bias being the difference $E\{R[l, m]\} - R_x[l]$. The bias is a measure of the error involved by using the estimator: the smaller the bias the better the estimator.
- $R[l, m]$ is asymptotically unbiased. From (3.56),

$$\lim_{N \to \infty} E\{R[l, m]\} = R_x[l]. \tag{3.57}$$

  Thus, $R[l, m]$ is asymptotically unbiased by definition.

See Exercises 3.10 and 3.11 for other versions of the nonrecursive estimator and their statistical properties.

Recursive Estimation Methods

For most speech coding applications, the frame length N is on the order of 200 samples, since it is roughly the time interval during which the signal remains stationary. An example is illustrated in Figure 3.6 (a), where a window of length 200 is used to calculate the autocorrelation values every 200 samples. In some applications it might be necessary to perform the calculation in an interval that is much shorter than 200, for instance, 40. By updating the estimates more frequently, delay associated with the buffering process (required to gather the input samples) is greatly reduced, which is highly desirable in practice. Figure 3.6(b) shows the solution

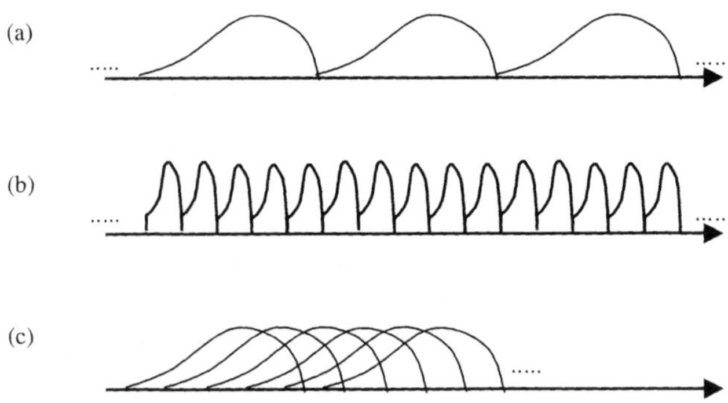

Figure 3.6 Illustration of nonrecursive autocorrelation estimation. (a) Estimation is performed every 200 samples; a window of length 200 is used. (b) Estimation is performed every 40 samples; a window of length 40 is used. (c) Estimation is performed every 40 samples; a window of length 200 is used. Note that the windows overlap each other.

where a window of length 40 is used. A short window, however, will increase the bias of the estimates, leading to inaccurate results. For higher precision, the situation depicted in Figure 3.6(c) can be applied, where a 200-sample window is employed every 40 samples, leading to overlapping. A disadvantage of this latter approach is in the computational aspect: for every short interval of time (40 samples in this case), 200 samples must be stored to compute the autocorrelation values, which is repeated for every 40-sample intervals. Since the windows are overlapping, some information should be reusable from interval to interval; in the present (nonrecursive) scheme, however, the procedure is not taking advantage of the situation, leading to a high degree of inefficiency.

To overcome these problems, a recursive approach is desirable. In this case, information from the past frames is used to update the estimates of the present frame so as to increase efficiency. Consider an estimator of autocorrelation based on the following relation:

$$R[l, m] = \sum_{n=-\infty}^{\infty} x[n]w[m-n]x[n-l]w[m-n+l], \tag{3.58}$$

which is essentially (3.54) without the scaling constant $1/N$. In most applications, only the relative magnitude between the autocorrelation values for different lag l is important; thus, the scaling constant can be omitted. The type of window that we consider here has the shape shown in Figure 3.7, where it is causal with a decaying amplitude ($w[n] \to 0$ as $n \to \infty$). Since the infinite-length window has very small amplitude outside a certain region, say, a region of length N, similar statistical properties can be drawn as in the nonrecursive case.

Barnwell Window

Barnwell (1981) proposed the following infinite-length sequence as the window used for autocorrelation estimation:

$$w[n] = (n+1)\alpha^n u[n], \tag{3.59}$$

with α a real positive constant smaller than one and $u[n]$ the unit step function. The window sequence $w[n]$ has z-transform

$$W(z) = \frac{1}{(1-\alpha z^{-1})^2}. \tag{3.60}$$

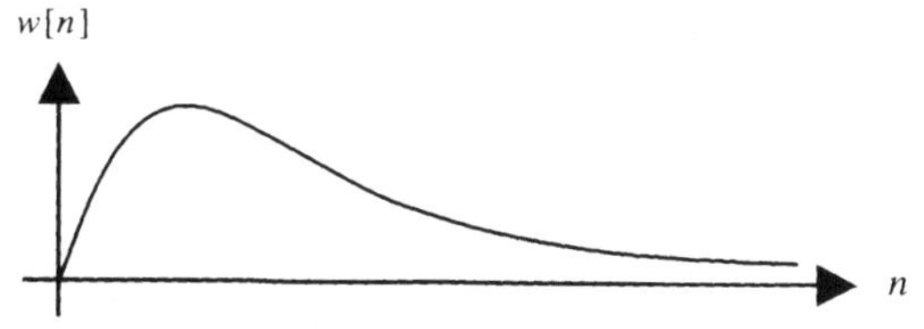

Figure 3.7 A causal window sequence of infinite length.

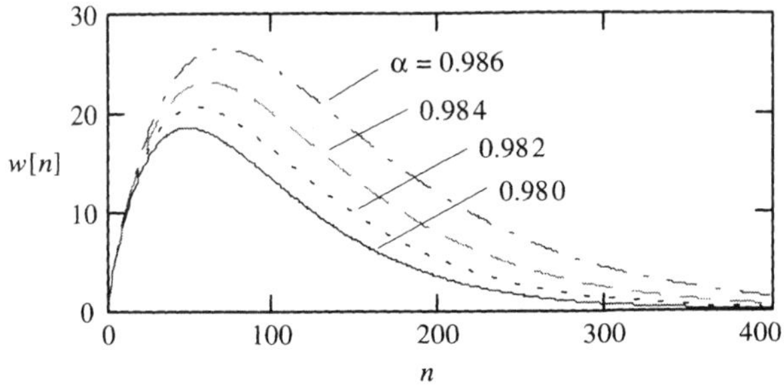

Figure 3.8 Barnwell window for four different values of α.

That is, it is an all-pole function with second-order pole located at $z = \alpha$. Figure 3.8 shows various window sequences for different values of α.

Note that the magnitude of the window is negligible outside a certain finite-length interval; the length of this interval is a function of the constant α. By choosing the right α, it is possible to include more or less data for the estimation process. Typical window lengths are on the order of 30 ms in speech coding or 240 samples for an 8-kHz sampling rate; a constant α near 0.98 is a good choice to meet the specification.

Let's define

$$x_l[n] = x[n]x[n-l] \tag{3.61}$$

and

$$w_l[n] = w[n]w[n+l]. \tag{3.62}$$

Equation (3.58) can be rewritten as

$$R[l,m] = \sum_{n=-\infty}^{\infty} x_l[n]w_l[m-n] = x_l[m] * w_l[m]. \tag{3.63}$$

That is, the autocorrelation estimate with lag l and at the frame end time m is given by the convolution between the sequence $x_l[m]$ and $w_l[m]$. $R[l,m]$ can therefore be considered as the output of an LTI filter with impulse response $w_l[m]$ and input $x_l[m]$. Recalling the fact that $w_l[m]$ is a causal sequence, (3.63) can be rewritten as

$$R[l,m] = \sum_{n=-\infty}^{m} x_l[n]w_l[m-n]. \tag{3.64}$$

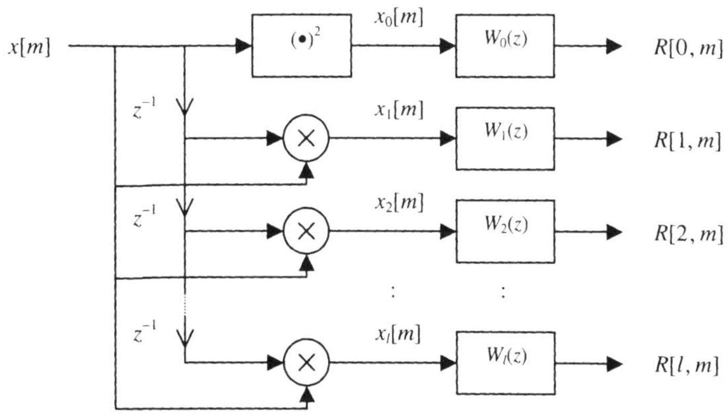

Figure 3.9 Block diagram of the system needed for recursive calculation of autocorrelation estimates.

Now, we seek the system function $W_l(z)$ of the LTI filter whose impulse response is $w_l[m]$. It can be shown (Exercise 3.12) that

$$W_l(z) = \frac{(l+1)\alpha^l - (l-1)\alpha^{l+2}z^{-1}}{1 - 3\alpha^2 z^{-1} + 3\alpha^4 z^{-2} - \alpha^6 z^{-3}}. \tag{3.65}$$

Equation (3.65) is the system function of the LTI filter needed for the estimation of autocorrelation. Figure 3.9 shows the system needed for recursive calculation of

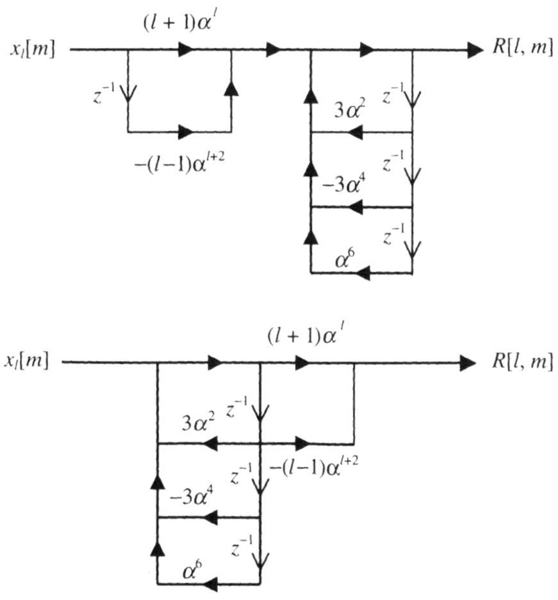

Figure 3.10 Direct form I (*top*) and direct form II (*bottom*) implementation of $W_l(z)$.

the autocorrelation estimates. The system function $W_l(z)$ can be implemented in direct form I or direct form II as shown in Figure 3.10. [Oppenheim and Schafer, 1989]. Direct form II realization has some important advantages: the filter can be separated into a recursive section where the multipliers $3\alpha^2$, $-3\alpha^4$, and α^6 are involved, and a nonrecursive section constructed with the multipliers $(l+1)\alpha$ and $-(l-1)\alpha^{l+2}$. The two products in the nonrecursive portion of the filters need only be carried out once on every frame interval and not on every sample, leading to substantial computational savings. In contrast, direct form I provides no such benefit.

To summarize, the Barnwell windowing method for recursive autocorrelation estimation presents the following differences when compared to nonrecursive techniques:

- Since the parameter α completely controls the window length, the same amount of computation is required regardless of the window length or frame size. In a nonrecursive approach using a finite-length window, the amount of computation is proportional to the window length.
- The scaling constants in the recursive sections of the linear filters for different lag l are all identical. This allows for less constant storage and simpler filter realizations.
- Since all the window information is contained in the linear filter coefficients, no extensive ROM storage is needed to support the window function. In contrast, nonrecursive methods often require samples of the window to be stored in memory, with the actual amount dependent on the window length.

Chen Window

In the aforementioned Barnwell windowing technique, the products of the current signal sample and previous samples are passed through a bank of third-order IIR filters, and the autocorrelation coefficients are obtained at the outputs of the filters. For fixed-point arithmetic, rounding is necessary and introduces errors that tend to accumulate as noise in the recursive structure of IIR filters. Since most target processors for speech coding applications are of the fixed-point type, use of the Barnwell window presents serious implementational problems. To avoid the problem associated with a recursive structure, a conventional blockwise nonrecursive window (such as the Hamming window) can be used. However, as mentioned earlier, with frequent updates and a high degree of window overlapping, the resulting scheme is inefficient, with excessive complexity.

Chen proposed a hybrid window consisting of a recursively decaying tail and a section of nonrecursive samples at the beginning [Chen, 1995]. The recursive part is exponentially decaying while the nonrecursive part is a section of the sine function. The overall shape is very similar to the Barnwell window. The purpose of the nonrecursive portion is to mimic the general shape of the Barnwell window, while the purpose of the recursive portion is to enable recursive calculation so as to reduce complexity (with respect to a nonrecursive approach). By using this window, numerical sensitivity is greatly reduced (with respect to Barnwell window), enabling the deployment of a hybrid window with sufficient accuracy using fixed-point arithmetic.

The window is defined by

$$w[n] = \begin{cases} 0, & n \leq 0, \\ \sin(cn), & 0 < n \leq L, \\ b\alpha^{n-L-1}, & n \geq L+1, \end{cases} \tag{3.66}$$

where L is the length of the nonrecursive section of the window and α, b, and c are constants that must be found for a particular window specification. To ensure a smooth junction between the sine function and the exponential function at $n = L + 1$, two conditions are imposed:

- Values of the two functions are equal at $n = L + 1$, which means

$$\sin(c(L+1)) = b. \tag{3.67}$$

- Slopes of the two functions (derivatives with respect to n) are equal at $n = L + 1$, implying

$$c\cos(cn) = b(\ln\alpha)\alpha^{n-L-1}. \tag{3.68}$$

  At $n = L + 1$,

$$\ln\alpha = c\,\mathrm{ctg}(c(L+1)) \tag{3.69}$$

  or

$$\alpha = \exp[c\,\mathrm{ctg}(c(L+1))]. \tag{3.70}$$

Summarizing, the following procedure is used for window design.

Step 1. The decaying factor α is first fixed; its choice depends on how long we want the effective length of the exponential tail to be.

Step 2. Length of the nonrecursive part is L. Its value is chosen based on how we want to shape the initial part of the window and how much computational complexity we are willing to have. Obviously, the larger the number L, the higher the complexity. Also, more storage is required for the longer nonrecursive window.

Step 3. Once α and L are known, (3.69) or (3.70) is solved for the constant c. This can be done with a graphical approach.

Step 4. Equation (3.67) is used to find b.

Example 3.3 The window design procedure is illustrated for the case when $\alpha = 0.5^{1/40} = 0.9828205985$ and $L = 30$. These specifications correspond to the

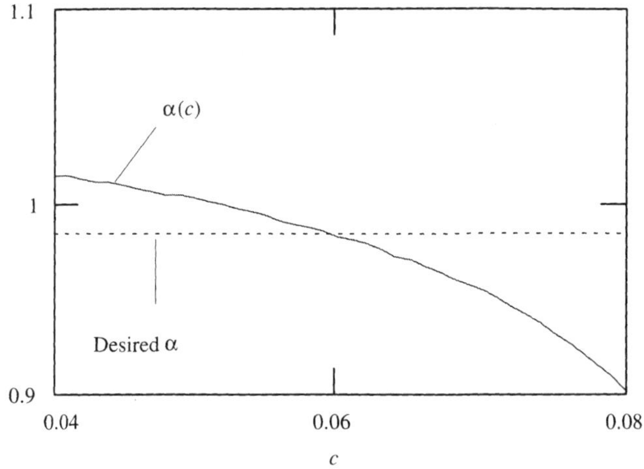

Figure 3.11 Graphical approach to finding the parameter c of the Chen window.

window used for the perceptual weighting filter of the ITU-T G.728 LD-CELP coder (Chapter 14). Thus, Steps 1 and 2 of the design procedure are already completed. For Step 3, we use a graphic method to solve c. In Figure 3.11, (3.70) is plotted as a function of c. From there we can see that when $c \approx 0.06$, the desired value of α is reached; the range of c used to search for the result is found experimentally. Fine tuning its value yields the final result of $c = 0.0597731696$, which is done manually on a trial-and-error basis. (Note that a simple computer program can be created to search for c.) In Step 4, the value of c is substituted in (3.67) to give $b = 0.96$. The window is therefore completely specified and is plotted in Figure 3.12.

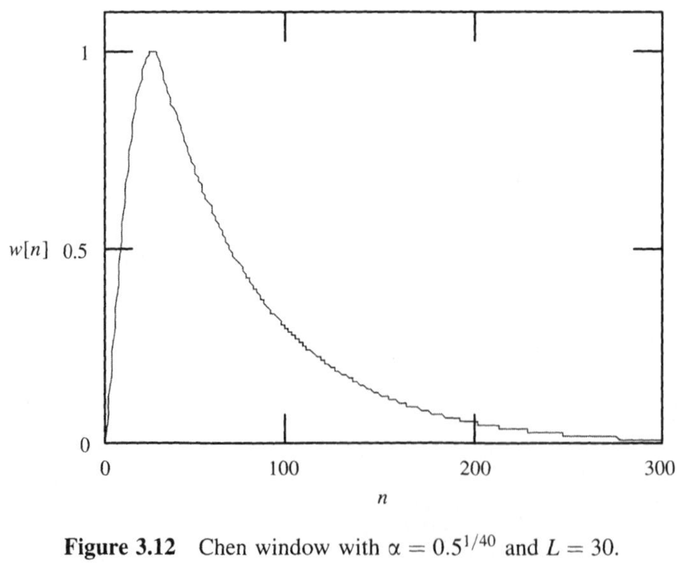

Figure 3.12 Chen window with $\alpha = 0.5^{1/40}$ and $L = 30$.

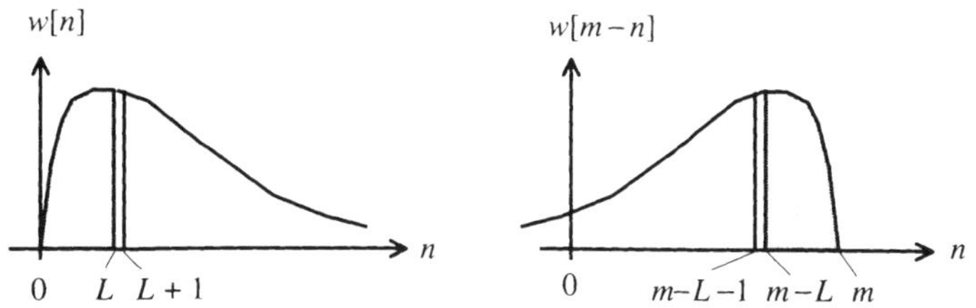

Figure 3.13 Chen window showing the nonrecursive and recursive portion (*left*), together with the time-reversed and -shifted version (*right*).

Computational Procedure for Chen Window

For causal windows, (3.58) reduces to

$$R[l, m] = \sum_{n=-\infty}^{m} x[n]w[m-n]x[n-l]w[m-n+l]. \tag{3.71}$$

The time-reversed and -shifted window sequence $w[m-n]$ is shown in Figure 3.13, where the limits separating the recursive and nonrecursive portions are shown. Equation (3.71) can be written as

$$\begin{aligned} R[l, m] = &\sum_{n=-\infty}^{m-L-1} x[n]w[m-n]x[n-l]w[m-n+l] \\ &+ \sum_{n=m-L}^{m} x[n]w[m-n]x[n-l]w[m-n+l]. \end{aligned} \tag{3.72}$$

Let's define

$$R_o[l, m] = \sum_{n=-\infty}^{m-L-1} x[n]w[m-n]x[n-l]w[m-n+l], \tag{3.73}$$

which represents the recursive part of the estimation equation. Assume now that we want to compute the recursive autocorrelation estimate at the frame end time $m + N$, with N being a positive integer. We can write

$$\begin{aligned} R_o[l, m+N] &= \sum_{n=-\infty}^{m+N-L-1} x[n]w[m+N-n]x[n-l]w[m+N-n+l] \\ &= \sum_{n=-\infty}^{m-L-1} x[n]w[m+N-n]x[n-l]w[m+N-n+l] \\ &\quad + \sum_{n=m-L}^{m+N-L-1} x[n]w[m+N-n]x[n-l]w[m+N-n+l]. \end{aligned} \tag{3.74}$$

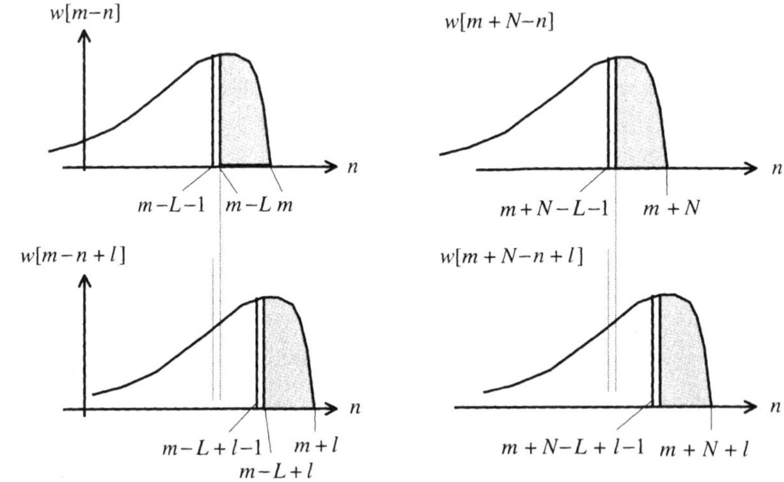

Figure 3.14 Relative positions of various time-reversed and -shifted window sequences.

Figure 3.14 shows the relative positions of the window sequences. We can see that the limits in the summations of (3.73) and (3.74) involve only the recursive part of the window. Substituting the actual expression for the window (3.66) in (3.73) leads to

$$R_o[l,m] = b^2\alpha^{2m-2L-2+l}\sum_{n=-\infty}^{m-L-1} x[n]x[n-l]\alpha^{-2n}. \tag{3.75}$$

Similarly, for (3.74) we have

$$R_o[l,m+N] = b^2\alpha^{2m-2L-2+l+2N}\sum_{n=-\infty}^{m-L-1} x[n]x[n-l]\alpha^{-2n} + \sum_{n=m-L}^{m+N-L-1} x[n]x[n-l]w[m+N-n]w[m+N-n+l]. \tag{3.76}$$

Comparing (3.75) and (3.76) gives

$$R_o[l,m+N] = \alpha^{2N}R_o[l,m] + \sum_{n=m-L}^{m+N-L-1} x[n]x[n-l]w[m+N-n]w[m+N-n+l]. \tag{3.77}$$

Therefore, $R_o[l, m+N]$ can be calculated recursively from $R_o[l, m]$ using (3.77). The autocorrelation estimate at $m+N$ is thus given by

$$R[l, m+N] = R_o[l, m+N] + \sum_{n=m+N-L}^{m+N} x[n]w[m+N-n]x[n-l]w[m+N-n+l]. \tag{3.78}$$

Note from Figure 3.14 that a total of $N + L + l_{\max} + 1$ values of the window must be stored. Discounting the first value of zero, a total of $N + L + l_{\max}$ values are needed. This number can also be found from the limits of the summations in (3.77) and (3.78).

The following pseudocode performs the calculations:

```
Ro[l, m] ← 0
temp ← 0
for n ← m − L to m + N − L − 1
        temp ← temp + x[n] x[n − l] w[m + N − n] w[m + N − n + l]
Ro[l, m + N] ← α^2N Ro[l, m] + temp
temp ← 0
for n ← m + N − L to m + N
        temp ← temp + x[n] x[n − l] w[m + N − n] w[m + N − n + l]
R[l, m + N] ← Ro[l, m + N] + temp // Results are here
m ← m + N
goto 2
```

3.5 OTHER SIGNAL MODELS

Besides the AR model presented in Section 3.3, there are other linear models that are encountered less frequently; however, they are sometimes applied for specific tasks. The models included in this section are the moving average (MA) model and the autoregressive–moving average (ARMA) model.

MA Model

The moving average process $x[n]$ of order K satisfies the difference equation

$$x[n] = v[n] + b_1 v[n-1] + \cdots + b_K v[n-K], \tag{3.79}$$

where the constants $b_1, b_2, \ldots, b_K$ are known as the MA parameters and $v[n]$ represents a white noise process. Thus, an MA process is formed by a linear combination of $(K+1)$ white noise samples. Figure 3.15 shows the MA process analyzer and synthesizer filters.

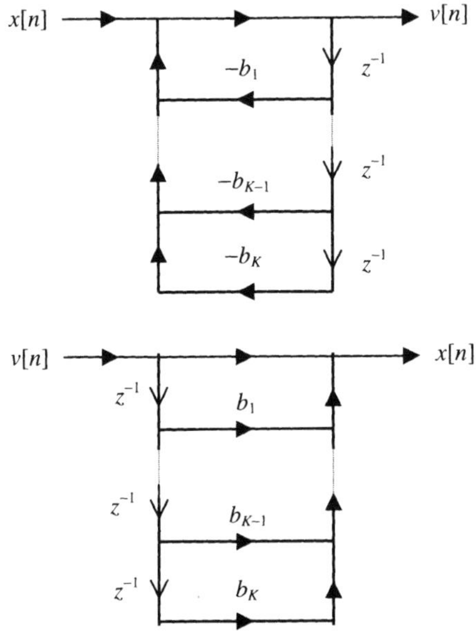

Figure 3.15 Direct form realization of the MA process analyzer filter (*top*) and the synthesizer filter (*bottom*).

ARMA Model

The autoregressive–moving average process $x[n]$ of orders (M, K) satisfies the difference equation

$$x[n] + a_1 x[n-1] + \cdots + a_M x[n-M] = v[n] + b_1 v[n-1] + \cdots + b_K v[n-K], \tag{3.80}$$

where the constants $a_1, \ldots, a_M, b_1, \ldots, b_K$ are the ARMA parameters, with $v[n]$ a white noise process. The ARMA model is the most flexible of all three linear models; however, its design and analysis are more difficult than the AR or the MA model.

3.6 SUMMARY AND REFERENCES

Important concepts in statistical signal processing are presented in this chapter, which form the foundations for many speech coding algorithms. As we will see in later chapters, many coding schemes attempt to estimate the parameters of a time-varying filter, used to capture the PSD of the original speech. Since the number of parameters needed to specify the time-varying filter is far less than the number of speech samples, a high compression ratio is achievable.

Further reading in spectrum estimation can be found in Stearns and Hush [1990] and Therrien [1992]. Foundation of stochastic processes are found in Papoulis [1991] and Peebles [1993]. A comprehensive discussion of finite-length effects in DSP systems, and noise performance of FIR and IIR filters are found in DeFatta et al. [1988].

EXERCISES

3.1 The Parseval theorem in the Fourier transform states that if

$$x[n] \overset{F}{\longleftrightarrow} X(e^{j\omega})$$

then

$$\sum_{n=-\infty}^{\infty} |x[n]|^2 = \frac{1}{2\pi}\int_{-\pi}^{\pi} |X(e^{j\omega})|^2 d\omega.$$

Use the theorem to derive (3.3).

3.2 Derive (3.20) from (3.13) based on the fact that $R[n, n+l] = R[l]$ for a WSS random signal. That is, the autocorrelation is constant with respect to the time variable n.

3.3 The zero mean condition for white noise can be verified as follows. Given the white noise signal $x[n]$ with zero mean ($E\{x[n]\} = 0$), form another signal with $y[n] = x[n] + \mu$, where $\mu \neq 0$ is a constant. Show that the autocorrelation function of $y[n]$ does not satisfy the white noise definition.

3.4 The random variables $\mathbf{x}$ and $\mathbf{y}$ are uncorrelated if the cross-covariance defined by

$$C_{\mathbf{xy}} = E\{(\mathbf{x} - \mu_{\mathbf{x}})(\mathbf{y} - \mu_{\mathbf{y}})\}$$

is zero, with $\mu_{\mathbf{x}}$ and $\mu_{\mathbf{y}}$ being the mean of $\mathbf{x}$ and $\mathbf{y}$, respectively. Argue why, for a white noise signal, samples from different time instances are uncorrelated.

3.5 Cross-covariance for the jointly WSS random processes $x[n]$ and $y[n]$ is defined by

$$C_{xy}[l] = E\{(x[n] - \mu_x)(y[n-l] - \mu_y)\};$$

that is, it is a correlation function for the random processes with the mean removed (μ_x is the mean of $x[n]$ and μ_y is the mean of $y[n]$). Derive the following equations:

$$\begin{aligned} C_{yx}[l] &= h[l] * C_x[l], \\ C_{xy}[l] &= h^*[-l] * C_x[l], \\ C_y[l] &= h[l] * C_{xy}[l], \\ C_y[l] &= h[l] * h^*[-l] * C_x[l], \end{aligned}$$

where similar conditions as for (3.25) to (3.28) apply.

3.6 We are given an LTI system with transfer function $H(e^{j\omega})$. The system input is the WSS process $x[n]$ with PSD $S_x(e^{j\omega})$; the output process is $y[n]$. Then

$$\begin{aligned} S_{yx}(e^{j\omega}) &= H(e^{j\omega})S_x(e^{j\omega}), \\ S_{xy}(e^{j\omega}) &= H^*(e^{j\omega})S_x(e^{j\omega}), \\ S_y(e^{j\omega}) &= H(e^{j\omega})S_{xy}(e^{j\omega}). \end{aligned}$$

S_{xy} and S_{yx} are the cross PSD between $x[n]$ and $y[n]$ and are defined as the Fourier transform of the respective cross-correlation functions.

3.7 Show that the normal equation can be written in the form

$$\begin{pmatrix} 1 & a_1 & \cdots & a_M \\ a_1 & 1+a_2 & \cdots & 0 \\ \vdots & \vdots & \ddots & \vdots \\ a_M & a_{M-1} & \cdots & 1 \end{pmatrix} \begin{pmatrix} R_x[0] \\ R_x[1] \\ \vdots \\ R_x[M] \end{pmatrix} = \begin{pmatrix} \sigma_v^2 \\ 0 \\ \vdots \\ 0 \end{pmatrix}.$$

This form of the equation can be used to solve for the autocorrelation sequence $R_x[l]$ given the model coefficients a_i and the white noise variance σ_v^2.

3.8 Ignoring the relation containing the white noise variance (σ_v^2), show that the normal equation can be written as

$$\begin{pmatrix} R_x[0] & R_x[1] & \cdots & R_x[M-1] \\ R_x[1] & R_x[0] & \cdots & R_x[M-2] \\ \vdots & \vdots & \ddots & \vdots \\ R_x[M-1] & R_x[M-2] & \cdots & R_x[0] \end{pmatrix} \begin{pmatrix} a_1 \\ a_2 \\ \vdots \\ a_M \end{pmatrix} = - \begin{pmatrix} R_x[1] \\ R_x[2] \\ \vdots \\ R_x[M] \end{pmatrix}.$$

Thus, the AR parameters a_i can be found from the experimentally accessible quantities $R_x[l]$; that is, $R_x[l]$ can be estimated from actual data samples of $x[n]$.

3.9 For the ten-order AR model presented in Example 3.2, use the alternative form of the normal equation as shown in Exercise 3.7 to solve for the variance of the AR signal, assuming that the input white noise has unit variance.

3.10 Consider the l-dependent rectangular window

$$w[n] = \begin{cases} \sqrt{\frac{N}{N-|l|}}, & n = 0, 1, \ldots, N-1, \\ 0, & \text{otherwise.} \end{cases}$$

Substituting in (3.54) yields the estimator

$$R[l, m] = \frac{1}{N - |l|} \sum_{n=m-N+1+|l|}^{m} x[n]x[n - |l|]$$

with $|l| < N$. Show that $R[l, m]$ is an unbiased autocorrelation estimator.

One problem with this estimator is that when $|l|$ approaches N, the denominator in the above estimation equation approaches zero, leading to numerical problems. Thus, even though the estimator is unbiased, it is seldom used in practice.

3.11 Consider the autocorrelation estimator

$$R[l, m] = \frac{1}{N} \sum_{n=-\infty}^{\infty} x[n]x[n - |l|]w[m - n].$$

In this estimator, the signal product is computed first; it is then multiplied by the window to calculate the autocorrelation. For a rectangular window of length N show that

$$R[l, m] = \frac{1}{N} \sum_{n=m-N+1}^{m} x[n]x[n - |l|]$$

is an unbiased estimator.

3.12 Given

$$w_l[n] = w[n]w[n + l],$$

where

$$w[n] = (n + 1)\alpha^n u[n],$$

prove that

$$W_l(z) = \frac{(l+1)\alpha^l - (l-1)\alpha^{l+2}z^{-1}}{1 - 3\alpha^2 z^{-1} + 3\alpha^4 z^{-2} - \alpha^6 z^{-3}},$$

where $w_l[n]$ and $W_l(z)$ form a z-transform pair. *Hint*: Use the time-shift property of the z-transform [Oppenheim and Schafer, 1989]

$$w[n + l] \stackrel{z}{\longleftrightarrow} z^l W(z).$$

For $\alpha < |z| < \infty$, apply the complex convolution theorem and solve the contour integral based on the residue theorem [Churchill and Brown, 1990].

3.13 Repeat the Chen window design procedure for (a) $\alpha = 0.75^{1/40}, L = 35$ and (b) $\alpha = 0.75^{1/8}, L = 20$. Find the values of b and c for both cases. These specifications are used by the ITU-T G.728 LD-CELP coder (Chapter 14), where the first one is for the synthesis filter while the second one is for the log-gain predictor.

3.14 Draw the signal flow graphs for the ARMA process analyzer/synthesizer filters.

CHAPTER 4

LINEAR PREDICTION

Linear prediction (LP) forms an integral part of almost all modern day speech coding algorithms. The fundamental idea is that a speech sample can be approximated as a linear combination of past samples. Within a signal frame, the weights used to compute the linear combination are found by minimizing the mean-squared prediction error; the resultant weights, or linear prediction coefficients (LPCs*), are used to represent the particular frame.

Within the core of the LP scheme lies the autoregressive model (Chapter 3). Indeed, linear prediction analysis is an estimation procedure to find the AR parameters, given samples of the signal. Thus, LP is an identification technique where parameters of a system are found from the observation. The basic assumption is that speech can be modeled as an AR signal, which in practice has been found to be appropriate.

Another interpretation of LP is as a spectrum estimation method. As explained earlier, LP analysis allows the computation of the AR parameters, which define the PSD of the signal itself (Chapter 3). By computing the LPCs of a signal frame, it is possible to generate another signal in such a way that the spectral contents are close to the original one.

LP can also be viewed as a redundancy removal procedure where information repeated in an event is eliminated. After all, there is no need for transmission if certain data can be predicted. By displacing the redundancy in a signal, the amount

*In some literature, the linear prediction coefficients are referred to as LPC parameters, with the acronym meaning "linear prediction coding," which is the name assigned to an early standardized coder covered in Chapter 9. Since linear prediction is a general tool that might not apply to coding applications, we take the simpler approach in this book by referring to the coefficient as LPC. Hence, the LPC acronym bears two different meanings, which is normally clear from the context.

of bits required to carry the information is lowered, therefore achieving the purpose of compression.

In this chapter, the basic problem of LP analysis is stated, followed by its adaptation toward nonstationary signals. Examples of processing on actual speech samples are provided. Two computationally efficient procedures, namely, the Levinson–Durbin algorithm and the Leroux–Gueguen algorithm, are explained. The concept of long-term linear prediction is described, followed by some LP-based speech synthesis models. Practical issues related to speech processing are explained, with an alternative prediction scheme based on the moving average (MA) model given at the end of the chapter. LP is by no means confined to the speech processing arena; in fact, it is widely applied to many diverse areas. Readers are encouraged to consult other sources for additional information on the topic.

4.1 THE PROBLEM OF LINEAR PREDICTION

Here, linear prediction is described as a system identification problem, where the parameters of an AR model are estimated from the signal itself. The situation is illustrated in Figure 4.1. The white noise signal $x[n]$ is filtered by the AR process synthesizer to obtain $s[n]$—the AR signal—with the AR parameters denoted by $\hat{a}_i$. A linear predictor is used to predict $s[n]$ based on the M past samples; this is done with

$$\hat{s}[n] = -\sum_{i=1}^{M} a_i s[n-i], \tag{4.1}$$

where the a_i are the estimates of the AR parameters and are referred to as the linear prediction coefficients (LPCs)*. The constant M is known as the prediction order. Therefore, prediction is based on a linear combination of the M past samples of the signal, and hence the prediction is linear. The prediction error is equal to

$$e[n] = s[n] - \hat{s}[n]. \tag{4.2}$$

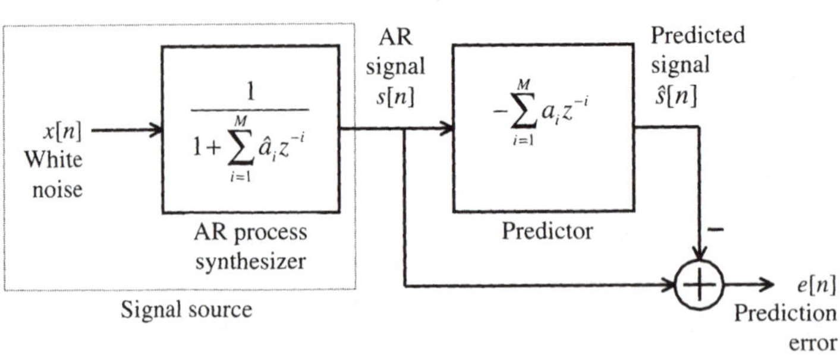

Figure 4.1 Linear prediction as system identification.

*In some literature, the sign convention for the LPC is reversed.

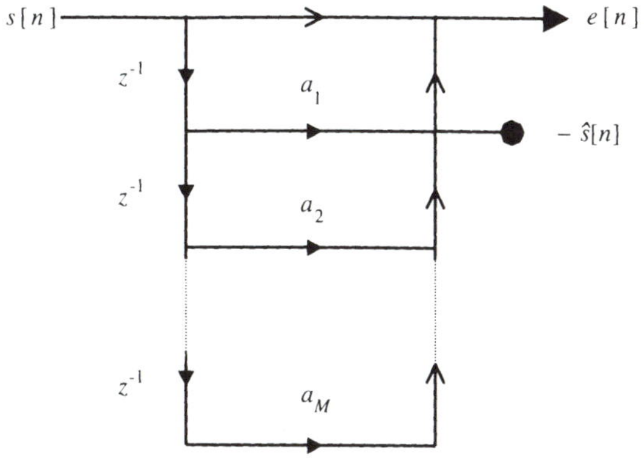

Figure 4.2 The prediction-error filter.

That is, it is the difference between the actual sample and the predicted one. Figure 4.2 shows the signal flow graph implementation of (4.2) and is known as the prediction-error filter: it takes an AR signal as input to produce the prediction-error signal at its output.

Error Minimization

The system identification problem consists of the estimation of the AR parameters $\hat{a}_i$ from $s[n]$, with the estimates being the LPCs. To perform the estimation, a criterion must be established. In the present case, the mean-squared prediction error

$$J = E\{e^2[n]\} = E\left\{\left(s[n] + \sum_{i=1}^{M} a_i s[n-i]\right)^2\right\} \tag{4.3}$$

is minimized by selecting the appropriate LPCs. Note that the cost function J is precisely a second-order function of the LPCs. Consequently, we may visualize the dependence of the cost function J on the estimates a_1, a_2, ..., a_M as a bowl-shaped $(M+1)$-dimensional surface with M degrees of freedom. This surface is characterized by a unique minimum. The optimal LPCs can be found by setting the partial derivatives of J with respect to a_i to zero; that is,

$$\frac{\partial J}{\partial a_k} = 2E\left\{\left(s[n] + \sum_{i=1}^{M} a_i s[n-i]\right)s[n-k]\right\} = 0 \tag{4.4}$$

for $k = 1, 2, \ldots, M$. At this point, it is maintained without proof that when (4.4) is satisfied, then $a_i = \hat{a}_i$; that is, the LPCs are equal to the AR parameters. Justification of this claim appears at the end of the section. Thus, when the LPCs are found, the system used to generate the AR signal (AR process synthesizer) is uniquely identified.

Normal Equation

Equation (4.4) can be rearranged to give

$$E\{s[n]s[n-k]\} + \sum_{i=1}^{M} a_i E\{s[n-i]s[n-k]\} = 0 \tag{4.5}$$

or

$$\sum_{i=1}^{M} a_i R_s[i-k] = -R_s[k] \tag{4.6}$$

for $k = 1, 2, \ldots, M$, where

$$R_s[i-k] = E\{s[n-i]s[n-k]\}, \tag{4.7}$$

$$R_s[k] = E\{s[n]s[n-k]\}. \tag{4.8}$$

Equation (4.6) defines the optimal LPCs in terms of the autocorrelation $R_s[l]$ of the signal $s[n]$. In matrix form, it can be written as

$$\mathbf{R_s a} = -\mathbf{r_s}, \tag{4.9}$$

where

$$\mathbf{R_s} = \begin{pmatrix} R_s[0] & R_s[1] & \cdots & R_s[M-1] \\ R_s[1] & R_s[0] & \cdots & R_s[M-2] \\ \vdots & \vdots & \ddots & \vdots \\ R_s[M-1] & R_s[M-2] & \cdots & R_s[0] \end{pmatrix}, \tag{4.10}$$

$$\mathbf{a} = [a_1 \quad a_2 \quad \cdots \quad a_M]^T, \tag{4.11}$$

$$\mathbf{r_s} = [R_s[1] \quad R_s[2] \quad \cdots \quad R_s[M]]^T. \tag{4.12}$$

Equation (4.9) is known as the normal equation. Assuming that the inverse of the correlation matrix $\mathbf{R_s}$ exists, the optimal LPC vector is obtained with

$$\mathbf{a} = -\mathbf{R_s}^{-1}\mathbf{r_s}. \tag{4.13}$$

Equation (4.13) allows the finding of the LPCs if the autocorrelation values of $s[n]$ are known from $l = 0$ to M.

Prediction Gain

The prediction gain of a predictor is given by

$$PG = 10\log_{10}\left(\frac{\sigma_s^2}{\sigma_e^2}\right) = 10\log_{10}\left(\frac{E\{s^2[n]\}}{E\{e^2[n]\}}\right) \tag{4.14}$$

and is the ratio between the variance of the input signal and the variance of the prediction error in decibels (dB). Prediction gain is a measure of the predictor's performance. A better predictor is capable of generating lower prediction error, leading to a higher gain.

Example 4.1: Predicting White Noise Consider the situation when $s[n]$ is a white noise signal; that is, $R_s[l] = \sigma_s^2\delta[l]$. From (4.12) we see that $\mathbf{r}_s$ is the zero vector and from (4.10) $\mathbf{R}_s$ is a diagonal matrix, leading to the LPC vector $\mathbf{a} = \mathbf{0}$. Hence, $e[n] = s[n]$ and the prediction gain is $PG = 0$ dB. The result means that white noise is *unpredictable*: nothing can be gained with a predictor. The unpredictability is due to the fact that no correlation exists between white noise samples. For most real-world signals, like speech, correlation exists and hence it is possible to obtain higher than zero gain with a linear predictor.

Minimum Mean-Squared Prediction Error

From Figure 4.1 we can see that when $a_i = \hat{a}_i$, $e[n] = x[n]$; that is, the prediction error is the same as the white noise used to generate the AR signal $s[n]$. Indeed, this is the optimal situation where the mean-squared error is minimized, with

$$J_{\min} = E\{e^2[n]\} = E\{x^2[n]\} = \sigma_x^2, \tag{4.15}$$

or equivalently, the prediction gain is maximized.

The optimal condition can be reached when the order of the predictor is equal to or higher than the order of the AR process synthesizer. In practice, M is usually unknown. A simple method to estimate M from a signal source is by plotting the prediction gain as a function of the prediction order. In this way it is possible to determine the prediction order for which the gain saturates; that is, further increasing the prediction order from a certain critical point will not provide additional gain. The value of the predictor order at the mentioned critical point represents a good estimate of the order of the AR signal under consideration.

As was explained before, the cost function J in (4.3) is characterized by a unique minimum. If the prediction order M is known, J is minimized when $a_i = \hat{a}_i$, leading to $e[n] = x[n]$; that is, prediction error is equal to the excitation signal of the AR process synthesizer. This is a reasonable result since the best that the prediction-error filter can do is to "whiten" the AR signal $s[n]$. Thus, the maximum prediction gain is given by the ratio between the variance of $s[n]$ and the variance of $x[n]$ in decibels.

Taking into account the AR parameters used to generate the signal $s[n]$, we have

$$J_{\min} = \sigma_x^2 = R_s[0] + \sum_{i=1}^{M} a_i R_s[i], \tag{4.16}$$

which was already derived in Chapter 3. The above equation can be combined with (4.9) to give

$$\begin{bmatrix} R_s[0] & \mathbf{r}_s^T \\ \mathbf{r}_s & \mathbf{R}_s \end{bmatrix} \begin{bmatrix} 1 \\ \mathbf{a} \end{bmatrix} = \begin{bmatrix} J_{\min} \\ \mathbf{0} \end{bmatrix} \tag{4.17}$$

and is known as the augmented normal equation, with $\mathbf{0}$ the $M \times 1$ zero vector. Equation (4.17) can also be written as

$$\sum_{i=0}^{M} a_i R_s[i-k] = \begin{cases} J_{\min}, & k = 0 \\ 0, & k = 1, 2, \ldots, M \end{cases} \tag{4.18}$$

where $a_0 = 1$.

4.2 LINEAR PREDICTION ANALYSIS OF NONSTATIONARY SIGNALS

So far the discussion is focused on a WSS stochastic process. Due to the dynamic nature of a speech signal, the LPCs must be calculated for every signal frame. Within a frame, one set of LPCs is determined and used to represent the signal's properties in that particular interval, with the underlying assumption that the statistics of the signal remain unchanged within the frame. The process of calculating the LPCs from signal data is called linear prediction analysis.

The problem of linear prediction is restated as follows. It is desired to calculate the LPCs on the N data points ending at time m: $s[m-N+1]$, $s[m-N+2], \ldots, s[m]$. The LPC vector is written as

$$\mathbf{a}[m] = [a_1[m] \quad a_2[m] \quad \cdots \quad a_M[m]]^T \tag{4.19}$$

with M being the prediction order. From the last section, we need to solve the normal equation, rewritten here in a time-adaptive form

$$\mathbf{R}[m]a[m] = -\mathbf{r}[m], \tag{4.20}$$

with

$$\mathbf{R}[m] = \begin{pmatrix} R[0,m] & R[1,m] & R[2,m] & \cdots & R[M-1,m] \\ R[1,m] & R[0,m] & R[1,m] & \cdots & R[M-2,m] \\ R[2,m] & R[1,m] & R[0,m] & \cdots & R[M-3,m] \\ \vdots & \vdots & \ddots & & \vdots \\ R[M-1,m] & R[M-2,m] & R[M-3,m] & \cdots & R[0,m] \end{pmatrix} \tag{4.21}$$

and

$$\mathbf{r}[m] = [R[1,m] \quad R[2,m] \quad \cdots \quad R[M,m]]^T. \tag{4.22}$$

Hence, for the case of nonstationary signals, LP analysis is performed for every signal frame ending at time m. The autocorrelation values $R[l,m]$ are estimated for each frame and the normal equation is solved to yield the set of LPCs associated with the particular frame. Methods of autocorrelation estimation are extensively discussed in Chapter 3.

Prediction Schemes

Different prediction schemes are used in various applications and are decided by system requirements. Generally, two main techniques are applied in speech coding: *internal prediction* and *external prediction*. Figure 4.3 illustrates the schemes. For internal prediction, the LPCs derived from the estimated autocorrelation values using the frame's data are applied to process the frame's data themselves. In external prediction, however, the derived LPCs are used in a future frame; that is, the

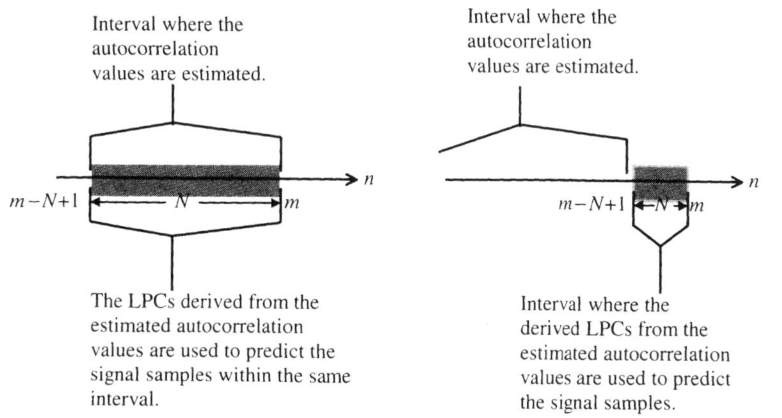

Figure 4.3 Illustration of internal prediction (*left*) and external prediction (*right*).

LPCs associated with the frame are not derived from the data residing within the frame, but from the signal's past. The reason why external prediction can be used is because the signal statistics change slowly with time. If the frame is not excessively long, its properties can be derived from the not so distant past.

Many speech coding algorithms use internal prediction, where the LPCs of a given frame are derived from the data pertaining to the frame. Thus, the resultant LPCs capture the statistics of the frame accurately. Typical length of the frame varies from 160 to 240 samples. A longer frame has the advantage of less computational complexity and lower bit-rate, since calculation and transmission of LPCs are done less frequently. However, a longer coding delay results from the fact that the system has to wait longer for sample collection. Also, due to the changing nature of a nonstationary environment, the LPCs derived from a long frame might not be able to produce good prediction gain. On the other hand, a shorter frame requires more frequent update of the LPCs, resulting in a more accurate representation of the signal statistics. Drawbacks include higher computational load and bit-rate. Most internal prediction schemes rely on nonrecursive autocorrelation estimation methods, where a finite-length window is used to extract the signal samples.

External prediction is prevalently used in those applications where low coding delay is the prime concern. In that case, a much shorter frame must be used (on the order of 20 samples, such as the LD-CELP standard—Chapter 14). A recursive autocorrelation estimation technique is normally applied so that the LPCs are derived from the samples before the time instant $n = m - N + 1$ (Figure 4.3). Note that the shape of the window associated with a recursive autocorrelation estimation technique puts more emphasis on recent samples. Thus, the statistics associated with the estimates are very close to the actual properties of the frame itself, even though the estimation is not based on the data internal to the frame.

In many instances, the notions of internal and external become fuzzy. As we will see later in the book, many LP analysis schemes adopted by standardized coders are based on estimating several (usually two) sets of LPCs from contiguous analysis intervals. These coefficients are combined in a specific way and applied to a given interval for the prediction task. We skip the details for now, which are covered thoroughly in Chapter 8, when interpolation of LPCs is introduced.

Prediction Gain

Prediction gain is given here using a similar definition as presented in the last section, with the expectations changed to summations

$$PG[m] = 10 \log_{10}\left(\frac{\sum_{n=m-N+1}^{m} s^2[n]}{\sum_{n=m-N+1}^{m} e^2[n]}\right), \tag{4.23}$$

where

$$e[n] = s[n] - \hat{s}[n] = s[n] + \sum_{i=1}^{M} a_i[m]s[n-i]; \quad n = m - N + 1, \ldots, m. \tag{4.24}$$

The LPCs $a_i[m]$ are found from the samples inside the interval $[m - N + 1, m]$ for internal prediction, and $n < m - N + 1$ for external prediction. Note that the prediction gain defined in (4.23) is a function of the time variable m. In practice, the average performance of a prediction scheme is often measured by the segmental prediction gain, defined with

$$SPG = A\{PG[m]\}, \tag{4.25}$$

which is the time average of the prediction gain for each frame in the decibel domain.

Example 4.2 White noise is generated using a random number generator with uniform distribution and unit variance. This signal is then filtered by an AR synthesizer with

$a_1 = 1.534$	$a_2 = 1$	$a_3 = 0.587$	$a_4 = 0.347$	$a_5 = 0.08$
$a_6 = -0.061$	$a_7 = -0.172$	$a_8 = -0.156$	$a_9 = -0.157$	$a_10 = -0.141$

The frame of the resultant AR signal is used for LP analysis, with a length of 240 samples. Nonrecursive autocorrelation estimation using a Hamming window is applied. LP analysis is performed with prediction order ranging from 2 to 20; prediction error and prediction gain are found for each case. Figure 4.4 summarizes the results, where we can see that the prediction gain grows initially from $M = 2$ and is maximized when $M = 10$. Further increasing the prediction order will not provide additional gain; in fact, it can even reduce it. This is an expected result since the AR model used to generate the signal has order ten.

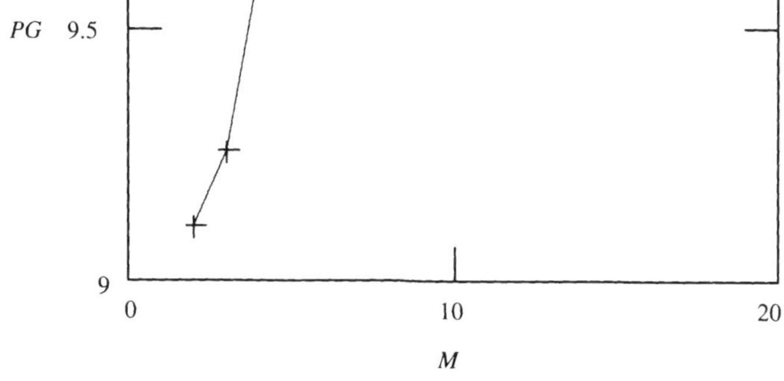

Figure 4.4 Prediction gain (PG) as a function of the prediction order (M) in an experiment.

Theoretically, the curve of prediction gain as a function of the prediction order should be monotonically increasing, meaning that $PG(M_1) \leq PG(M_2)$ if $M_1 \leq M_2$. In the present experiment, however, only one sample realization of the random process is utilized; thus, the general behavior of the linear predictor is not fully revealed. For a more accurate study on the behavior of the signal, a higher number of sample realizations for the random signal are needed.

Figure 4.5 compares the theoretical PSD (defined with the original AR parameters) with the spectrum estimates found with the LPCs computed from the signal frame using $M = 2$, 10, and 20. For low prediction order, the resultant spectrum is not capable of fitting the original PSD. An excessively high order, on the other hand, leads to overfitting, where undesirable errors are introduced. In the present case, a prediction order of 10 is optimal. Note how the spectrum of the original signal is captured by the estimated LPCs. This is the reason why LP analysis is known as a spectrum estimation technique, specifically a parametric spectrum estimation method since the process is done through a set of parameters or coefficients.

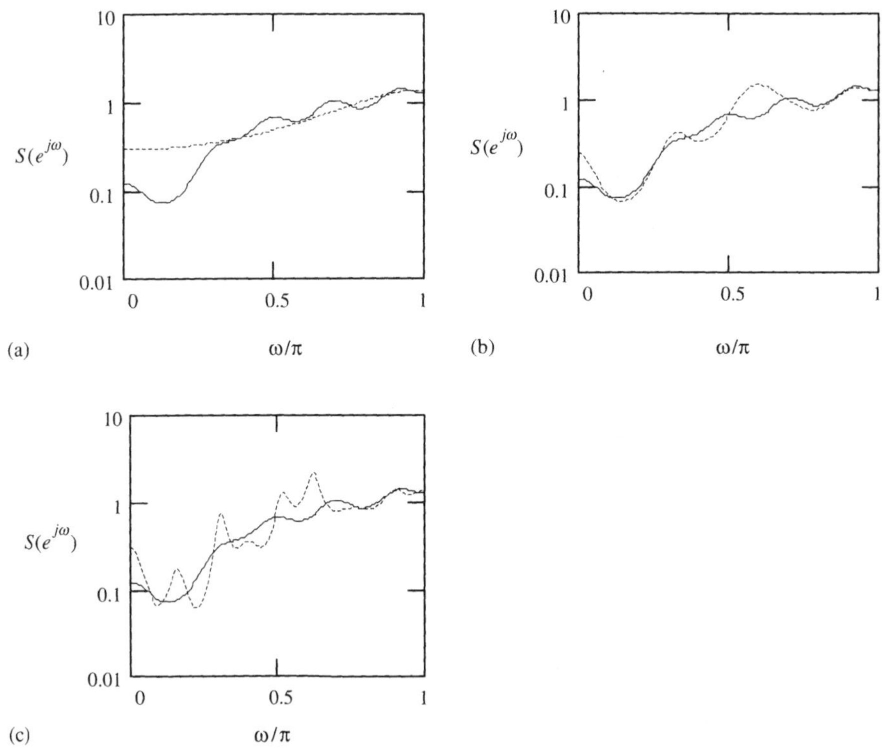

Figure 4.5 Plots of PSD (solid trace) together with several estimates (dot trace) using the LPC found with (a) $M = 2$, (b) $M = 10$, and (c) $M = 20$.

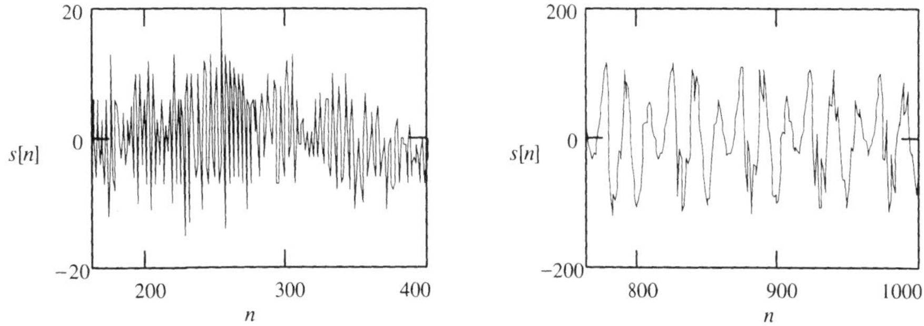

Figure 4.6 The speech frames used in the experiment. *Left*: Unvoiced ($m = 400$). *Right*: Voiced ($m = 1000$).

4.3 EXAMPLES OF LINEAR PREDICTION ANALYSIS OF SPEECH

So far the linear prediction analysis technique was described in a general context. For applications involving a speech signal, the signal itself is often assumed to satisfy the AR model. Facts related to LP analysis of speech are derived in this section from real speech samples, where accuracy of the AR assumption is evaluated. The observations made are used to tailor the scheme of LP as applied to speech coding.

Example 4.3 Speech samples of a male subject are used in the experiment. Figure 4.6 shows the speech frames considered. As we can see, the frame ending at $m = 400$ is unvoiced, and the frame ending at $m = 1000$ is voiced, with a pitch period approximately equal to 49 time-units. Also note that the unvoiced frame has amplitude far lower than the voiced frame, which is commonly the case in practice. Length of each frame is equal to 240 samples—a popular value used in speech coding. Periodograms of the two frames are plotted in Figure 4.7. Note that the

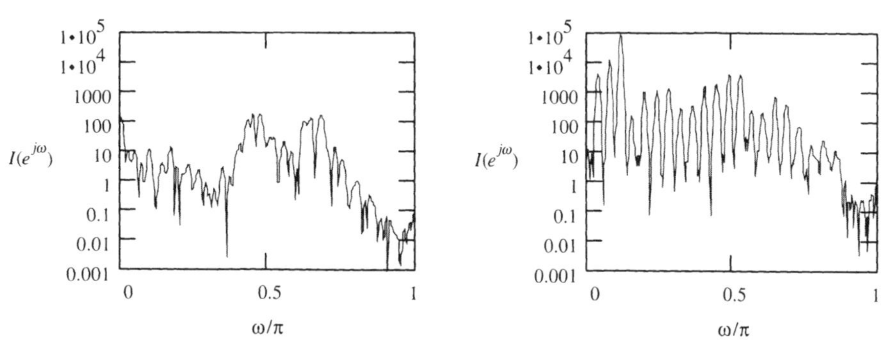

Figure 4.7 Periodograms of the signal frames in Figure 4.6. *Left*: Unvoiced. *Right*: Voiced.

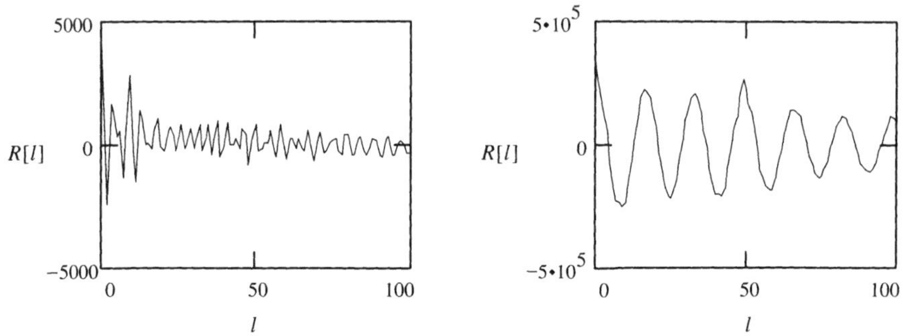

Figure 4.8 Autocorrelation values for the signal frames in Figure 4.6. *Left*: Unvoiced. *Right*: Voiced.

spectrum of the unvoiced frame is relatively smooth, while for the voiced frame a harmonic structure is present, indicating a strong fundamental component in the signal. Obviously the harmonics are associated with periodicity in the time domain.

Periodicity can also be detected or measured from the autocorrelation values shown in Figure 4.8, where the lag ranges from 0 to 100. For the noise-like unvoiced frame, the values of the autocorrelation have low magnitude when the lag is higher than ten, indicating the fact that correlation between distant samples is low. For the voiced frame, on the other hand, high correlation exists between samples, which is particularly strong when the lag is equal to the pitch period and is around 49 in the present case. As expected, the value of the autocorrelation gradually decreases with increasing lag in both cases since correlation between samples tends to weaken. These results show that it is possible to classify a frame as unvoiced or voiced by calculating its autocorrelation, and pitch period can be determined by locating the peaks of the autocorrelation.

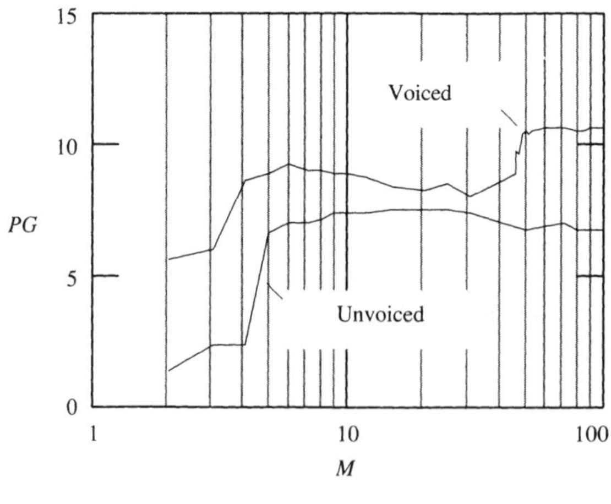

Figure 4.9 Plot of prediction gain (PG) as a function of the prediction order (M) for the signal frames in Figure 4.6.

The selected frames are used for LP analysis, where the derived LPCs are employed to predict the samples within the frame (internal prediction). The prediction gain results are found for prediction order ranging from 2 to 100 and are plotted in Figure 4.9.

STATEMENT 1: *For a given prediction order, the average prediction gain obtainable for voiced frames is higher than for unvoiced frames.*

The above statement is partially reflected in Figure 4.9 and is true in general when a large number of frames are analyzed. This can be understood from the nature of the signal itself. An unvoiced frame is highly random, with low correlation between samples, and therefore less predictable than a voiced frame.

Back to Figure 4.9, we observe that for the unvoiced frame, prediction gain increases abruptly when the prediction order goes from 2 to 5. Further increasing the prediction order provides additional gain increase, but at a milder pace. For $M > 10$, prediction gain remains essentially constant, implying the fact that correlation between far separated samples is low.

For the voiced frame, prediction gain is low for $M \leq 3$, it remains almost constant for $4 < M < 49$, and it reaches a peak for $M \geq 49$. The phenomenon is due to the fact that, for the voiced frame under consideration, the pitch period is approximately equal to 49. For $M < 49$, the number of LPCs is not enough to remove the correlation between samples one pitch period apart. For $M \geq 49$, however, the linear predictor is capable of modeling the correlation between samples one pitch period apart, leading therefore to a substantial improvement in prediction gain. Further note that the change in prediction gain is abrupt: between $M = 48$ and 49, for instance, a jump of nearly 3 dB in prediction gain is observed.

STATEMENT 2: *For a voiced frame, the prediction gain associated with a predictor having a prediction order large enough to cover one pitch period is substantially higher than the prediction gain associated with a predictor having a prediction order lower than one pitch period.*

The above statement is a key observation to develop the concept of long-term prediction (described later), which is an efficient modeling strategy for voiced signals.

The effectiveness of the predictor at different prediction orders can be studied further by observing the level of "whiteness" in the prediction-error sequence. The prediction-error filter associated with a good predictor is capable of removing as much correlation as possible from the signal samples, leading to a prediction-error sequence with a flat PSD. Figure 4.10 illustrates the prediction-error sequence of the unvoiced frame and the corresponding periodogram for different prediction order. Note that $M = 4$ is not enough to "whiten" the original signal frame, where we can see that the periodogram of the prediction error does not mimic the flat spectrum of a white noise signal. For $M = 10$, however, flatness is achieved in the periodogram and, hence, the prediction error becomes "roughly" white.

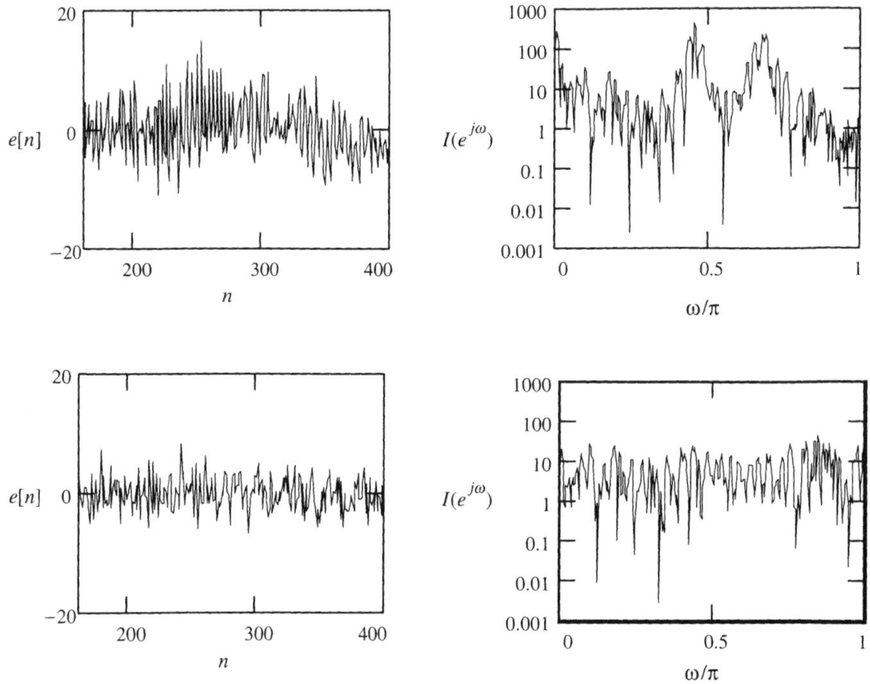

Figure 4.10 Plots of prediction error and periodograms for the unvoiced frame in Figure 4.6. *Top*: $M = 4$. *Bottom*: $M = 10$.

Figure 4.11 shows the prediction-error sequences and the corresponding periodograms of the voiced frame. We can see that for $M = 3$, a high level of periodicity is still present in the prediction-error sequence and a harmonic structure is observed in the corresponding periodogram. When $M = 10$, the amplitude of the prediction-error sequence becomes lower. However, the periodic components remain. As we can see, the periodogram develops a flatter appearance, but the harmonic structure is still present. For $M = 50$, periodicity in the time domain and frequency domain is reduced to a minimum. Hence, in order to effectively "whiten" the voiced frame, a minimum prediction order equal to 50 is required.

STATEMENT 3: *To remove the correlation between samples using a linear predictor, a much higher prediction order is required for a voiced frame than for an unvoiced frame. For effective whitening of a voiced frame, the prediction order should be greater than or equal to the underlying pitch period of the signal.*

For many speech coding algorithms where the LPCs are quantized and transmitted as information on the frame, a prediction order of ten is normally used. In general, an order of ten can describe quite well the PSD of an unvoiced frame, but it

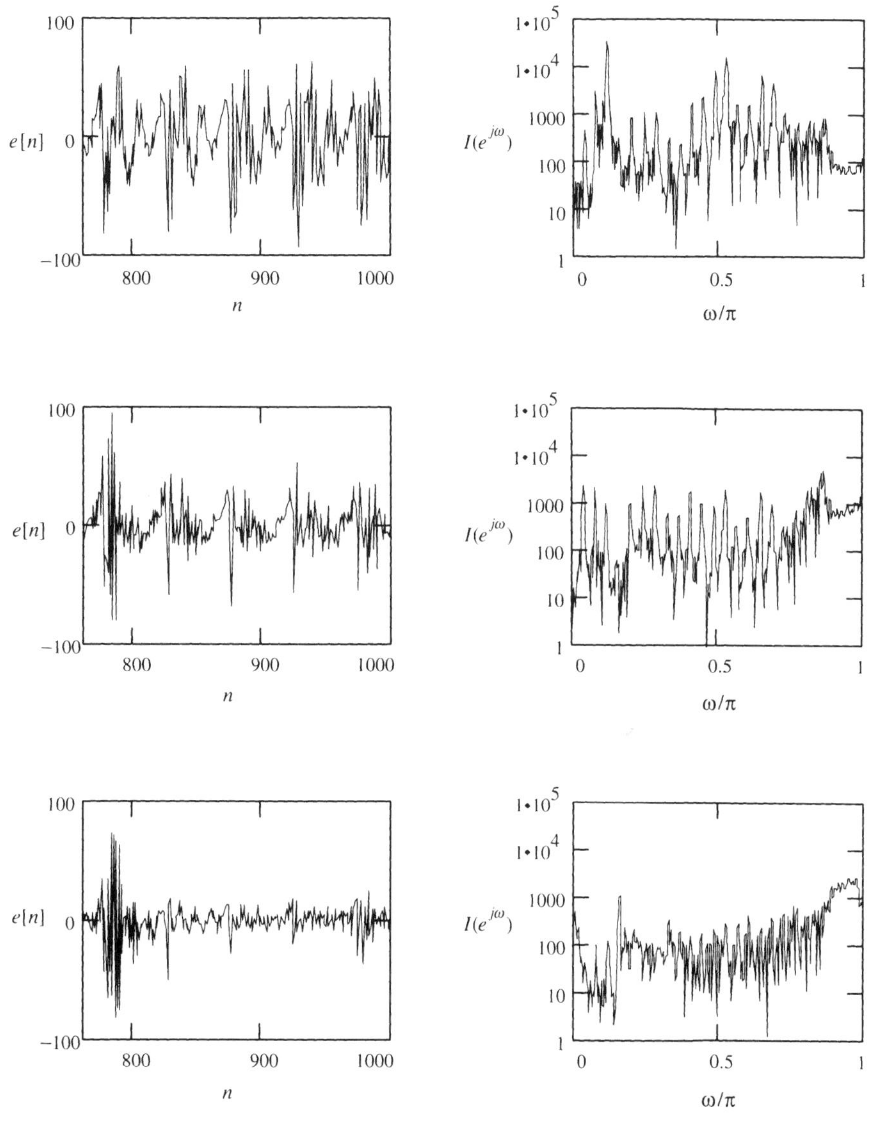

Figure 4.11 Plots of prediction error and periodograms for the voiced frame in Figure 4.6. *Top*: $M = 3$. *Middle*: $M = 10$. *Bottom*: $M = 50$.

is definitely inadequate for a voiced frame. As we will see later in this book, different coding algorithms use different strategies to recreate the spectrum of a voiced frame. Most of these algorithms rely on a predictor of order ten to capture the "envelope" of the PSD. The idea of the envelope of the spectrum is illustrated in Figure 4.12, where the PSD of the voiced frame using the derived LPCs is superimposed with the periodogram of the speech signal. We can see that the spectrum

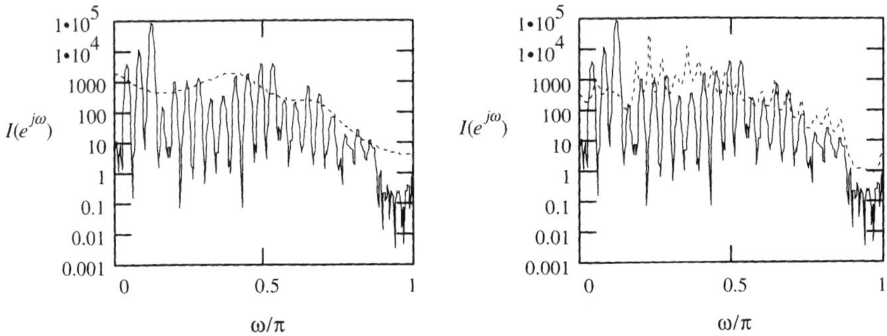

Figure 4.12 LPC-based spectrum estimate (dotted line) and periodogram (solid line) for a voiced speech frame. *Left*: $M = 10$. *Right*: $M = 50$.

estimate with a prediction order of ten represents a smoothed version, or the envelope of the signal spectrum. When $M = 50$, the spectrum estimate becomes much closer to the periodogram.

Example 4.4: External Prediction Using the Chen Window The Chen window (Chapter 3) with $\alpha = 0.5^{1/40}$ and $L = 30$ is used in autocorrelation estimation. With a prediction order of 50, a linear predictor is derived from the autocorrelation values and used in external prediction, where the frame length is varied from 10 to 50. Segmental prediction gain is measured using roughly 40 seconds of speech material. Figure 4.13 summarizes the results, where the segmental prediction gain is plotted as

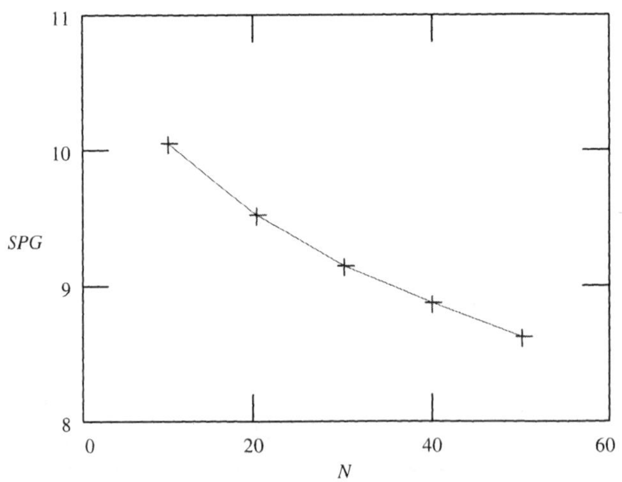

Figure 4.13 Segmental prediction gain (*SPG*) as a function of the frame's length (*N*) in an external prediction experiment.

a function of the frame's length. As we can see, prediction gain is the highest for short frames, which is expected, since as the frame's length increases, the statistics derived from the signal's past become less and less accurate for representing the frame itself; therefore prediction gain drops. In the ITU-T G.728 LD-CELP coder (Chapter 14), the frame's length is equal to 20 samples and is a trade-off between computational complexity, coding delay, and quality of the synthesized speech.

4.4 THE LEVINSON–DURBIN ALGORITHM

The normal equation as given in (4.9) can be solved by finding the matrix inverse for $\mathbf{R}_s$, with the solution provided in (4.13). In general, inverting a matrix is quite computationally demanding. Fortunately, efficient algorithms are available to solve the equation, which take advantage of the special structure of the correlation matrix. This section discusses the Levinson–Durbin algorithm while the next one is concerned with the Leroux–Gueguen algorithm, both highly suitable for practical implementation of LP analysis. Consider the augmented normal equation of form

$$\begin{bmatrix} R[0] & R[1] & \cdots & R[M] \\ R[1] & R[0] & \cdots & R[M-1] \\ \vdots & \vdots & \ddots & \vdots \\ R[M] & R[M-1] & \cdots & R[0] \end{bmatrix} \begin{bmatrix} 1 \\ a_1 \\ \vdots \\ a_M \end{bmatrix} = \begin{bmatrix} J \\ 0 \\ \vdots \\ 0 \end{bmatrix} \qquad (4.26)$$

with the objective being the solution for the LPCs $a_i, i = 1, \ldots, M$, given the autocorrelation values $R[l], l = 0, 1, \ldots, M$. J represents the minimum mean-squared prediction error or the variance of the input white noise for the AR process synthesizer. In a practical situation, the autocorrelation values are estimated from the signal samples and J is usually unknown; however, the Levinson–Durbin solution is formulated to solve for this quantity as well.

The Levinson–Durbin approach finds the solution to the Mth-order predictor from that of the $(M-1)$th-order predictor. It is an iterative–recursive process where the solution of the zero-order predictor is first found, which is then used to find the solution of the first-order predictor; this process is repeated one step at a time until the desired order is reached. The algorithm relies on two key properties of the correlation matrix:

- The correlation matrix of a given size contains as subblocks all the lower-order correlation matrices.
- If

$$\begin{bmatrix} R[0] & R[1] & \cdots & R[M] \\ R[1] & R[0] & \cdots & R[M-1] \\ \vdots & \vdots & \ddots & \vdots \\ R[M] & R[M-1] & \cdots & R[0] \end{bmatrix} \begin{bmatrix} a_0 \\ a_1 \\ \vdots \\ a_M \end{bmatrix} = \begin{bmatrix} b_0 \\ b_1 \\ \vdots \\ b_M \end{bmatrix}, \qquad (4.27)$$

then

$$\begin{bmatrix} R[0] & R[1] & \cdots & R[M] \\ R[1] & R[0] & \cdots & R[M-1] \\ \vdots & \vdots & \ddots & \vdots \\ R[M] & R[M-1] & \cdots & R[0] \end{bmatrix} \begin{bmatrix} a_M \\ a_{M-1} \\ \vdots \\ a_0 \end{bmatrix} = \begin{bmatrix} b_M \\ b_{M-1} \\ \vdots \\ b_0 \end{bmatrix}; \tag{4.28}$$

that is, the correlation matrix is invariant under interchange of its columns and then its rows. The mentioned properties are direct consequences of the fact that the correlation matrix is Toeplitz. We say that a square matrix is Toeplitz if all the elements on its main diagonal are equal, and if the elements on any other diagonal parallel to the main diagonal are also equal.

We consider the solution to the augmented normal equation starting from zero prediction order. It is shown that the solution for a certain order can be obtained from the lower prediction order results.

Predictor of Order Zero

In this case we consider the equation

$$R[0] = J_0, \tag{4.29}$$

which is already solved. The above relation states basically that the minimum mean-squared prediction error achievable with a zero-order predictor is given by the autocorrelation of the signal at lag zero, or the variance of the signal itself. For zero order the prediction is always equal to zero; hence, the prediction error is equal to the signal itself. Expanding (4.29) to the next dimension, we have

$$\begin{bmatrix} R[0] & R[1] \\ R[1] & R[0] \end{bmatrix} \begin{bmatrix} 1 \\ 0 \end{bmatrix} = \begin{bmatrix} J_0 \\ \Delta_0 \end{bmatrix}, \tag{4.30}$$

which is the two-dimensional (2-D) version of (4.26) with $a_1 = 0$. Since $a_1 = 0$, the optimal condition cannot be achieved in general, and the term Δ_0 is introduced on the right-hand side to balance the equation. This quantity is found from the equation as

$$\Delta_0 = R[1]. \tag{4.31}$$

From the property of the correlation matrix, (4.30) is equivalent to

$$\begin{bmatrix} R[0] & R[1] \\ R[1] & R[0] \end{bmatrix} \begin{bmatrix} 0 \\ 1 \end{bmatrix} = \begin{bmatrix} \Delta_0 \\ J_0 \end{bmatrix} \tag{4.32}$$

Equations (4.30) and (4.32) are used in the next step.

Predictor of Order One

We seek to solve

$$\begin{bmatrix} R[0] & R[1] \\ R[1] & R[0] \end{bmatrix} \begin{bmatrix} 1 \\ a_1^{(1)} \end{bmatrix} = \begin{bmatrix} J_1 \\ 0 \end{bmatrix}, \tag{4.33}$$

where $a_1^{(1)}$ is the LPC of the predictor; the superscript denotes the prediction order of one. J_1 represents the minimum mean-squared prediction error achievable using a first-order predictor. Thus, we have two unknowns in (4.33): $a_1^{(1)}$ and J_1. Consider a solution of the form

$$\begin{bmatrix} 1 \\ a_1^{(1)} \end{bmatrix} = \begin{bmatrix} 1 \\ 0 \end{bmatrix} - k_1 \begin{bmatrix} 0 \\ 1 \end{bmatrix}, \tag{4.34}$$

with k_1 being a constant. Multiplying both sides by the correlation matrix, we have

$$\begin{bmatrix} R[0] & R[1] \\ R[1] & R[0] \end{bmatrix} \begin{bmatrix} 1 \\ a_1^{(1)} \end{bmatrix} = \begin{bmatrix} R[0] & R[1] \\ R[1] & R[0] \end{bmatrix} \begin{bmatrix} 1 \\ 0 \end{bmatrix} - k_1 \begin{bmatrix} R[0] & R[1] \\ R[1] & R[0] \end{bmatrix} \begin{bmatrix} 0 \\ 1 \end{bmatrix} \tag{4.35}$$

Substituting (4.30), (4.32), and (4.33) gives

$$\begin{bmatrix} J_1 \\ 0 \end{bmatrix} = \begin{bmatrix} J_0 \\ \Delta_0 \end{bmatrix} - k_1 \begin{bmatrix} \Delta_0 \\ J_0 \end{bmatrix}. \tag{4.36}$$

Then

$$k_1 = \frac{\Delta_0}{J_0} = \frac{R[1]}{J_0}, \tag{4.37}$$

where (4.31) is used. The LPC of this predictor is readily found from (4.34) to be

$$a_1^{(1)} = -k_1. \tag{4.38}$$

Using (4.36) and (4.37), we find

$$J_1 = J_0\left(1 - k_1^2\right). \tag{4.39}$$

Thus, the first-order predictor is completely specified. The parameter k_1 is known as the reflection coefficient (RC), representing an alternative form of LPC. Note that k_1 (and therefore $a_1^{(1)}$ and J_1) is derived from the results of the previous step: the zero-order predictor. In a similar manner, we can expand (4.33) to dimension three:

$$\begin{bmatrix} R[0] & R[1] & R[2] \\ R[1] & R[0] & R[1] \\ R[2] & R[1] & R[0] \end{bmatrix} \begin{bmatrix} 1 \\ a_1^{(1)} \\ 0 \end{bmatrix} = \begin{bmatrix} J_1 \\ 0 \\ \Delta_1 \end{bmatrix} \quad (4.40)$$

or

$$\begin{bmatrix} R[0] & R[1] & R[2] \\ R[1] & R[0] & R[1] \\ R[2] & R[1] & R[0] \end{bmatrix} \begin{bmatrix} 0 \\ a_1^{(1)} \\ 1 \end{bmatrix} = \begin{bmatrix} \Delta_1 \\ 0 \\ J_1 \end{bmatrix}, \quad (4.41)$$

where Δ_1 represents the additional term necessary to balance the equation when a first-order predictor is used and $R[2] \neq 0$. This quantity is solved as

$$\Delta_1 = R[2] + a_1^{(1)} R[1]. \quad (4.42)$$

Predictor of Order Two

We go one step further by solving

$$\begin{bmatrix} R[0] & R[1] & R[2] \\ R[1] & R[0] & R[1] \\ R[2] & R[1] & R[0] \end{bmatrix} \begin{bmatrix} 1 \\ a_1^{(2)} \\ a_2^{(2)} \end{bmatrix} = \begin{bmatrix} J_2 \\ 0 \\ 0 \end{bmatrix}. \quad (4.43)$$

The unknowns in this case are the LPCs $a_1^{(2)}$ and $a_2^{(2)}$ and the minimum mean-squared prediction error J_2. Consider a solution of the form

$$\begin{bmatrix} 1 \\ a_1^{(2)} \\ a_2^{(2)} \end{bmatrix} = \begin{bmatrix} 1 \\ a_1^{(1)} \\ 0 \end{bmatrix} - k_2 \begin{bmatrix} 0 \\ a_1^{(1)} \\ 1 \end{bmatrix}, \quad (4.44)$$

with k_2 as the RC. Multiplying both sides by the correlation matrix leads to

$$\begin{bmatrix} J_2 \\ 0 \\ 0 \end{bmatrix} = \begin{bmatrix} J_1 \\ 0 \\ \Delta_1 \end{bmatrix} - k_2 \begin{bmatrix} \Delta_1 \\ 0 \\ J_1 \end{bmatrix}, \quad (4.45)$$

where (4.40), (4.41), and (4.43) are used to derive the above relation. The RC k_2 can be found from (4.45) and using (4.42) for Δ_1:

$$k_2 = \frac{1}{J_1}\left(R[2] + a_1^{(1)} R[1]\right). \tag{4.46}$$

From (4.44), we find

$$a_2^{(2)} = -k_2, \tag{4.47}$$

$$a_1^{(2)} = a_1^{(1)} - k_2 a_1^{(1)}. \tag{4.48}$$

Finally, J_2 is found from (4.45) and (4.46) as

$$J_2 = J_1\left(1 - k_2^2\right). \tag{4.49}$$

For the next step, (4.43) is expanded according to

$$\begin{bmatrix} R[0] & R[1] & R[2] & R[3] \\ R[1] & R[0] & R[1] & R[2] \\ R[2] & R[1] & R[0] & R[1] \\ R[3] & R[2] & R[1] & R[0] \end{bmatrix} \begin{bmatrix} 1 \\ a_1^{(2)} \\ a_2^{(2)} \\ 0 \end{bmatrix} = \begin{bmatrix} J_2 \\ 0 \\ 0 \\ \Delta_2 \end{bmatrix} \tag{4.50}$$

or

$$\begin{bmatrix} R[0] & R[1] & R[2] & R[3] \\ R[1] & R[0] & R[1] & R[2] \\ R[2] & R[1] & R[0] & R[1] \\ R[3] & R[2] & R[1] & R[0] \end{bmatrix} \begin{bmatrix} 0 \\ a_2^{(2)} \\ a_1^{(2)} \\ 1 \end{bmatrix} = \begin{bmatrix} \Delta_2 \\ 0 \\ 0 \\ J_2 \end{bmatrix}. \tag{4.51}$$

Note that

$$\Delta_2 = R[3] + a_1^{(2)} R[2] + a_2^{(2)} R[1]. \tag{4.52}$$

Predictor of Order Three

In this case, the solution to be considered has the form

$$\begin{bmatrix} 1 \\ a_1^{(3)} \\ a_2^{(3)} \\ a_3^{(3)} \end{bmatrix} = \begin{bmatrix} 1 \\ a_1^{(2)} \\ a_2^{(2)} \\ 0 \end{bmatrix} - k_3 \begin{bmatrix} 0 \\ a_2^{(2)} \\ a_1^{(2)} \\ 1 \end{bmatrix}. \tag{4.53}$$

Proceeding in a similar manner, one arrives at the solution

$$k_3 = \frac{1}{J_2}\left(R[3] + a_1^{(2)}R[2] + a_2^{(2)}R[1]\right), \tag{4.54}$$

$$a_3^{(3)} = -k_3, \tag{4.55}$$

$$a_2^{(3)} = a_2^{(2)} - k_3 a_1^{(2)}, \tag{4.56}$$

$$a_1^{(3)} = a_1^{(2)} - k_3 a_2^{(2)}, \tag{4.57}$$

$$J_3 = J_2\left(1 - k_3^2\right). \tag{4.58}$$

The procedure continues until the desired prediction order is reached.

A Summary

The Levinson–Durbin algorithm is summarized as follows. Inputs to the algorithm are the autocorrelation coefficients $R[l]$, with the LPCs and RCs the outputs.

- Initialization: $l = 0$, set

$$J_0 = R[0].$$

- Recursion: for $l = 1, 2, \ldots, M$

Step 1. Compute the lth RC

$$k_l = \frac{1}{J_{l-1}}\left(R[l] + \sum_{i=1}^{l-1} a_i^{(l-1)} R[l-i]\right). \tag{4.59}$$

Step 2. Calculate LPCs for the lth-order predictor:

$$a_l^{(l)} = -k_l, \tag{4.60}$$

$$a_i^{(l)} = a_i^{(l-1)} - k_l a_{l-i}^{(l-1)}; \quad i = 1, 2, \ldots, l-1. \tag{4.61}$$

Stop if $l = M$.

Step 3. Compute the minimum mean-squared prediction error associated with the lth-order solution

$$J_l = J_{l-1}\left(1 - k_l^2\right). \tag{4.62}$$

Set $l \leftarrow l + 1$; return to Step 1.

The final LPCs are

$$a_i = a_i^{(M)}; \quad i = 1, 2, \ldots, M. \tag{4.63}$$

Note that in the process of solving the LPCs, the set of RCs (k_i, $i = 1, 2, \ldots, M$) is also found.

A virtue of the Levinson–Durbin algorithm lies in its computational efficiency. Its use results in a huge saving in the number of operations and storage locations compared to standard methods for matrix inversion. Another benefit of its use is in the set of RCs, which can be used for the verification of the *minimum phase* property of the resultant prediction-error filter. A system is minimum phase when its poles and zeros are inside the unit circle. Thus, a minimum phase system has a stable and causal inverse [Oppenheim and Schafer, 1989]. Dependence of the minimum phase condition on RCs is stated in the following theorem.

Theorem 4.1. The prediction-error filter with system function

$$A(z) = 1 + \sum_{i=1}^{M} a_i z^{-i}, \tag{4.64}$$

where the a_i are the LPCs found by solving the normal equation, is a minimum phase system if and only if the associated RCs k_i satisfy the condition

$$|k_i| < 1; \quad i = 1, 2, \ldots, M. \tag{4.65}$$

See Appendix A for a proof of the theorem.

The fact that $A(z)$ represents a minimum phase system implies that the zeros of $A(z)$ are inside the unit circle of the z-plane. Thus, the poles of the inverse system $1/A(z)$ are also inside the unit circle. Hence, the inverse system is guaranteed to be stable if the RCs satisfy condition (4.65). Since the inverse system is used to synthesize the output signal in an LP-based speech coding algorithm, stability is mandatory with all the poles located inside the unit circle. Therefore, by using the Levinson–Durbin algorithm to solve for the LPCs, it is straightforward to verify the stability of the resultant synthesis filter by inspecting the RCs. If the magnitudes of the RCs are less than one, the filter is stable.

What should we do in the case where the filter is unstable? A simple heuristic is commonly applied to fix the situation. For instance, the LPC from the last frame (representing a stable filter) can be taken and used in the present frame. Since adjacent frames often share similar statistical properties, the distortion introduced is perceptually small.

Conversion of Reflection Coefficients to Linear Prediction Coefficients

As mentioned earlier, the RC represents an alternative form of the LPC. Indeed, a one-to-one correspondence exists between the two sets of parameters. RCs possess several desirable properties, making them the preferred parameters to deal with in many practical situations. Here we consider the problem of finding the LPCs given the set of RCs. Consider the set of RCs $k_i, i = 1, \ldots, M$. It is desired to find the corresponding LPCs a_i. The problem can be solved directly from the equations in the Levinson–Durbin algorithm, summarized as follows:

For $l = 1, 2, \ldots, M$,

$$a_l^{(l)} = -k_l, \tag{4.66}$$

$$a_i^{(l)} = a_i^{(l-1)} - k_l a_{l-i}^{(l-1)}; \quad i = 1, 2, \ldots, l-1. \tag{4.67}$$

At the end of the loop, the desired result is $a_i = a_i^{(M)}$.

Conversion of Linear Prediction Coefficients to Reflection Coefficients

Given the set of LPCs $a_i, i = 1, \ldots, M$, it is desired to find the corresponding RCs k_i. This problem can again be solved using the Levinson–Durbin algorithm, working in a reversed fashion. By changing the index in (4.67) to

$$a_{l-i}^{(l)} = a_{l-i}^{(l-1)} - k_l a_i^{(l-1)} \tag{4.68}$$

and substituting (4.68) in (4.67) to eliminate $a_{l-i}^{(l-1)}$ leads to

$$a_i^{(l)} = a_i^{(l-1)} - k_l a_{l-i}^{(l)} - k_l^2 a_i^{(l-1)},$$

or

$$a_i^{(l-1)} = \frac{a_i^{(l)} + k_l a_{l-i}^{(l)}}{1 - k_l^2}. \tag{4.69}$$

The above equation is used to find the RCs based on the following loop, with $a_i^{(M)} = a_i$:

For $l = M, M-1, \ldots, 1$,

$$k_l = -a_l^{(l)}, \tag{4.70}$$

$$a_i^{(l-1)} = \frac{a_i^{(l)} + k_l a_{l-i}^{(l)}}{1 - k_l^2}; \quad i = 1, 2, \ldots, l-1. \tag{4.71}$$

4.5 THE LEROUX–GUEGUEN ALGORITHM

A potential problem with the Levinson–Durbin algorithm lies in the values of the LPCs, since they possess a large dynamic range and a bound on their magnitudes cannot be found on a theoretical basis. The issue is of little concern if the algorithm is implemented under a floating-point environment. However, it could present some difficulties for fixed-point implementation.

Example 4.5 A total of 1300 frames having 240 samples each are used to demonstrate the typical distribution of LPCs and RCs. These frames are LP analyzed with a prediction order of ten. Figure 4.14 shows the histogram plots for the LPCs a_1, a_2, a_3, a_6, a_7, and a_{10}. In general, we observe that the low-order coefficients tend to have a wider dynamic range. For high-order coefficients, the magnitudes tend to

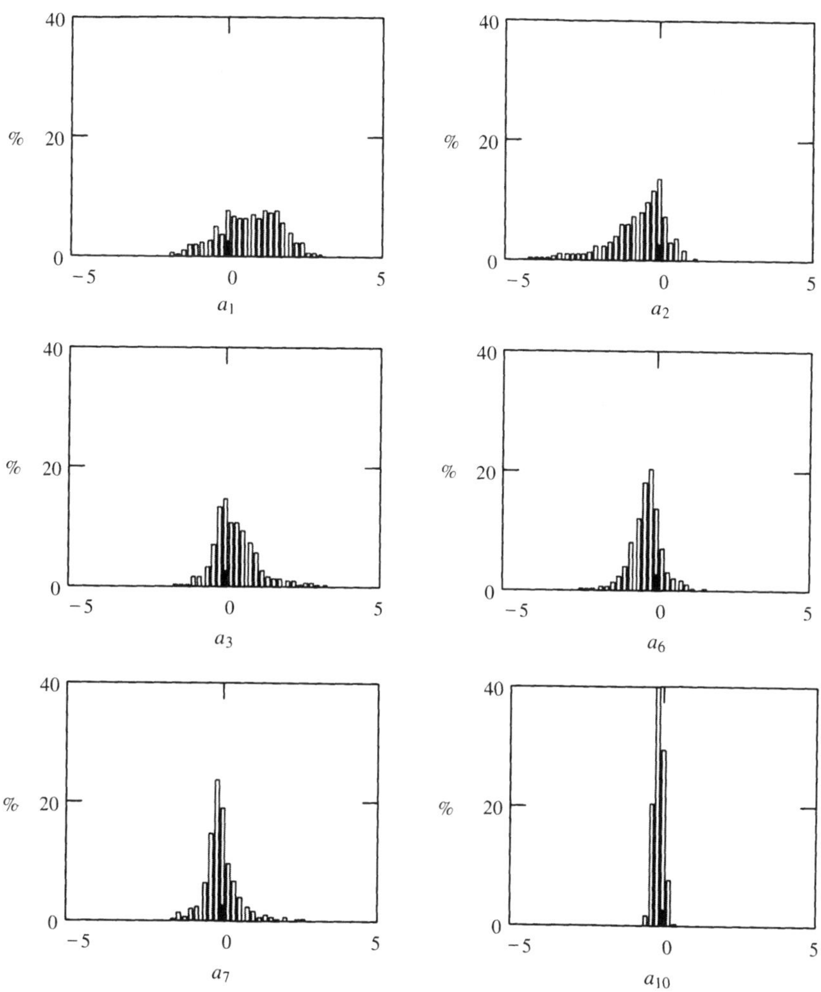

Figure 4.14 Histogram plots of some LPCs, obtained from 1300 frames of speech material. Vertical axis is the percentage of occurrence, while the horizontal axis is the value of the coefficients.

be small and most coefficients are gathered around the origin. Even though high-magnitude (>4) coefficients are scarce, no clear bounds exist, leading to problems in fixed-point implementation of the Levinson–Durbin algorithm.

Figure 4.15 shows the histogram plots of the corresponding RCs, where k_1, k_2, k_3, k_6, k_7, and k_{10} are displayed. Note that in all cases they are bounded so that $|k_i| \leq 1$. This bound is important for efficient design of algorithms under a fixed-point environment since usage of the finite numerical range can be maximized.

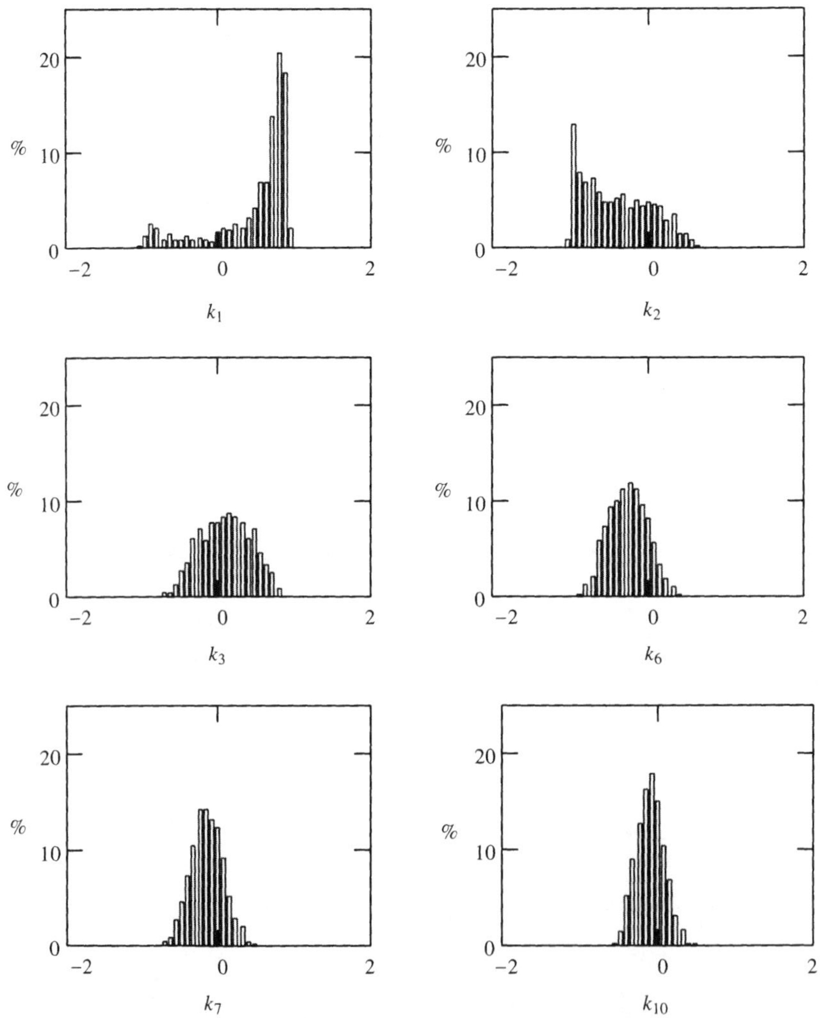

Figure 4.15 Histogram plots of some reflection coefficients, obtained from 1300 frames of speech material. Vertical axis is the percentage of occurrence, while the horizontal axis is the value of the coefficients.

The Leroux–Gueguen Solution

Leroux and Gueguen [1979] proposed a method to compute the RCs from the autocorrelation values without dealing directly with the LPCs. Hence, problems related to dynamic range in a fixed-point environment are eliminated. Consider the parameter

$$\varepsilon^{(l)}[k] = E\{e^{(l)}[n]s[n-k]\} = \sum_{i=0}^{l} a_i^{(l)} R[i-k], \tag{4.72}$$

where

$$\begin{aligned} e^{(l)}[n] &= \text{prediction error using an } l\text{th-order prediction-error filter,} \\ a_i^{(l)} &= \text{LPC of an } l\text{th-order predictor,} \\ R[k] &= \text{autocorrelation value of the signal } s[n]. \end{aligned}$$

A fixed-point implementation arises from the fact that the parameters ε, as defined in (4.72), have a fixed dynamic range. This is stated in the following theorem.

Theorem 4.2. Given the parameters ε as defined in (4.72), then

$$\left|\varepsilon^{(l)}[k]\right| \leq R[0]. \tag{4.73}$$

Proof. Given any two random variables $\mathbf{x}$ and $\mathbf{y}$, the condition

$$(E\{\mathbf{xy}\})^2 \leq E\{\mathbf{x}^2\}E\{\mathbf{y}^2\}, \tag{4.74}$$

known as the Schwarz inequality [Papoulis, 1991], is satisfied. Applying the inequality to our problem leads to

$$\left|\varepsilon^{(l)}[k]\right|^2 = \left|E\{e^{(l)}[n]s[n-k]\}\right|^2 \leq E\{(e^{(l)}[n])^2\}E\{s^2[n-k]\} = J_l R[0],$$

with J_l denoting the variance of the prediction error. Since the power of the prediction error (J_l) is in general less than or equal to the power of the signal $s[n]$ ($R[0]$), we have $|\varepsilon^{(l)}[k]|^2 \leq R[0]^2$ and the theorem is proved.

To derive the Leroux–Gueguen algorithm, relations between the ε parameters and variables in the Levinson–Durbin algorithm are found. First, note that

$$k_l = \frac{\varepsilon^{(l-1)}[l]}{\varepsilon^{(l-1)}[0]}; \quad l = 1, 2, \ldots, M, \tag{4.75}$$

and is found directly from (4.72) and (4.59). Substituting the recursive relation for the LPCs, (4.61) gives

$$\begin{aligned}\varepsilon^{(l)}[k] &= \sum_{i=0}^{l-1}\left(a_i^{(l-1)} - k_l a_{l-i}^{(l-1)}\right)R[i-k] - k_l R[l-k]\\ &= \sum_{i=0}^{l-1} a_i^{(l-1)} R[i-k] - k_l \sum_{i=0}^{l-1} a_i^{(l-1)} R[i+k-l].\end{aligned}$$

Note that the relation $R[l] = R[-l]$ is used. Comparing with (4.72) leads to

$$\varepsilon^{(l)}[k] = \varepsilon^{(l-1)}[k] - k_l \varepsilon^{(l-1)}[l-k]. \tag{4.76}$$

The above equation relates the ε parameters at the lth order with the parameters at order $l-1$. From (4.72) we observe that

$$\varepsilon^{(0)}[k] = R[k]. \tag{4.77}$$

Hence, (4.75) and (4.76) can be used to solve the RCs recursively starting from $l = 1$. Higher-order solutions are built on solutions from a previous order. The question to ask next is how many ε parameters need to be computed at each order l or what is the range of k. The answer can be found by deriving the range of k at each l, descending from $l = M$, with M being the desired prediction order. At $l = M$,

$$k_M = \frac{\varepsilon^{(M-1)}[M]}{\varepsilon^{(M-1)}[0]}.$$

Thus, we only need

$$\varepsilon^{(M-1)}[0], \quad \varepsilon^{(M-1)}[M]$$

from order $M-1$. At order $M-1$, we need to solve

$$\begin{aligned}k_{M-1} &= \frac{\varepsilon^{(M-2)}[M-1]}{\varepsilon^{(M-2)}[0]},\\ \varepsilon^{(M-1)}[M] &= \varepsilon^{(M-2)}[M] - k_{M-1}\varepsilon^{(M-2)}[-1],\\ \varepsilon^{(M-1)}[0] &= \varepsilon^{(M-2)}[0] - k_{M-1}\varepsilon^{(M-2)}[M-1].\end{aligned}$$

Hence, the parameters

$$\varepsilon^{(M-2)}[-1], \quad \varepsilon^{(M-2)}[0], \quad \varepsilon^{(M-2)}[M-1], \quad \varepsilon^{(M-2)}[M]$$

TABLE 4.1 The ε Parameters Required at Each Order *l* in the Leroux–Gueguen Algorithm

l	Parameters Required
M	
$M-1$	$\varepsilon^{(M-1)}[0], \varepsilon^{(M-1)}[M]$
$M-2$	$\varepsilon^{(M-2)}[-1], \varepsilon^{(M-2)}[0], \varepsilon^{(M-2)}[M-1], \varepsilon^{(M-2)}[M]$
$M-3$	$\varepsilon^{(M-3)}[-2], \ldots, \varepsilon^{(M-3)}[0], \varepsilon^{(M-3)}[M-2], \ldots, \varepsilon^{(M-3)}[M]$
$M-4$	$\varepsilon^{(M-4)}[-3], \ldots, \varepsilon^{(M-4)}[0], \varepsilon^{(M-4)}[M-3], \ldots, \varepsilon^{(M-4)}[M]$
$\vdots$	
1	$\varepsilon^{(1)}[-M+2], \ldots, \varepsilon^{(1)}[0], \varepsilon^{(1)}[2], \ldots, \varepsilon^{(1)}[M]$
0	$\varepsilon^{(0)}[-M+1], \ldots, \varepsilon^{(0)}[0], \varepsilon^{(0)}[1], \ldots, \varepsilon^{(0)}[M]$

are needed at order $M-2$. Proceeding in this way we can find the parameters needed at each order l. Table 4.1 summarizes the results.

The Leroux–Gueguen algorithm can now be summarized as follows.

- Initialization: $l = 0$, set

$$\varepsilon^{(0)}[k] = R[k], \quad k = -M+1, \ldots, M. \tag{4.78}$$

- Recursion: for $l = 1, 2, \ldots, M$

Step 1. Compute the lth RC

$$k_l = \frac{\varepsilon^{(l-1)}[l]}{\varepsilon^{(l-1)}[0]}. \tag{4.79}$$

Stop if $l = M$.

Step 2. Calculate the ε parameters:

$$\varepsilon^{(l)}[k] = \varepsilon^{(l-1)}[k] - k_l \varepsilon^{(l-1)}[l-k]; \quad k = -M+l+1, \ldots, 0, l+1, \ldots, M. \tag{4.80}$$

Set $l \leftarrow l+1$; return to Step 1.

Leroux–Gueguen Versus Levinson–Durbin

The Leroux–Gueguen algorithm is better suited for fixed-point implementation since all intermediate variables have known bound. A drawback is the fact that only RCs are returned as the result, which is not a problem if the associated filter is in lattice form. LPCs are required for the direct-form filter and can be obtained either using the Levinson–Durbin method or the Leroux–Gueguen algorithm followed by the conversion procedure explained in Section 4.4.

Use of a lattice filter is often uninviting due to the increased amount of computation. In addition, for a time-varying situation, coefficient update from frame to frame introduces a stronger undesirable transient in the lattice structure. On the other hand, the Leroux–Gueguen approach followed by RC-to-LPC conversion does not provide significant computational saving, if any, when compared to the Levinson–Durbin algorithm. All these factors combined make the Levinson–Durbin approach more popular in practice, even though it is known to have numerical problems.

In the practical implementation of the Levinson–Durbin algorithm under a fixed-point environment, careful planning is necessary to ensure that all variables are within the allowed range. See Chen et al. [1991] for a discussion on the selection between the two algorithms within the context of LD-CELP coder design, with the Levinson–Durbin method choosen at the end.

4.6 LONG-TERM LINEAR PREDICTION

Experiments in Section 4.3 using real speech data have shown that the prediction order must be high enough to include at least one pitch period in order to model adequately the voiced signal under consideration. A linear predictor with an order of ten, for instance, is not capable of accurately modeling the periodicity of the voiced signal having a pitch period of 50. The problem is evident when the prediction error is examined: a lack of fit is indicated by the remaining periodic component. By increasing the prediction order to include one pitch period, the periodicity in the prediction error has largely disappeared, leading to a rise in prediction gain. High prediction order leads to excessive bit-rate and implementational cost since more bits are required to represent the LPCs, and extra computation is needed during analysis. Thus, it is desirable to come up with a scheme that is simple and yet able to model the signal with sufficient accuracy.

Important observation is derived from the experimental results of Section 4.3 (Figure 4.9). An increase in prediction gain is due mainly to the first 8 to 10 coefficients, plus the coefficient at the pitch period, equal to 49 in that particular case. The LPCs at orders between 11 and 48 and at orders greater than 49 provide essentially no contribution toward improving the prediction gain. This can be seen from the flat segments from 10 to 49, and beyond 50. Therefore, in principle, the coefficients that are not contributing toward elevating the prediction gain can be eliminated, leading to a more compact and efficient scheme. This is exactly the idea of long-term linear prediction, where a short-term predictor is connected in cascade with a long-term predictor, as shown in Figure 4.16. The short-term predictor is basically the one we have studied so far, with a relatively low prediction order M in the range of 8 to 12. This predictor eliminates the correlation between nearby

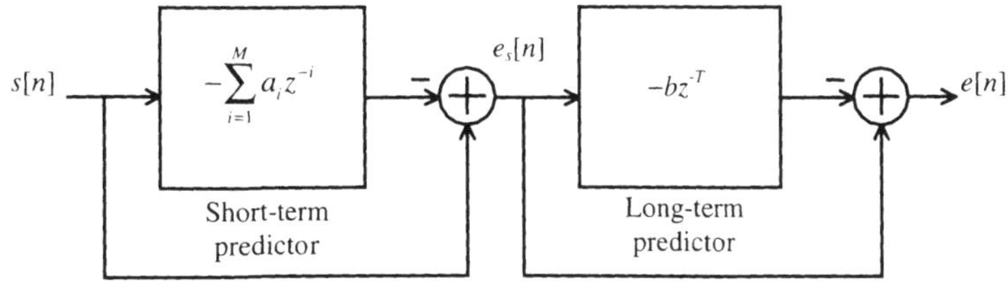

Figure 4.16 Short-term prediction-error filter connected in cascade to a long-term prediction-error filter.

samples or is short-term in the temporal sense. The long-term predictor, on the other hand, targets correlation between samples one pitch period apart.

The long-term prediction-error filter with input $e_s[n]$ and output $e[n]$ has system function

$$H(z) = 1 + bz^{-T}. \tag{4.81}$$

Note that two parameters are required to specify the filter: pitch period T and long-term gain b (also known as long-term LPC or pitch gain). The procedure to determine b and T is referred to as long-term LP analysis. Positions of the predictors in Figure 4.16 can actually be exchanged. However, experimentally it was found that the shown configuration achieves on average a higher prediction gain [Ramachandran and Kabal, 1989]. Thus, it is adopted by most speech coding applications.

Long-Term LP Analysis

A long-term predictor predicts the current signal sample from a past sample that is one or more pitch periods apart, using the notation of Figure 4.16:

$$\hat{e}_s[n] = -be_s[n-T], \tag{4.82}$$

where b is the long-term LPC, while T is the pitch period or lag. Within a given time interval of interest, we seek to find b and T so that the sum of squared error

$$J = \sum_n (e_s[n] - \hat{e}_s[n])^2 = \sum_n (e_s[n] + be_s[n-T])^2 \tag{4.83}$$

is minimized. Differentiating the above equation with respect to b and equating to zero, one can show that

$$b = -\frac{\sum_n e_s[n]e_s[n-T]}{\sum_n e_s^2[n-T]}, \tag{4.84}$$

which gives the optimal long-term gain as a function of two correlation quantities of the signal, with the correlation quantities a function of the pitch period T. An exhaustive search procedure can now be applied to find the optimal T. Substituting (4.84) back into (4.83) leads to

$$J = \sum_n e_s^2[n] - \frac{\left(\sum_n e_s[n]e_s[n-T]\right)^2}{\sum_n e_s^2[n-T]}. \tag{4.85}$$

The step-by-step procedure of long-term LP analysis is summarized with the following pseudocode:

```
1.  J_min ← ∞
2.  for T ← T_min to T_max
3.        (Use (4.84) to compute b)
4.        (Use (4.83) or (4.85) to compute J)
5.        if J < J_min
6.              J_min ← J
7.              b_opt ← b
8.              T_opt ← T
9.  return b_opt, T_opt
```

The parameters T_{min} and T_{max} in Line 2 define the search range within which the pitch period is determined. The reader must be aware that the pseudocode is not optimized in terms of execution speed. In fact, computation cost can be reduced substantially by exploring the redundancy within the procedure (Exercise 4.8).

Example 4.6 The same speech frame as in Example 4.3 is used here, where long-term LP analysis is applied to the 240-sample frame ending at time $m = 1000$. As was explained earlier in this section, analysis is done on the prediction-error sequence obtained at the output of the short-term prediction-error filter (with a

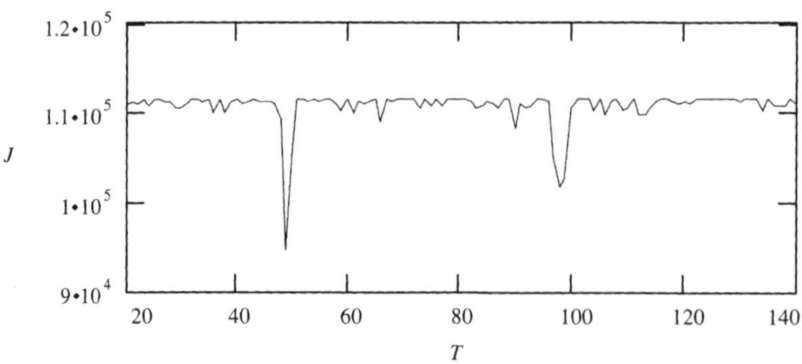

Figure 4.17 Example of the sum of squared error (J) as a function of the pitch period (T) in long-term LP analysis.

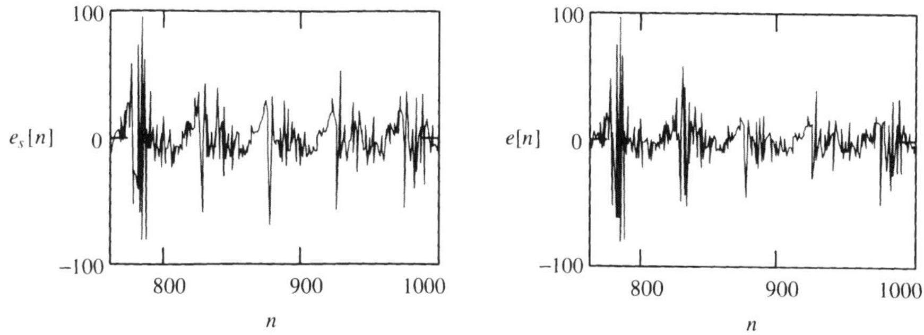

Figure 4.18 *Left*: Input to long-term prediction-error filter (short-term prediction error). *Right*: Output of the long-term prediction-error filter (overall prediction error).

prediction order of 10). That is, short-term LP analysis is applied to the frame at $m = 1000$, and short-term prediction error is calculated using the LPC found. Of course, short-term prediction error prior to the frame under consideration is available so that long-term LP analysis can be completed.

The sum of the squared error as a function of the pitch period ($20 \leq T \leq 140$) is plotted in Figure 4.17. The overall minimum is located at $T = 49$ and coincides roughly with the period of the waveform in the time domain. Figure 4.18 shows the short-term prediction error and the overall prediction error, with the latter slightly lower in amplitude. In this case, prediction gain of the long-term prediction-error filter is found to be 0.712 dB.

The Frame/Subframe Structure

Results of Example 4.6 show that the effectiveness of the long-term predictor on removing long-term correlation is limited. In fact, the overall prediction-error sequence is very much like the short-term prediction-error sequence, containing a strong periodic component whose period is close to the pitch period itself.

The crux of the problem is that the parameters of the long-term predictor need to be updated more frequently than the parameters of the short-term predictor. That is, it loses its effectiveness when the time interval used for estimation becomes too long, which is due to the dynamic nature of the pitch period as well as long-term LPCs. Experiments using an extensive amount of speech samples revealed that by shortening the time interval in which the long-term parameters were estimated from 20 to 5 ms, an increase in prediction gain of 2.2 dB was achievable [Ramachandran and Kabal, 1989].

The idea of frame and subframe was born as a result of applying short-term LP analysis to a relatively long interval, known as the frame. Inside the frame, it is divided into several smaller intervals, known as subframes. Long-term LP analysis is applied to each subframe separately. The scheme is depicted in Figure 4.19. Typical numbers as used by the FS1016 CELP coder (Chapter 12) are 240 samples for the frame, which is comprised of four subframes having 60 samples each.

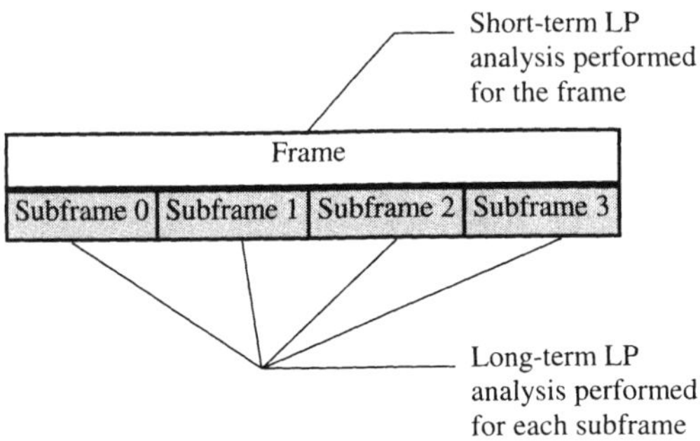

Figure 4.19 The frame/subframe structure.

More frequent update of the long-term predictor obviously requires a higher bit-rate. However, the resultant scheme is still more economical than the one using 50 or more LPCs to capture the signal's statistics.

Example 4.7 The same experimental setup as in Example 4.6 is considered, with the exception that long-term LP analysis is applied to the four subframes within the frame defined by $n \in [761, 1000]$. Intervals of the subframes are $n \in [761, 820]$, $[821, 880]$, $[881, 940]$, and [941, 1000]. Figure 4.20 shows the error curves for the

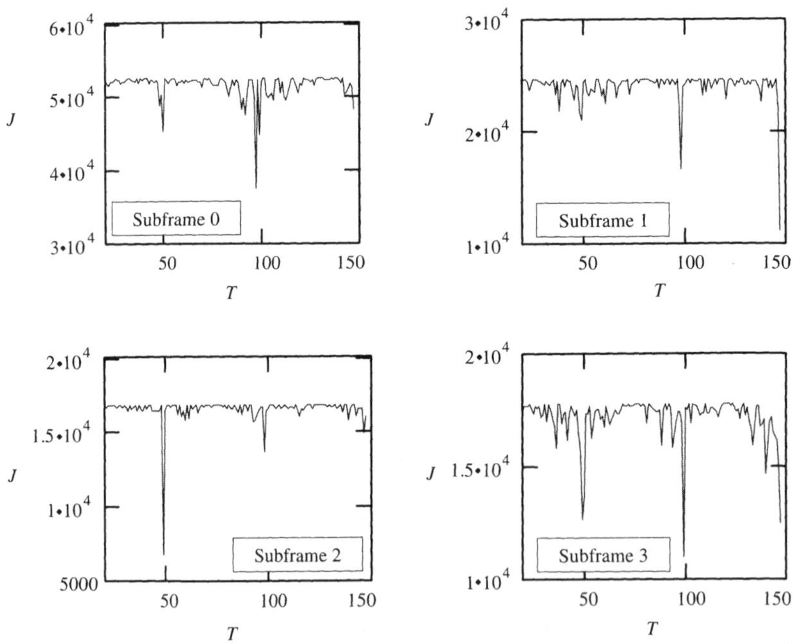

Figure 4.20 Example of the sum of squared error (J) as a function of the pitch period (T) for the four subframes.

TABLE 4.2 Results Summary for an Example of Long-Term LP Analysis

Subframe Number	b	T
0	−0.833	97
1	−0.638	147
2	−0.735	49
3	−0.627	99

four subframes, where the minimums indicate the optimal pitch periods ($20 \leq T \leq 147$). Parameters of the long-term predictor are summarized in Table 4.2. Note that both the long-term gain and pitch period change drastically from subframe to subframe.

Figure 4.21 shows the final prediction-error sequence. Compared to the outcome of Example 4.6 (Figure 4.18), it is clear that in the present case the sequence is "whiter," with lower amplitude and periodicity largely removed. A prediction gain of 2.26 dB is registered, which is a substantial improvement with respect to the 0.712 dB obtained in Example 4.6.

Fractional Pitch Period

One error source that limits the resolution and hence the accuracy of the long-term predictor is time discretization, or quantization of the estimates of pitch period. The problem is introduced by the sampling process, where a continuous-time signal is

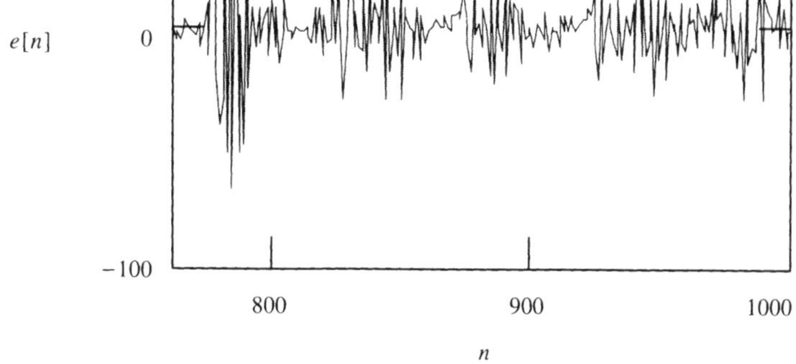

Figure 4.21 Example of long-term prediction-error filter's output.

transformed to discrete time. A pitch estimate, expressed as an integer multiple of the sampling interval, contains a time quantization error, which may lead to audible distortions.

Another problem with integer pitch period is the phenomenon of pitch multiplication. For periodic signals the current period is not only similar to the previous one but also to periods that occurred multiple periods ago. Quantization error of the continuous-valued pitch period can lead to mismatch during correlation calculation, resulting in the exhaustive search procedure producing the best delay value to be equal to a multiple of the "true" pitch period. Pitch multiplication is disadvantageous for coding since a smooth pitch contour can be encoded more efficiently. In addition, the sudden jump of pitch might lead to artifacts in the synthesized speech. This effect is clearly observed in Example 4.7 (Figure 4.20 and Table 4.2), where the four values of the pitch period are clearly multiples of a "true" value located between 48 and 50.

Fractional pitch period is introduced as a mean to increase temporal resolution. In this case, the pitch period is allowed to have a fractional part plus the integer part. Analysis is performed using short-term prediction error delayed by a fractional quantity and is obtained via interpolation. In general, long-term predictors with high temporal resolution suffer less from pitch multiplication than low-resolution ones. Experimental results have shown that by using fractional delay, the average prediction gain is increased by 1.5 to 2.5 dB [Kroon and Atal, 1991]. The improvement is more notorious for female speech since the shorter pitch period makes it more susceptible to quantization errors.

Example 4.8 The same experimental setup as in Example 4.7 is considered with the difference that long-term LP analysis is performed using fractional pitch period. Only 1 bit is used to code the fractional part of the pitch period. Thus, the fraction can only take two values: 0 or 0.5. Fractional delay is calculated using a simple linear interpolation approach, where the fractionally delayed version of the short-term prediction error (input to long-term prediction-error filter) is calculated with

$$e_s[n + 0.5] = \tfrac{1}{2}\left(e_s[n] + e_s[n + 1]\right). \tag{4.86}$$

The same exhaustive search procedure can be applied to find the optimal long-term gain and pitch period. In this particular case, a difference is observed only for the last subframe, where the error curve as a function of the pitch period is shown in Figure 4.22. The optimal pitch period for the subframe is found to be $T = 49.5$ with long-term gain $b = -1.086$. Compared to the results of Example 4.7, we can see that pitch multiplication of the last subframe is eliminated. Overall prediction gain due to long-term prediction is 2.41 dB, a 0.15-dB improvement with respect to the case of integer pitch period.

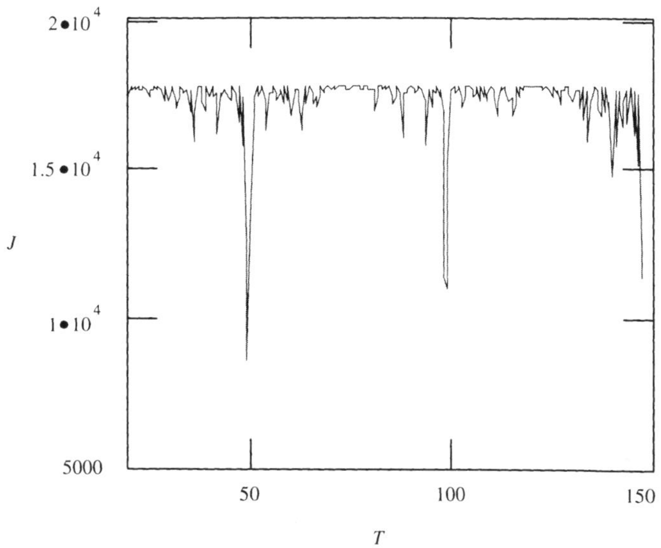

Figure 4.22 Error curve for subframe 3 when fractional delay is applied (compare to Figure 4.20).

4.7 SYNTHESIS FILTERS

So far we have focused on analyzing the signal with the purpose of identifying the parameters of a system, based on the assumption that the system itself satisfies the AR constraint. The identification process is done by minimizing the prediction error. If the prediction error is "white" enough, we know that the estimated system is a good fit; therefore, it can be used to synthesize signals having similar statistical properties as the original one. In fact, by exciting the synthesis filter with the system function

$$H(z) = \frac{1}{1 + \sum_{i=1}^{M} a_i z^{-i}} \tag{4.87}$$

using a white noise signal, the filter's output will have a PSD close to the original signal as long as the prediction order M is adequate. In (4.87), the a_i are the LPCs found from the original signal. Figure 4.23 shows the block diagram of the synthesis filter, where a unit-variance white noise is generated and scaled by the gain g and is input to the synthesis filter to generate the synthesized speech at the output. Since $x[n]$ has unit variance, $gx[n]$ has variance equal to g^2. From (4.16) we can readily write

$$g = \gamma\sqrt{R_s[0] + \sum_{i=1}^{M} a_i R_s[i]}. \tag{4.88}$$

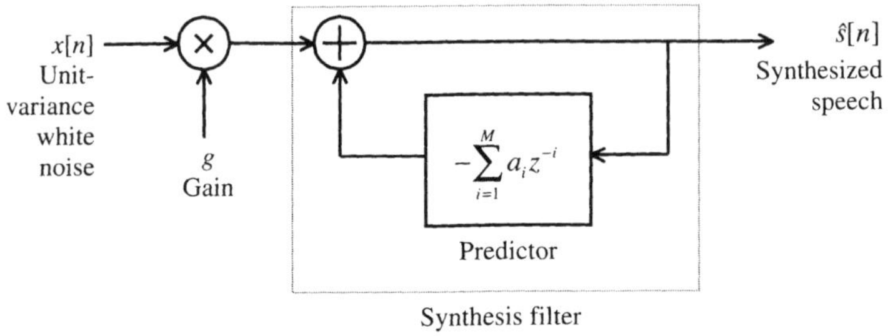

Figure 4.23 The synthesis filter.

Thus, the gain can be found by knowing the LPCs and the autocorrelation values of the original signal. In (4.88), γ is a scaling constant. A scaling constant is needed because the autocorrelation values are normally estimated using a window that weakens the signal's power. The value of γ depends on the type of window selected and can be found experimentally. Typical values of γ range from 1 to 2. In addition, it is important to note that the autocorrelation values in (4.88) must be the time-averaged ones, instead of merely the sum of products.

Example 4.9 The same voiced frame as in Example 4.3 is analyzed to give a set of 50 LPCs, corresponding to a predictor of order 50. The derived predictor is used in synthesis where white noise with uniform distribution and unit variance is used as input to the synthesis filter. The gain g is found from (4.88) with $\gamma = 1.3$. The synthesized speech and periodogram are displayed in Figure 4.24. Compared to the original signal (Figures 4.6 and 4.7), we can see that the two signals share many common attributes in both the time and frequency domains. In fact, sounds generated by the two waveforms are very much alike.

As discussed earlier in the chapter, using high prediction order (> 12) is computationally expensive and in most instances inefficient. Thus, many LP-based speech

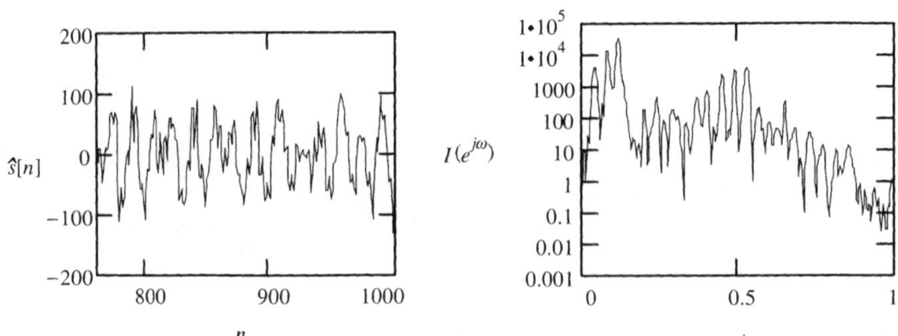

Figure 4.24 Plots of waveform and periodogram of the synthesized speech signal and periodogram in an experiment.

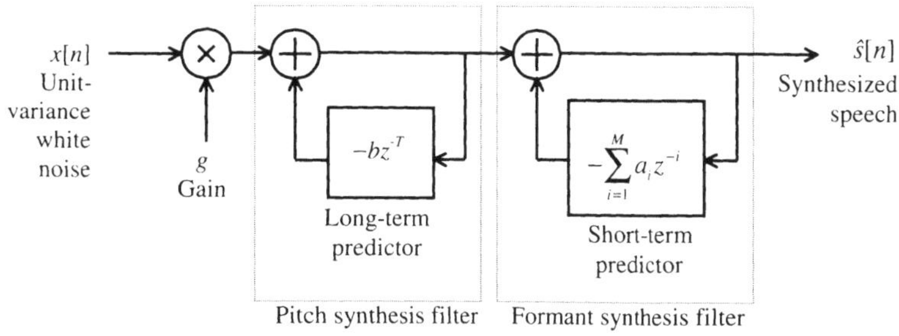

Figure 4.25 Long-term and short-term linear prediction model for speech production.

coding algorithms rely on a prediction order between 8 and 12, with order ten being the most widely employed. Since this low prediction order is not sufficient to recreate the PSD for voiced signal, a non-white-noise excitation is utilized as input to the synthesis filter. The choice of excitation is a trade-off between complexity and quality of synthesized speech. Different algorithms use different approaches to target the problem and the details are given in subsequent chapters.

Long-Term and Short-Term Linear Prediction Model for Speech Synthesis

The long-term predictor studied in Section 4.6 is considered for synthesis purposes. A block diagram is shown in Figure 4.25, known as the long-term and short-term linear prediction model for speech production. The parameters of the two predictors are again estimated from the original speech signal. The long-term predictor is responsible for generating correlation between samples that are one pitch period apart. The filter with system function

$$H_P(z) = \frac{1}{1 + bz^{-T}}, \tag{4.89}$$

describing the effect of the long-term predictor in synthesis, is known as the long-term synthesis filter or pitch synthesis filter. On the other hand, the short-term predictor recreates the correlation present between nearby samples, with a typical prediction order equal to ten. The synthesis filter associated with the short-term predictor, with system function given by (4.87), is also known as the formant synthesis filter since it generates the envelope of the spectrum in a way similar to the vocal track tube, with resonant frequencies known simply as formants. The gain g in Figure 4.25 is usually found by comparing the power level of the synthesized speech signal to the original level.

Example 4.10 The magnitude of the transfer functions for the pitch synthesis filter and formant synthesis filter obtained from Example 4.7 are plotted in Figure 4.26. In the same figure, the product of the transfer functions is also plotted. Since

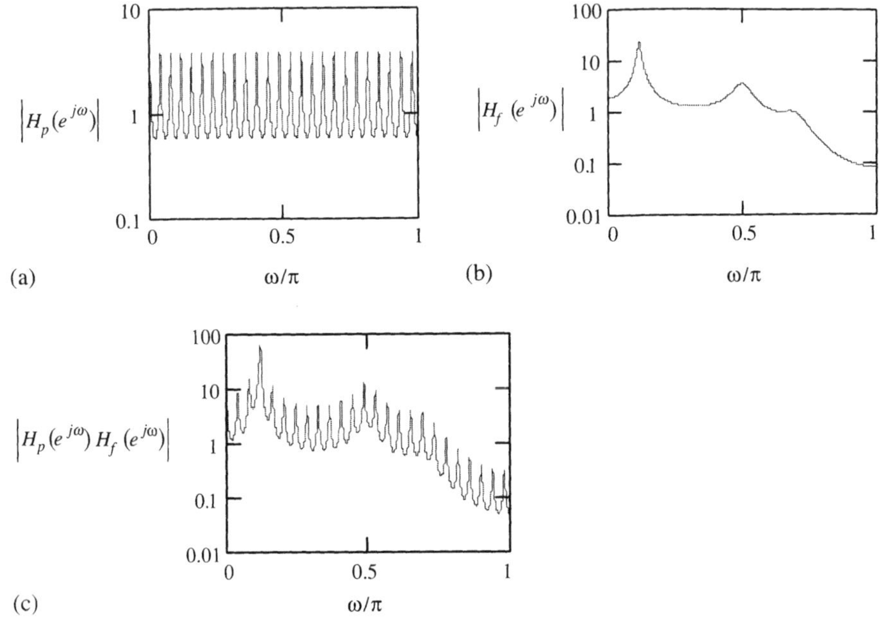

Figure 4.26 Magnitude plots of the transfer functions for (a) a pitch synthesis filter, (b) a formant synthesis filter, and (c) a cascade connection between pitch synthesis filter and formant synthesis filter.

the two filters are in cascade, the overall transfer function is equal to their product. Parameters of the filters are

$$b = -0.735 \quad T = 49$$
$$a_1 = -1.502 \quad a_2 = 1.738 \quad a_3 = -2.029 \quad a_4 = 1.789 \quad a_5 = -1.376$$
$$a_6 = 1.255 \quad a_7 = -0.693 \quad a_8 = 0.376 \quad a_9 = -0.08 \quad a_{10} = 0.033$$

Note that the pitch synthesis filter generates the harmonic structure due to the fundamental pitch frequency, while the formant synthesis filter captures the spectrum envelope. Product of the two recreates the original signal spectrum. Compared to Figure 4.7, we can see that the spectrum due to the synthesis filters has a shape that closely matches the PSD of the original signal.

Stability Issues

In many coding situations, the synthesis filters are excited by a random noise sequence; stability of the filters is a prime concern. For the formant synthesis filter with system function (4.87), we already know from Theorem 4.1 that it is a minimum-phase system as long as the RCs have magnitude less than one, which can be verified while solving the normal equation during LP analysis.

For the pitch synthesis filter with system function (4.89), the system poles are found by solving

$$1 + bz^{-T} = 0$$

or

$$z^{-T} = -b. \tag{4.90}$$

There are a total of T different solutions for z, and hence the system has T different poles. These poles lie at the vertices of a regular polygon of T sides that is inscribed in a circle of radius $|b|^{1/T}$. Thus, in order for the filter to be stable, the following condition must be satisfied:

$$|b| < 1. \tag{4.91}$$

An unstable pitch synthesis filter arises when the absolute value of the numerator is greater than the denominator as in (4.84), resulting in $|b| > 1$. This usually arises when a transition from an unvoiced to a voiced segment takes place and is marked by a rapid surge in signal energy. When processing a voiced frame that occurs just after an unvoiced frame, the denominator quantity $\Sigma e_s^2[n - T]$ involves the sum of the squares of amplitudes in the unvoiced segment, which is normally weak. On the other hand, the numerator quantity $\Sigma e_s[n]e_s[n - T]$ involves the sum of the products of the higher amplitudes from the voiced frame and the lower amplitudes from the unvoiced frame. Under these circumstances, the numerator can be larger in magnitude than the denominator, leading to $|b| > 1$. Therefore, an unstable pitch synthesis filter can arrive when the signal energy shows a sudden increase. To ensure stability, the long-term gain is often truncated so that its magnitude is always less than one.

Maintaining the long-term gain to have a magnitude strictly less than one is often not a good strategy, since subjective quality could be adversely affected. This is true for various kinds of speech sounds generated by a sudden release of pressure, such as the stop consonants b and d. By easing the constraint on the long-term gain, sounds of a transient, noncontinuant nature can be captured more accurately by the underlying model, leading to an increase in subjective quality. Thus, it is common for various coding algorithms to tolerate short-term instability in the pitch synthesis filter. A popular choice for the upper bound of the long-term gain is between 1.2 and 2.

4.8 PRACTICAL IMPLEMENTATION

In general, LP analysis is a well-behaved procedure in the sense that the resultant synthesis filter is guaranteed to be stable as long as the magnitudes of the RCs are less than one (Section 4.4). In practice, however, there are situations under which stability can be threatened. For instance, under marginally stable conditions, the limited precision of the computational environment can lead to errors high enough to produce an unstable filter; this could happen for signals with sustained oscillation, where the spectrum is associated with poles close to the unit circle. In this section we study several techniques employed in speech coding to fix the described problem, all of them aimed at alleviating ill-conditioning during LP analysis and, at the same time, improving the stability of the resultant synthesis filter, as well as the quality of the synthetic speech. These techniques can be used in an isolated fashion or combined together.

Pre-emphasis of the Speech Waveform

The typical spectral envelope of the speech signal has a high frequency roll-off due to radiation effects of the sound from the lips. Hence, high-frequency components have relatively low amplitude, which increases the dynamic range of the speech spectrum. As a result, LP analysis requires high computational precision to capture the features at the high end of the spectrum. More importantly, when these features are very small, the correlation matrix can become ill-conditioned and even singular, leading to computational problems. One simple solution is to process the speech signal using the filter with system function

$$H(z) = 1 - \alpha z^{-1}, \tag{4.92}$$

which is highpass in nature. The purpose is to augment the energy of the high-frequency spectrum. The effect of the filter can also be thought of as a flattening process, where the spectrum is "whitened." Denoting $x[n]$ as the input to the filter and $y[n]$ as the output, the following difference equation applies:

$$y[n] = x[n] - \alpha x[n-1]. \tag{4.93}$$

The filter described in (4.92) is known as the pre-emphasis filter. By pre-emphasizing, the dynamic range of the power spectrum is reduced. This process substantially reduces numerical problems during LP analysis, especially for low-precision devices. A value of α near 0.9 is usually selected.

It is common to find in a typical speech coding scheme that the input speech is first pre-emphasized using (4.92). To keep a similar spectral shape for the synthetic speech, it is filtered by the de-emphasis filter with system function

$$G(z) = \frac{1}{1 - \alpha z^{-1}} \tag{4.94}$$

at the decoder side, which is the inverse filter with respect to pre-emphasis. Figure 4.27 shows the magnitude plots of the filter's transfer functions.

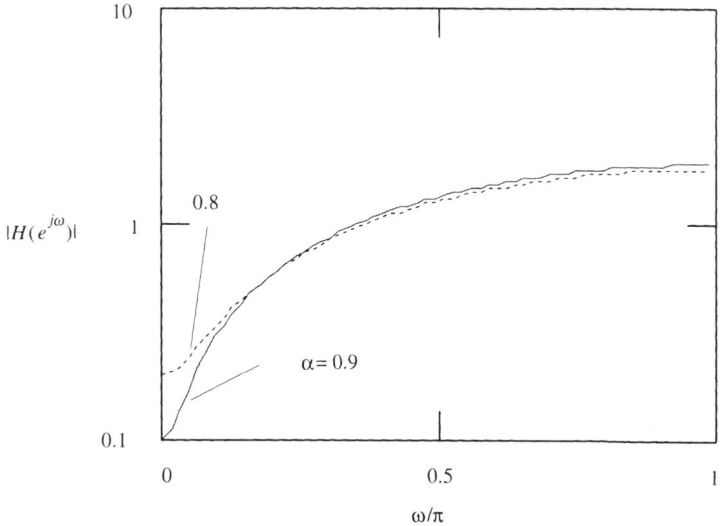

Figure 4.27 Magnitude plots of the transfer functions of the pre-emphasis filter.

Bandwidth Expansion Through Modification of the LPC

In the application of linear prediction, the resultant synthesis filter might become marginally stable due to the poles located too close to the unit circle. The problem is aggravated in fixed-point implementation, where a marginally stable filter can actually become unstable (with the poles located outside the unit circle) after quantization and loss of precision during processing. This problem creates occasional "chirps" or oscillations in the synthesized signal.

Stability can be improved by modifying the LPCs according to

$$a_{new_i} = \gamma^i a_i; \quad i = 1, 2, \ldots, M, \tag{4.95}$$

with $\gamma < 1$ a positive constant. The operation moves all the poles of the synthesis filter radially toward the origin, leading to improved stability. By doing so, the original spectrum is bandwidth expanded, in the sense that the spectrum becomes flatter, especially around the peaks, where the width is widened. Typical values for γ are between 0.988 and 0.996.

Another advantage of the bandwidth expansion technique is the shortening of the duration of the impulse response, which improves robustness against channel errors. This is because the excitation signal (in some speech coders the excitation signal is coded and transmitted) distorted by channel errors is filtered by the synthesis filter, and a shorter impulse response limits the propagation of channel error effects to a shorter duration.

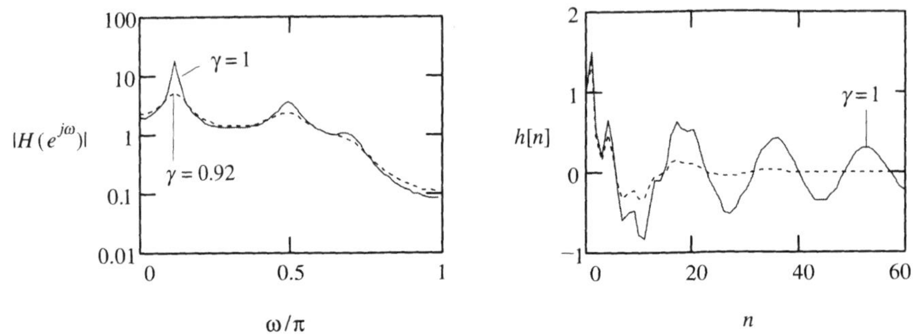

Figure 4.28 Magnitude of the transfer function (*left*) and impulse response (*right*) of the original (solid line) and bandwidth-expanded (dotted line) synthesis filters.

Example 4.11 The LPCs from Example 4.10 are modified for bandwidth expansion, using a constant γ of 0.92. Figure 4.28 shows a comparison between the original and modified magnitude response and impulse response. Note how the bandwidth-expanded version has a smoother, flatter frequency response; in addition, the impulse response decays faster toward zero. Poles of the system function are plotted in Figure 4.29, where, after bandwidth expansion, they are pulled toward the origin.

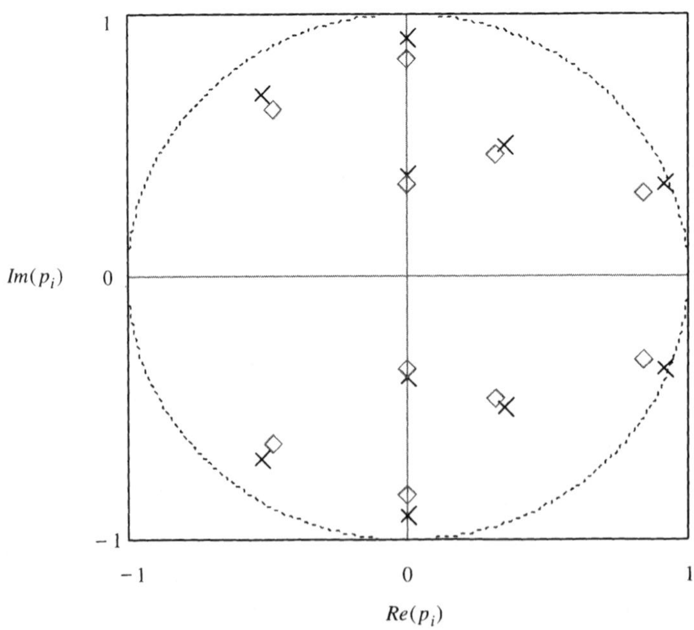

Figure 4.29 Plot of poles for the original (×) and bandwidth-expanded (◇) synthesis filters.

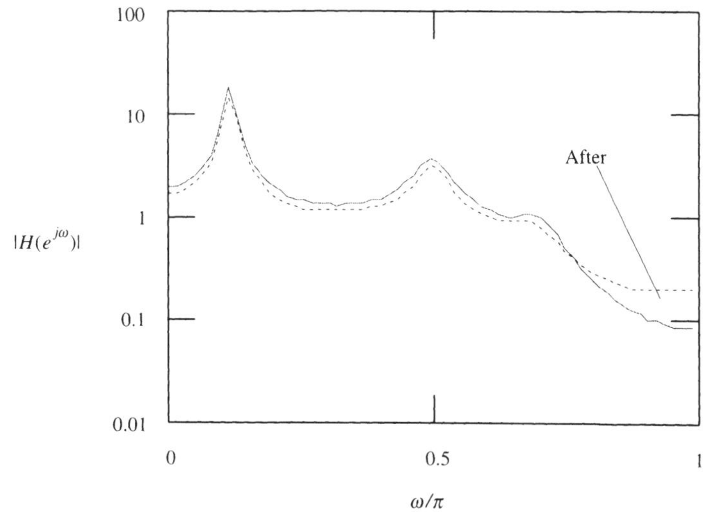

Figure 4.30 Comparison between the magnitude plots of the synthesis filter's transfer functions before and after white noise correction.

White Noise Correction

White noise correction mitigates ill-conditioning in LP analysis by directly reducing the spectral dynamic range and is accomplished by increasing the autocorrelation coefficient at zero lag by a small amount. The procedure is described by

$$R[0] \leftarrow \lambda . R[0]$$

with $\lambda > 1$ a constant. The constant λ is usually selected to be slightly above one. For the G.728 LD-CELP coder (Chapter 14), $\lambda = 257/256 = 1.00390625$, an increase of 0.39%. The process is equivalent to adding a white noise component to the original signal with a power that is 24 dB below the original average power. This directly reduces the spectral dynamic range and reduces the possibility of ill-conditioning in LP analysis. The drawback is that such an operation elevates the spectral valleys. By carefully choosing the constant λ, the degradation in speech quality can be made imperceptible.

Example 4.12 Figure 4.30 compares the magnitude plots of the synthesis filter before and after white noise correction, where the LPCs are the same as in Example 4.10 and $\lambda = 257/256$. Note that the dynamic range of the original function is reduced, where the lowest portion is elevated significantly.

Spectral Smoothing by Autocorrelation Windowing

In the bandwidth expansion method described earlier, the spectrum represented by the LPCs is smoothed by manipulating the values of the coefficients. The technique is applied after the LPCs are obtained.

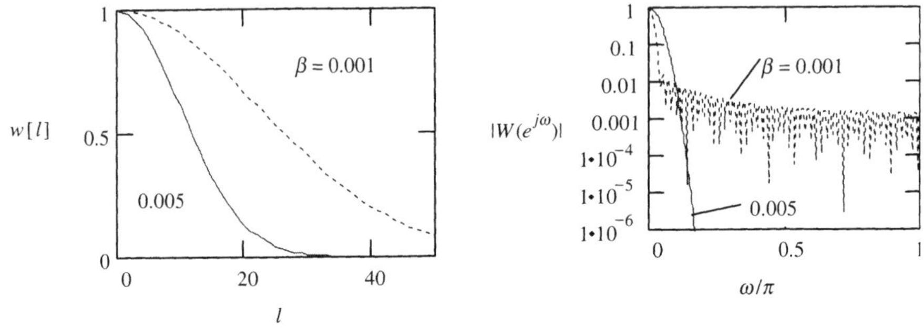

Figure 4.31 Gaussian windows and their Fourier transforms (magnitude normalized).

On some occasions, it is desirable to introduce some smoothing before obtaining the LPCs, since the solution algorithms (Levinson–Durbin or Leroux–Gueguen) require many computational steps leading to error accumulation. This can be done by windowing the autocorrelation function. Since the autocorrelation function and the power spectral density form a Fourier transform pair (Chapter 3), multiplying the autocorrelation values with a window (in lag domain) has the effect of convolving the power spectral density with the Fourier transform of the window (in frequency domain) [Oppenheim and Schafer, 1989]. By selecting an appropriate window, the desired effect of spectral smoothing is achieved. Given the autocorrelation function $R[l]$, windowing is performed with

$$R_{\text{new}}[l] = R[l] \cdot w[l]; \quad l = 0, 1, \ldots, M; \tag{4.96}$$

a suitable choice for $w[l]$ is the Gaussian window defined by

$$w[l] = e^{-\beta l^2}, \tag{4.97}$$

where β is a constant. Figure 4.31 shows some plots of a Gaussian window for various values of β.

The described technique can be used to alleviate ill-conditioning of the normal equation before it is solved; after convolving in the frequency domain, all sharp spectral peaks are smoothed out. The spectral dynamic range is reduced with the poles of the associated synthesis filter farther away from the unit circle.

Example 4.13 The autocorrelation values corresponding to the LPCs of Example 4.10 are Gaussian windowed with $\beta = 0.01$. Figure 4.32 compares the original spectrum with the one obtained after smoothing: note how the sharp peaks are lowered and widened. The net effect is similar to a bandwidth expansion procedure with direct manipulation of the LPCs.

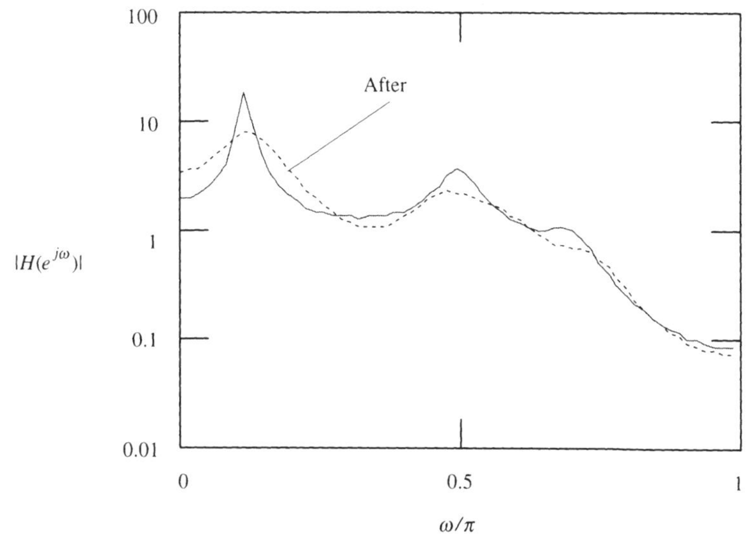

Figure 4.32 Comparison between the magnitude plots of the synthesis filter's transfer functions before and after spectral smoothing.

4.9 MOVING AVERAGE PREDICTION

The discussion so far is based on the AR model. Figure 4.33 shows the block diagrams of the AR process analyzer and synthesizer filters, where a predictor with the difference equation given by (4.1) is utilized. It is straightforward to verify that these block diagrams generate the exact same equations for the AR model. In practical coding applications, parameters of the predictor are often found from the signal itself since a computationally efficient procedure is available, enabling real-time adaptation.

The MA model, as explained in Chapter 3, is in a sense the dual of the AR model. Figure 4.34 shows the predictor-based block diagrams of the analyzer and synthesizer filters. In this case, however, the difference equation of the predictor is given by

$$\hat{s}[n] = -\sum_{i=1}^{K} b_i x[n-i], \tag{4.98}$$

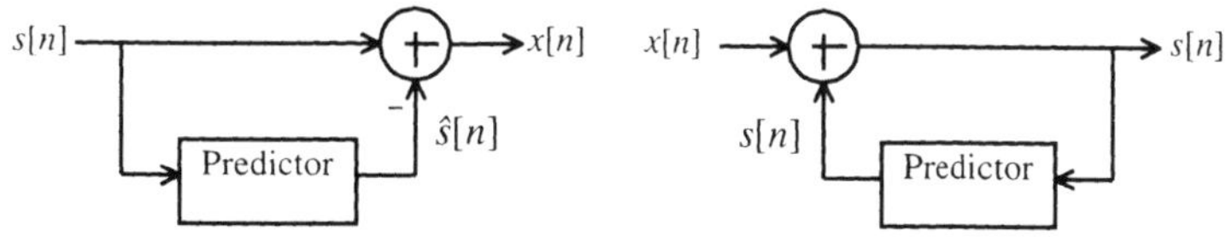

Figure 4.33 Block diagram of the AR analyzer filter (*left*) and synthesizer filter (*right*).

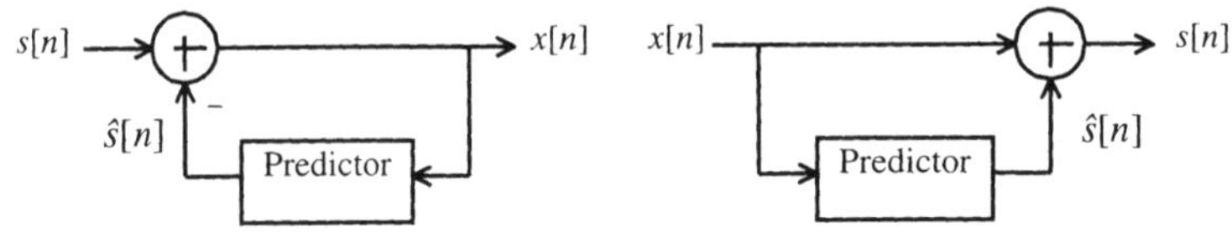

Figure 4.34 Block diagram of the MA analyzer filter (*left*) and synthesizer filter (*right*).

with K the order of the model and b_i the MA parameters. When compared with (4.1) we can see that "prediction" is now based on a linear combination of excitation or samples of prediction error $x[n]$, which in theory are white noise.

Unlike the AR model, where the optimal parameters can be found by solving a set of linear equations based on the statistics of the observed signal, the MA parameters can only be found using a set of nonlinear equations, and in practice highly computationally demanding. Hence, other approaches are normally applied to find the model parameters; these include spectral factorization [Therrien, 1992] and adaptive filtering techniques such as the least-mean-square (LMS) algorithm [Haykin, 1991], as well as other iterative methods.

Even though (4.98) is a sort of "linear prediction" scheme, where the prediction is based on a linear combination of samples, the name of LP is traditionally associated with AR modeling. When prediction is based on the MA model, it is explicitly referred to as "MA prediction" in the literature. Why do we bother with MA prediction? The technique offers some unique advantages and will be explained in Chapter 6, where differential pulse code modulation (DPCM) is introduced, and also in Chapter 7 with the introduction of predictive vector quantization (PVQ). Finally, in Chapter 15, MA prediction is applied to the design of a predictive quantizer for linear prediction coefficients.

4.10 SUMMARY AND REFERENCES

In this chapter, a theoretical foundation and practical implementation of linear prediction are thoroughly explained. Linear prediction is described as a system identification problem, where the parameters of an underlying autoregressive model are estimated from the signal. To find these parameters, autocorrelation values are obtained from the signal and a set of linear equations is solved. The resultant estimation is optimal in the sense that the variance of the prediction error is minimized. For nonstationary signals such as speech, the LP analysis procedure is applied to each short interval of time, known as a frame. The extracted LPCs from each frame result in a time-varying filter representing the activity of the human speech production organs. LP is often associated with the acoustic tubes model for speech production. Details can be found in Rabiner and Schafer [1978]. Efficient algorithms to solve the normal equation were introduced. Two such procedures—the Levinson–Durbin algorithm and the Leroux–Gueguen algorithm—can be used, with the latter more suitable for fixed-point implementation since all intermediate quantities of the procedure are bounded.

The method of LP analysis presented in this chapter is known in the literature as the autocorrelation method. Other schemes exist for LP analysis. The covariance method, for instance, formulates the problem in a different way, with the sum of squared error minimized inside the frame. This method has not received wide acceptance mainly because it cannot be solved as efficiently as the autocorrelation method. Also, no simple procedure allows a stability check. For additional information readers are referred to classical textbooks such as Markel and Gray [1976] and Rabiner and Schafer [1978]. A discussion of the computational cost for various LP analysis procedures is found in Deller et al. [1993].

Long-term linear prediction is an efficient scheme where correlation of the speech signal is modeled by two predictors. The short-term predictor is in charge of correlation between nearby samples, while the long-term predictor is in charge of correlation located one or multiple pitch periods away. The method described in this chapter is known as the one-tap predictor; that is, prediction is based on one single sample from the distant past. For a multitap long-term predictor, see Ramachandran and Kabal [1989]. However, the extra complexity and slight performance improvement limit the application of the multitap long-term predictor in practice [Kroon and Atal, 1991]. See Veeneman and Mazor [1993] for additional insight.

Several techniques to alleviate ill-conditioning, improve stability, and increase quality of the synthetic speech are presented. In a typical speech coding algorithm, these methods are used separately or combined together, and they are often included as standard computational steps. These procedures are cited in subsequent chapters, where different standard coders are studied. Autocorrelation windowing was introduced in Tohkura et al. [1978], developed originally to combat bandwidth underestimation. See Chen [1995] for a discussion of the incorporation of white noise correction, autocorrelation windowing, and bandwidth expansion to the framework of the LD-CELP coder.

Prediction can also be defined within the context of other signal models, such as MA. A good coverage of various statistical models can be found in Therrien [1992], as well as other textbooks such as Haykin [1991] and Picinbono [1993].

One of the criticisms about the application of LP in speech modeling is the fact that no zeros are incorporated in the system function of the synthesis filter, which introduces inaccuracies when representing certain classes of signals, such as nasal sounds. Difficulties related to a pole-zero type of system function, or ARMA model, are mainly due to the lack of efficient computational procedure to locate the parameters of the model. See Lim and Lee [1993] for pole-zero modeling of speech signals.

EXERCISES

4.1 Within the context of linear prediction, let $e[n]$ denote the prediction error under optimal conditions. Show that

$$E\{e[n]s[n-k]\} = 0$$

for $k = 1, 2, \ldots, M$. That is, $e[n]$ is orthogonal to $s[n-k]$. The relation is known as the principle of orthogonality.

4.2 An alternative way to obtain

$$J_{\min} = R_s[0] + \sum_{i=1}^{M} a_i R_s[i]$$

is by substituting (4.6), the condition required to minimize the cost function J(4.3), into J itself. Show the details of this alternative derivation.

4.3 In internal prediction where the analysis interval (for autocorrelation estimation) is the same as the prediction interval (the derived LPCs are used to predict the signal samples), find out the prediction gain when different windows are involved in the LP analysis procedure. Using a prediction order of ten and a frame length of 240 samples, calculate the segmental prediction gain by averaging the prediction gain results for a large number of signal frames for the two cases where the rectangular window or Hamming window is involved. Which window provides higher performance?

4.4 Consider the situation of external prediction where the autocorrelation values are estimated using a recursive method based on the Barnwell window. Using a prediction order of 50 and a frame length of 20 samples, measure the prediction gain for a high number of frames. Repeat the experiment using various values of the parameter α of the window (Chapter 3). Plot the resultant segmental prediction gain as a function of α. Based on the experiment, what is the optimal value of the parameter α?

4.5 From the system function of the pitch synthesis filter, find the analytical expression of the impulse response. Plot the impulse response of the pitch synthesis filter for the following two cases:

(a) $b = 0.5, T = 50$.
(b) $b = 1.5, T = 50$.

What conclusions can be reached about the stability of the filter?

4.6 Within the context of the Levinson–Durbin algorithm, (a) prove that

(a)

$$J_l = J_0 \prod_{i=1}^{l} \left(1 - k_i^2\right),$$

which is the minimum mean-squared prediction error achievable with an lth-order predictor.

(b) Prove that the prediction gain of the lth-order linear predictor is

$$PG_l = -10 \log_{10}\left(\prod_{i=1}^{l} \left(1 - k_i^2\right)\right).$$

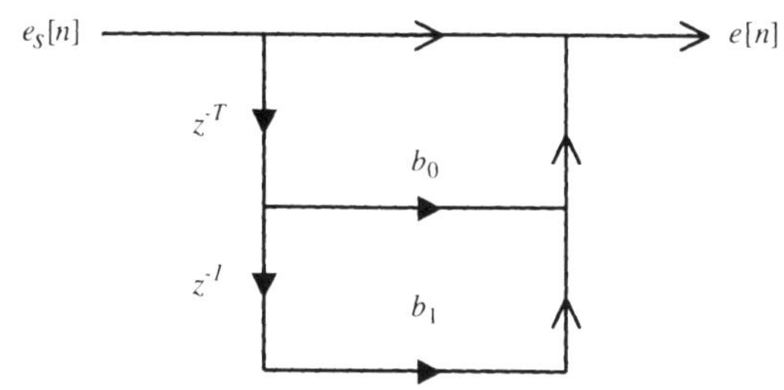

Figure 4.35 Equivalent signal flow graph of a long-term prediction-error filter with fractional delay.

4.7 In Example 4.8, where the simple linear interpolation procedure is applied to create fractional delay, show that the long-term prediction-error filter can be implemented as in Figure 4.35, with the long-term LPC summarized in Table 4.3, where b is the long-term gain given by (4.84). Thus, the considered long-term predictor with fractional delay is indeed a two-tap long-term predictor. What happens with the cases when two or more bits are used to encode the fractions?

4.8 In the long-term LP analysis procedure, minimization of J is equivalent to maximizing the quantity

$$\frac{\left(\sum_n e_s[n]e_s[n-T]\right)^2}{\sum_n e_s^2[n-T]}.$$

Justify the above statement. Develop a more efficient pseudocode to perform the task.

4.9 One suboptimal way to perform long-term LP analysis is by determining the pitch period T based on maximizing the autocorrelation

$$R[T] = \sum_n e_s[n]e_s[n-T].$$

Note that the sum of squared error J is not necessarily minimized. An advantage of the method is the obvious computation saving. Write down the pseudocode to perform long-term LP analysis based on the described approach. Mention the property of the resultant long-term gain b. *Hint*: The maximum autocorrelation value is in general greater than or equal to zero.

TABLE 4.3 Long-Term LPC for a Prediction Error Filter with Two Fractional Values: 0 or 0.5

Fraction	b_0	b_1
0	b	0
1/2	$b/2$	$b/2$

4.10 Use some speech signal to obtain a set of autocorrelation values for a 10th order predictor. Find the corresponding LPCs and plot the magnitude of the response for the associated synthesis filter. Also, plot the poles of the system function. Repeat using the LPCs obtained by first applying a white noise correction ($\lambda = 257/256$), followed by a Gaussian windowing ($\beta = 0.001$), and finally apply bandwidth expansion with $\gamma = 0.98$ to the resultant LPCs.

4.11 Within the context of AR modeling, where the prediction error is $e[n]$ and the prediction is $\hat{s}[n]$, derive the difference equation relating $e[n]$ to $\hat{s}[n]$ and show that the system function of the filter with $e[n]$ as input and $\hat{s}[n]$ as output is

$$H(z) = \frac{-\sum_{i=1}^{M} a_i z^{-i}}{1 + \sum_{i=1}^{M} a_i z^{-i}}.$$

4.12 Develop the pseudocode to perform long-term LP analysis using the fractional delay scheme described in Example 4.8. Consider two cases: an exhaustive search approach, where all possible delay values are evaluated, and a two-step suboptimal scheme, where the integer pitch period is located first followed by a fractional refinement near the integer result found.

4.13 In the long-term and short-term linear prediction model for speech production, the long-term predictor has a delay of T, while the short-term predictor has an order of M, with $T > M$. Is it functionally equivalent to replace the cascade connection of pitch synthesis filter and formant synthesis filter with a single filter composed of a predictor of order T with system function

$$-\sum_{i=1}^{T} a_i z^{-i},$$

where $a_i = 0$ for $i = M+1, M+2, \ldots, T-1$? Why or why not?

CHAPTER 5

SCALAR QUANTIZATION

Representation of a large set of elements with a much smaller set is called quantization. The number of elements in the original set in many practical situations is infinite, like the set of real numbers. In speech coding, prior to storage or transmission of a given parameter, it must be quantized. Quantization is needed to reduce storage space or transmission bandwidth so that a cost-effective solution is deployed. In the process, some quality loss is introduced, which is undesirable. How to minimize loss for a given amount of available resources is the central problem of quantization.

In this chapter, the basic definitions involved with scalar quantization are given, followed by an explanation of uniform quantizers—a common type of quantization method widely used in practice. Conditions to achieve optimal quantization are included with the results applied toward the development of algorithms used for quantizer design. Algorithmic implementation is discussed in the last section, where computational cost is addressed. The presentation of the material is intended to be rigorous mathematically. However, the main goal is to understand the practical aspects of scalar quantization, so as to incorporate the techniques in the coding of speech.

5.1 INTRODUCTION

In this section the focus is on the basic issues of scalar quantization.

Definition 5.1: Scalar Quantizer. A scalar quantizer Q of size N is a mappin from the real number $x \in \boldsymbol{R}$ into a finite set $\boldsymbol{Y}$ containing N output values (also known as reproduction points or codewords) y_i. Thus,

$$Q: \boldsymbol{R} \to \boldsymbol{Y}$$

where

$$(y_1, y_2, \ldots, y_N) \in \boldsymbol{Y}.$$

$\boldsymbol{Y}$ is known as the codebook of the quantizer. The mapping action is written as

$$Q(x) = y_i;\ x \in \boldsymbol{R};\quad i = 1, \ldots, N. \tag{5.1}$$

In all cases of practical interest, N is finite so that a finite number of binary digits is sufficient to specify the output value. We further assume that the indexing of output values is chosen so that

$$y_1 < y_2 < \cdots < y_N.$$

Definition 5.2: Resolution. We define the resolution r of a scalar quantizer as

$$r = \log_2 N \equiv \lg N, \tag{5.2}$$

which measures the number of bits needed to uniquely specify the quantized value.

Definition 5.3: Cell. Associated with every N point quantizer is a partition of the real line $\boldsymbol{R}$ into N cells $\boldsymbol{R}_i$, $i = 1, \ldots, N$. The ith cell is defined by

$$\boldsymbol{R}_i = \{x \in \boldsymbol{R}: Q(x) = y_i\} = Q^{-1}(y_i). \tag{5.3}$$

It follows that

$$\bigcup_i \boldsymbol{R}_i = \boldsymbol{R}, \tag{5.4}$$

and if $i \neq j$

$$\boldsymbol{R}_i \cap \boldsymbol{R}_j = \emptyset. \tag{5.5}$$

Definition 5.4: Granular Cell and Overload Cell. A cell that is unbounded is called an overload cell. The collection of all overload cells is called the overload region. A cell that is bounded is called a granular cell. The collection of all granular cells is called the granular region.

The set of numbers

$$x_0 < x_1 < x_2 < \cdots < x_N,$$

known as the boundary points, are used to define the cells for the quantizer, which are given by

$$R_i = (x_{i-1}, x_i]; \quad i = 1, \ldots, N, \tag{5.6}$$

where $x \in (a, b]$ implies $a < x \leq b$. Based on this definition, we have $x_0 = -\infty$ and $x_N = \infty$. Two overload cells exist: $\boldsymbol{R}_1 = (x_0, x_1] = (-\infty, x_1]$ and $\boldsymbol{R}_N = (x_{N-1}, x_N] = (x_{N-1}, \infty]$. The number of granular cells is equal to $N-2$, defined by (5.6) with $i = 2$ to $N-1$.

Definition 5.5: Regular Quantizer. A quantizer is defined to be regular if each cell $\boldsymbol{R}_i$ is an interval such that $y_i \in (x_{i-1}, x_i)$.

Since most quantizers for coding applications are regular, only regular quantizers are considered in this book. Figure 5.1 shows an example of the transfer characteristic of a regular quantizer.

Definition 5.6: Encoder and Decoder. Every quantizer can be viewed as the combined effect of two successive operations—an encoder E and a decoder D:

$$E: \boldsymbol{R} \to \boldsymbol{I},$$

$$D: \boldsymbol{I} \to \boldsymbol{R},$$

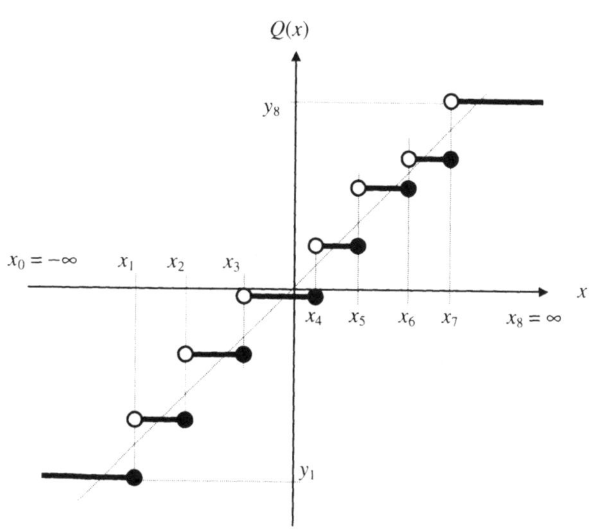

Figure 5.1 Example of the transfer characteristic for a regular quantizer with eight output levels.

with I the index set $\{1, 2, \ldots, N\}$. Thus, if $Q(x) = y_i$, then $E(x) = i$ and $D(i) = y_i$. Furthermore,

$$\hat{x} = Q(x) = D(E(x)) = y_i. \tag{5.7}$$

Definition 5.7: Distance or Distortion Measure. A distance or distortion measure is an assignment of a nonnegative cost $d(x, Q(x))$ associated with quantizing any input value x with a reproduction point $Q(x)$:

$$d(x, Q(x)) = \begin{cases} 0, & Q(x) = x, \\ >0, & \text{otherwise.} \end{cases} \tag{5.8}$$

Given a distortion measure we can quantify the performance of a system by the expected value of d. Let $\mathbf{x}$ denote a continuously distributed random variable with a specified PDF $f_{\mathbf{x}}(x)$. Then the expected value of the distortion can be expressed as

$$D = E\{d(\mathbf{x}, Q(\mathbf{x}))\} = \int_{-\infty}^{\infty} d(x, Q(x)) f_{\mathbf{x}}(x)\, dx. \tag{5.9}$$

The performance of a quantizer is often specified in terms of a signal-to-noise (SNR) ratio, given by

$$\text{SNR} = 10 \log_{10}\left(\frac{\sigma_{\mathbf{x}}^2}{D}\right) \tag{5.10}$$

measured in decibels, with $\sigma_{\mathbf{x}}^2$ denoting the variance of $\mathbf{x}$. Taking into account the partition into cells, (5.9) can be rewritten as

$$D = \sum_{i=1}^{N} \int_{R_i} d(x, Q(x)) f_{\mathbf{x}}(x)\, dx. \tag{5.11}$$

This is equivalent to

$$D = \sum_{i=1}^{N} P\{\mathbf{x} \in R_i\} E\{d(\mathbf{x}, y_i) | \mathbf{x} \in R_i\}, \tag{5.12}$$

with $P\{\cdot\}$ denoting the probability of an event. A conditional expectation term is incorporated.

Example 5.1: Mean-Squared Error Criterion. Due to its simplicity and analytical elegance, the mean-squared error is widely used in many practical situations. Consider the distortion measure defined by the squared error:

$$d(x, \hat{x}) = (x - \hat{x})^2. \tag{5.13}$$

Then from (5.11), the expected value of the distortion, or mean-squared error, is given by

$$D = E\{(\mathbf{x} - Q(\mathbf{x}))^2\} = \sum_{i=1}^{N} \int_{R_i} (x - y_i)^2 f_{\mathbf{x}}(x)\, dx = \sum_{i=1}^{N} \int_{x_{i-1}}^{x_i} (x - y_i)^2 f_{\mathbf{x}}(x)\, dx. \tag{5.14}$$

5.2 UNIFORM QUANTIZER

A uniform quantizer is one of the simplest to design and widely used in practice. For a uniform quantizer, the transfer characteristic $Q(x)$ is such that

$$y_{i+1} - y_i = \Delta; \quad i = 1, 2, \ldots, N-1, \tag{5.15a}$$

$$x_{i+1} - x_i = \Delta; \quad x_i,\ x_{i+1} \text{ finite} \tag{5.15b}$$

with Δ a constant known as the step size. The output levels for a uniform quantizer are given by

$$y_i = x_i - \Delta/2; \quad i = 1, \ldots, N-1, \tag{5.16a}$$

$$y_N = x_{N-1} + \Delta/2. \tag{5.16b}$$

Figure 5.2 shows an example of a uniform quantizer. Quantization error is defined by

$$e(x) = x - Q(x). \tag{5.17}$$

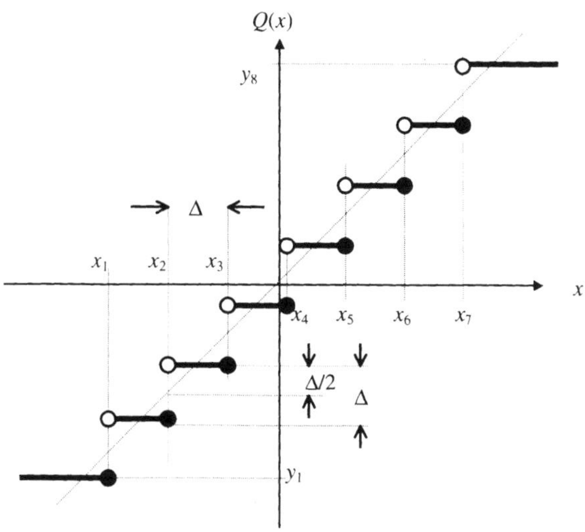

Figure 5.2 Example of the transfer characteristic for a uniform quantizer with eight output levels.

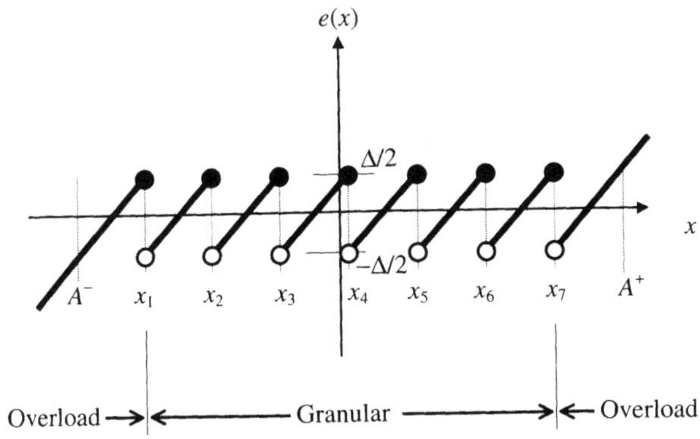

Figure 5.3 Example of quantization error for a uniform quantizer with eight output levels.

Figure 5.3 plots the error $e(x)$ corresponding to the quantizer of Figure 5.2. Note that

$$|e(x)| \leq \Delta/2, \quad A^- \leq x \leq A^+, \tag{5.18}$$

with

$$A^+ = x_{N-1} + \Delta \tag{5.19a}$$

and

$$A^- = x_1 - \Delta. \tag{5.19b}$$

One design technique for uniform quantizers is to assign A^+ and A^- to be equal to the maximum and minimum of the input value, respectively. Hence, excessive overload error is eliminated. Once the values of A^+ and A^- are known, the step size Δ can be found from (5.19) and Figure 5.3, given by

$$\Delta = \frac{A^+ - A^-}{N}. \tag{5.20}$$

Uniform Input Distribution

Consider the case of a uniform quantizer, where the input is bounded with values lying in the range $[A^-, A^+]$. Further assume that the input is uniformly distributed within that range. It is readily seen from Figure 5.3 that the quantizer error considered as the continuous random variable **e** has a uniform PDF in $[-\Delta/2, \Delta/2]$. This is shown in Figure 5.4. Thus, the mean of the quantization error is

$$E\{\mathbf{e}\} = 0 \tag{5.21}$$

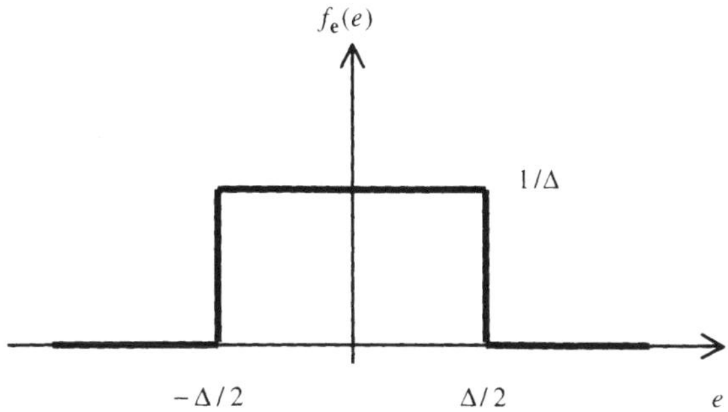

Figure 5.4 Probability density function of the quantization error for uniform input distribution.

with variance

$$var\{\mathbf{e}\} = E\{\mathbf{e}^2\} = \Delta^2/12. \tag{5.22}$$

Equation (5.22) is equal to the expected value of the distortion if the mean-squared error criterion is adopted. Therefore, to reduce the expected distortion, the step size must be decreased, which is accomplished by increasing the quantizer size N (see (5.20)). An excessively high N, however, requires a large amount of bits, translating directly to higher coding cost.

5.3 OPTIMAL QUANTIZER

The primary goal of quantizer design is to select the reproduction levels and the partition regions or cells so as to provide the minimum possible average distortion for a fixed number of levels N, or equivalently a fixed resolution r. Here, conditions for a quantizer to achieve optimality are given. These conditions will serve as references to develop the optimization procedure required for quantizer design.

Definition 5.8: Optimal Quantizer. A quantizer Q of size N is said to be optimum if it minimizes the expected value of the distortion

$$D = E\{d(\mathbf{x}, Q(\mathbf{x}))\} = \sum_{i=1}^{N} \int_{\mathbf{R}_i} d(x, y_i) f_{\mathbf{x}}(x)\, dx, \tag{5.23}$$

with $\mathbf{R}_i$ the cells of the quantizer and $f_{\mathbf{x}}(x)$ the PDF of the input random variable $\mathbf{x}$.

Therefore, for optimal operation, it is necessary to specify the output points y_i and partition cells $\mathbf{R}_i$ for a given PDF of $\mathbf{x}$ so as to minimize D.

The Nearest-Neighbor Condition for Optimality

For a given codebook $\mathbf{Y}$ of size N, the optimal partition cells satisfy

$$\mathbf{R}_i = \{x : d(x, y_i) \leq d(x, y_j)\} \tag{5.24}$$

for all $i \neq j$. That is, $Q(x) = y_i$ only if $d(x,y_i) \leq d(x,y_j)$. Hence,

$$d(x, Q(x)) = \min_i d(x, y_i). \tag{5.25}$$

To show that (5.24) yields a minimum expected distortion D, consider the relation

$$D = \int d(x, Q(x)) f_{\mathbf{x}}(x)\, dx \geq \int (\min_i d(x, y_i)) f_{\mathbf{x}}(x)\, dx \tag{5.26}$$

and this lower bound is indeed attained when $Q(x)$ performs the nearest-neighbor mapping with the given codebook $\boldsymbol{Y}$.

Definition 5.9: Centroid. We define the centroid cent($\boldsymbol{R}_o$), of any nonempty set $\boldsymbol{R}_o \in \boldsymbol{R}$, as the value y_o (if it exists) that minimizes the expected distortion between $\mathbf{x}$ and y_o, given that $\mathbf{x}$ lies in $\boldsymbol{R}_o$. Thus,

$$\text{cent}(\boldsymbol{R}_o) = \{y_o : \boldsymbol{E}\{\boldsymbol{d}(\mathbf{x}, y_o)|\mathbf{x} \in \boldsymbol{R}_o\} \leq \boldsymbol{E}\{\boldsymbol{d}(\mathbf{x}, y)|\mathbf{x} \in \boldsymbol{R}_o\}\}; \quad \forall y \in \boldsymbol{R}. \tag{5.27}$$

The Centroid Condition for Optimality

Given a partition $\{\boldsymbol{R}_i;\ i = 1, \ldots, N\}$, the optimal codewords satisfy

$$y_i = \text{cent}(\boldsymbol{R}_i). \tag{5.28}$$

The statement can be proved in the following manner. From (5.12), the average distortion is written as

$$D = \sum_{i=1}^{N} P\{\mathbf{x} \in \boldsymbol{R}_i\} E\{d(\mathbf{x}, y_i)|\mathbf{x} \in \boldsymbol{R}_i\} \geq \sum_{i=1}^{N} P\{\mathbf{x} \in \boldsymbol{R}_i\} \min_y E\{d(\mathbf{x}, y)|\mathbf{x} \in \boldsymbol{R}_i\} \tag{5.29}$$

and the inequality becomes an equality if the y_i are the centroids.

The Centroid Optimality Condition for Squared Error Distortion Measure

Given a partition $\{\boldsymbol{R}_i:\ i = 1, \ldots, N\}$, if the distortion measure is the squared error defined by

$$d(x, y_i) = (x - y_i)^2, \tag{5.30}$$

then the optimal codewords satisfy

$$y_i = \text{cent}(\boldsymbol{R}_i) = E\{\mathbf{x}|\mathbf{x} \in \boldsymbol{R}_i\} = \frac{\int_{R_i} x f_{\mathbf{x}}(x)\, dx}{\int_{R_i} f_{\mathbf{x}}(x)\, dx}. \tag{5.31}$$

The result is obtained by writing the expected value of the distortion as

$$D = \sum_{i=1}^{N} P\{\mathbf{x} \in \boldsymbol{R}_i\} E\{(\mathbf{x} - y_i)^2 | \mathbf{x} \in \boldsymbol{R}_i\}, \tag{5.32}$$

which can be minimized by choosing the appropriate codeword y_i for each cell. The value of y_i that minimizes (5.32) is precisely given by the conditional mean indicated in (5.31). See Exercise 5.6 for a proof.

5.4 QUANTIZER DESIGN ALGORITHMS

Given the codebook size N, it is desired to find the input partition cells and codewords such that the average distortion $D = E\{d(\mathbf{x}, Q(\mathbf{x}))\}$ is minimized, with $\mathbf{x}$ being the input random variable or the source with a given PDF.

The Lloyd Iteration for Codebook Improvement

It is possible to improve the quantizer by following the two iteration steps indicated below, known as the Lloyd iteration.

Step 1. Given the codebook $\boldsymbol{Y}_m = \{y_{m,i};\ i = 1, 2, \ldots, N\}$, find the optimal partition into quantization cells; that is, use the nearest-neighbor condition to form the nearest-neighbor cells:

$$\boldsymbol{R}_{m,i} = \{x : d(x, y_{m,i}) \leq d(x, y_{m,j})\}$$

for all $j \neq i$.

Step 2. Using the centroid condition, find $\boldsymbol{Y}_{m+1}$, the optimal reproduction codewords for the cells just found. Note that the input PDF must be known in order to compute the centroids.

The Lloyd Algorithm

The Lloyd iteration can be used to improve a quantizer starting from an initial codebook. If the input PDF is not mathematically tractable, a sample distribution based on empirical observations is used instead. The actual algorithm of Lloyd for quantizer design is stated as follow:

Step 1. Begin with an initial codebook $\boldsymbol{Y}_1$. Set $m = 1$.

Step 2. Given the codebook $\boldsymbol{Y}_m$, perform the Lloyd iteration to generate the improved codebook $\boldsymbol{Y}_{m+1}$.

Step 3. Compute the average distortion for $\boldsymbol{Y}_{m+1}$ (D_{m+1}). If it has changed by a small enough amount since the last iteration, stop. Otherwise, set $m + 1 \rightarrow m$ and go to Step 2.

One reasonable stopping criterion is to use the fractional drop in distortion, $(D_m - D_{m+1})/D_m$. The algorithm stops when the ratio is below a suitable threshold.

It can easily be shown that the algorithm will necessarily produce a sequence of codebooks with monotone nonincreasing values of average distortion. If the algorithm converges to a codebook in the sense that further iterations no longer produce any changes in the set of reproduction values, then the resulting codebook must simultaneously satisfy both necessary conditions for optimality.

Quantizer Design Based on Squared Error

Here, the Lloyd algorithm is adapted to squared error measure applied to a set of training data. The theory presented so far is based on knowledge of the PDF of the source to be quantized. In practice, the PDF of a given source is either unknown or difficult to deal with analytically. Therefore, the common practice is to collect a finite number of samples from the source of interest and use this set of data, known as the training data set, to "train" the quantizer so as to reduce the average distortion over the data set.

Consider the case where we want to design the quantizer for the random variable $\mathbf{x}$. Further assume that a total of N_t samples are available, denoted by

$$x_k; \quad k = 0, 1, \ldots, N_t - 1,$$

where N_t is the size of the training data set. Based on the squared distortion measure and assuming that the training data set is partitioned into N cells, then from (5.31) the centroid of the ith cell is found to be

$$y_i \approx \frac{\sum\limits_{x_k \in R_i} x_k \frac{1}{N_t}}{\sum\limits_{x_k \in R_i} \frac{1}{N_t}} = \frac{1}{N_i} \sum_{x_k \in R_i} x_k \tag{5.33}$$

with N_i the number of elements in the ith cell. In (5.33) the integrals are approximated by the sums and each element in the training data set has the same probability of occurrence equal to $1/N_t$. Thus, the centroid of $\boldsymbol{R}_i$ is given by the mean of the elements pertaining to the cell.

The adapted version of Lloyd algorithm (also known as the k-means algorithm) is specified as follows:

Step 1. Initialization: Elements of the initial codebook are selected randomly from the training data set.

Step 2. Partition into nearest-neighbor cells: Given the codebook, partition the training data set into cells based on the squared distortion criteria.

Step 3. Centroid calculation: Update the codebook by calculating the centroids found from the cells obtained in Step 2. The centroid is the arithmetic mean of the elements pertaining to a particular cell.

Step 4. Compute the distortion sum for the new codebook. If it has changed by a small enough amount since the last iteration, stop. Otherwise go to Step 2.

The distortion sum in Step 4 is concerned with the total sum of differences between the elements of the training set with respect to their quantized versions. The average distortion sum is given by the distortion sum divided by N_t. Both are valid indicators for the quality of the quantizer.

Example 5.2 Behavior of the Lloyd algorithm is illustrated here, where a quantizer with a codebook size of $N = 8$ is designed for a Gaussian source with zero mean and unit variance. Size of the training data set is $N_t = 600$. A histogram of the training data is shown in Figure 5.5, where we can see that it assumes the shape of the PDF corresponding to a normally distributed random variable. Figure 5.6 shows the distortion sum as a function of the iteration number; note how it rapidly drops as the algorithm progresses. After ten iterations, it almost stabilizes to a minimum value. Figure 5.7 displays the trajectories of the eight codewords, where

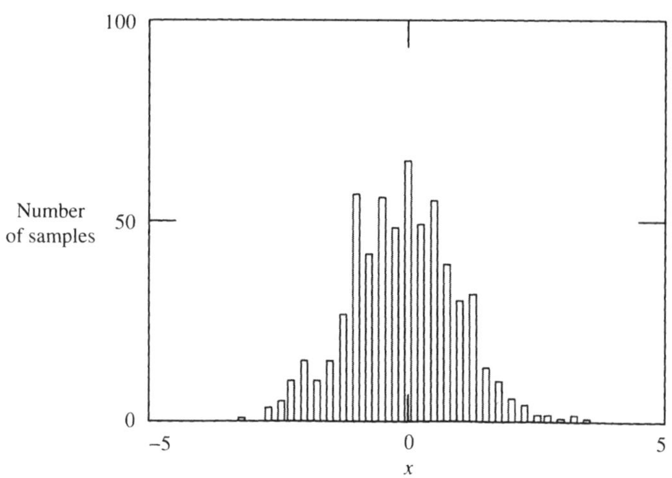

Figure 5.5 Histogram of the training data to be used in an experiment.

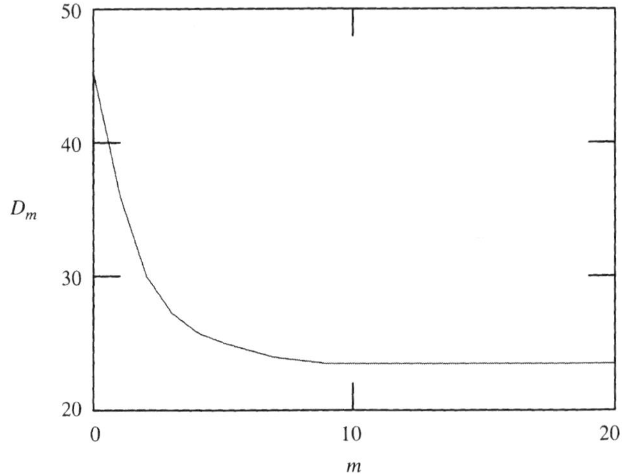

Figure 5.6 Distortion sum as a function of the number of iterations in a quantizer design experiment.

the initial values are selected arbitrarily from the training data set. Like the distortion sum, the codewords converge rapidly toward a stable value after ten iterations.

One fundamental issue for the Lloyd algorithm is the problem of the local minimum, where there is no guarantee that the codewords after convergence can provide the best possible solution, or global minimum. In practice, the algorithm converges and gives a solution representing a local minimum. In general, the convergence points of the codewords, and thus the final distortion sum, depend on the initial codewords selected to start the training process. The situation is illustrated for the case of Example 5.2, where a total of ten random initializations are performed

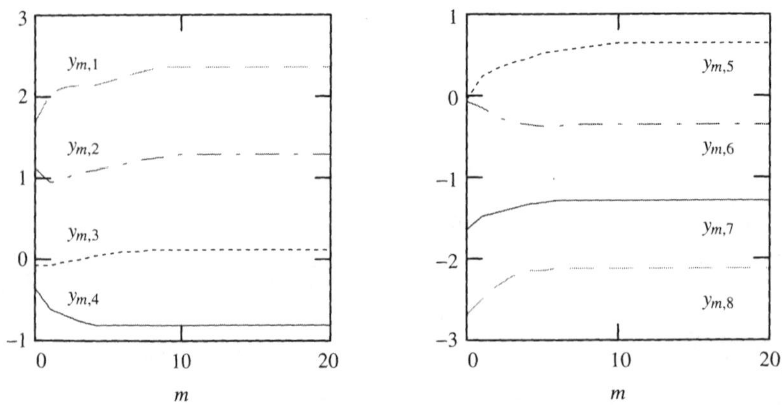

Figure 5.7 Codeword trajectories in a quantizer design experiment.

TABLE 5.1 Results for Ten Random Initializations in a Quantizer Design Experiment

Initialization Number	Distortion Sum	SNR – Training (dB)	SNR – Testing (dB)
0	23.6	14.5	14.5
1	24.1	14.4	13.8
2	21.7	14.9	13.8
3	23.6	14.5	14.5
4	21.6	14.9	13.9
5	23.6	14.5	14.5
6	21.6	14.9	13.9
7	21.6	14.9	13.9
8	24.5	14.4	14.0
9	23.6	14.5	14.5

with the results summarized in Table 5.1. The results include the distortion sum, the signal-to-noise ratio for the training data set, and the signal-to-noise ratio for the testing data set. The distortion sum reflects the total sum of squared error between the training data set and the final codewords. The testing data set is derived from the same source but is different from the training data set; its purpose is to verify the generalization ability of the solution found by the algorithm: that is, to see whether the quantizer can provide good performance for data outside the training data set having the same statistical property as the source. In the present case, size of the testing data set is the same as the size of the training data set.

Note that different initializations result in distinct solutions at convergence with different performance. Also note that the SNR for the testing data set is consistently below the SNR for the training data set, which is an expected result since the sum of distortion is optimized for the training data set. It is not possible in general to predict which initialization results in better solution. One simple approach in practice is to run the algorithm for a number of random initializations and pick the best one as the final solution. A similar problem is associated with vector quantization, and additional techniques to overcome this problem are explained in Chapter 7.

5.5 ALGORITHMIC IMPLEMENTATION

Traditionally, a scalar quantizer is implemented as electronic circuits using amplifiers, comparators, and digital logic gates. The analog-to-digital converter, for instance, is one form of scalar quantizer. With the availability of programmable processors, software implementation becomes feasible. In software, parameters of the quantizer can easily be modified, and the system is inherently more stable against changing operating conditions, like temperature and supply voltage. Since most programmable processors work in a sequential fashion, where instructions contained in a program are executed one by one, we will consider two algorithmic implementations suitable to any sequential machine.

In the following we will assume that all cell boundary values $-\infty < x_1 < x_2 < \cdots < x_{N-1} < \infty$ and the output levels $y_1 < y_2 < \cdots < y_N$ are known.

Linear Search

One straightforward implementation is to compare the input value to the boundary values starting from the lowest point x_1. The idea is summarized with the following pseudocode:

```
QUANTIZE(x)
for i ← 1 to N-1
        if x ≤ x_i return y_i
return y_N
```

Execution time of QUANTIZE(x) can be quantified with the number of comparison operations required to return the output level. From the pseudocode we can see that the shortest possible time is one single comparison operation and it happens when $x \leq x_1$. In the worst case with $x > x_{N-2}$, the number of comparison operations is equal to $N - 1$. With the assumption that the input value x has an equal probability of falling within any particular cell, then, on average, the number of comparison operations is given by

$$\begin{aligned}\text{Number of comparisons} &= \frac{1}{N} + \frac{2}{N} + \cdots + \frac{N-1}{N} + \frac{N-1}{N} \\ &= \frac{1}{N}\sum_{i=1}^{N-1} i + \frac{N-1}{N} \\ &= \frac{(N-1)(N+2)}{2N}. \end{aligned} \tag{5.34}$$

As $N \to \infty$, the number of comparisons approaches $N/2$. Thus, the execution time grows asymptotically with N.

Tree Search

Alternatively, the cell boundary values can be organized in a tree structure for comparison purposes. Figure 5.8 shows a tree structure corresponding to the case of $N = 8$. To see how it works, assume that the input value x is such that $x_3 < x \leq x_4$. First x is compared to x_4; since $x \leq x_4$, the path on the tree is to move to the left and compare x with x_2. Now, $x > x_2$ and the algorithm moves down one more level to compare x with x_3. Since $x > x_3$ and the bottom of the tree is reached, output level y_4 is returned as the quantized version of x.

The idea can be extended to any quantizer size. If the size is a power of two, then regardless of the input x, the number of comparison operations is always equal to

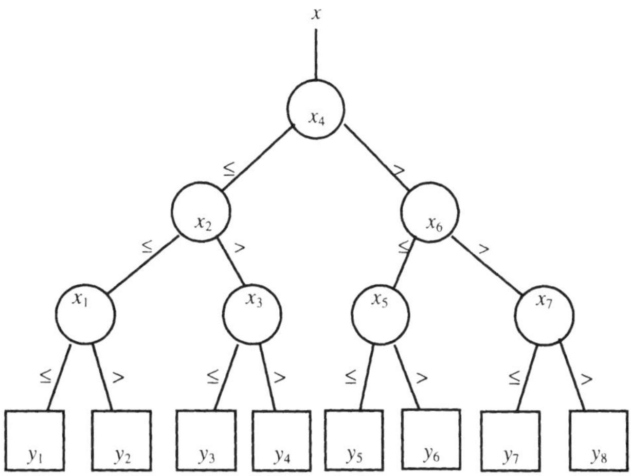

Figure 5.8 A tree structure for scalar quantization with $N = 8$.

lg N. Hence, the execution time of this quantizer is proportional to lg N. A comparison between the number of operations required by the two implementations is shown in Figure 5.9. Only sizes that are a power of two are considered. The quantizer size ranges from 4 to 1024. The number of operations is taken to be the average number of comparisons for the two methods. As we can see, a tree search requires far less operations to complete its task, which is especially true when the size of the quantizer increases. At $N = 1024$, or a resolution of 10 bits, the

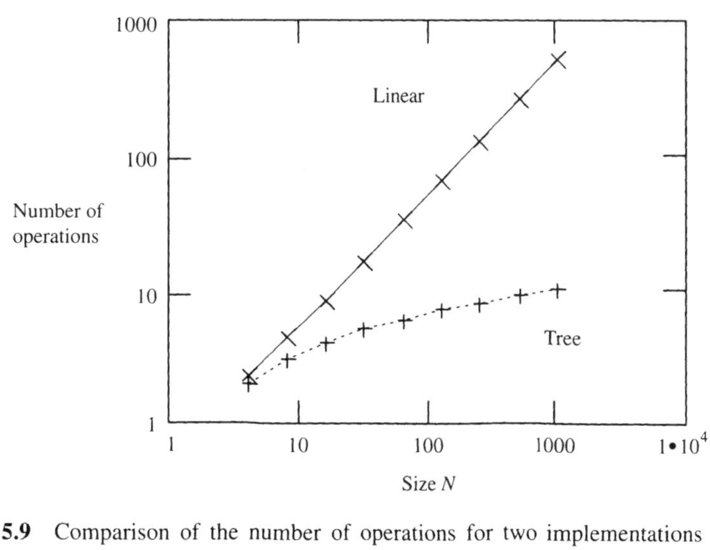

Figure 5.9 Comparison of the number of operations for two implementations of scalar quantizer.

tree search technique requires 10 operations while a linear search method needs almost 513 operations! This case illustrates the fact that algorithms, like computer hardware, are indispensable for the improvement of the overall system performance.

5.6 SUMMARY AND REFERENCES

In this chapter, the fundamental concepts of scalar quantization are introduced, followed by an explanation of a uniform quantizer, which is widely used in practice due to its simplicity. Conditions for optimal operation are given. These conditions form the foundation for the derivation of algorithms for quantizer design. Software implementation of a scalar quantizer is exemplified with two types of search; performance is analyzed and compared in terms of the number of operations required. The studied cases are based on the assumption that the quantizer is implemented in a sequential machine, where instructions are executed one at a time. Many modern processors provide a certain degree of parallelism, allowing an algorithm to execute at a much faster pace. See Intel [1997] for an example of such a processor.

The objective of scalar quantizer design is to specify the transfer characteristic, which is completely determined by the boundaries of the cells and the corresponding output levels. For uniform quantizers, input boundaries and output levels are found directly from the minimum and maximum values of the source. In order to minimize the average distortion, an optimization method, called the Lloyd algorithm, is used to find the parameters of the quantizer. The algorithm is capable of producing locally optimal solutions adapted to the statistical property of the source.

Many theoretical issues concerning scalar quantization can be found in Gersho and Gray [1995]. Also see Sayood [1996] for a comprehensive discussion of related topics. For random variables and probability, see Papoulis [1991] and Peebles [1993]. For a thorough discussion of algorithmic structure and computational cost, see Cormen et al. [1990].

Vector quantization, the subject of Chapter 7, contains generalization of the theory presented in this chapter.

EXERCISES

5.1 A symmetric quantizer satisfies

$$Q(x) = -Q(-x).$$

A quantizer is midtread if it is regular and one of the output levels is zero; a quantizer is midrise if it is regular and no output level is equal to zero. What is the constraint for N, the number of output values of the quantizer, so that the quantizer is midtread symmetric? Repeat for midrise symmetric.

5.2 Given a signal source x such that $-1 \leq x \leq 3$, design a uniform quantizer with $N = 7$ by specifying all input boundaries x_i and output levels y_i.

5.3 Prove that if Q is a quantizer whose codebook satisfies the centroid condition and the distortion is measured with squared error, then:

(a) The mean of the quantizer output is the same as the mean of the input

$$E\{Q(\mathbf{x})\} = E\{\mathbf{x}\}.$$

(b) The quantizer output is uncorrelated with the quantizer error

$$E\{Q(\mathbf{x})(Q(\mathbf{x}) - \mathbf{x})\} = 0.$$

(c) The expected squared quantizer error is the difference of the second moments of the input signal and the quantized output

$$E\{(\mathbf{x} - Q(\mathbf{x}))^2\} = E\{\mathbf{x}^2\} - E\{(Q(\mathbf{x}))^2\}.$$

(d) The second moment of the quantizer output is less than or equal to the second moment of the input

$$E\{(Q(\mathbf{x}))^2\} \leq E\{\mathbf{x}^2\}.$$

5.4 Obtain a certain number of speech samples (say, 2000) and use them to train scalar quantizers having sizes of 4, 8, 16, 32, 64, and 128. Plot the resultant quantizer transfer characteristic in each case. Measure the distortion sum with respect to the training data set and plot it as a function of the quantizer size. Using another set of speech samples as the testing data set, measure the performance of the resultant quantizers.

5.5 The following pseudocode implements a scalar quantizer based on the *full search* method, where the distance between the input value x and each codeword is calculated. The codeword producing the smallest distance is returned as the quantized result.

```
QUANTIZE(x)
min ← ∞
for i ← 1 to N
        distance ← d(x, y_i)
        if distance < min
                min ← distance
                index ← i
return y_index
```

Without counting the operations necessary for distance computation, how many comparisons (Line 4) are required? Compare this approach with respect to linear search and tree search. Comment on any advantage and/or disadvantage.

5.6 To find the optimal codewords under squared-error distortion measure, one can differentiate the conditional expected distortion

$$E\{(\mathbf{x} - y_i)^2 | \mathbf{x} \in \boldsymbol{R}_i\}$$

with respect to y_i and equating to zero. Show that the optimal codeword is indeed the centroid of the cell $\boldsymbol{R}_i$.

5.7 The linear search technique as explained in Section 5.5 can also be implemented as a tree. Draw the corresponding tree for the case of $N = 8$ and find the average number of comparisons assuming equal probability for the input value to fall within any particular cell.

5.8 The adaptive quantizer design algorithm can be used as an alternative for quantizer design under squared error.

Step 1. Initialization: Elements of the initial codebook are selected randomly from the training data set.

Step 2. Centroid update: For each training data x_k, find the associated codeword y_i, then modify the codeword according to

$$y_i \leftarrow \begin{cases} y_i - \alpha|y_i - x_k|, & y_i > x_k, \\ y_i + \alpha|y_i - x_k|, & y_i < x_k, \end{cases}$$

with $\alpha < 1$ a constant. The purpose of this step is to move the codeword a little closer to a certain training data.

Step 3. Compute the distortion sum for the new codebook. If it has changed by a small enough amount since the last iteration, stop. Otherwise go back to Step 2.

The logic of the algorithm is that by adjusting the codewords repeatedly toward the associated training data (Step 2), the codebook should eventually converge to a local optimum. In practice, the constant α is often reduced gradually as training progresses.

Implement the adaptive algorithm and compare its performance with the Lloyd algorithm using a set of randomly generated training data.

5.9 Design the tree structure for a quantizer of size $N = 7$. Assuming that the input variable has an equal probability of falling within any particular cell, what is the average number of comparison operations? Repeat for $N = 6$.

CHAPTER 6

PULSE CODE MODULATION AND ITS VARIANTS

Essentially, pulse code modulation (PCM) refers to the process of quantizing the samples of a discrete-time signal, so that both time and amplitude are represented in a discrete form. PCM is the most obvious method developed for the digital coding of waveforms. It is conceptually straightforward to understand and broadly accepted as a comparison standard for other coders.

This chapter can be considered as an extension to Chapter 5, where scalar quantization is discussed in detail with the material applied to any signal type. In this chapter, however, the focus is on speech samples. The chapter is organized as follows: performance of uniform quantizers is first found, followed by standardized nonuniform quantizers; the idea of differential PCM is given, and several adaptive methods are spelled out at the end.

6.1 UNIFORM QUANTIZATION

Due to the convenience of binary code representation, a quantizer with an even number of output values is often used for the quantization of speech samples. Table 6.1 illustrates the case of 3-bit resolution, or a size of 8, where the binary indices and output values are shown. For the output values, Δ represents the step size. Negative indices are written in second-complement form, with the most significant bit containing sign information. Second-complement representation is widely used in most microprocessors, due to its convenience for arithmetic operations [Mano, 1993].

Figure 6.1 shows the transfer characteristic of the quantizer. Note that the zero output level is present, and the number of negative output levels is equal to the

TABLE 6.1 Binary Index and Quantized Values for a Uniform Quantizer with 3-bit Resolution

Binary Index	Quantized Value
011	3Δ
010	2Δ
001	Δ
000	0
111	$-\Delta$
110	-2Δ
101	-3Δ
100	-4Δ

number of positive output levels plus one. All uniform quantizers considered in this section adopt the described structure.

Loading Factor

Given a quantizer, the loading factor γ is defined by

$$\gamma = A/\sigma, \tag{6.1}$$

where σ is the standard deviation of the input signal, and A is the peak signal magnitude that can be quantized without incurring an excessive overload error. For our uniform quantizer, A is defined to have the value of $-x_1$. The loading factor indicates how well the input is matched to the quantizer.

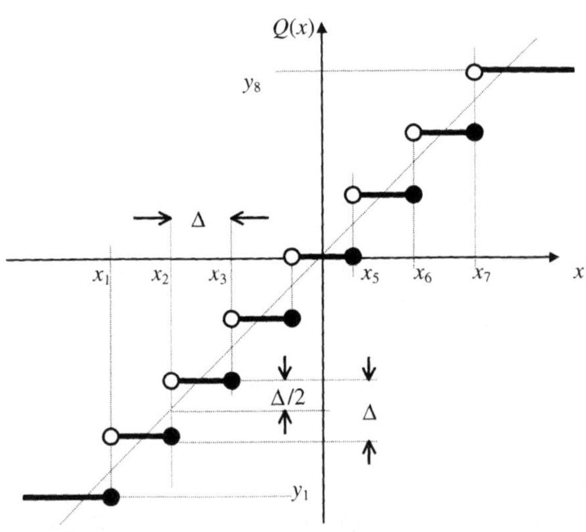

Figure 6.1 Transfer characteristic of a uniform quantizer with eight output levels.

Uniform Input Distribution

For a uniformly distributed input variable **x** with standard deviation σ, the PDF is given by

$$f_{\mathbf{x}}(x) = \begin{cases} \frac{1}{2\sqrt{3}\sigma}, & |x| \leq \sqrt{3}\sigma, \\ 0, & \text{otherwise}, \end{cases} \tag{6.2}$$

where zero mean is assumed. Expected distortion incurred with quantizing the variable **x** using an N-point quantizer is given by (Chapter 5)

$$D = \sum_{i=1}^{N} \int_{x_{i-1}}^{x_i} (x - y_i)^2 f_{\mathbf{x}}(x)\, dx, \tag{6.3}$$

where the x_i are the boundary points and the y_i are the codewords. In (6.3), the distortion measure considered is the squared error. The signal-to-noise ratio (SNR) is

$$\text{SNR} = 10 \log_{10}(\sigma^2/D) \tag{6.4}$$

and is the basic performance measure for the quantizer.

For uniform quantizers, separations between the (finite) boundary points are constant and are also true for the codewords. Expected distortion can be solved directly from (6.3) using numerical methods for the integral. By selecting a set of input standard deviation values σ, one can solve D and hence SNR for each σ. The resultant curves (D or SNR versus σ) allow us to assess the performance and behavior of the quantizer.

Example 6.1 A uniform quantizer with $N = 16$ (resolution equal to 4 bits) is designed with $x_1 = -1$, leading to a step size Δ of 0.133. Figure 6.2 shows the expected distortion and signal-to-noise ratio as a function of σ, where σ is varied between 0.01 and 2. The input variable is uniformly distributed.

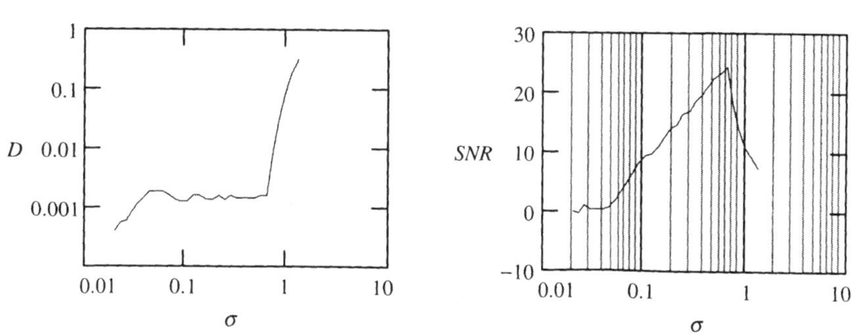

Figure 6.2 Expected distortion (*left*) and signal-to-noise ratio (*right*) as a function of the input standard deviation. Uniform quantization with 4-bit resolution and uniformly distributed input.

For small values of input standard deviation, the expected distortion is relatively low and remains almost constant inside the range $0.1 < \sigma < 0.7$. It then grows abruptly for $\sigma > 0.7$. This behavior is explained as follows: for small σ the quantizer output is closed to zero, generating a small distortion value. As σ increases, the expected distortion remains almost constant, due to the fact that the quantizer's cells are uniform, and so does the input PDF. However, increasing the value of σ further drives the quantizer to its overload regions (Chapter 5), increasing greatly the total distortion. From the SNR plot we can see that SNR $\approx$ 0 dB for $\sigma < 0.05$; that is, the signal energy is near the noise energy. This is expected since the output level is zero for low input, leading to an error level close to the input level. For $0.05 < \sigma < 0.7$, SNR grows almost 6 dB per octave until the point where overload noise is so severe that SNR decreases sharply.

Effects of Resolution

In analyzing the behavior of the quantizer, it is preferable to use relative quantities, like signal-to-noise ratio and loading factor, instead of absolute quantities, such as expected distortion and standard deviation. Relative parameters portray the behavior of the system in a way that is independent of the signal level and hence is more general. Figure 6.3 plots the SNR as a function of the loading factor for four resolution values: $r = 3$, 4, 5, and 6 bits. As we can see, SNR is elevated in

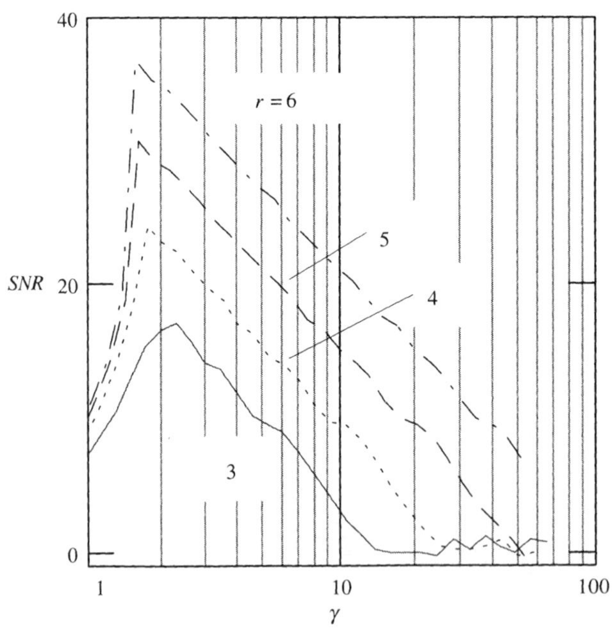

Figure 6.3 Signal-to-noise ratio as a function of the loading factor for four resolution values. Uniform quantization and uniformly distributed input.

general for higher resolution. Roughly speaking, the peak SNR increases by 6 dB for each additional bit. See Exercise 6.1 for a theoretical justification of this behavior. A loading factor between 1.5 and 2 seems to be optimal since SNR is maximized.

Probability Distribution Function of Speech Signals' Amplitudes

As discussed in Chapter 5, the characteristic of a quantizer must adapt to the statistical distribution of the input itself in order to achieve optimality, with optimality referring to the minimization of the average distortion. The PDF of speech samples can be estimated by determining a histogram for a large number of samples. It is commonly accepted that the Laplacian density

$$f_{\mathbf{x}}(x) = \frac{1}{\sqrt{2}\sigma} e^{-\sqrt{2}|x|/\sigma} \tag{6.5}$$

is a good approximation to the actual distribution of speech samples [Rabiner and Schafer, 1978]. Figure 6.4 shows the PDF plots for several values of the variance. As we can see, the number of low-amplitude samples is higher than the number of high-amplitude samples. Due to this distribution shape, use of uniform quantization is suboptimal.

Figure 6.5 shows several SNR curves for Laplacian input PDF. In comparison to the results of uniform input PDF (Figure 6.3), the curves are much smoother, with no abrupt transition. This is due to the fact that the Laplacian PDF is a smooth continuous function (except at the origin). In addition, the SNR values in the present case are in general lower for a particular resolution, as is evidenced by comparing Figure 6.3 with Figure 6.5. Since uniform quantization is optimal for uniformly distributed input, the SNR decreases for other input distribution functions at the same resolution. From the shape of the Laplacian PDF, the SNR can be improved if the quantizer is modified in such a way that a higher number of output levels are assigned to low-amplitude input, effectively reducing the average distortion. This technique is studied in the next section.

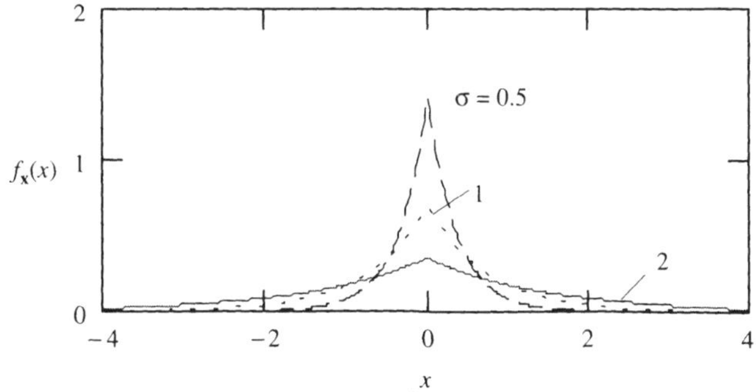

Figure 6.4 Plot of PDF for random variables with Laplacian distribution.

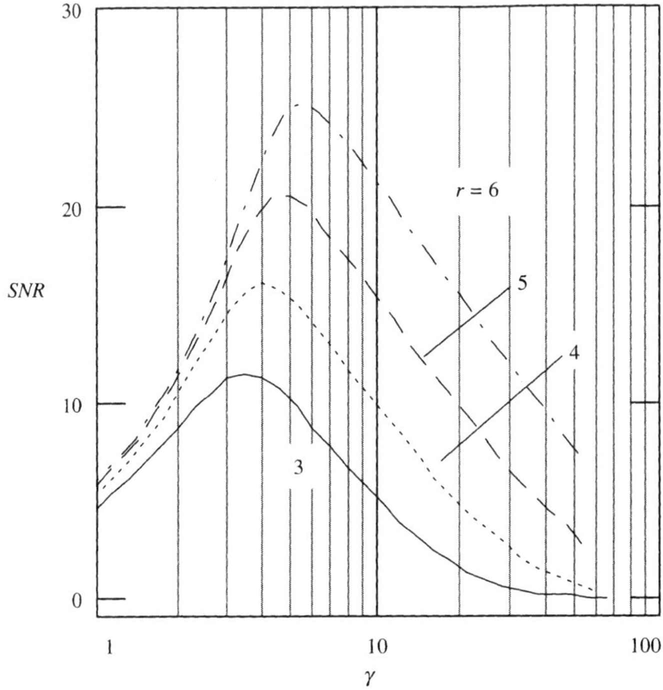

Figure 6.5 Signal-to-noise ratio as a function of the loading factor for four resolution values (input with Laplacian distribution).

6.2 NONUNIFORM QUANTIZATION

As shown in the last section, use of uniform quantization is suboptimal for speech samples, mainly due to their statistical distribution, where the percentage of occurrence of a sample having a certain amplitude diminishes with an increase in magnitude. Due to this property, a quantizer with a nonuniform transfer characteristic yields higher performance. In this section, commonly used nonlinear quantization schemes are analyzed; these schemes form the core of the ITU-T G.711 PCM standard.

For a nonuniform quantizer, the input variable x is first transformed with a memoryless monotonic nonlinearity $f(\cdot)$ to produce an output $f(x)$. Then it is quantized uniformly with the quantized values transformed by the inverse nonlinearity $f^{-1}(\cdot)$. Denoting the transfer characteristic of the uniform quantizer as $Q_u(x)$ (see Figure 6.6), the transfer characteristic of the nonuniform quantizer, denoted by $Q(x)$, is given by

$$Q(x) = f^{-1}(Q_u(f(x))). \tag{6.6}$$

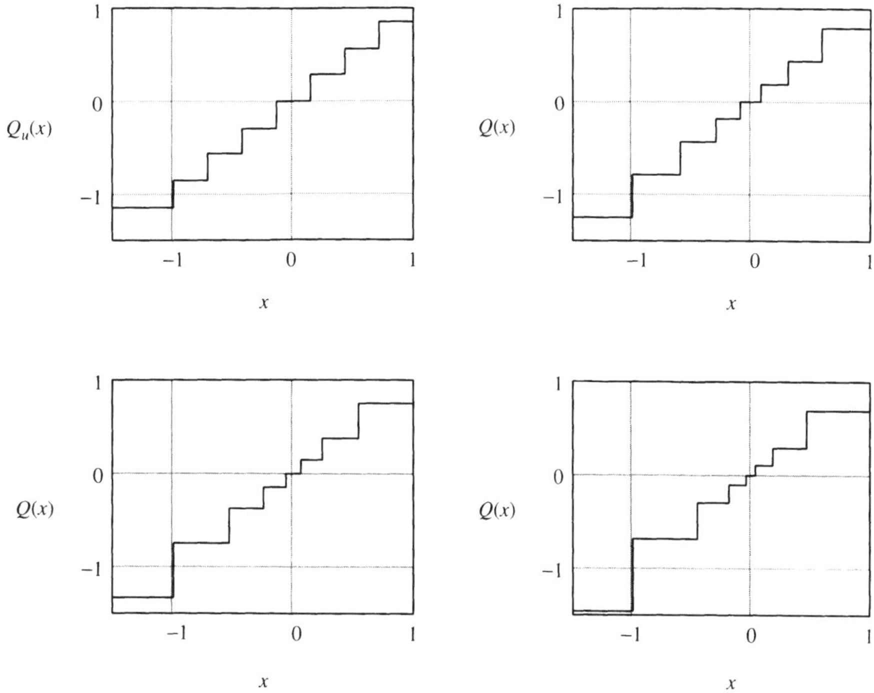

Figure 6.6 Transfer characteristic of a uniform quantizer of size 8 (*top left*). Transfer characteristics of nonuniform quantizers designed based on the described uniform quantizer and the μ-law function, with $\mu = 2$ (*top right*), $\mu = 4$ (*bottom left*), and $\mu = 10$ (*bottom right*).

The nonlinearity $f(\cdot)$ is usually designed to have a reduced slope for large input amplitudes and vice versa, leading to compression effects.

Logarithmic Nonlinear Functions

Two logarithmic functions have become widely used as design guidelines for nonuniform quantization of speech in digital telephony. The μ-law characteristic is given by

$$f(x) = A\frac{\ln(1 + \mu|x|/A)}{\ln(1 + \mu)}\operatorname{sgn}(x), \quad |x| \leq A, \tag{6.7}$$

where A is the peak-input magnitude and μ is a constant that controls the degree of compression. The prevailing value used in practice is $\mu = 255$. Figure 6.7 shows

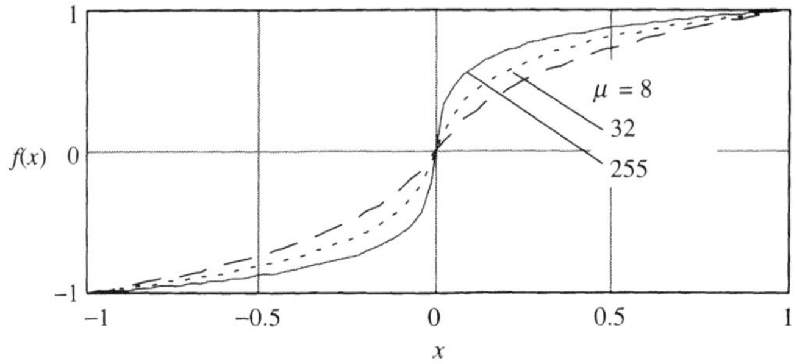

Figure 6.7 Plots of μ-law characteristics with $\mu = 255$, 32, and 8. $A = 1$ for all cases.

some plots of the μ-law characteristic. It is easy to show that the inverse function for (6.7) is

$$f^{-1}(y) = \frac{A}{\mu}\left[\exp\left(\frac{\ln(1+\mu)\cdot|y|}{A}\right) - 1\right]\operatorname{sgn}(y), \quad |y| \le A. \tag{6.8}$$

The A-law characteristic is given by

$$f(x) = \begin{cases} \dfrac{A_o|x|}{1+\ln A_o}\operatorname{sgn}(x), & |x| \le \dfrac{A}{A_o}, \\ \dfrac{A(1+\ln(A_o|x|/A))}{1+\ln A_o}\operatorname{sgn}(x), & \dfrac{A}{A_o} \le |x| \le A, \end{cases} \tag{6.9}$$

with A_o a constant that controls the degree of compression. In practice, $A_o = 87.6$ is a commonly used value. Some plots of A-law characteristic are shown in Figure 6.8. The inverse of (6.9) is given by

$$f^{-1}(y) = \begin{cases} \left(\dfrac{1+\ln A_o}{A_o}\right)y, & |y| \le \dfrac{A}{1+\ln A_o}, \\ \dfrac{A}{A_o}\exp\left[\left(\dfrac{1+\ln A_o}{A}\right)\cdot|y| - 1\right]\operatorname{sgn}(y), & \dfrac{A}{1+\ln A_o} \le y < A. \end{cases} \tag{6.10}$$

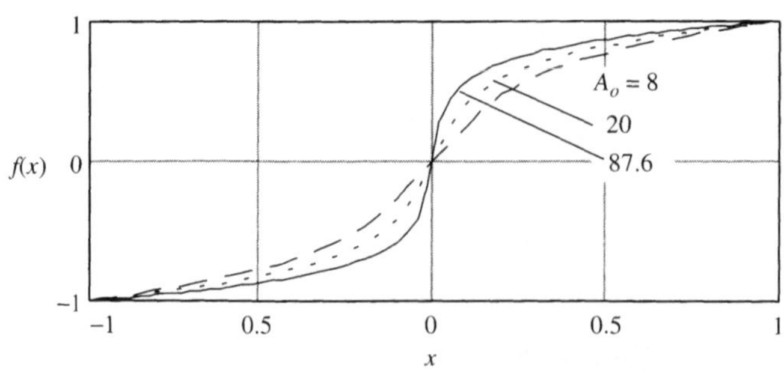

Figure 6.8 Plots of A-law characteristics with $A_o = 87.6$, 20, and 8. $A = 1$ for all cases.

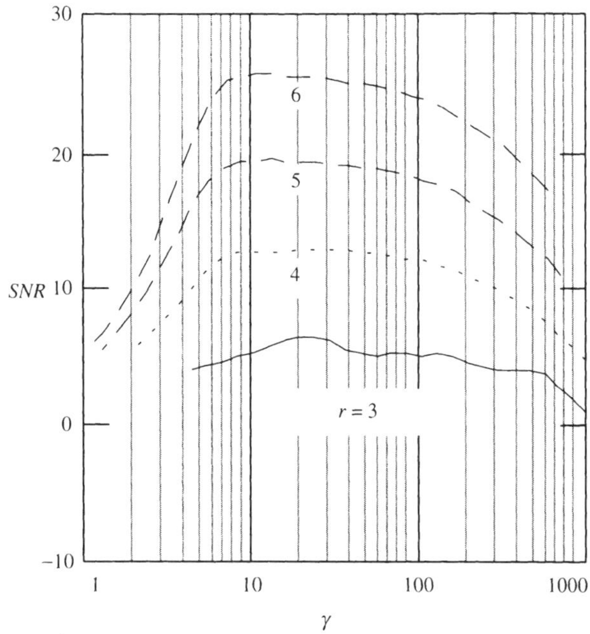

Figure 6.9 Signal-to-noise ratio as a function of the loading factor for four resolution values (input with Laplacian distribution) for the case of nonuniform quantization using μ-law ($\mu = 255$).

Example 6.2 A uniform quantizer with $N = 8$ is designed with $x_1 = -1$; the transfer characteristic $Q_u(x)$ is shown in Figure 6.6. Based on this design, several nonuniform quantizers are found from (6.6) by combining with the μ-law characteristic, where $A = 1$ and $\mu = 2$, 4, and 10. Note that as μ increases, the output levels tend to gather more and more toward the low-amplitude zone, leading to a better representation of low-level inputs; this is due to the fact that the amount of nonlinearity augments with μ.

Example 6.3 Figure 6.9 shows the SNR curves when the nonuniform quantizer uses μ-law with $\mu = 255$; the input has Laplacian distribution. From the SNR curves, we can see that the nonuniform quantizer is capable of maintaining an almost constant SNR for large variations in input signal variance, something not achievable with uniform quantization. Figure 6.10 compares the SNR curves for the uniform and nonuniform quantizers, at a resolution of 6 bits. Even though the peak SNR values for the two cases are similar, the nonuniform quantizer can sustain a high SNR value for a far wider input dynamic range. As we can see, while the SNR of the uniform quantizer drops with increase of loading factor (decrease in input variance), the SNR of the nonuniform quantizer remains near the peak. This property allows the nonuniform quantizer to represent low-amplitude samples with superior accuracy.

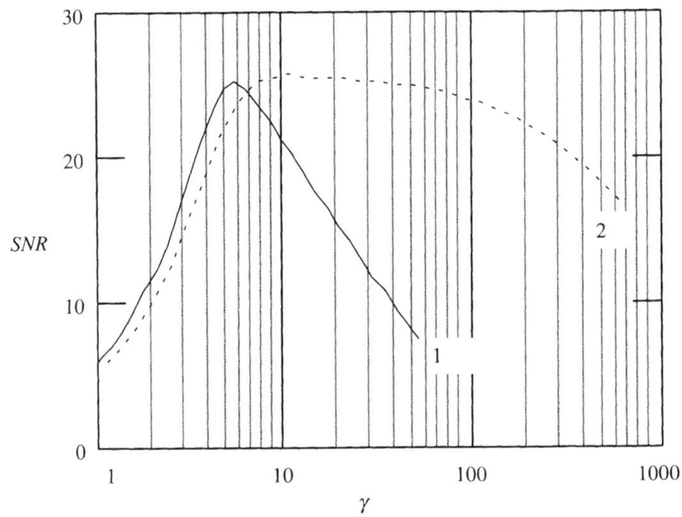

Figure 6.10 Comparison between (1) uniform and (2) nonuniform quantization at a resolution of 6 bits.

For speech signals the energy may vary as much as 40 dB among speakers and depending on transmission environment. In addition, for a given speaking environment, the amplitude of the speech signal varies considerably from voiced speech to unvoiced speech, and even within voiced sounds. If a uniform quantizer is used, a relatively high resolution must be employed in order to maintain good quality. High resolution automatically translates into elevated bit-rate and cost. To reduce the number of bits per sample, nonuniform schemes are desirable due to the fact that high SNR can be sustained for large variations of input signal level. Figure 6.11 compares the peak SNR values for the quantization cases considered. Note that they all increase approximately 6 dB per additional bit. For Laplacian distributed input, peak SNR is higher for a nonuniform quantizer when the resolution is above 6 bits.

ITU-T Recommendation G.711

This is a nonuniform PCM standard recommended for encoding speech signals. The recommendation is based on digitally linearizable companding, which permits a precise control of quantization characteristics. The compression and expansion characteristics are piecewise linear approximations to μ- and A-law, with $\mu = 255$ and $A_o = 87.56$; 8 bits/sample is adopted, leading to a bit-rate of 64 kbps at 8 kHz of sampling frequency. Within the standard [ITU, 1993], the nonlinear characteristics, or input-to-output mappings, are explicitly expressed in the forms of a table, where the inputs are 13- or 14-bit uniform PCM samples.

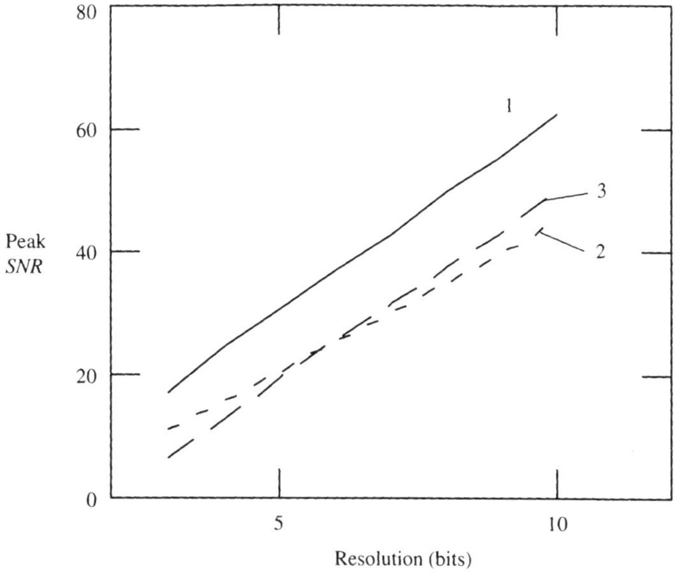

Figure 6.11 Comparison between the peak SNR achievable for a certain resolution for three quantization situations: (1) uniform quantization of uniformly distributed input, (2) uniform quantization of Laplacian distributed input, and (3) nonuniform quantization (μ-law with μ = 255) of Laplacian distributed input.

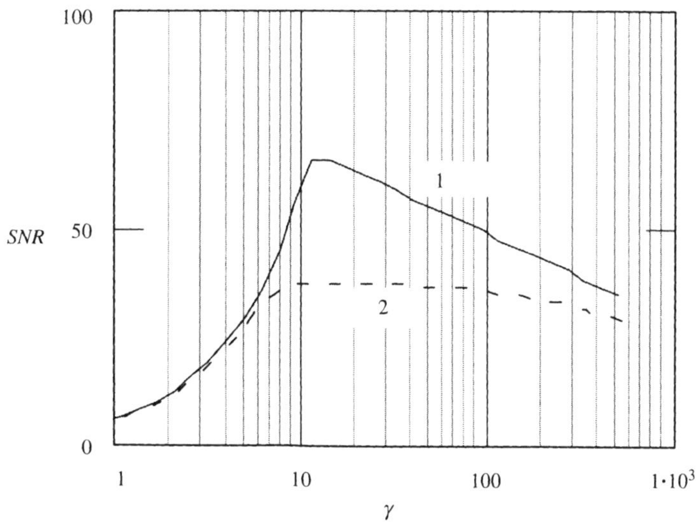

Figure 6.12 Performance comparison between (1) uniform quantization at 14-bit resolution and (2) nonuniform quantization at 8-bit resolution.

It is important to note that the quality of 14-bit uniform quantization is higher than 8-bit nonuniform quantization; therefore, by encoding through the nonlinear quantizer, irreversible losses are introduced. Figure 6.12 compares the two schemes, where the nonuniform quantizer is based on μ-law with $\mu = 255$. Peak SNR for the uniform quantizer is around 66 dB, while for the nonuniform quantizer, it is equal to 37.8 dB. Subjectively, G.711 provides toll quality acceptable for telephone communication.

6.3 DIFFERENTIAL PULSE CODE MODULATION

Differential PCM (DPCM) is based on the notion of quantizing the prediction-error signal. In many signal sources of interest, such as speech, the samples do not change a great deal from one sample to the next; in other words, the samples are correlated with their neighbors. If we can predict the current sample from the past history, it is possible to form the prediction-error signal, with significantly lower variance and dynamic range. By quantizing the prediction error, a higher signal-to-noise ratio can be achieved for a given resolution. The technique is shown in Figure 6.13, where the prediction error $e[n]$ — obtained by subtracting the input $x[n]$ from the prediction $x_p[n]$ — is quantized. The indices at the output of the quantizer's encoder represent the DPCM bit-stream. The indices are entered into the quantizer's decoder to obtain the quantized prediction error, which is combined with the prediction $x_p[n]$ to form the quantized input. The predictor takes the quantized input samples to calculate the prediction. The DPCM decoder works in a similar fashion to obtain the quantized samples from the indices.

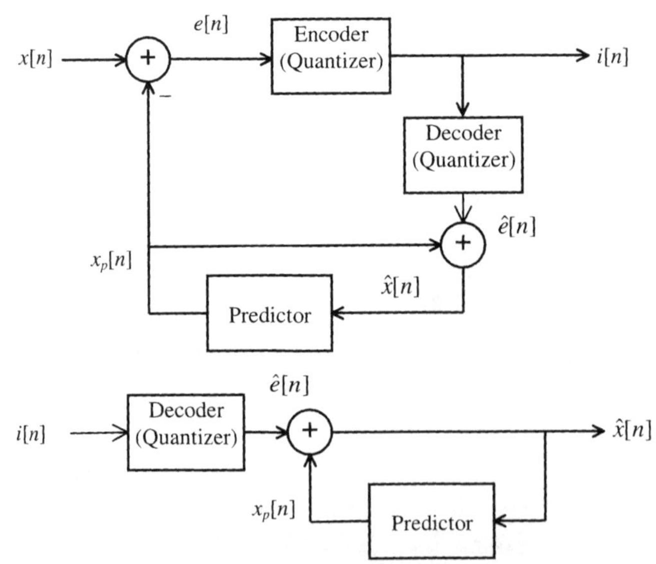

Figure 6.13 DPCM encoder (*top*) and decoder (*bottom*).

Note that prediction is based on the quantized signal samples, which is indeed suboptimal since higher performance can be achieved using the original, unquantized signal samples. The approach is utilized mainly because the decoder has no access to the original input, and synchronization must be maintained between encoder and decoder.

Example 6.4 Consider the predictor defined with

$$x_p[n] = \hat{x}[n-1]; \tag{6.11}$$

that is, the prediction is the past sample. As an illustration we will measure the performance of PCM and compare it with DPCM using the mentioned prediction scheme.

The input signal is the sine wave shown in Figure 6.14. To design the uniform PCM scheme for this signal, we only have to take into account the range of amplitude. A size eight uniform quantizer is designed with codewords $\{-1, -0.75, -0.5, -0.25, 0, 0.25, 0.5, 0.75\}$. Figure 6.15 shows the quantized signal and quantization error, leading to a SNR of 16.1 dB.

For DPCM, we need to design the quantizer for the prediction error. Computing the pseudo-prediction error as $x[n] - x[n-1]$ yields the conclusion that the signal is roughly limited in $(-0.25, 0.25)$. A size four uniform quantizer is designed with codewords $\{-0.5, -0.25, 0, 0.25\}$. Figure 6.16 shows the quantized signal, quantization error, and prediction error (quantizer input). The resultant SNR is 20.3 dB.

As we can see in the above example, DPCM beats PCM by ~4 dB, using half the bit-rate. The performance is achieved by the use of a reduced-range quantizer, with smaller step size leading to lower error. The reduction in range is possible because the predictor diminishes the amplitude of the signal entering the quantizer by

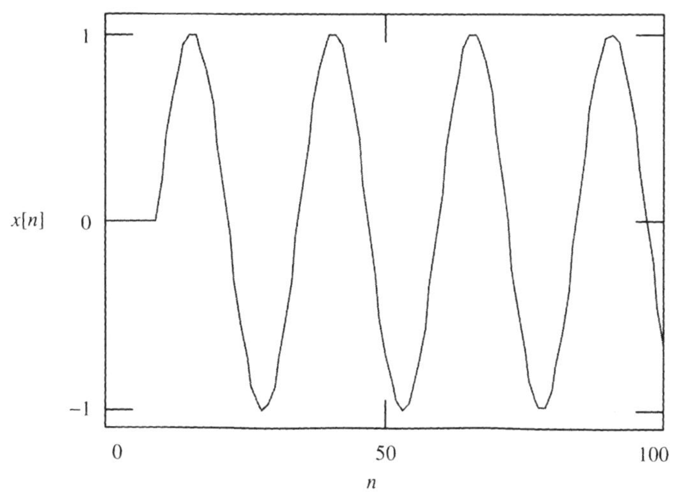

Figure 6.14 An example of input signal to be quantized.

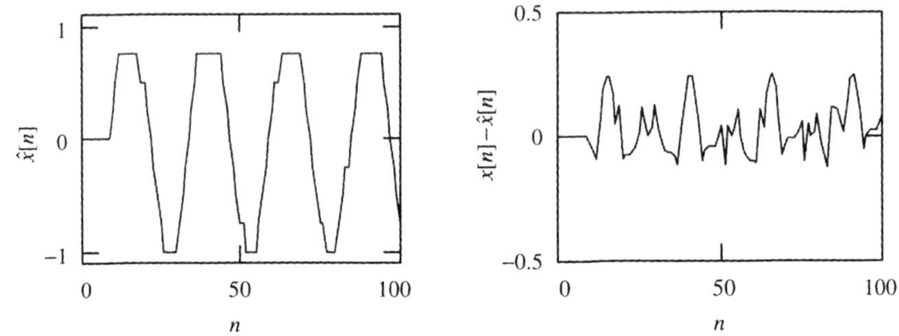

Figure 6.15 PCM quantized signal (*left*) with input from Figure 6.14, and quantization error (*right*).

removing redundancy. After all, there is no need to transmit if one can predict from the past.

It is important to point out that the above example only serves the purpose of illustration. One can rely on a fixed predictor only if the signal source is stationary. Otherwise, the predictor must change with time to adapt to the input signal

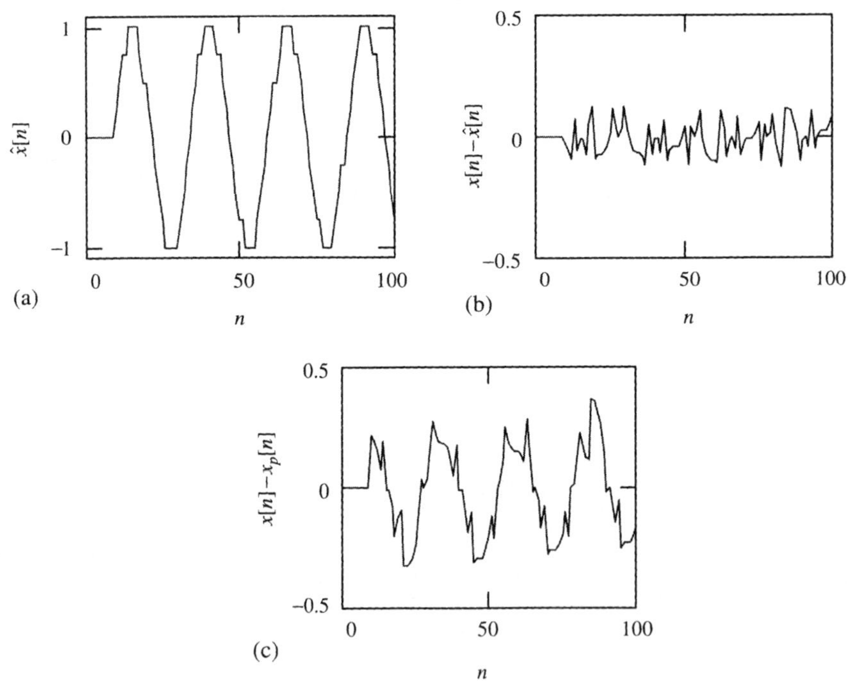

Figure 6.16 DPCM quantized signal with (a) input from Figure 6.14, (b) quantization error, and (c) prediction error.

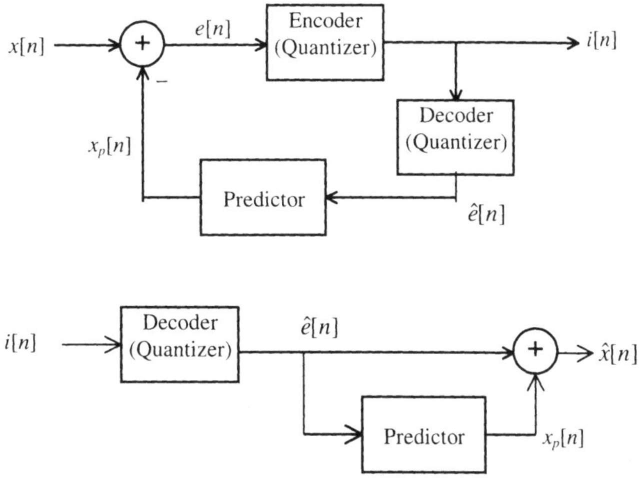

Figure 6.17 Encoder (*top*) and decoder (*bottom*) of DPCM with MA prediction.

properties. Principles of DPCM are applied not only to speech coding, but many other signal compression applications as well.

DPCM with MA Prediction

The predictor in Figure 6.13 utilizes the past quantized input samples, therefore obeying the AR model. An alternative is to base the prediction on the MA model (Chapter 3), where input to the predictor is the quantized prediction-error signal, as shown in Figure 6.17. Performance of the MA predictor is in general inferior; however, it provides the advantage of being more robust against channel errors.

Consider what happens in the DPCM decoder of Figure 6.13, when channel error is present; the error not only affects the current sample but will propagate indefinitely toward the future due to the involved loop. For the DPCM decoder of Figure 6.17, however, a single error will affect the current decoded sample, plus a finite number of future samples, with the number determined by the order of the predictor. Thus, DPCM with MA prediction behaves better under noisy channel conditions.

Often, in practice, the predictor combines the quantized input and quantized prediction error to form the prediction. Hence, high prediction gain of the AR model is mixed with high robustness of the MA model, resulting in an ARMA model-based predictor (Chapter 3).

6.4 ADAPTIVE SCHEMES

In scalar quantization, adaptation is necessary for optimal performance when dealing with nonstationary signals like speech, where properties of the signal change rapidly with time. These schemes are often referred to as adaptive PCM (APCM) and are the topics of this section.

Forward Gain-Adaptive Quantizer

Forward adaptation can accurately control the gain level of the input sequence to be quantized, but side information must be transmitted to the decoder. The general structure of a forward gain-adaptive quantizer is shown in Figure 6.18.

A finite number N of input samples (frame) are used for gain computation, where $N \geq 1$ is a finite number, known as the frame length. The estimated gain is quantized and used to scale the input signal frame; that is, $x[n]/\hat{g}[m]$ is calculated for all samples pertaining to a particular frame. Note that a different index m is used for the gain sequence, with m being the index of the frame. The scaled input is quantized with the index $i_a[n]$ and $i_g[m]$ transmitted to the decoder. These two indices represent the encoded bit-stream. Thus, for each frame, N indices $i_a[n]$ and one index $i_g[m]$ are transmitted. If transmission errors occur at a given moment, distortions take place in one frame or a group of frames; however, subsequent frames will be unaltered. With sufficiently low error rates, the problem is not serious.

Many choices are applicable for gain computation. Some popular schemes are

$$g[m] = k_1 \max_n\{|x[n]|\} + k_2, \tag{6.12}$$

$$g[m] = k_1 \sum_n x^2[n] + k_2, \tag{6.13}$$

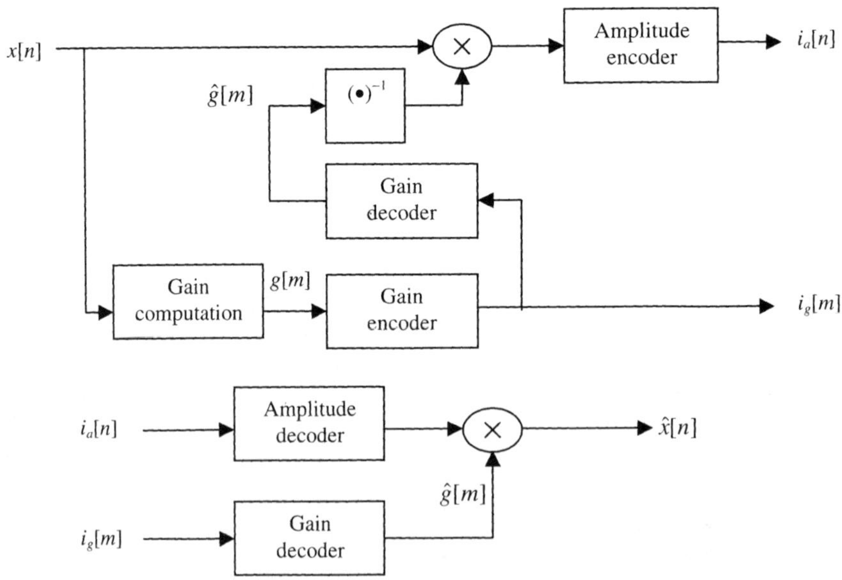

Figure 6.18 Encoder (*top*) and decoder (*bottom*) of the forward gain-adaptive quantizer.

with the range of n pertaining to the frame associated with index m, and (k_1, k_2) positive constants. As we can see, the purpose of the gain is to normalize the amplitude of the samples inside the frame, so that high-amplitude frames and low-amplitude frames are quantized optimally with a fixed quantizer. To avoid numerical problems with low-amplitude frames, k_2 is incorporated so that divisions by zero are avoided.

For nonstationary signals like speech having a wide dynamic range, use of APCM is far more efficient than a fixed quantizer. At a given bit-rate, the SNR and especially the SSNR are greatly improved with respect to PCM (see Chapter 19 for the definition of SSNR).

Backward Gain-Adaptive Quantizer

In a backward gain-adaptive quantizer, gain is estimated on the basis of the quantizer's output. The general structure is shown in Figure 6.19. Such schemes have the distinct advantage that the gain need not be explicitly retained or transmitted since it can be derived from the output sequence of the quantizer. A major disadvantage of backward gain adaptation is that a transmission error not only causes the current sample to be incorrectly decoded but also affects the memory of the gain estimator, leading to forward error propagation.

Similar to the case of the forward gain-adaptive quantizer, gain is estimated so as to normalize the input samples. In this way, the use of a fixed amplitude quantizer is adequate to process signals with wide dynamic range. One simple implementation consists of setting the gain $g[n]$ proportional to the recursive estimate of variance

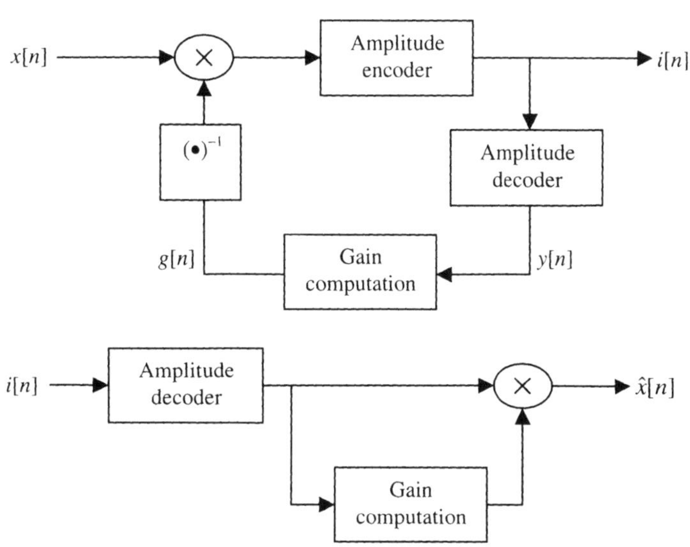

Figure 6.19 Encoder (*top*) and decoder (*bottom*) of the backward gain-adaptive quantizer.

for the normalized-quantized samples, where the variance is estimated recursively with

$$\sigma^2[n] = \alpha\sigma^2[n-1] + (1-\alpha)y^2[n], \tag{6.14}$$

where $\alpha < 1$ is a positive constant. This constant determines the update rate of the variance estimate. For faster adaptation, set α close to zero. The gain is computed with

$$g[n] = k_1\sigma^2[n] + k_2, \tag{6.15}$$

where k_1 and k_2 are positive constants. The constant k_1 fixes the amount of gain per unit variance. The constant k_2 is incorporated to avoid division by zero. Hence, the minimum gain is equal to k_2.

In general, it is very difficult to analytically determine the impact of various parameters (α, k_1, k_2) on the performance of the quantizer. In practice, these parameters are determined experimentally depending on the signal source.

Adaptive Differential Pulse Code Modulation

The DPCM system described in Section 6.3 has a fixed predictor and a fixed quantizer; much can be gained by adapting the system to track the time-varying behavior of the input. Adaptation can be performed on the quantizer, on the predictor, or on both. The resulting system is called adaptive differential PCM (ADPCM).

Figure 6.20 shows the encoder and decoder of an ADPCM system with forward adaptation. As for the forward APCM scheme, side information is transmitted, including gain and predictor information. In the encoder, a certain number of samples (frame) are collected and used to calculate the predictor's parameters. For the case of the linear predictor, a set of LPCs is determined through LP analysis (Chapter 4). The predictor is quantized with the index $i_p[m]$ transmitted.

As in DPCM, prediction error is calculated by subtracting $x[n]$ from $x_p[n]$. A frame of the resultant prediction-error samples is used in gain computation, with the resultant value quantized and transmitted. The gain is used to normalize the prediction-error samples, which are then quantized and transmitted. Note that the quantized quantities (samples of normalized prediction error, gain, and the predictor's parameters) are used in the encoder to compute the quantized input $\hat{x}[n]$, and the prediction $x_p[n]$ is derived from the quantized input. This is done because, on the decoder side, it is only possible to access the quantized quantities; in this way, synchronization is maintained between encoder and decoder since both are handling the same variables.

As we will see, many speech coding algorithms use a scheme similar to the forward-adaptive ADPCM. In many such algorithms, LP analysis is performed with the resultant coefficients quantized and transmitted. Thus, a good understanding of ADPCM allows a better digestion of the material presented in subsequent chapters.

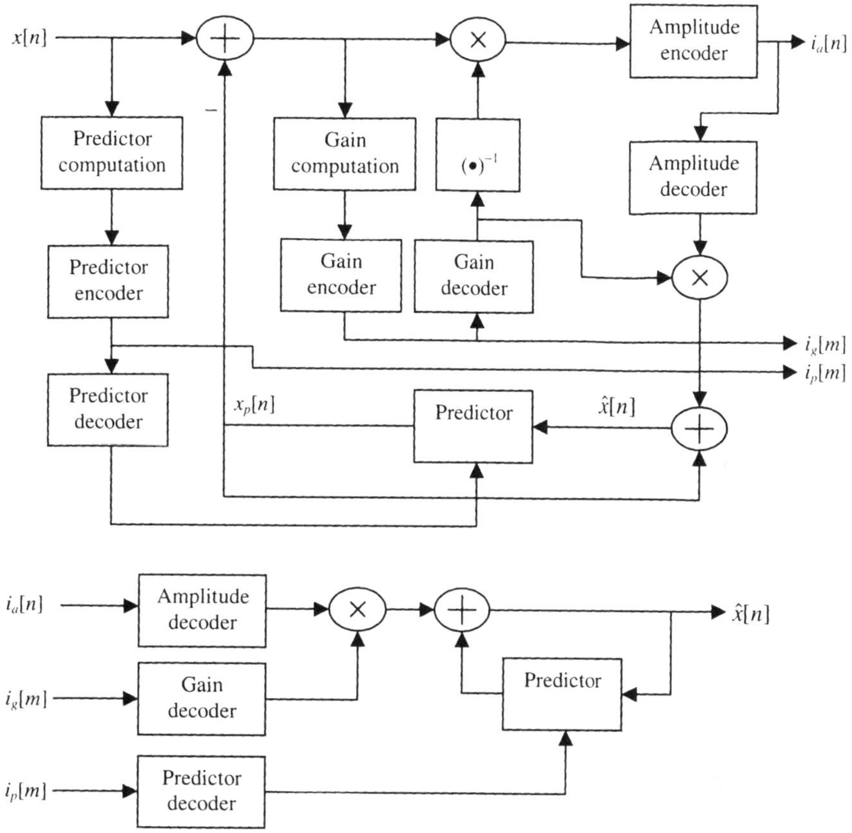

Figure 6.20 Encoder (*top*) and decoder (*bottom*) of a forward-adaptive ADPCM quantizer.

One shortcoming of the forward-adaptation scheme is the delay introduced due to the necessity of collecting a given number of samples before processing can start. The amount of delay is proportional to the length of the frame. This delay can be critical in certain applications since echo and annoying artifacts can be generated.

Backward adaptation is often preferred in those applications where delay is critical. Figure 6.21 shows an alternative ADPCM scheme with backward adaptation. Note that the gain and predictor are derived from the quantized-normalized prediction-error samples; hence, there is no need to transmit any additional parameters except the index of the quantized-normalized samples.

Similar to DPCM, the input is subtracted from the prediction to obtain the prediction error, which is normalized, quantized, and transmitted. The quantized-normalized prediction error is used for gain computation. The derived gain is used in denormalization of the quantized samples; these prediction-error samples are added with the predictions to produce the quantized input samples. The

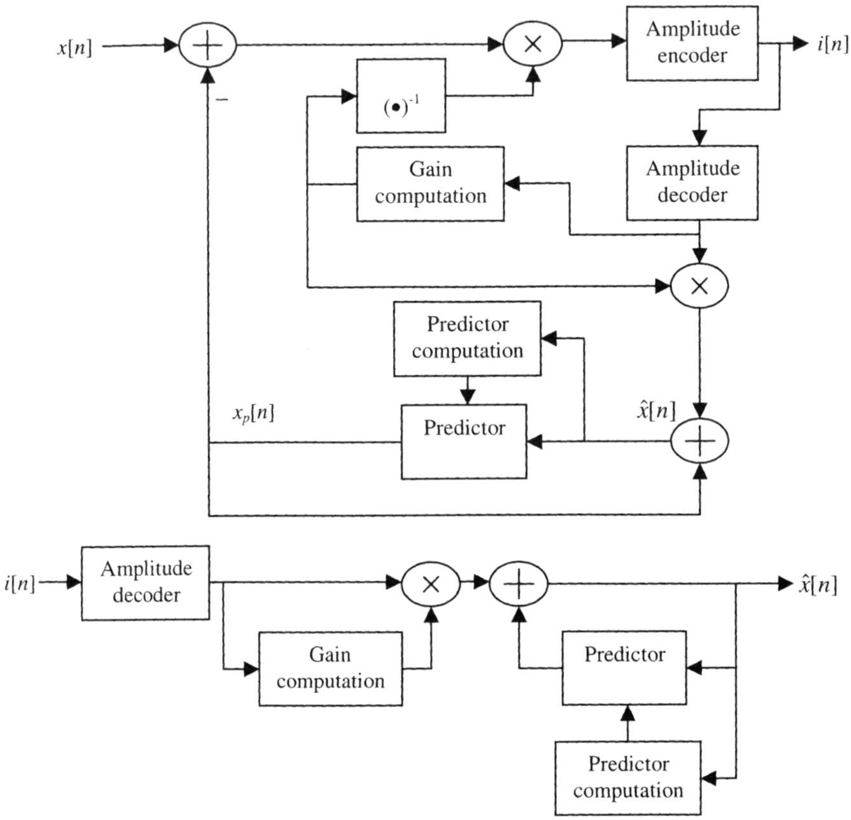

Figure 6.21 Encoder (*top*) and decoder (*bottom*) of a backward-adaptive ADPCM quantizer.

predictor is determined from the quantized input $\hat{x}[n]$. Techniques for linear predictor calculation are given in Chapter 4. Using recursive relations for the gain calculation ((6.14) and (6.15)) and linear prediction analysis, the amount of delay is minimized since a sample can be encoded and decoded with little delay. This advantage is due mainly to the fact that the system does not need to collect samples of a whole frame before processing. However, the reader must be aware that backward schemes are far more sensitive to transmission errors, since they not only affect the present sample but all future samples due to the recursive nature of the technique.

6.5 SUMMARY AND REFERENCES

The major facts about PCM are presented in this chapter, where performance of uniform quantization as a function of resolution is found. For resolution higher than 4 bits, increasing the resolution by 1 bit roughly increases the SNR by 6 dB. Since the speech samples have a statistical distribution close to the Laplacian type, nonuniform quantization with more low-amplitude output levels is better suited for the quantization task. The μ-law and A-law nonlinearities are introduced as commonly accepted rules for the design of nonuniform quantizers. Unlike uniform quantizers, where good quality output is maintainable only for a relatively narrow range of input level, nonuniform quantizers can sustain high SNR for a wider range of input level. Therefore, it is more suitable for speech coding since the energy level of the signal varies drastically with speakers, environments, equipment, and so on.

For nonstationary signals like speech, use of some kind of adaptation can achieve higher coding efficiency. Several schemes of APCM and ADPCM are presented. Forward adaptation is less sensitive to transmission errors, with higher delay and lower coding efficiency when compared to backward schemes. In subsequent chapters, the principles of adaptive methods are applied to speech coding algorithm design.

Discussion of APCM and ADPCM has been limited to high-level concepts and system structures, without entering into the details of implementation of any coding standard. For ADPCM a well-known standard exists and is due to ITU. This specifies a 64-kbps nonuniform PCM (G.711) input, and four possible bit-rates are available: 40, 32, 24, and 16 kbps. It is known as recommendation G.726 and is based on backward adaptation. It utilizes a pole-zero or ARMA predictor with its parameters updated through a gradient-based adaptive algorithm. Readers are referred to ITU [1990] for details. Also, Jayant and Noll [1984] contains ample descriptions of many PCM-related schemes, as well as early developments in the field.

EXERCISES

6.1 In uniform quantization, consider the case when the size is large (high resolution).

(a) Given the input is in a particular cell, argue why the conditional PDF of the quantization error is uniform over the interval $(-\Delta/2, \Delta/2)$. Show that the conditional variance of the quantization error is $\Delta^2/12$, with Δ the quantizer's step size.

(b) Argue why the unconditional average distortion is given approximately by

$$D = \Delta^2/12.$$

(c) Show that

$$\Delta \approx 2A/N$$

and

$$D = \sigma^2\gamma^2 2^{-2r}/3$$

with N the quantizer's size.

(d) Show that

$$\text{SNR} = 6.02r + 10\log_{10}\left(\frac{3}{\gamma^2}\right);$$

that is, SNR increases approximately 6 dB for each additional bit used to quantize an input sample.

6.2 Using a large amount of speech samples, construct the histogram and verify the validity of the proposed Laplacian distribution.

6.3 Generate SNR curves (as a function of γ) for nonuniform μ-law quantization using uniformly distributed input. Compare with uniform quantization in terms of peak SNR.

6.4 Generate SNR curves (as a function of γ) for A-law quantization. Compare with μ-law results.

6.5 In DPCM, assuming stationary input, show that

$$\text{SNR} = \frac{\sigma_x^2}{E\{e^2[n]\}} = \text{PG} \cdot \text{QG}$$

where

$$\sigma^2 = \text{input variance},$$
$$e[n] = \text{quantization error},$$
$$\text{PG} = \text{prediction gain},$$
$$\text{QG} = \text{quantizer gain},$$

and

$$\text{PG} = \frac{\sigma_x^2}{\sigma_e^2},$$

$$\text{QG} = \frac{\sigma_e^2}{E\{(e[n] - \hat{e}[n])^2\}}.$$

6.6 Delta modulation is the 1-bit (or two-level) version of DPCM. In this scheme, the prediction is given by the past quantized sample

$$x_p[n] = \hat{x}[n-1]$$

and the quantizer has only two levels

$$\hat{e}[n] = \Delta\text{sgn}(e[n]),$$

where sgn(·) is the sign function, while Δ is the quantizer's step size. Show that

$$x_p[n] = \Delta \sum_{i=1}^{n} \text{sgn}(e[i]),$$

assuming that the system is initialized at instant $n = 0$ with $x_p[0] = 0$.

6.7 In Figure 6.5, the peaks of the SNR curves tend to shift toward higher values of γ for increasing resolution. Explain this phenomenon. Why isn't the situation occurring for uniformly distributed input (Figure 6.3)?

6.8 Draw the block diagrams of a DPCM system based on an ARMA predictor; that is, the predictor takes as inputs the samples of the quantized input signal and the quantized prediction-error signal.

6.9 Draw the block diagrams of the encoder and decoder for an ADPCM system with forward gain adaptation and backward predictor adaptation. Repeat for the case of backward gain adaptation and forward predictor adaptation.

6.10 Similar to Example 6.4, design a 2-bit DPCM-MA scheme using a first-order predictor and compare its performance to 3-bit PCM. The predictor coefficient can be determined by trial-and-error. Develop a systematic way to find the optimal predictor coefficient, when the quantizer is fixed.

CHAPTER 7

VECTOR QUANTIZATION

Vector quantization (VQ) concerns the mapping in multidimensional space from a (possibly continuous-amplitude) source ensemble to a discrete ensemble. The mapping function proceeds according to some distortion criterion or metric employed to measure the performance of VQ. VQ offers several unique advantages over scalar quantization, including the ability to exploit the linear and nonlinear dependencies among the vector components, and is highly versatile in the selection of multidimensional quantizer cell shapes. Due to these reasons, for a given resolution (measured in bits), use of VQ typically results in lower distortion than scalar quantization.

In VQ, vectors of a certain dimension form the input to the vector quantizer. At both the encoder and decoder of the quantizer there is a set of vectors, having the same dimension as the input vector, called the codebook. The vectors in this codebook, known as codevectors, are selected to be representative of the population of input vectors. At the encoder, the input vector is compared to each codevector in order to find the closest match. The elements of this codevector represent the quantized vector. A binary index is transmitted to the decoder in order to inform about the selected codevector. Because the decoder has exactly the same codebook, it can retrieve the codevector given its binary index.

Some materials in this chapter are natural generalizations of results in scalar quantization. After all, scalar quantization is VQ with unit dimension. VQ has become more and more significant for signal coding applications, mainly due to its high performance. In subsequent chapters we will see how the technique is applied to various speech coding standards; hence, it is imperative to acquire proficiency in the subject. In this chapter, the basic definitions involved with vector quantization are given, followed by the conditions required for optimal quantization; algorithms for quantizer design are described in detail. It is shown that optimal

VQ is highly costly to implement in practice. Therefore, several suboptimal schemes with certain structure are explained in detail; these include multistage VQ (MSVQ), split VQ, and conjugate VQ. These suboptimal schemes provide good performance at a reasonable implementational cost, enabling the deployment of VQ to many speech coding applications. Similar to DPCM in scalar quantization, incorporation of prediction into the VQ framework leads to improved performance in most cases; these schemes are discussed in a separate section, where predictive VQ (PVQ) and PVQ-MA are introduced.

7.1 INTRODUCTION

The basic issues of vector quantization are introduced in this section. Many topics are mere extensions of scalar quantization from one dimension to multiple dimensions.

Definition 7.1: Vector Quantizer. A vector quantizer Q of dimension M and size N is a mapping from a vector $\mathbf{x}$ in M-dimensional Euclidean space $\boldsymbol{R}^M$ into a finite set $\boldsymbol{Y}$ containing N M-dimensional outputs or reproduction points, called codevectors or codewords. Thus,

$$Q: \boldsymbol{R}^M \rightarrow \boldsymbol{Y},$$

where

$$\mathbf{x} = [x_1, x_2, \ldots, x_M]^T,$$
$$(\mathbf{y}_1, \mathbf{y}_2, \ldots, \mathbf{y}_N) \in \boldsymbol{Y},$$
$$\mathbf{y}_i = [y_{i1}, y_{i2}, \ldots, y_{iM}]^T; \quad i = 1, \ldots, N.$$

$\boldsymbol{Y}$ is known as the codebook of the quantizer. The mapping action is written as

$$Q(\mathbf{x}) = \mathbf{y}_i; \quad i = 1, \ldots, N. \tag{7.1}$$

Definition 7.2: Resolution. We define the resolution of a vector quantizer as

$$r = \lg N, \tag{7.2}$$

which measures the number of bits needed to uniquely address a specific codeword.

Definition 7.3: Cell. Associated with every N-point M-dimensional vector quantizer is a partition of $\boldsymbol{R}^M$ into N regions or cells, $\boldsymbol{R}_i, i = 1, \ldots, N$. The ith cell is defined by

$$\boldsymbol{R}_i = \{\mathbf{x} \in \boldsymbol{R}^M : Q(\mathbf{x}) = \mathbf{y}_i\} = Q^{-1}(\mathbf{y}_i). \tag{7.3}$$

Definitions of granular cell and overload cell found in scalar quantization apply directly to VQ.

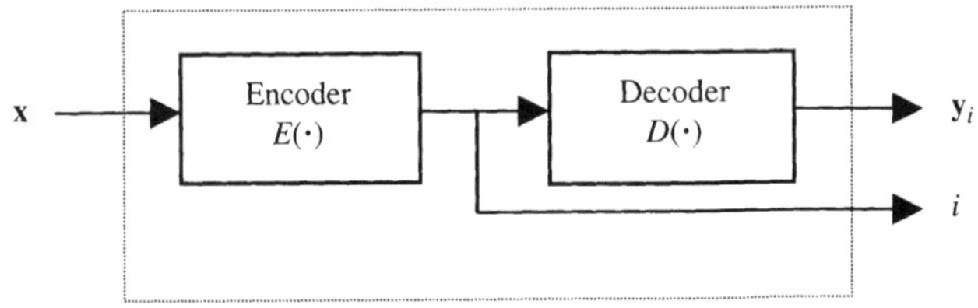

Figure 7.1 Vector quantizer as the encoder followed by the decoder.

Definition 7.4: Encoder and Decoder. A vector quantizer can be decomposed into two component operations, the vector encoder and the vector decoder. The encoder E is the mapping from $\boldsymbol{R}^M$ to the index set $\boldsymbol{I} = \{1, 2, \ldots, N\}$, and the decoder D maps the index set $\boldsymbol{I}$ into the reproduction set $\boldsymbol{Y}$. Thus,

$$E{:}\,\boldsymbol{R}^M \rightarrow \boldsymbol{I},$$
$$D{:}\,\boldsymbol{I} \rightarrow \boldsymbol{R}^M.$$

The task of the encoder is to identify in which of N geometrically specified regions the input vector lies. On the other hand, the decoder is simply a table lookup and is fully determined by specifying the codebook. The overall operation of VQ can be regarded as the cascade or composition of two operations:

$$\hat{\mathbf{x}} = Q(\mathbf{x}) = D(E(\mathbf{x})) = \mathbf{y}_i. \tag{7.4}$$

Figure 7.1 illustrates the concepts of encoder and decoder.

Definition 7.5: Distance or Distortion Measure. A distance or distortion measure is an assignment of a nonnegative cost $d(\mathbf{x},Q(\mathbf{x}))$ associated with quantizing any input vector $\mathbf{x}$ with a reproduction vector $Q(\mathbf{x})$:

$$d(\mathbf{x}, Q(\mathbf{x})) = \begin{cases} 0, & Q(\mathbf{x}) = \mathbf{x}, \\ >0, & \text{otherwise}. \end{cases} \tag{7.5}$$

Given such a measure we can quantify the performance of a system by the expected value of d. Let $\mathbf{X}$ denote a continuously distributed random vector* in $\boldsymbol{R}^M$ with a specified PDF $f_{\mathbf{X}}(x)$. Then the expected value of the distortion can be expressed as

$$D = E\{d(\mathbf{X}, Q(\mathbf{X}))\} = \int_{R^M} d(\mathbf{x}, Q(\mathbf{x})) f_{\mathbf{X}}(\mathbf{x})\, d\mathbf{x}. \tag{7.6}$$

*Here, bold capital letters represent random vectors. In M-dimension, $\mathbf{X} = [\mathbf{x}_1, \ldots, \mathbf{x}_M]^T$, where the $\mathbf{x}_i$ are random variables. The sample outcome of $\mathbf{X}$ is the vector $\mathbf{x} = [x_1, \ldots, x_M]^T$. As we can see, bold lowercase letters represent both vectors or random variables. The meaning of the symbol is normally explicit from the context.

When the input vector has a discrete distribution, the probablility mass function (PMF) $p_{\mathbf{X}}(\mathbf{x})$ can be used instead:

$$D = E\{d(\mathbf{X}, Q(\mathbf{X}))\} = \sum_{k} d(\mathbf{x}_k, Q(\mathbf{x}_k)) p_{\mathbf{X}}(\mathbf{x}_k), \tag{7.7}$$

with the $\mathbf{x}_k$ being the outcomes of $\mathbf{X}$ with nonzero probability.

Assuming the input range to be partitioned into N cells, the expected distortion can be expressed using probability terms coupled with conditional expectation as follows:

$$D = \sum_{i=1}^{N} P\{\mathbf{X} \in \boldsymbol{R}_i\} E\{d(\mathbf{X}, \mathbf{y}_i) | \mathbf{X} \in \boldsymbol{R}_i\}. \tag{7.8}$$

A widely used distortion measure is the squared error, given by

$$d(\mathbf{x}, \hat{\mathbf{x}}) = \|\mathbf{x} - \hat{\mathbf{x}}\|^2 = \sum_{j=1}^{M} (x_j - \hat{x}_j)^2, \tag{7.9}$$

which is the squared Euclidean distance between two vectors.

Definition 7.6: Nearest-Neighbor Quantizer. Suppose that $d(\mathbf{x}, \mathbf{y})$ is a distortion measure on the input/output vector space. We define a nearest-neighbor vector quantizer as one whose partition cells are given by

$$\boldsymbol{R}_i = \{\mathbf{x} : d(\mathbf{x}, \mathbf{y}_i) \leq d(\mathbf{x}, \mathbf{y}_j);\ \forall j \in I\}, \quad i \in \boldsymbol{I}, \tag{7.10}$$

where $\boldsymbol{I} = \{1, 2, \ldots, N\}$ is the index set. Thus, for a nearest-neighbor encoder each cell $\boldsymbol{R}_i$ consists of all points $\mathbf{x}$ that have less distortion when reproduced with codevector $\mathbf{y}_i$ than with any other codevector.

Since most quantizers for coding applications are of the nearest-neighbor type, only nearest-neighbor quantizers are considered in this book. The following pseudocode is a direct implementation of a nearest-neighbor quantizer:

```
QUANTIZE(x)
min ← ∞
for i ← 1 to N
        distance = d(x,y_i)
        if distance < min
                min ← distance
                index ← i
return y_index
```

Running time of the algorithm specified before is proportional to N, the size of the quantizer. Also note that a total of $M \times N$ locations are required to store the codebook, with M being the dimension of the quantizer. It is assumed that one location is required to store one component of a vector.

7.2 OPTIMAL QUANTIZER

In this section, the optimal operation of a vector quantizer is defined. Conditions to achieve optimality are explained. These conditions serve as the foundation in developing the optimization procedure used for quantizer design.

Definition 7.7: Optimal Quantizer. Assume that a particular distortion measure d has been selected. The vector quantizer is said to be optimal if it minimizes the expected value of the distortion ((7.6) or (7.7)).

Optimality can be achieved by selecting the codewords $\mathbf{y}_i$ and partition cells $\boldsymbol{R}_i$ for a given PDF of the source $\mathbf{X}$ so as to minimize the expected distortion.

Nearest-Neighbor Condition for Optimality

For a given codebook $\boldsymbol{Y}$ of size N, the optimal partition cells satisfy

$$\boldsymbol{R}_i = \{\mathbf{x}: d(\mathbf{x}, \mathbf{y}_i) \leq d(\mathbf{x}, \mathbf{y}_j)\}, \tag{7.11}$$

where $i, j = 1, \ldots, N$ and $i \neq j$. That is, $Q(\mathbf{x}) = \mathbf{y}_i$ only if $d(\mathbf{x}, \mathbf{y}_i) \leq d(\mathbf{x}, \mathbf{y}_j)$. Thus, given the codebook $\boldsymbol{Y}$, the encoder contains a minimum distortion or nearest-neighbor mapping, with

$$d(\mathbf{x}, Q(\mathbf{x})) = \min_i d(\mathbf{x}, \mathbf{y}_i). \tag{7.12}$$

See Exercise 7.2 for a proof of the condition.

Definition 7.8: Centroid. We define the centroid cent($\boldsymbol{R}_o$), of any nonempty set $\boldsymbol{R}_o \in \boldsymbol{R}^M$ as the vector $\mathbf{y}_o$ (if it exists) that minimizes the distortion between a point $\mathbf{X} \in \boldsymbol{R}_o$ and $\mathbf{y}_o$, averaged over the probability distribution of $\mathbf{X}$ given $\mathbf{X} \in \boldsymbol{R}_o$. Thus,

$$\text{cent}(\boldsymbol{R}_o) = \{\mathbf{y}_o : E\{d(\mathbf{X}, \mathbf{y}_o)|\mathbf{X} \in \boldsymbol{R}_o\} \leq E\{d(\mathbf{X}, \mathbf{y})|\mathbf{X} \in \boldsymbol{R}_o\}\}, \quad \forall \mathbf{y} \in \boldsymbol{R}^M. \tag{7.13}$$

Centroid Condition for Optimality

For a given partition $\{\boldsymbol{R}_i; i = 1, \ldots, N\}$, the optimal codevectors satisfy

$$\mathbf{y}_i = \text{cent}(\boldsymbol{R}_i). \tag{7.14}$$

See Exercise 7.3 for a proof of the condition.

Definition 7.9: Boundary Set. Given the partition cells $\boldsymbol{R}_i, i = 1, \ldots, N$, the boundary set is defined as

$$\boldsymbol{B} = \{\mathbf{x}: d(\mathbf{x}, \mathbf{y}_j) = d(\mathbf{x}, \mathbf{y}_i), \quad i \neq j\} \tag{7.15}$$

for all $i, j = 1, \ldots, N$. Thus, the boundary consists of points that are equally close to both $\mathbf{y}_j$ and to some other $\mathbf{y}_i$ and hence do not have a unique nearest neighbor.

Zero Probability Boundary Condition for Optimality

A necessary condition for a codebook to be optimal for a given source distribution is

$$\boldsymbol{B} = \emptyset. \tag{7.16}$$

That is, the boundary set must be empty. Alternatively, we can write

$$P\{\mathbf{x}: d(\mathbf{x}, \mathbf{y}_i) = d(\mathbf{x}, \mathbf{y}_j), \quad i \neq j\} = 0 \tag{7.17}$$

for all $i, j = 1, \ldots, N$.

Suppose that the boundary set is not empty and hence there is at least one $\mathbf{x}$ that is equidistant to the codevectors $\mathbf{y}_i$ and $\mathbf{y}_j$. Mapping $\mathbf{x}$ to $\mathbf{y}_i$ or $\mathbf{y}_j$ will yield two encoding schemes with the same average distortion. By including the nonzero probability input point $\mathbf{x}$ into either cell ($\boldsymbol{R}_i$ and $\boldsymbol{R}_j$ associated with $\mathbf{y}_i$ and $\mathbf{y}_j$, respectively) will necessarily modify the centroids of $\boldsymbol{R}_i$ and $\boldsymbol{R}_j$, meaning that the codebook is no longer optimal for the new partition.

7.3 QUANTIZER DESIGN ALGORITHMS

Given a codebook $\boldsymbol{Y}$ of size N, it is desired to find the input partition cells $\boldsymbol{R}_i$ and codewords such that the average distortion $D = E\{d(\mathbf{X}, Q(\mathbf{X}))\}$ is minimized, where $\mathbf{X}$ is the input random vector with a given PDF.

The Lloyd Iteration for Codebook Improvement

The Lloyd iteration as explained in Chapter 5 can be applied directly to VQ after some notational changes. For the case of vectors, the centroid computation requires the evaluation of multiple integrals (assume continuously distributed input) over a complicated region and is generally impossible by analytical means. In practice, a training vector set consisting of N_t vectors

$$\mathbf{x}_k; \quad k = 0, 1, \ldots, N_t - 1 \tag{7.18}$$

is used to optimize the vector quantizer. These vectors are sample outcomes of the input random vector $\mathbf{X}$. It is assumed that each sample vector has a probability mass of $1/N_t$. If the value of N_t is sufficiently high, the statistical property of the training set approaches the continuously distributed random vector $\mathbf{X}$.

From the training set, the Lloyd iteration is described below.

Step 1. Given the codebook $\boldsymbol{Y}_m = \{\mathbf{y}_{m,i}; i = 1, 2, \ldots, N\}$, partition the training set into cluster sets using the nearest-neighbor condition

$$\boldsymbol{R}_{m,i} = \{\mathbf{x}_k : d(\mathbf{x}_k, \mathbf{y}_{m,i}) \leq d(\mathbf{x}_k, \mathbf{y}_{m,j})\} \tag{7.19}$$

for all $i \neq j, k = 0, \ldots, N_t - 1$. To ensure zero probability boundary condition, a tie-breaking rule is necessary. For instance, if $d(\mathbf{x}_k, \mathbf{y}_i) = d(\mathbf{x}_k, \mathbf{y}_i)$, then a random assignment can be arranged where $\mathbf{x}_k$ is assigned arbitrarily to $\boldsymbol{R}_i$ or $\boldsymbol{R}_j$.

Step 2. Using the centroid condition, compute the centroids for the cells just found to obtain the new codebook $\boldsymbol{Y}_{m+1}$. If an empty cell was generated in Step 1, an alternative codevector assignment is made for that cell.

An empty cell is defined as the one where no training vector is assigned to it. This situation can happen when the particular codeword of the cell is very far away from the training vectors. Empty cells do not contribute toward lowering the final distortion sum and therefore must be eliminated. A variety of heuristic solutions have been proposed to handle the empty cell problem. One simple approach is to split the biggest cell into two cells by adding the corresponding codeword with random numbers having small variance, generating in this way two close versions of the original codeword.

The Generalized Lloyd Algorithm

The generalized Lloyd algorithm (GLA) is a generalization of the Lloyd algorithm discussed in Chapter 5 to the design of vector quantizers. In essence it follows the exact same steps except for some notational changes. The GLA is also known in literature as the LBG algorithm after Linde, Buzo, and Gray [1980].

Quantizer Design Based on Squared Euclidean Distance Measure

Consider the distance measure between two M-dimensional vectors given by

$$d(\mathbf{x}, \mathbf{y}) = \|\mathbf{x} - \mathbf{y}\|^2 = \sum_{m=1}^{M} (x_m - y_m)^2. \tag{7.20}$$

Using this measure, it can be shown (Exercise 7.4) that for $\mathbf{x}_k \in \boldsymbol{R}_i$, the centroid of $\boldsymbol{R}_i$, denoted by $\mathbf{y}_i$, is given by

$$\mathbf{y}_i = \frac{1}{N_i} \sum_{\mathbf{x}_k \in \boldsymbol{R}_i} \mathbf{x}_k \tag{7.21}$$

with N_i the number of elements in the ith cell. Thus, the centroid of $\boldsymbol{R}_i$ is given by the mean of the elements pertaining to the cell.

Example 7.1 In this example, the GLA is applied to VQ design for 2-D vectors. The quantizer size is $N = 8$ and a total of $N_t = 500$ training vectors are used. The training vectors are obtained from random number generators. Figure 7.2 shows the locations of the training vectors in the two-dimensional plane. Initial codewords are selected randomly from the training vectors; that is, N vectors are selected randomly from the training vector set to use as the initial codebook. Trajectories of the codewords during training and codewords after convergence are recorded. Note how the final codewords tend toward locations of the plane where density of training vectors is higher; this is an expected result since the algorithm tries to minimize the average distortion over the training data set. Figure 7.3 shows the distance sum as a function of the number of iterations, where convergence occurs approximately at iteration number 40. Note that the curve decreases monotonically with the number of iterations.

Similar to scalar quantization, the solution at convergence depends on the initial codebook. The algorithm converges at local minimums, leading to differing levels of performance. As was discussed in Chapter 5, a number of random initializations can be employed so that the best codebook (with the minimum distortion sum) is

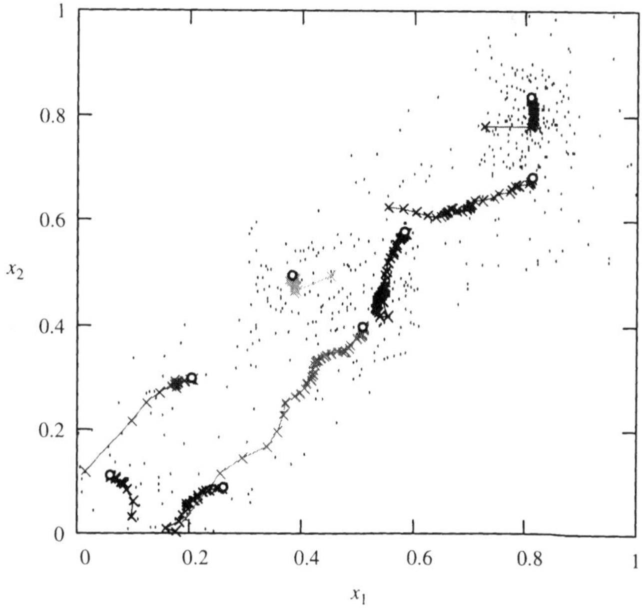

Figure 7.2 Training vectors and codeword trajectories. Positions of training vectors are marked with dots. Trajectories of codewords are joined with the updated codeword marked with x. Positions of codewords after convergence are marked with o.

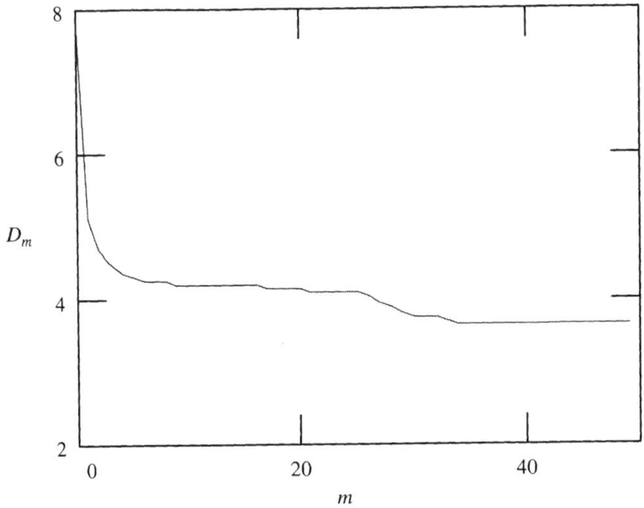

Figure 7.3 Distance sum as a function of the number of iterations.

selected as the final solution. An alternative technique is discussed next, which can improve the performance after convergence.

Stochastic Relaxation

Stochastic relaxation is a method applied to vector quantizer design which is able to avoid poor local minimums. The basic idea is to simultaneously perturb either all encoder or all decoder parameters with an independent white noise process, whose variance starts from a maximum and decreases gradually with the number of iterations. After each perturbation, a Lloyd iteration is performed.

The heuristic justification of the approach is that by adding high levels of noise at the beginning of the training process, the state of the quantizer is essentially randomized. As the noise is reduced, the algorithm finds it more difficult to escape a deep minimum in a single step, but a shallow local minimum will not confine the state. Since the added noise goes to zero, it is much more probable that the final state is in a deep minimum of the total distortion sum.

One proposed perturbation noise variance as a function of the iteration number m is given by

$$\sigma_m^2 = \begin{cases} A(1-(m-1)/I)^3, & m = 1, \ldots, I, \\ 0, & m > I, \end{cases} \tag{7.22}$$

where I is a constant defining the number of noisy iterations to run. In general, I should be large enough so that random perturbation becomes effective. A is the initial noise variance. According to (7.22), noise variance for perturbation is high at the beginning and gradually decreases toward zero. The choice of A depends on the magnitude of the parameters to be disturbed. One version of the stochastic relaxation algorithm, based on perturbing the training vector set, is outlined below.

Step 1. Begin with an initial codebook $\boldsymbol{Y}_1$. Set $m = 1$. The initial codebook can be obtained by randomly selecting some training vectors as codewords.

Step 2. Perturb the training vector set. Generate noise samples with variance given by (7.22). Add these noise samples to the training vector set.

Step 3. Perform a nearest-neighbor search with the perturbed training vector set to form the cells $\boldsymbol{R}_{m,i}, i = 1, \ldots, N$. With N being the quantizer size, or the number of codewords.

Step 4. Check for an empty cell: if some of the cells found in Step 3 are empty, they should be removed to avoid problems with the centroid calculation in the next step. One simple approach is to eliminate the codeword associated with the empty cell and split the cell with the highest number of members into two codewords according to

$$\mathbf{y}_e \leftarrow \mathbf{y}_b + \mathbf{u},$$
$$\mathbf{y}_b \leftarrow \mathbf{y}_b - \mathbf{u},$$

where $\mathbf{y}_e$ is the codeword associated with the empty cell, $\mathbf{y}_b$ is the codeword associated with the biggest cell, and $\mathbf{u}$ is a low-norm vector. The vector $\mathbf{u}$ can be fixed or generated using a low-variance random number generator. The low-norm requirement is to ensure that the two resultant codewords are not far from the original $\mathbf{y}_b$, and hence the new $\mathbf{y}_e$ and $\mathbf{y}_b$ must take some of the members of the biggest cell, splitting effectively into two nonempty cells. If none of the cells from Step 3 is empty, the algorithm proceeds to the next step. If some of the cells are empty, a cell-splitting procedure is performed and the algorithm has to go back to Step 3 again for one nearest-neighbor search. Then Step 4 is executed again to eliminate empty cells. Unless all cells are nonempty, the algorithm will bounce between Steps 3 and 4.

Step 5. Compute the centroids using the uncorrupted training vectors for the cell just found to obtain the new codebook $\boldsymbol{Y}_{m+1}$.

Step 6. Compute the distortion sum for $\boldsymbol{Y}_{m+1}$ (D_{m+1}). If it has changed by a small enough amount since the last iteration, stop. Otherwise set $m \leftarrow m + 1$ and go to Step 2.

Example 7.2 A similar design situation as in Example 7.1 is considered here. A total of 100 2-D training vectors are used with the quantizer size equal to 10. The elements of the training vectors are between 0 and 1. Figure 7.4 shows the distance sum curves for the GLA and stochastic relaxation (SR), after 500 iterations. In stochastic relaxation, $I = 300$ and $A = 1$. At the end of the training section, the GLA produces a distance sum of 1.576, while 1.407 is recorded for stochastic relaxation.

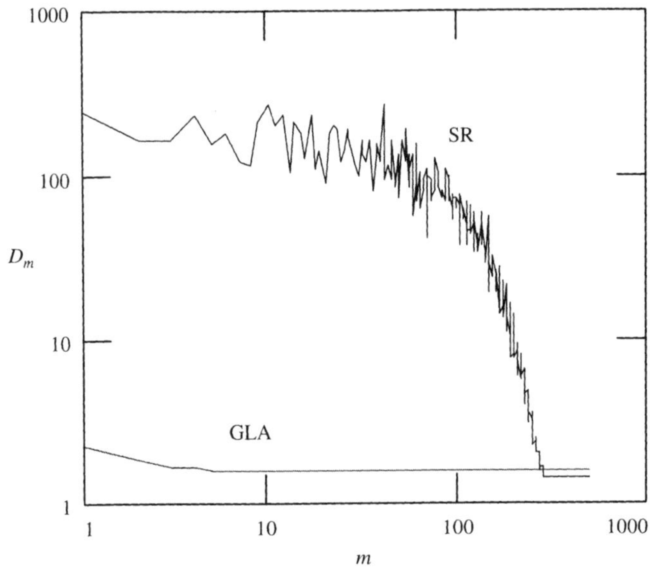

Figure 7.4 Comparison between the distance sum curves for stochastic relaxation and the GLA. Note that both horizontal and vertical scales are logarithmic.

Therefore, for this particular experiment, stochastic relaxation produces a better result.

The price to pay for using stochastic relaxation is obvious: more training time. In general, the constant I must be high enough so that random perturbation becomes effective. The result at convergence also depends on the initial codebook as well as the noise sequence used for perturbation. Therefore, stochastic relaxation can only be viewed as an alternative training method, and there is no guarantee of achieving global optimality. In practice, the amount of computation incurred in stochastic relaxation might be comparable to running several GLA using different random initializations, and the outcomes of the two approaches might be close. Thus, the advantage of one method over the other is hard to determine or nonexistent.

7.4 MULTISTAGE VQ

A direct implementation of VQ having a reasonable resolution as presented in the last sections requires a great deal of implementational costs—costs in terms of space (storage needed for codebook, referred to as memory cost) and time (operations required to search for the best codeword, referred to as computational cost). In order to lower the implementational costs, a certain structure can be imposed to reduce either the space requirement, the time requirement, or both. In this section we consider a structured scheme, known as multistage vector quantization (MSVQ), which provides substantial savings in both space and time. The price paid for these advantages is the performance degradation when compared with an equivalent unconstrained VQ. However, cost reduction often compensates for the drop in performance in many practical applications.

Figure 7.5 shows the block diagrams of the encoder and decoder for MSVQ with K stages. In the encoder, the input vector $\mathbf{x}$ is compared with

$$\hat{\mathbf{x}} = \mathbf{y}_{i_1}^{(1)} + \mathbf{y}_{i_2}^{(2)} + \cdots + \mathbf{y}_{i_K}^{(K)}, \tag{7.23}$$

where $\mathbf{y}_i^{(l)}$ is the ith codevector from the lth stage codebook. That is, the first-stage codebook $\boldsymbol{Y}_1$ of size N_1 submits the codevector $\mathbf{y}_{i_1}^{(1)}$; the second-stage codebook $\boldsymbol{Y}_2$ of size N_2 submits the codevector $\mathbf{y}_{i_2}^{(2)}$; and so on. Note that all codevectors have the same dimension as the input vector. The decoders D_1 to D_K of the different stages merely output the codevector using the input index.

By choosing different indices, the encoder seeks to minimize the distance between $\mathbf{x}$ and $\hat{\mathbf{x}}$. The index set $\{i_1, i_2, \ldots, i_K\}$ that minimizes the distance is

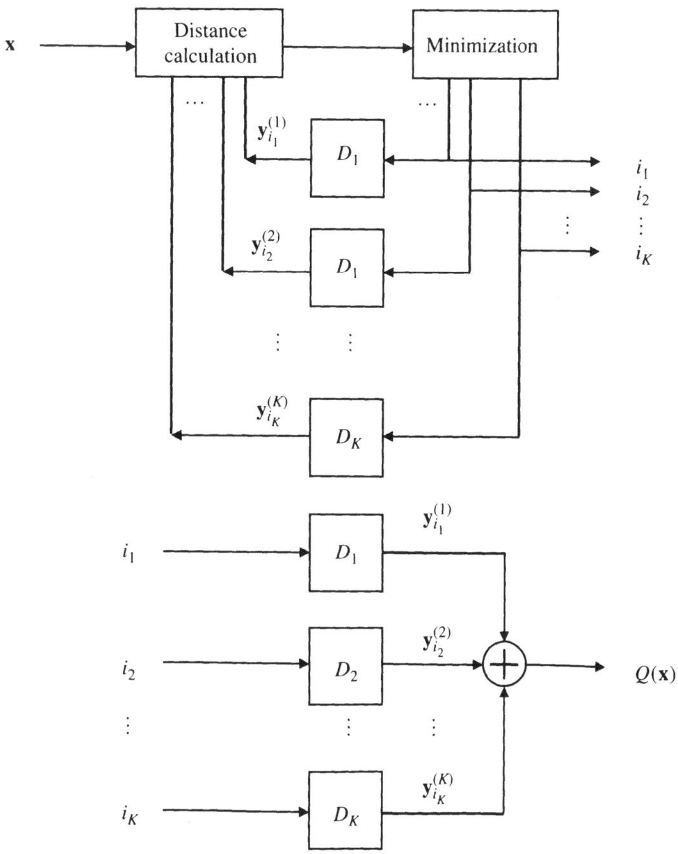

Figure 7.5 MSVQ encoder (*top*) and decoder (*bottom*).

transmitted to the MSVQ decoder, where codevectors from different stages are added together to form the quantized version of the input.

Resolution

We are given a K-stage MSVQ with codebook sizes $N_1, N_2, \ldots, N_K$. The resolution of each stage is given by

$$r_1 = \lg N_1, r_2 = \lg N_2, \ldots, r_K = \lg N_K$$

with an overall resolution of

$$r = \sum_{l=1}^{K} r_l = \sum_{l=1}^{K} \lg N_l = \lg\left(\prod_{l=1}^{K} N_l\right) \tag{7.24}$$

bits.

Memory Cost

Assuming that one memory location is needed to store one codevector, the memory cost (MC) for a K-stage MSVQ is given by

$$MC = \sum_{l=1}^{K} N_l = \sum_{l=1}^{K} 2^{r_l}. \tag{7.25}$$

That is, MC memory locations are needed for codebook storage.

Example 7.3 Here, comparison is made between the memory cost of unconstrained VQ and MSVQ, for typical values of resolution. Table 7.1 shows some numerical results. In the first column, resolution values are given. The second column contains the memory cost for an unconstrained VQ, calculated as 2^r. The Third column and fourth column contain the memory cost for an MSVQ having two stages and three stages, respectively, together with the selected configuration within parenthesis. For MSVQ, the memory cost depends on the resolution as well as the

TABLE 7.1 Comparison of Memory Cost for Typical Values of Resolution

Resolution (in bits)	Unconstrained VQ	MSVQ: Two Stages	MSVQ: Three Stages
6	64	16 (3, 3)	12 (2, 2, 2)
8	256	32 (4, 4)	20 (2, 3, 3)
10	1024	64 (5, 5)	32 (3, 3, 4)
15	32768	384 (7, 8)	96 (5, 5, 5)
20	1048576	2048 (10, 10)	320 (6, 7, 7)

particular codebook configuration; for instance, in two stages, 10-bit resolution can be configured as $(1, 9)$ $(r_1 = 1, r_2 = 9)$, $(3, 7)$ $(r_1 = 3, r_2 = 7)$, and so on; that is, as long as the sum of the resolutions for the two codebooks is equal to the overall resolution. A different codebook configuration, in general, produces different memory cost.

Several observations can be drawn from Table 7.1. First, memory cost is far lower for MSVQ, especially when the resolution increases. At 10-bit resolution, for instance, unconstrained VQ needs 16 times the amount of memory required by a two-stage MSVQ. On the other hand, a higher number of stages in an MSVQ configuration usually need less memory.

In general, a more costly system requiring more memory provides higher performance, mainly because more parameters are available for tuning. In many instances, however, slight performance gain is obtained at a huge cost increase. Thus, from a practical point of view, the small gain in system quality does not justify the excessive cost.

Example 7.4: Minimizing Memory Cost for a Fixed Resolution and Number of Stages It is clear from (7.25) that different configurations of MSVQ having the same overall resolution r (given by (7.24)) produce different costs. Table 7.2 shows an example of 6-bit resolution, where all possible costs for two-stage and three-stage MSVQ are summarized. In this example, when a particular configuration has the same size for all stages, the memory cost seems to be minimized ((3, 3) vs. (1, 5) and (2, 4); (2, 2, 2) vs. (1, 1, 4) and (1, 2, 3)). Is the memory cost minimized when the sizes of different stages are the same? The answer to the question is yes and the reader is referred to Exercise 7.7 for a proof. Since one of the main advantages in using MSVQ is space saving, it is common in practice to use stages having the same size in order to minimize MC. The ultimate configuration, however, is decided by the required performance level and the amount of available resources.

Search Procedure

Consider a K-stage MSVQ with codebook sizes $N_1, N_2, \ldots, N_K$. The task of the encoder is to search for the best indices, grouped as the vector $\mathbf{i} = [i_1, i_2, \ldots, i_K]$ so as to minimize the quantization error. Various search strategies exist leading

TABLE 7.2 Memory Cost for MSVQ with 6-bit Resolution

Two Stages ($K = 2$)	Three Stages ($K = 3$)
34 (1, 5)	20 (1, 1, 4)
20 (2, 4)	14 (1, 2, 3)
16 (3, 3)	12 (2, 2, 2)

to different design procedures, implementational complexities, and overall performance. The determination of the optimal set of indices is a complex combinatorial optimization problem that can only be solved in general by trying all sets of indices (full search). Suboptimal sequential solutions are typically used in practice to reduce the complexity.

Full Search

This is the optimal procedure and evaluates all possible combinations of indices to find the best codeword. That is, $i_1, i_2, \ldots, i_K$ are exhaustively searched to yield the optimal index vector $\mathbf{i}_o = [i_{o1}, i_{o2}, \ldots, i_{oK}]$ such that

$$d\left(\mathbf{x}, \mathbf{y}_{i_{o1}}^{(1)} + \mathbf{y}_{i_{o2}}^{(2)} + \cdots + \mathbf{y}_{i_{oK}}^{(K)}\right) \leq d\left(\mathbf{x}, \mathbf{y}_{i_1}^{(1)} + \mathbf{y}_{i_2}^{(2)} + \cdots + \mathbf{y}_{i_K}^{(K)}\right); \quad \forall \mathbf{i} \neq \mathbf{i}_o, \tag{7.26}$$

where $d(.,.)$ is the chosen distance or distortion measure.

Sequential Search

The sequential search procedure determines $\mathbf{i}_o = [i_{o1}, i_{o2}, \ldots, i_{oK}]$ in a sequential manner. Starting from Stage 1, i_{o1} is found so that

$$d\left(\mathbf{x}, \mathbf{y}_{i_{o1}}^{(1)}\right) \leq d\left(\mathbf{x}, \mathbf{y}_{i_1}^{(1)}\right); \quad \forall\, i_1 \neq i_{o1}. \tag{7.27}$$

Next, for Stage 2, i_{o1} is fixed and i_{o2} is searched to reach

$$d\left(\mathbf{x}, \mathbf{y}_{i_{o1}}^{(1)} + \mathbf{y}_{i_{o2}}^{(2)}\right) \leq d\left(\mathbf{x}, \mathbf{y}_{i_{o1}}^{(1)} + \mathbf{y}_{i_2}^{(2)}\right); \quad \forall\, i_2 \neq i_{o2}. \tag{7.28}$$

In general, for the jth stage,

$$d\left(\mathbf{x}, \mathbf{y}_{i_{o1}}^{(1)} + \mathbf{y}_{i_{o2}}^{(2)} + \cdots + \mathbf{y}_{i_{oj}}^{(j)}\right) \leq d\left(\mathbf{x}, \mathbf{y}_{i_{o1}}^{(1)} + \mathbf{y}_{i_{o2}}^{(2)} + \cdots + \mathbf{y}_{i_j}^{(j)}\right); \quad \forall\, i_j \neq i_{oj}. \tag{7.29}$$

Tree Search

Tree search is a generalization of the sequential search procedure in which more than one index is passed on from one stage to the next. In the first stage, the set of M_a (an integer) minimum distortion indices is determined. Denoting this set by $\boldsymbol{I}_1$, we have

$$i_{o1} \in \boldsymbol{I}_1 \text{ if and only if } d\left(\mathbf{x}, \mathbf{y}_{i_{o1}}^{(1)}\right) \leq d\left(\mathbf{x}, \mathbf{y}_{i_1}^{(1)}\right); \quad \forall\, i_1 \notin \boldsymbol{I}_1. \tag{7.30}$$

In the second stage, a joint search is performed with the added constraint that the first index is from $\boldsymbol{I}_1$. The M_a minimum distortion sets of indices are maintained from this search, and this set of (pairs of) indices is denoted by $\boldsymbol{I}_2$,

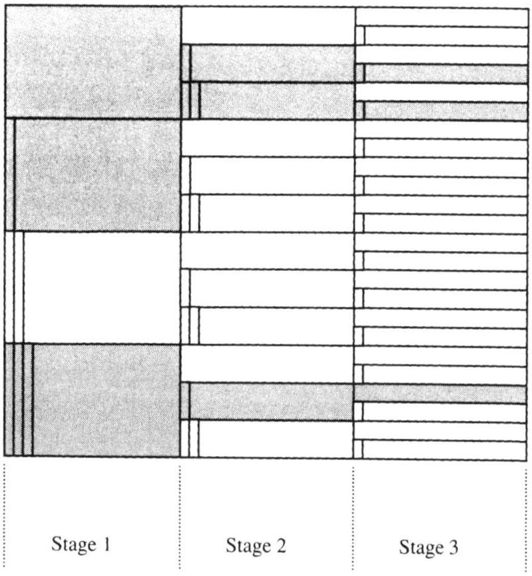

Figure 7.6 Illustration of a tree search; the best paths are indicated by filled blocks.

which satisfies

$$(i_{o1}, i_{o2}) \in \boldsymbol{I}_2 \text{ if and only if } d\left(\mathbf{x}, \mathbf{y}_{i_{o1}}^{(1)} + \mathbf{y}_{i_{o2}}^{(2)}\right) \le d\left(\mathbf{x}, \mathbf{y}_{i_{o1}}^{(1)} + \mathbf{y}_{i_2}^{(2)}\right); \\ \forall\, i_{o1} \in \boldsymbol{I}_1, \quad (i_{o1}, i_2) \notin \boldsymbol{I}_2. \tag{7.31}$$

Similarly, at the jth stage, the set of M_a indices $\boldsymbol{I}_j$ are defined by

$$(i_{o1}, \ldots, i_{oj}) \in \boldsymbol{I}_j \text{ if and only if } d\left(\mathbf{x}, \sum_{l=1}^{j-1} \mathbf{y}_{i_{ol}}^{(l)} + \mathbf{y}_{i_{oj}}^{(j)}\right) \le d\left(\mathbf{x}, \sum_{l=1}^{j-1} \mathbf{y}_{i_{ol}}^{(l)} + \mathbf{y}_{i_j}^{(j)}\right); \\ \forall\ (i_{o1}, \ldots, i_{o,j-1}) \in \boldsymbol{I}_{j-1}, (i_{o1}, \ldots, i_j) \notin \boldsymbol{I}_j. \tag{7.32}$$

At the final stage, the minimum distortion set of indices from the set $\boldsymbol{I}_K$ is chosen as the overall optimum. It is important to emphasize that only M_a elements exist in the index set $\boldsymbol{I}_j$, $j = 1$ to K.

Example 7.5 Tree search is illustrated here for a three-stage MSVQ. Codebook sizes are $N_1 = 4$, $N_2 = 3$, and $N_3 = 2$. Thus, $i_1 = 1, 2, 3, 4$; $i_2 = 1, 2, 3$; and $i_3 = 1, 2$. A hypothetical search situation is shown in Figure 7.6 with $M_a = 3$. Search starts with the first stage, where the input vector is compared with the four first-stage codevectors; that is, these distances are calculated as

$$d_1 = d\left(\mathbf{x}, \mathbf{y}_1^{(1)}\right), \qquad d_2 = d\left(\mathbf{x}, \mathbf{y}_2^{(1)}\right),$$
$$d_3 = d\left(\mathbf{x}, \mathbf{y}_3^{(1)}\right), \qquad d_4 = d\left(\mathbf{x}, \mathbf{y}_4^{(1)}\right).$$

Assume that d_1, d_2, and d_4 are all smaller than d_3; then

$$\boldsymbol{I}_1 = \{1, 2, 4\}.$$

Proceeding with the second stage the following pairs of indices (i_1, i_2) are considered:

$$\begin{array}{ccc} (1,1) & (1,2) & (1,3) \\ (2,1) & (2,2) & (2,3) \\ (4,1) & (4,2) & (4,3) \end{array}$$

and the best M_a pairs of indices are maintained. Assume that (1, 2), (1, 3), and (4, 2) are the best (lowest distance among the nine pairs of indices); then

$$\boldsymbol{I}_2 = \{(1,2), (1,3), (4,2)\}.$$

The final stage tests the following triplets of indices:

$$\begin{array}{cc} (1,2,1) & (1,2,2) \\ (1,3,1) & (1,3,2) \\ (4,2,1) & (4,2,2) \end{array}$$

In this example, we assume

$$\boldsymbol{I}_3 = \{(1,2,2), (1,3,2), (4,2,1)\}.$$

That is, the index triplets in $\boldsymbol{I}_3$ give the lowest distance among the six alternatives. Finally, the set of indices from $\boldsymbol{I}_3$ that provides the lowest distortion is the desired encoder output. The search process is illustrated graphically in Figure 7.6.

Computational Cost

Here, computational costs associated with various search procedures in MSVQ are analyzed. For simplicity the cost is expressed as the number of distance computations required to complete the search.

Full Search

Since all possible combinations of indices among all stages are evaluated, a total of

$$CC = N_1 N_2 \cdots N_K = \prod_{l=1}^{K} N_l \tag{7.33}$$

distance computations are required.

Sequential Search

In this case, each stage is searched only once; thus, the number of distance computations is given by

$$CC = N_1 + N_2 + \cdots + N_K = \sum_{l=1}^{K} N_l. \tag{7.34}$$

Tree Search

In the general case, the number of distance computations is given by

$$CC = N_1 + \min(N_1, M_a)N_2 + \min(\min(N_1, M_a)N_2, M_a)N_3 + \cdots. \tag{7.35}$$

For the first stage, N_1 distance computations are needed to locate M_a codevectors. For the second stage, N_2 distance computations are required for each of the M_a, or N_1 (depending which number is smaller), first-stage codevectors. Counting in this way up to the Kth stage gives (7.35). The minimum operator min(.,.) compares two numbers and returns the smallest.

Example 7.6 Figure 7.7 shows an example of computational cost where the MSVQ consists of four stages having the same size of 4. That is, $N_l = 4$ for $l = 1$

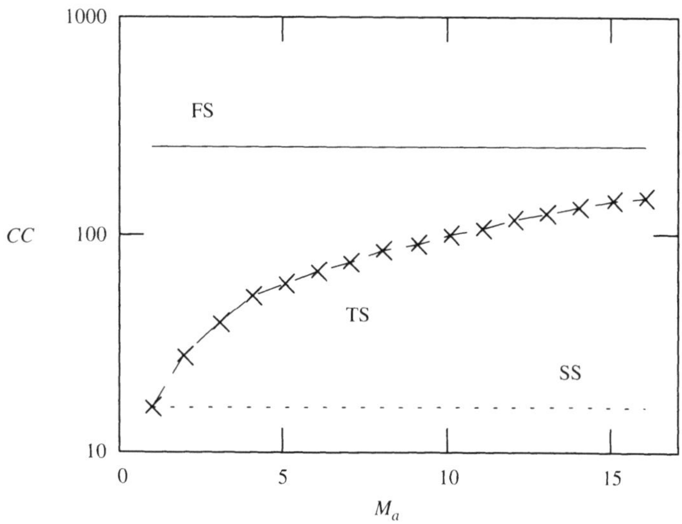

Figure 7.7 An example of computational cost comparison between three codebook search methods under MSVQ: full search (FS), tree search (TS), and sequential search (SS), with M_a being the number of indices preserved at each stage. Note that the vertical scale is logarithmic.

to 4. The cases considered are full search, sequential search, and tree search with different M_a. Note that the maximum value of M_a is equal to 16 since a maximum of $4 \times 4 = 16$ index sets are passed to the next stage. From this example we can see that sequential search requires far fewer distance computations with respect to full search. The cost involved with tree search is the same as with sequential search when $M_a = 1$; indeed, the two methods are equivalent under that condition. For higher values of M_a, the cost increases proportionately.

Since the full search procedure verifies all combinations of codevectors, its performance is the best. However, it is demonstrated later that the performance of the tree search technique with a moderate value of M_a can be close to a full search approach, with the accompanying benefit of substantial cost reduction.

MSVQ Design Algorithms

Several algorithms for MSVQ codebook design are presented next. It is important to note that they are only applicable to the case of squared Euclidean distance measure. We will consider exclusively this particular distance measure since it is widely employed in speech coding applications. Differing strategies must be used for other distance measures.

Sequential Codebook Design Algorithm

This method optimizes the current stage codebook under the assumption that all subsequent stages are populated by zero vectors only. Hence, the training vector set (7.18) is used to generate the first-stage codebook using the GLA. For the second stage, a new training vector set is generated with

$$\mathbf{x}_k^{(2)} = \mathbf{x}_k - Q_1(\mathbf{x}_k); \quad k = 0, 1, \ldots, N_t - 1, \tag{7.36}$$

where $Q_1(\cdot)$ is the first-stage quantization function. This training set is now used to create the second-stage codebook. In general, the lth-stage codebook is generated with the training set

$$\begin{aligned} \mathbf{x}_k^{(l)} &= \mathbf{x}_k^{(l-1)} - Q_{l-1}(\mathbf{x}_k^{(l-1)}) \\ &= \mathbf{x}_k - Q_1(\mathbf{x}_k) - \cdots - Q_{l-1}(\mathbf{x}_k^{(l-1)}) \\ &= \mathbf{x}_k - \sum_{i=1}^{l-1} Q_i(\mathbf{x}_k^{(i)}), \end{aligned} \tag{7.37}$$

where $\mathbf{x}_k^{(1)} = \mathbf{x}_k$.

This codebook design procedure is suboptimal in the sense that it does not find the best set of codebooks for sequentially encoding a vector stage-by-stage with the multistage structure. It is, however, greedy in the sense that it finds the codebook for the first stage that would be optimal if there were only one stage. Then it finds the best codebook for the second stage given the first-stage codebook, assuming there are only two stages. Similarly, each successive stage is optimal given all previous stage codebooks and assuming that it is the last stage.

Example 7.7 Some properties of the sequential algorithm are illustrated here through an actual design case. A group of 2-D vectors is selected for codebook design. The group consists of 3000 vectors (Figure 7.8, *top left*) and a four-stage MSVQ is designed, where all stages have the same size of four (overall resolution = 8 bits). Initial codebooks are randomly initialized. As explained before, the first stage with four codevectors is designed using the GLA with the 3000 input vectors. Then, the training data for the second stage is formed by computing the difference between the input vectors and the corresponding quantized versions from the first stage. These difference vectors form the training data for the second stage. The process is repeated for each stage. Training vectors for the four stages are shown in Figure 7.8. It is possible to point out two key observations from this example:

- The arithmetic mean of the training vectors for Stages 2, 3, and 4 tend to be the zero vector. This is evidenced from the fact that they gather around the

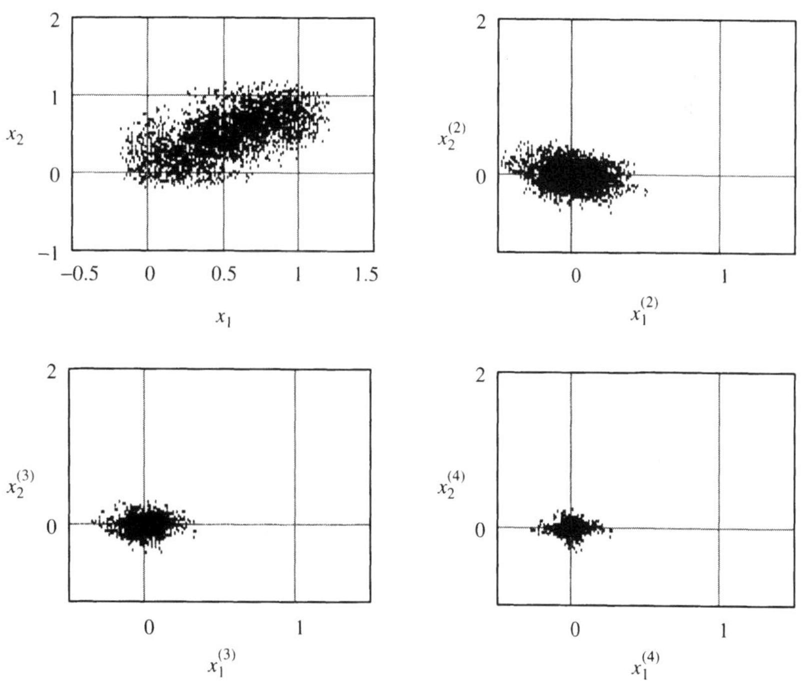

Figure 7.8 Example training data for a four-stage MSVQ. Original (*top left*), second stage (*top right*), third stage (*bottom left*), and fourth stage (*bottom right*).

origin (Figure 7.8, *top right*, *bottom left*, and *bottom right*). Note that the mean of the original training vector set is not zero (Figure 7.8, *top left*).

- The amplitude of the training vectors for successive stages tends to diminish. In other words, the energy of the training vectors for successive stages is smaller.

These two phenomena result from the operation of the VQ: training vectors for successive stages are essentially quantization errors of previous stages. If the VQ is well designed, the amplitudes of training vectors for successive stages should become lower and lower, with an arithmetic mean approaching zero. Later in the section, these properties are applied to develop the joint codebook design algorithm for a tree search.

The way that the quantizer is designed also has important consequences regarding performance under different search procedures. In the present case, the distance sum for the input training set under full search is 5.28, while the minimum distance sum under the tree search is 5.48. This represents a 3.8% performance degradation by changing from full search to tree search. The computational cost for full search is 256, while for the tree search ($M_a = 4$) it is 52. Thus, full search requires roughly five times the amount of computation done by a tree search. As we can see in this example, computation saving is huge with relatively low performance degradation when full search is replaced by tree search.

Figure 7.9 shows the distance sum for the input training set for the full search and tree search. Note how the tree search performance saturates rapidly as M_a increases ($M_a \geq 4$). This characteristic is compared with other design approaches later in the chapter.

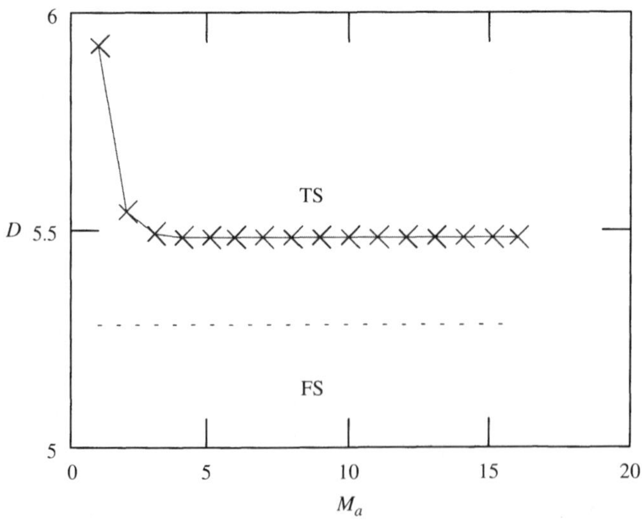

Figure 7.9 Distance sum (D) for the training data set for full search (FS) and tree search (TS) as a function of M_a.

Stacked Codebook Description of MSVQ

Notations are introduced here to describe the operation of MSVQ. The codebook of the lth stage can be represented by the stacked stage codebook:

$$\mathbf{y}^{(l)} = \begin{bmatrix} \mathbf{y}_1^{(l)} \\ \mathbf{y}_2^{(l)} \\ \vdots \\ \mathbf{y}_{N_l}^{(l)} \end{bmatrix}; \quad l = 1, \ldots, K, \tag{7.38}$$

formed by placing all N_l codevectors as a single $MN_l \times 1$ vector. Furthermore, by stacking all K stacked stage codebooks, we form the overall stacked codebook:

$$\mathbf{y} = \begin{bmatrix} \mathbf{y}^{(1)} \\ \mathbf{y}^{(2)} \\ \vdots \\ \mathbf{y}^{(K)} \end{bmatrix}, \tag{7.39}$$

which is an $M(N_1 + N_2 + \cdots + N_K) \times 1$ vector.

To select a particular codeword of the lth-stage codebook, we multiply the corresponding stacked codebook with the selection matrix:

$$\mathbf{y}_i^{(l)} = \mathbf{B}_i^{(l)}\mathbf{y}^{(l)}; \qquad i = 1, \ldots, N_l, \quad l = 1, \ldots, K, \tag{7.40}$$

with $\mathbf{B}_i^{(l)}$ an $M \times (MN_l)$ matrix whose elements are one or zero. Figure 7.10 shows the structure of the matrix. The multistage selection matrix is formed with

$$\mathbf{B} = \begin{bmatrix} \mathbf{B}_{i_1}^{(1)} & \mathbf{B}_{i_2}^{(2)} & \cdots & \mathbf{B}_{i_K}^{(K)} \end{bmatrix}, \tag{7.41}$$

which is an $M \times M(N_1 + \cdots + N_K)$ matrix. Equation (7.23) can now be expressed as

$$\hat{\mathbf{x}} = \mathbf{B} \cdot \mathbf{y}, \tag{7.42}$$

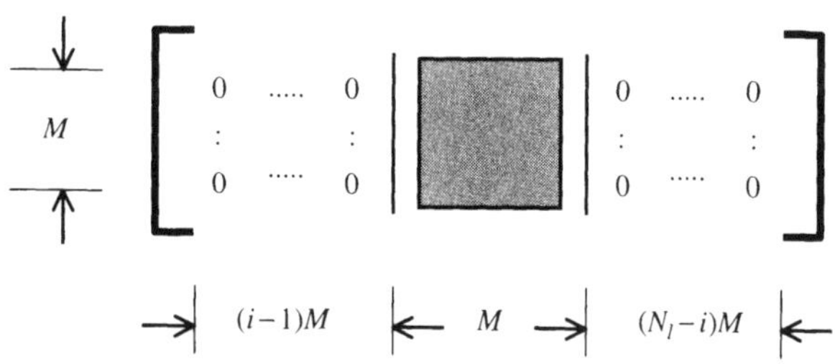

Figure 7.10 Structure of the selection matrix $\mathbf{B}_i^{(l)}$. The shaded area represents an $M \times M$ identity matrix.

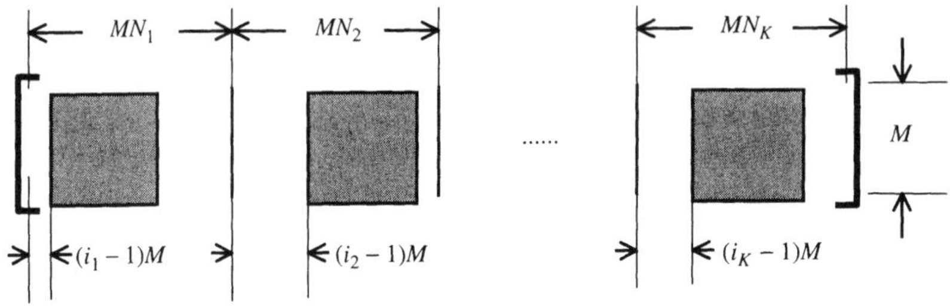

Figure 7.11 General form of the multistage selection matrix **B**. The shaded area represents an $M \times M$ identity matrix.

where **B** is the multistage selection matrix while **y** is the quantizer's stacked codebook. Given the indices $i_1, \ldots, i_K$, the matrix **B** selects the corresponding codewords from the stage codebooks $\mathbf{y}^{(1)}, \ldots, \mathbf{y}^{(K)}$. Figure 7.11 shows the structure of the multistage selection matrix.

Derivation of the Joint Codebook Design Algorithm

The goal of joint codebook design is to jointly optimize all codevectors over all stages after each pass through the training set. The codebook design procedure seeks to minimize the distance sum:

$$D = \sum_k \|\mathbf{x}_k - \hat{\mathbf{x}}_k\|^2. \tag{7.43}$$

Using the notation in (7.42),

$$\begin{aligned} D &= \sum_k (\mathbf{x}_k - \mathbf{B}_k\mathbf{y})^T(\mathbf{x}_k - \mathbf{B}_k\mathbf{y}) \\ &= \sum_k \mathbf{x}_k^T\mathbf{x}_k - 2\mathbf{y}^T\sum_k \mathbf{B}_k^T\mathbf{x}_k + \mathbf{y}^T\left(\sum_k \mathbf{B}_k^T\mathbf{B}_k\right)\mathbf{y}. \end{aligned} \tag{7.44}$$

Let's define

$$\mathbf{v} = \sum_k \mathbf{B}_k^T\mathbf{x}_k, \tag{7.45}$$

which is an $M(N_1 + \cdots + N_K) \times 1$ vector, and

$$\mathbf{Q} = \sum_k \mathbf{B}_k^T\mathbf{B}_k. \tag{7.46}$$

which is an $M(N_1 + \cdots + N_K) \times M(N_1 + \cdots + N_K)$ symmetric matrix. Also define

$$D_o = \sum_k \mathbf{x}_k^T\mathbf{x}_k. \tag{7.47}$$

Substituting (7.45), (7.46), and (7.47) in (7.44) leads to

$$D = D_o - 2\mathbf{y}^T\mathbf{v} + \mathbf{y}^T\mathbf{Q}\mathbf{y}. \tag{7.48}$$

Differentiating the above equation with respect to $\mathbf{y}$ and equating the result to zero gives

$$\frac{\partial D}{\partial \mathbf{y}} = -2\mathbf{v} + 2\mathbf{Q}\mathbf{y} = 0. \tag{7.49}$$

Thus,

$$\mathbf{v} = \mathbf{Q} \cdot \mathbf{y} \tag{7.50}$$

or

$$\mathbf{y} = \mathbf{Q}^{-1}\mathbf{v}. \tag{7.51}$$

In general, the inverse of $\mathbf{Q}$ does not exist. Indeed, an infinite number of solutions for the joint codebook exist since adding a constant vector to one stage while subtracting the same constant vector from any other stage leads to the same ensemble of possible codewords. Alternative techniques must be used to solve (7.50). Before continuing with the solution of (7.50), let's consider the general forms of the various matrices involved in the derivation. The product $\mathbf{B}^T\mathbf{B}$ has the general form shown in Figure 7.12. From Figure 7.12 we can understand the structure of the $\mathbf{Q}$ matrix (7.46), which is a symmetric matrix. Note that the rows or columns

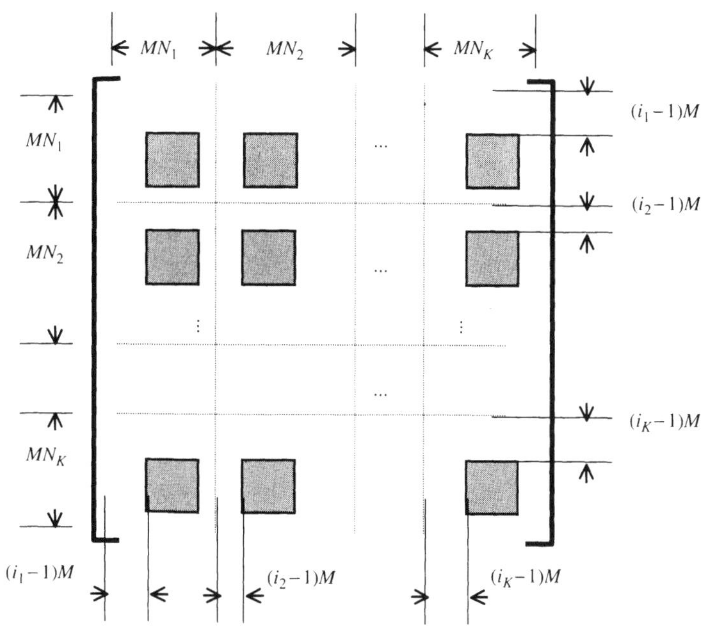

Figure 7.12 General form of the product $\mathbf{B}^T\mathbf{B}$. The shaded areas represent an $M \times M$ identity matrix.

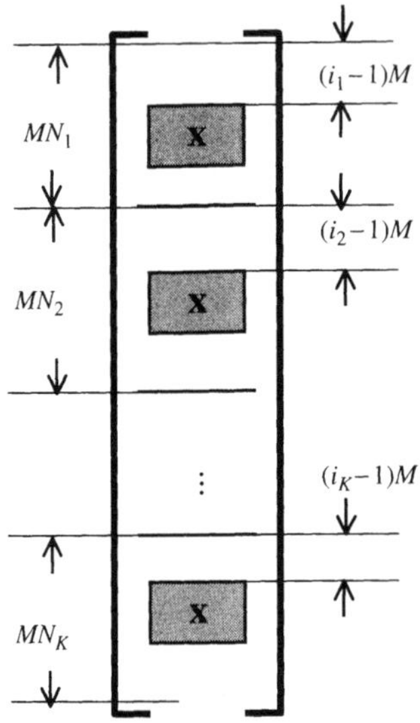

Figure 7.13 General form of the vector $\mathbf{B}^{\mathrm{T}}\mathbf{x}$. The selection matrix $\mathbf{B}$ puts copies of the vector $\mathbf{x}$ in the final product according to the indices $i_1, \ldots, i_K$.

of $\mathbf{Q}$ can be linearly dependent. Therefore, $\mathbf{Q}$ is usually not a full rank matrix; in other words, the inverse does not exist. In addition, moving along the main diagonal we can find that the $\mathbf{Q}$ matrix is comprised of diagonal submatrices of dimensions MN_1, MN_2, $\ldots$, MN_K.

Another important observation about the structure of $\mathbf{Q}$ is that the main diagonal will not contain any zero element if and only if all codewords from the stage codebooks are utilized at least once. That is, after passing the whole training set through the system, all codewords from all stage codebooks must have been involved.

Multiplying $\mathbf{B}^{\mathrm{T}}$ by $\mathbf{x}$ creates the $M(N_1 + \cdots + N_K) \times 1$ vector whose structure is illustrated in Figure 7.13.

For simplicity, a projection method is utilized for the solution of (7.50). Note that (7.39) can be written as

$$\mathbf{y} = \mathbf{y}_o^{(l)} + \mathbf{S}_l \mathbf{y}^{(l)}; \quad l = 1, \ldots, K, \tag{7.52}$$

where

$$\mathbf{y}_o^{(l)} = \begin{bmatrix} \mathbf{y}^{(1)} \\ \vdots \\ \mathbf{y}^{(l-1)} \\ \mathbf{0}^{(l)} \\ \mathbf{y}^{(l+1)} \\ \vdots \\ \mathbf{y}^{(K)} \end{bmatrix} \tag{7.53}$$

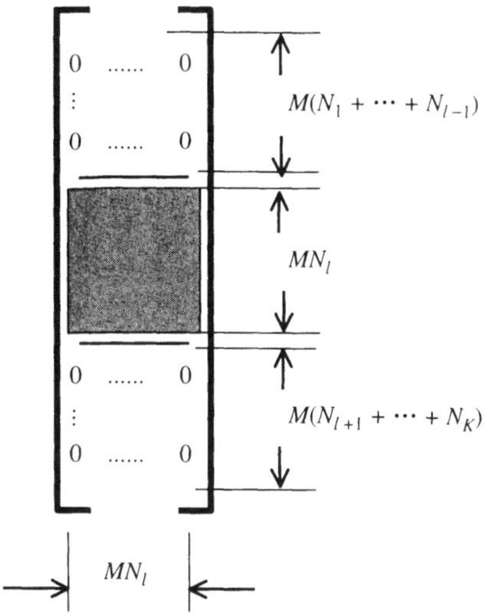

Figure 7.14 General form of the shifting matrix $\mathbf{S}_l$. The shaded area represents an $M \cdot N_l \times M \cdot N_l$ identity matrix.

with $\mathbf{0}^{(l)}$ a zero vector of length MN_l. $\mathbf{S}_l$ is a shifting matrix of dimension $M(N_1 + \cdots + N_K) \times MN_l$, and has the general form shown in Figure 7.14. Substituting (7.53) in (7.48) gives:

$$D = D_o - 2\left(\mathbf{y}_o^{(l)} + \mathbf{S}_l\mathbf{y}^{(l)}\right)^T \mathbf{v} + \left(\mathbf{y}_o^{(l)} + \mathbf{S}_l\mathbf{y}^{(l)}\right)^T \mathbf{Q}\left(\mathbf{y}_o^{(l)} + \mathbf{S}_l\mathbf{y}^{(l)}\right) \tag{7.54}$$

or

$$D = D_o - 2\mathbf{y}_o^{(l)T}\mathbf{v} + \mathbf{y}_o^{(l)T}\mathbf{Q}\mathbf{y}_o^{(l)} - 2\mathbf{y}^{(l)T}\mathbf{S}_l^T\mathbf{v} + 2\mathbf{y}^{(l)T}\mathbf{S}_l^T\mathbf{Q}\mathbf{y}_o^{(l)} + \mathbf{y}^{(l)T}\mathbf{S}_l^T\mathbf{Q}\mathbf{S}_l\mathbf{y}^{(l)}. \tag{7.55}$$

Let's define

$$D_o^{(l)} = D_o - 2\mathbf{y}_o^{(l)T}\mathbf{v} + \mathbf{y}_o^{(l)T}\mathbf{Q}\mathbf{y}_o^{(l)}, \tag{7.56}$$

$$\mathbf{v}^{(l)} = \mathbf{S}_l^T\left(\mathbf{v} - \mathbf{Q}\mathbf{y}_o^{(l)}\right), \tag{7.57}$$

$$\mathbf{Q}^{(l)} = \mathbf{S}_l^T\mathbf{Q}\mathbf{S}_l. \tag{7.58}$$

Substituting (7.56), (7.57), and (7.58) in (7.55) gives

$$D = D_o^{(l)} - 2\mathbf{y}^{(l)T}\mathbf{v}^{(l)} + \mathbf{y}^{(l)T}\mathbf{Q}^{(l)}\mathbf{y}^{(l)}. \tag{7.59}$$

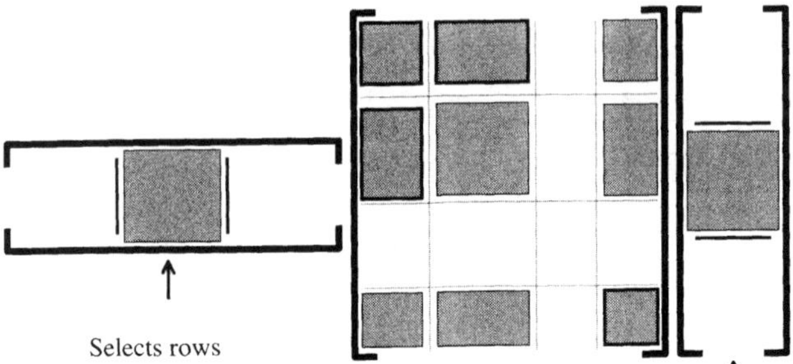

Figure 7.15 Illustration of the product $\mathbf{S}_l^T\mathbf{Q}\mathbf{S}_l$.

Thus, to minimize D, we differentiate (7.59) with respect to $\mathbf{y}^{(l)}$ and equate the result to zero, leading to

$$\mathbf{v}^{(l)} = \mathbf{Q}^{(l)}\mathbf{y}^{(l)}, \tag{7.60}$$

or

$$\mathbf{y}^{(l)} = \left[\mathbf{Q}^{(l)}\right]^{-1}\mathbf{v}^{(l)}. \tag{7.61}$$

The existence of $[\mathbf{Q}^{(l)}]^{-1}$ is investigated next. From (7.58) the effect of the two selection matrices is to "truncate" the matrix $\mathbf{Q}$ into the $MN_l \times MN_l$ matrix $\mathbf{Q}^{(l)}$. The truncation is symmetrical along the main diagonal. This process is illustrated in Figure 7.15. From Figure 7.12 we can see that $\mathbf{Q}$ is comprised of K^2 diagonal submatrices. Since $\mathbf{S}_l^T$ selects the rows* $M(N_1 + \cdots + N_{l-1}) + 1$ to $M(N_1 + \cdots + N_l)$ while $\mathbf{S}_l$ selects the columns $M(N_1 + \cdots + N_{l-1}) + 1$ to $M(N_1 + \cdots + N_l)$, we conclude that $\mathbf{Q}^{(l)}$ is a diagonal matrix of dimension $MN_l \times MN_l$.

As discussed earlier, if all codewords in the lth stage have been utilized during training, the elements on the main diagonal of $\mathbf{Q}^{(l)}$ are nonzero. Therefore, the inverse of $\mathbf{Q}^{(l)}$ exists and is simple to find.

Once the stacked codebook $\mathbf{y}$ has been reoptimized, the training sequence is again repartitioned using the new codebooks. This method is referred to as simultaneous joint design, since all codebooks are reoptimized simultaneously and jointly after each pass over the training sequence. A step-by-step description of the joint codebook design algorithm is presented at the end of the section.

*For an $M \times N$ matrix, the rows are counted from 1 to M, while the columns are counted from 1 to N.

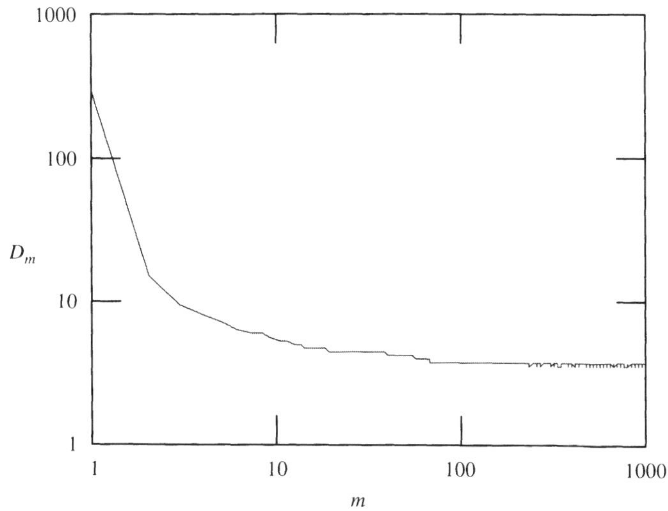

Figure 7.16 Distance sum as a function of the number of iterations in an example of joint codebook design.

Example 7.8 The same training vector set as in Example 7.7 is used to illustrate the properties of the joint codebook design algorithm. The same MSVQ configuration is considered. The initial codebooks are randomly initialized. As in the case of the GLA, the final performance depends on the initial codebook values. Unlike the GLA, however, the distortion sum after each pass through the training data might not decrease. Figure 7.16 shows an example distortion sum curve after 1000 passes through the training data set, where convergence to a minimum is obtained. Small oscillations are observed after roughly 200 passes through the training data set.

Figure 7.17 plots the final codevectors together with the training vectors. The 256 codevectors are obtained by evaluating all possible combinations between the codebooks from the four stages (full search). Unlike the GLA, where the final codevectors upon convergence usually stay as centroids of the training data, in the present case, some codevectors are clearly located far away from the training data, implying therefore a loss in performance. Performance degradation is due to the structure imposed on the MSVQ, a trade-off with respect to implementational costs.

Joint Codebook Design Algorithm and Tree Search

The joint design algorithm described before is done under the assumption that a full search is used. If a sequential or tree search procedure is employed, the performance may degrade. Theoretically, joint design combined with sequential search may even result in worse performance than sequentially designed codebooks. Note that if a full search is used, the ordering of the codebooks is unimportant since

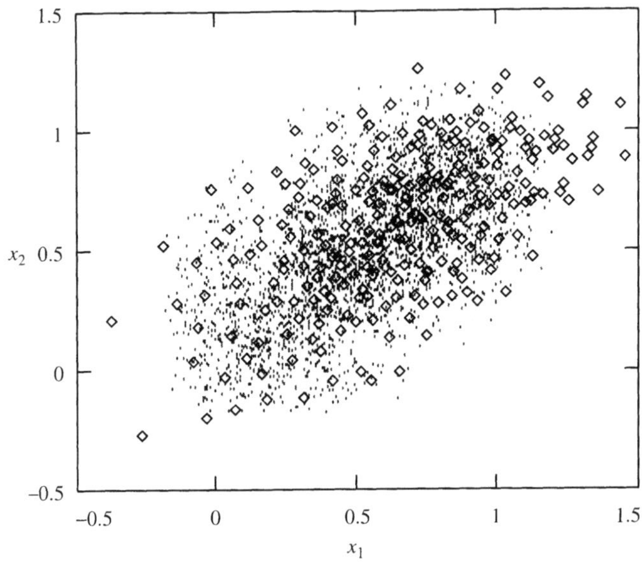

Figure 7.17 Plot of the input training vectors (dots) and full-search codevectors (diamonds) of the four-stage MSVQ.

it does not change the ensemble of reconstruction values. If a sequential or tree search is used, the ordering may be very important.

Some empirical procedures are proposed next that can improve the performance of MSVQ under a tree search designed using the joint algorithm. These procedures are based on observations of the characteristics of the codebooks optimized using the sequential algorithm.

Improvement 1. Begin with an initial codebook sequentially designed. Since the final solution depends on the initial codevectors, by using the sequential algorithm first and further optimizing with the joint codebook design algorithm, it is expected that the solution at convergence will be better adapted for a sequential or tree search.

Improvement 2. Remove the mean of all codewords of each codebook (except the first codebook) so that

$$\sum_{i=1}^{N_l} \mathbf{y}_i^{(l)} = \mathbf{0} \tag{7.62}$$

for $l = 2$ to K, with $\mathbf{0}$ the zero vector. This step should be performed after each update of the codewords in the joint codebook design algorithm. Since for both sequential search and tree search, the codevectors from stage 2 to K contain quantized versions of difference vectors, it is expected that the mean should be close to zero. By removing the mean, the resultant codebooks are better adapted to the natural structure required for a sequential or tree search. Indeed, this is a property of sequentially designed codebooks, as shown in Example 7.7.

Improvement 3. Codebook reordering: Modify the order of the codebooks such that the energy in $\mathbf{y}^{(j)}$ is less than the energy in $\mathbf{y}^{(l)}$ for all $j > l$. For all energy computation, the mean is first removed. Thus,

$$[\mathbf{y}^{(j)}]^{\mathrm{T}}\mathbf{y}^{(j)} \leq [\mathbf{y}^{(l)}]^{\mathrm{T}}\mathbf{y}^{(l)}; \quad \forall j > l. \tag{7.63}$$

This step should be performed after each update of the codewords in the joint codebook design algorithm.

Justification for the improvement propositions are as follows. During sequential search or tree search each additional stage (except the first one) provides the difference vector that, when added with the codewords from previous stages, produces a better approximation to the input vector. Hence, it is expected that the codewords for successive codebooks in a MSVQ have lower and lower amplitude, or contain less and less energy. By incorporating this reordering step, the resultant codebooks are better suited for a sequential or tree search. This is another property observed in sequentially designed codebooks, as shown in Example 7.7.

Example 7.9 This example provides an overview of the properties of MSVQ designed with various algorithms. The same training data and quantizer configuration as in Example 7.7 are considered (four stages, 2 bits per stage). Comparison is made for the following schemes:

1. Design based on a sequential algorithm, same as Example 7.7.
2. Design based on a joint algorithm with randomly initialized codebook, same as Example 7.8.
3. Design based on a joint algorithm, where the improvement propositions for tree search as explained previously are incorporated. Initial codebooks are randomly generated.
4. Same as 3 with the exception that the initial codebooks are designed using the sequential algorithm.

Performance under full search is considered first. Table 7.3 shows the distance sum under full search for the codebooks designed using the four schemes. The input training data set is used to measure the distance sum. As we can see, Scheme 1 provides the worst performance since it is suboptimal for a full search. Schemes 2 and 3 give almost the same distance sum, implying that the optimization procedure for a tree search has no significant effect on a full search performance. The best result is obtained for Scheme 4, where the sequentially designed codebooks are jointly optimized.

TABLE 7.3 Performance Comparison for Four Design Schemes

Scheme	Distance Sum (Full-Search)	Distance Sum (Tree-Search)
1	5.282	5.483 ($M_a = 4$)
2	3.651	12.91 ($M_a = 16$)
3	3.817	9.407 ($M_a = 16$)
4	2.933	3.556 ($M_a = 5$)

In the same table we can appreciate the performance of the four schemes under a tree search, where the minimum distance sum, together with the minimum M_a to achieve that value of distance sum, is recorded. Note that Scheme 4 again provides the highest performance. Scheme 2 is the worst in this case since no optimization for a tree search is incorporated.

Behavior under a tree search is better revealed by plotting the distance sum as a function of M_a, shown in Figure 7.18. As we can see, Scheme 1 provides good performance for a tree search; however, it is not as good as Scheme 4. Both these approaches are relatively insensitive to changes in M_a, especially when $M_a \geq 4$. Scheme 2 is the worst; its poor performance is due to the fact that no optimization is performed for a tree search. For Schemes 1 and 4, the distance sum essentially remains constant when $M_a \geq 4$. That is, no improvement is obtainable for additional computation. Thus, good performance can be achieved at low cost.

In summary, Scheme 4 provides the best performance in both a full search and tree search.

Summary of the Joint Codebook Design Algorithm

The following is a step-by-step description of the joint codebook design algorithm. Optimization procedures for a tree search are incorporated.

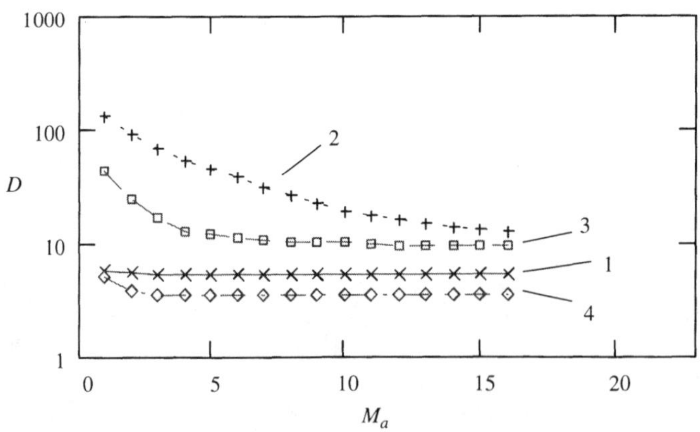

Figure 7.18 Performance comparison in a tree search for four MSVQ design schemes in an experiment.

Step 1. Begin with an initial stacked codebook $\mathbf{y}$. Set $m = 1$. The initial codebook is designed using a sequential algorithm.

Step 2. Partition: For each vector $\mathbf{x}_k$, $k = 0, \ldots, N_t - 1$, in the training set, determine the codebook indices $i_{1k}, i_{2k}, \ldots, i_{Kk}$ that minimize the distortion (full search). Keep a count on the number of times that a specific codeword is found to minimize the quantization error.

Step 3. Check for underutilized codewords. If a particular codeword has not been used after presenting the whole training set, it should be eliminated. The most frequently used codeword in the same stage is split into two codewords by addition with random numbers of small variance. Go back to Step 2 after codeword splitting. If all codewords have been utilized at least once after Step 2, proceed with Step 4.

Step 4. Compute:

$$\mathbf{v} = \sum_k \mathbf{B}_k^T \mathbf{x}_k,$$

$$\mathbf{Q} = \sum_k \mathbf{B}_k^T \mathbf{B}_k.$$

Then for $l = 1$ to K,

$$\begin{aligned}
\mathbf{Q}^{(l)} &= \mathbf{S}_l^T \mathbf{Q} \mathbf{S}_l, \\
\mathbf{v}^{(l)} &= \mathbf{S}_l^T \left(\mathbf{v} - \mathbf{Q}\mathbf{y}_o^{(l)}\right), \\
\mathbf{y}^{(l)} &\leftarrow \left[\mathbf{Q}^{(l)}\right]^{-1} \mathbf{v}^{(l)} && \text{(Update stacked stage codebook)}, \\
\mathbf{u}^{(l)} &= \frac{1}{N_l} \sum_{i=1}^{N_l} \mathbf{y}_i^{(l)} && \text{(Mean computation)}, \\
\mathbf{y}_i^{(l)} &\leftarrow \mathbf{y}_i^{(l)} - \mathbf{u}^{(l)}, \quad i = 1, \ldots, N_l && \text{(Mean removal)}, \\
e^{(l)} &= \left[\mathbf{y}^{(l)}\right]^T \mathbf{y}^{(l)} && \text{(Energy computation)}.
\end{aligned}$$

Step 5. Codebook reordering: Sort the energy $e^{(l)}$ and reorder the codebooks by reassigning the index l in such a way that

$$e^{(1)} > e^{(2)} > \cdots > e^{(K)}.$$

During reordering, the orders of the mean vectors $\mathbf{u}^{(l)}$ and stacked stage codebooks $\mathbf{y}^{(l)}$ are changed accordingly.

Step 6. Mean recovery for first stage codebook:

$$\mathbf{y}_i^{(1)} \leftarrow \mathbf{y}_i^{(1)} + \mathbf{u}^{(1)}; \quad i = 1, 2, \ldots, N_1.$$

Step 7. Compute the distance sum for $\mathbf{y}$. If it has changed by a small enough amount since the last iteration, stop. Otherwise set $m \leftarrow m + 1$ and go to Step 2. Note that distance sum is calculated using the full search procedure.

7.5 PREDICTIVE VQ

A predictive scheme can be deployed to take advantage of the correlation between consecutive vectors entering the quantizer. The basic scheme is a straightforward extension to DPCM or predictive quantization in the scalar case (Chapter 6).

The Basics of PVQ

Figure 7.19 shows the block diagrams of the encoder and decoder for PVQ and is typically applied to those sequences of vectors with a high degree of redundancy. That is, between the current vector and the past vectors, for instance, the vector elements have some degree of correlation. During encoding, the difference vector or prediction-error vector

$$\mathbf{e}[n] = \mathbf{x}[n] - \mathbf{x}_p[n] \tag{7.64}$$

is quantized. The quantized version of the input vector is given by

$$\hat{\mathbf{x}}[n] = \hat{\mathbf{e}}[n] + \mathbf{x}_p[n], \tag{7.65}$$

where $\hat{\mathbf{e}}[n]$ is the quantized difference vector and $\mathbf{x}_p[n]$ the prediction vector. Prediction is usually based on a small number of past quantized results. On the decoder side, the index $i[n]$ is used to recover the quantized difference vector, which is combined with the prediction vector to form the quantization result.

Design Methodologies with Known Predictor

In PVQ design, it is necessary to determine the predictor as well as the VQ codebook for the population of prediction-error vectors. Several techniques for VQ

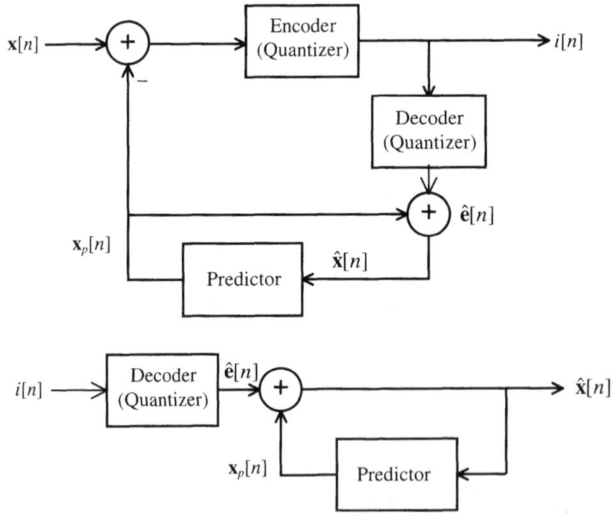

Figure 7.19 PVQ encoder (*top*) and decoder (*bottom*).

codebook design are explained here; in all these techniques, the predictor is assumed to be known and fixed.

Open-Loop Method

Given the set of training vectors, and by knowing the predictor, a corresponding set of prediction-error vectors can be generated, with the assumption of no quantization. These vectors, in turn, can be used to design a VQ using the GLA. Combining the predictor with the VQ yields a complete PVQ system.

Closed-Loop Method

It is possible to modify the GLA so as to produce a PVQ design technique as follows:

Step 1. Begin with the initial codebook $\boldsymbol{Y}_1$. Set $m = 1$. The initial codebook can be designed using the open-loop method explained before.

Step 2. Encode the set of training vectors using $\boldsymbol{Y}_m$; perform Lloyd iteration to generate the improved codebook $\boldsymbol{Y}_{m+1}$.

Step 3. Compute the average distortion for $\boldsymbol{Y}_{m+1}$. If it has changed by a small enough amount since the last iteration, stop. Otherwise set $m + 1 \rightarrow m$ and go to Step 2.

For the specified procedure, it is assumed that a set of training vectors is available; also, an initial VQ codebook exists ($\boldsymbol{Y}_1$), and the predictor is known. Note that the training sequence $\mathbf{x}[n]$ is used in each iteration to generate a prediction-error sequence $\mathbf{e}[n]$, which depends on the previous codebook. The sequence $\mathbf{e}[n]$ is used in the design of the next codebook. That is, an attempt is made to design a codebook from a prediction-error sequence that depends on the previous codebook; implying that the training sequence on which the Lloyd iteration is based changes constantly; therefore, the procedure is suboptimal and the distortion is not guaranteed to decrease monotonically. In practice, however, it has been found to be an effective algorithm and a substantial coding gain can be achieved over the open-loop design approach.

It is assumed in the described procedure that a single Lloyd iteration step is applied after completing each encoding step. In actual implementation, multiple iterations can be executed at Step 2 before encoding the training vectors again. This latter approach might lead to faster convergence.

PVQ-MA

By utilizing the principle of MA prediction, higher robustness against channel error is achieved (see Chapter 6). Figure 7.20 shows the block diagrams of the system, referred to as PVQ-MA. We consider the design of PVQ-MA, which consists of finding the optimal VQ codebook and the predictor, with the predictor being the linear type with prediction order equal to M.

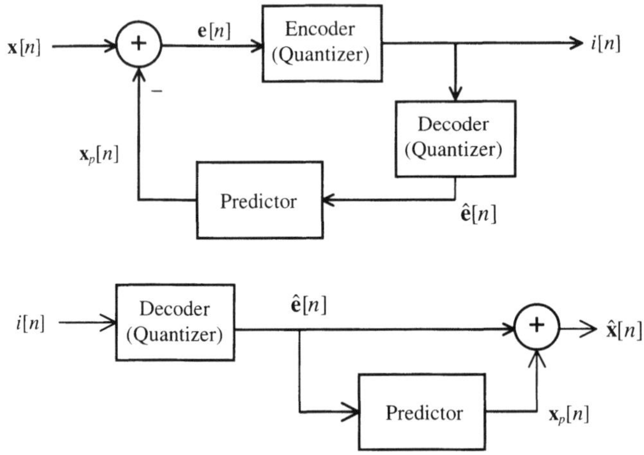

Figure 7.20 PVQ-MA encoder (*top*) and decoder (*bottom*).

The design procedure can be separated into two parts: VQ codebook and predictor. When the VQ codebook is updated, the predictor is left intact and vice versa. The actual procedure is specified below.

Step 1. Start with the initial codebook $\boldsymbol{Y}_1$, and the initial predictor $\mathbf{b}_1 = [b_{1,1}, b_{1,2}, \ldots, b_{1,M}]^T$, with $b_{1,i}$, $i = 1$ to M, being the MA coefficients. Set $m = 1$.

Step 2. Encode the set of training vectors using $\boldsymbol{Y}_m$ and $\mathbf{b}_m$; perform Lloyd iteration to generate the improved codebook $\boldsymbol{Y}_{m+1}$.

Step 3. Encode the set of training vectors using $\boldsymbol{Y}_{m+1}$ and $\mathbf{b}_m$; find the new set of MA coefficients $\mathbf{b}_{m+1}$.

Step 4. Compute the average distortion for $\boldsymbol{Y}_{m+1}$ and $\mathbf{b}_{m+1}$. If it has changed by a small enough amount since the last iteration, stop. Otherwise set $m + 1 \rightarrow m$ and go to Step 2.

Both the VQ codebook and the predictor can be randomly initialized. In fact, the initial MA coefficients can be set to zero; in that case, the prediction is always zero and hence the prediction error is equal to the input. In Step 2, after encoding the input training vectors, a set of prediction-error vectors is generated, used by the GLA for codebook generation.

After updating the VQ codebook, the set of input vectors is encoded using the new codebook and optimization is performed on the MA coefficients. The objective is to minimize

$$\begin{aligned} D &= \sum_n \|\mathbf{x}[n] - \hat{\mathbf{x}}[n]\|^2 \\ &= \sum_n \left\| \mathbf{x}[n] - \left(\hat{\mathbf{e}}[n] + \sum_{i=1}^{M} b_i \, \hat{\mathbf{e}}[n-i] \right) \right\|^2 \end{aligned} \tag{7.66}$$

by selecting the appropriate MA coefficients $\mathbf{b}$, with the range of n covering the whole training data set. Differentiating (7.66) with respect to b_k, $k = 1, 2, \ldots, M$, gives

$$\frac{\partial D}{\partial b_k} = -2\sum_n \left(\hat{\mathbf{e}}[n-k]^T(\mathbf{x}[n] - \hat{\mathbf{e}}[n]) - \sum_{i=1}^{M} b_i\, \hat{\mathbf{e}}[n-k]^T \hat{\mathbf{e}}[n-i] \right). \tag{7.67}$$

Equating (7.67) to zero leads to

$$\sum_{i=1}^{M} b_i \sum_n \hat{\mathbf{e}}[n-k]^T \hat{\mathbf{e}}[n-i] = \sum_n \hat{\mathbf{e}}[n-k]^T(\mathbf{x}[n] - \hat{\mathbf{e}}[n]); \quad k = 1, 2, \ldots, M, \tag{7.68}$$

which is a set of M linear equations. Written in matrix form, we have

$$\mathbf{A} \cdot \mathbf{b} = \mathbf{c}, \tag{7.69}$$

where

$$\mathbf{A} = \begin{pmatrix} \sum_n \|\hat{\mathbf{e}}[n-1]\|^2 & \sum_n \hat{\mathbf{e}}[n-1]^T \hat{\mathbf{e}}[n-2] & \cdots & \sum_n \hat{\mathbf{e}}[n-1]^T \hat{\mathbf{e}}[n-M] \\ \sum_n \hat{\mathbf{e}}[n-2]^T \hat{\mathbf{e}}[n-1] & \sum_n \|\hat{\mathbf{e}}[n-2]\|^2 & \cdots & \sum_n \hat{\mathbf{e}}[n-2]^T \hat{\mathbf{e}}[n-M] \\ \vdots & \vdots & \ddots & \vdots \\ \sum_n \hat{\mathbf{e}}[n-M]^T \hat{\mathbf{e}}[n-1] & \sum_n \hat{\mathbf{e}}[n-M]^T \hat{\mathbf{e}}[n-2] & \cdots & \sum_n \|\hat{\mathbf{e}}[n-M]\|^2 \end{pmatrix}, \tag{7.70}$$

$$\mathbf{b} = [b_1 b_2 \ldots b_M]^T, \tag{7.71}$$

$$\mathbf{c} = \Bigg[\sum_n \hat{\mathbf{e}}[n-1]^T(\mathbf{x}[n] - \hat{\mathbf{e}}[n]) \quad \sum_n \hat{\mathbf{e}}[n-2]^T(\mathbf{x}[n] - \hat{\mathbf{e}}[n]) \quad \cdots \\ \times \sum_n \hat{\mathbf{e}}[n-M]^T(\mathbf{x}[n] - \hat{\mathbf{e}}[n]) \Bigg]^T. \tag{7.72}$$

Solving (7.69) leads to the set of optimal MA coefficients.

7.6 OTHER STRUCTURED SCHEMES

As shown in previous sections, by imposing a certain structure on the basic VQ framework, it is possible to obtain schemes that are highly efficient, with some quality degradation. Deterioration in quality is often acceptable in practice since VQ schemes are in general far superior to scalar quantization at similar resolution. In this section, we study two additional structured techniques widely applied in practice that offer similar advantages as MSVQ.

Split VQ

This method is also known as partitioned VQ. The simplest and most direct way to reduce the search and storage complexity in coding a high-dimensional vector is simply to split the vector into two or more subvectors. In the case of a two splitting, the input vector

$$\mathbf{x} = [x_1, x_2, \ldots, x_M]^T$$

is partitioned into

$$\mathbf{x}_a = [x_1, x_2, \ldots, x_K]^T,$$

a K-dimensional vector with $K < M$ and

$$\mathbf{x}_b = [x_{K+1}, x_{K+2}, \ldots, x_M]^T$$

a vector of dimension $M - K$.

Assuming squared-error measure, we have

$$\|\mathbf{x} - \hat{\mathbf{x}}\|^2 = \|\mathbf{x}_a - \hat{\mathbf{x}}_a\|^2 + \|\mathbf{x}_b - \hat{\mathbf{x}}_b\|^2. \tag{7.73}$$

Thus, the encoder must simply find the nearest codevector to $\mathbf{x}_a$ in the codebook Y_a and, as a separate and independent task, the nearest codevector to $\mathbf{x}_b$ in the codebook Y_b. Furthermore, the optimal codebook design procedure is to derive two separate training vector sets of K- and $(M - K)$-dimensional vectors to generate Y_a and Y_b, respectively.

Conjugate VQ

In the transmission of VQ indices through noisy channels, a bit error can cause severe distortion to the decoded output since a completely different output vector is decoded. One remedy is to use several codebooks for quantization purposes, where the decoded codevector is obtained by a combination of vectors from various codebooks, addressed by different indices. In this way, if one index is contaminated, the final combination of vectors from various codebooks might still be near to the original one, alleviating the damage caused by a mistaken index.

The use of two codebooks having the same dimensions as the input vector is an effective way to reduce the quantization distortion for a noisy channel. The scheme is named conjugate VQ since the relationship of the two codebooks is "conjugate" in some sense. Assume, for instance, that the output vector is obtained as the sum of the codevectors from the two codebooks. If one of the codevectors has channel errors, then although it may be quite different from the intended codevector, the sum of the two codevectors is not so different from the desired output vector. Hence, the effects of bit errors in the complete vector are reduced.

The conjugate VQ scheme is essentially a two-stage MSVQ under full search. Thus, it is possible to utilize the same training algorithm to find the codebook elements for the case of Euclidean distance measure. The conjugate structure is a suboptimal scheme; its advantages are robustness against channel errors and lower memory cost. However, under error-free conditions, its output has lower quality than a single codebook at the same equivalent size. This is expected since at the same size of N, a single codebook requires N codevectors, while for two codebooks the total number of codevectors is $2\sqrt{N}$ (each has the same size of $\sqrt{N}$), leading to a reduced capability for information storage.

7.7 SUMMARY AND REFERENCES

Vector quantization is introduced in this chapter, where many basic concepts are mere extensions of the scalar counterpart. Conditions for optimal quantization are given, which serve as the foundation for the development of quantizer design algorithms. The generalized Lloyd algorithm is then derived for quantizer optimization; stochastic relaxation is introduced as an alternative to avoid poor local minimums. The price to pay for this benefit is additional computational burden.

Unconstrained VQ is relatively complex to implement in practice due to the required memory and computational costs. For cost reduction, structure is added to the codebook, with MSVQ being one of the most versatile strategies. It is shown that substantial saving in memory and computation can be obtained, with some performance degradation that can be tolerated in many practical circumstances. Several design algorithms and search procedures for MSVQ are then presented and their properties are illustrated through actual design situations. As demonstrated, performance level of MSVQ under a tree search with a relatively small M_a can actually be close to a full search when the codebooks are designed using the appropriate scheme; that is, by applying a sequential algorithm first and then optimizing the resultant codebooks using a joint algorithm. Thus, low computational complexity is obtainable with a tree search, with a small penalty in performance.

PVQ explores the redundancy present among the input vectors so as to improve the performance of the quantization system. The technique has been applied successfully for the quantization of line spectral frequencies, covered in Chapter 15. See Chang and Gray [1986] for gradient algorithms in PVQ design. Two additional structured schemes are introduced: the split VQ method separates a high-dimensional vector into two or more lower-dimensional vectors and quantizes them separately, and the conjugate VQ technique utilizes two codebooks with the goal of increasing robustness against channel errors.

Many theoretical issues concerning VQ can be found in Gersho and Gray [1995], which also contains discussions of many other structured VQ schemes. Additional theoretical framework for MSVQ can be found in Chan et al. [1992]. Application of MSVQ to the quantization of linear prediction coefficients is found in LeBlanc et al. [1993]. See Paliwal and Atal [1993] for split VQ of linear prediction coefficients. The idea of conjugate VQ was proposed by Moriya [1992]; it was used later for the design of a conjugate-structure code-excited linear prediction (CS-CELP) speech coder [Kataoka et al., 1996] and ultimately led to the ITU-T G.729 CS-ACELP standard [Salami et al., 1998], covered in Chapter 16.

Unlike scalar quantization, a transmission error in VQ can cause serious errors at the decoder side, since a totally different vector is recovered. It is possible to optimize the index assignment so as to improve performance under channel errors. See Hedelin et al. [1995a, 1995b] for the theoretical framework and practical solutions. The problem of achieving global optimality in VQ design is still a challenging problem. The reader is referred to the latest literature for additional information.

EXERCISES

7.1 Consider a 2-D VQ, where each component of the input vector $\mathbf{X} = [\mathbf{x}_1, \mathbf{x}_2]^T$ is quantized using an identical uniform scalar quantizer. If $0 \leq \mathbf{x}_i \leq 1$, plot the cells and codewords in the 2-D plane when the size of the scalar quantizers is equal to four.

7.2 Prove that when the nearest-neighbor condition for optimality (7.12) is satisfied, the expected distortion is minimized. *Hint*: The proof is similar to the scalar quantization case.

7.3 Prove that when the centroid condition for optimality (7.14) is satisfied, the expected distortion is minimized. *Hint*: The proof is similar to the scalar quantization case.

7.4 Prove the centroid relation under squared Euclidean distance measure (7.21). *Hint*: From the expression of average distortion differentiate with respect to the codeword.

7.5 *Split VQ Design Algorithm*. This is an alternative method for quantizer design and is outlined as follows:

Step 1. Begin with a codebook of size one. The codeword is found as the centroid of the training data set.

Step 2. Codeword splitting: Each codeword in the codebook is split into two by addition with a low-norm vector. That is, if $\mathbf{y}$ is the original codeword, two codewords are created with $\mathbf{y} + \mathbf{u}$ and $\mathbf{y} - \mathbf{u}$, with $\mathbf{u}$ a low-norm vector ($\mathbf{u}$ can be fixed or randomly generated). Size of the quantizer is now doubled.

Step 3. Run the GLA over the newly found codebook until convergence.

Step 4. Go back to Step 2 if the target codebook size has not been reached; otherwise stop.

Implement the split algorithm and compare its performance with the GLA using a set of randomly generated 2-D training vectors.

7.6 The stochastic relaxation algorithm presented was based on random perturbation of the training vector set. Without perturbing the training vector set, develop a scheme where the codebook is perturbed. Implement the method and make a performance comparison. Is there any advantage on using this scheme?

7.7 For a two-stage MSVQ, memory cost is given by

$$MC = N_1 + N_2 = 2^{r1} + 2^{r2}.$$

Assume that

$$r_1 + r_2 = r$$

with r constant. Prove that MC is minimum when

$$r_1 = r_2 = r/2.$$

Extend the proof to a higher number of stages.

7.8 We are given the following configurations of vector quantizers, all having the same resolution of 10 bits:

Unconstrained VQ	$r = 10$
Two-stage MSVQ	$(r_1, r_2) = (3, 7)$
Two-stage MSVQ	$(r_1, r_2) = (5, 5)$
Three-stage MSVQ	$(r_1, r_2, r_3) = (2, 2, 6)$
Three-stage MSVQ	$(r_1, r_2, r_3) = (3, 3, 4)$
Four-stage MSVQ	$(r_1, r_2, r_3, r_4) = (1, 1, 1, 7)$
Four-stage MSVQ	$(r_1, r_2, r_3, r_4) = (2, 2, 3, 3)$

Calculate the memory cost for each configuration. Sort the vector quantizers in terms of relative performance under a full search. Explain your results.

7.9 In the expression for the computational cost (CC) for a tree search, assume that the minimum operator min(.,.) always returns M_a. How can we rewrite CC?

7.10 *Iterative Sequential Search.* The iterative sequential search procedure begins with the indices provided by the sequential search method and reoptimizes each index assuming all other indices are fixed. Index i_{oj} is determined to satisfy

$$d\left(\mathbf{x}, \sum_{l=1,l\neq j}^{K} \mathbf{y}_{i_{ol}}^{(l)} + \mathbf{y}_{i_{oj}}^{(j)}\right) \leq d\left(\mathbf{x}, \sum_{l=1,l\neq j}^{K} \mathbf{y}_{i_{ol}}^{(l)} + \mathbf{y}_{i_j}^{(j)}\right); \quad \forall i_j \neq i_{oj}, \quad j = 1, \ldots, K.$$

Design an MSVQ codebook using a training vector set. Compare the performance of the sequential search and iterative sequential search. Is there any significant improvement between the two methods?

7.11 *Iterative Sequential Algorithm for Codebook Design.* For the iterative sequential algorithm, the training set for the lth stage of a K-stage MSVQ is given by

$$\mathbf{x}_k^{(l)} = \mathbf{x}_k - \sum_{i=1, i \neq l}^{K} Q_i(\mathbf{x}_k^{(i)}).$$

In this approach, all other stages are assumed fixed and known, and the technique is to reoptimize the lth stage. Once an initial set of codebooks is obtained, each stage can be optimized given the other stages. Thus, the final error is minimized rather than the error after each stage.

Implement the algorithm and compare its performance with respect to the sequential algorithm.

7.12 *Squared Euclidean Distance Measure with Weighting.* Consider the distance measure given by

$$d(\mathbf{x}, \hat{\mathbf{x}}) = (\mathbf{x} - \hat{\mathbf{x}})^T \mathbf{W} (\mathbf{x} - \hat{\mathbf{x}}),$$

where $\mathbf{W}$ is a diagonal matrix that may depend on $\mathbf{x}$. $\mathbf{W}$ is a weighting matrix that is incorporated so as to control the contribution of each vector component to the final measure. If $\mathbf{W} = \mathbf{I}$ (identity matrix), it reduces to the squared Euclidean distance measure.

Derive the joint codebook design algorithm using this distance measure by minimizing the distance sum

$$D = \sum_k (\mathbf{x}_k - \hat{\mathbf{x}}_k)^T \mathbf{W}_k (\mathbf{x}_k - \hat{\mathbf{x}}_k).$$

7.13 Translate the design algorithm for PVQ-MA, where the codebook and the predictor are updated sequentially, to the case of PVQ. Write down the steps and equations involved.

7.14 Consider a split VQ scheme where two codebooks are employed. The first codebook has size N_1 with K-dimensional codevectors and the second codebook has size N_2 with $(M - K)$-dimensional codevectors. What is the overall size of the resultant scheme? Find the memory cost in terms of the required storage space for each vector element and compare to that of the optimal case.

7.15 Consider the same split VQ scheme as described in Exercise 7.14, where each subvector (after splitting) is quantized by searching each codebook exhaustively in an independent manner. Show that the corresponding computational complexity expressions for distance computation over all codevectors are

$$\#\text{sums} = N_1(2K - 1) + N_2(2(M - K) - 1),$$
$$\#\text{squares} = N_1 K + N_2(M - K)$$

where #squares is the total number of square operations $(\cdot)^2$, and #sums includes addition and subtraction. For $N_1 = 2^{12}$, $N_2 = 2^{12}$, $M = 10$, and $K = 4$, evaluate #sums and #squares. Compare the results with the case of a full search with a codebook size of $N_1 N_2$.

7.16 Consider the conjugate VQ scheme where the codebooks Y_1 and Y_2 contain the codevectors $\mathbf{y}_i^{(1)}$ and $\mathbf{y}_i^{(2)}$, respectively, with $i = 1, \ldots, N_o$; that is, the two codebooks have the same size of N_o. Under Euclidean distance measure, the quantized vector is

$$\hat{\mathbf{x}} = \mathbf{y}_i^{(1)} + \mathbf{y}_j^{(2)},$$

where i and j are selected in such a way that

$$d = \|\mathbf{x} - \hat{\mathbf{x}}\|^2 = \|\mathbf{x} - (\mathbf{y}_i^{(1)} + \mathbf{y}_j^{(2)})\|^2 \tag{7.74}$$

is minimized for all combinations of i and j.

(a) What are the computational costs when (7.74) is evaluated directly? The answers should be expressed as #sums and #squares (see Exercise 7.15), and in terms of N_o (codebook size) and M (vector dimension).

(b) Find the memory cost involved in storing the two codebooks. Give the answer in terms of the number of vector elements.

(c) Show that (7.74) is equivalent to

$$d = \|\mathbf{x}\|^2 - 2\mathbf{x}^T\left(\mathbf{y}_i^{(1)} + \mathbf{y}_j^{(2)}\right) + \left\|\mathbf{y}_i^{(1)} + \mathbf{y}_j^{(2)}\right\|^2. \tag{7.75}$$

Consider the alternative distance expression

$$d' = -2\mathbf{x}^T(\mathbf{y}_i^{(1)} + \mathbf{y}_j^{(2)}) + \|\mathbf{y}_i^{(1)} + \mathbf{y}_j^{(2)}\|^2. \tag{7.76}$$

Can we use (7.76) for codebook search? Justify your answer.

(d) Consider an encoding scheme based on (7.76), where $\|\mathbf{y}_i^{(1)} + \mathbf{y}_j^{(2)}\|^2$, $i,j = 1$ to N_o, are precomputed and stored. Find the memory cost and computational cost for this scheme.

(e) Consider an encoding scheme based on (7.76), where $\|\mathbf{y}_i^{(1)} + \mathbf{y}_j^{(2)}\|^2$ and $\mathbf{y}_i^{(1)} + \mathbf{y}_j^{(2)}$, $i,j = 1$ to N_o, are precomputed and stored. Find the memory cost and computational cost for this scheme. Note that there is no need to store $\mathbf{y}_i^{(1)}$ and $\mathbf{y}_j^{(2)}$.

(f) With $N_o = 2^6$ and $M = 10$, compare the implementational costs of (a) and (b), with (d) and (e). Which scheme would you choose?

7.17 Moriya (1992) proposed the use of the distance measure

$$d = \mu \left\| \mathbf{x} - \left(\mathbf{y}_i^{(1)} + \mathbf{y}_j^{(2)} \right) \right\|^2 + (1 - \mu) \left\{ \left\| \mathbf{x} - \frac{\mathbf{y}_i^{(1)}}{2} \right\|^2 + \left\| \mathbf{x} - \frac{\mathbf{y}_j^{(2)}}{2} \right\|^2 \right\} \tag{7.77}$$

for codebook design in conjugate VQ, where $0 \leq \mu \leq 1$ is a constant known as the redundancy control parameter. What happens when $\mu = 1$? And $\mu = 0$? How do we choose μ if higher robustness against channel errors is desired? How do we choose μ if lower distortion under error-free conditions is preferred?

CHAPTER 8

SCALAR QUANTIZATION OF LINEAR PREDICTION COEFFICIENT

Most LP-based speech coding algorithms transmit the LPC as information on the signal frames. Before transmission can take place, the LPC must be quantized. As in any quantization scheme, it is a requirement to reduce the distortion as much as possible. In the case of the LPC, it is also required that the synthesis filter remain stable after quantization of the respective coefficients. Direct scalar quantization of the LPC is usually not done since small quantization errors in the individual coefficients can produce relatively large spectral errors and can lead to instability for the synthesis filter. Because of these problems, the LPC is usually transformed to another representation so that stability of the synthesis filter is ensured after quantization.

In this chapter, the distance measure known as spectral distortion is defined, which is a widely used parameter to quantify the performance of a given LPC quantizer. Alternative representations of LPC, including reflection coefficient (RC), log area ratio (LAR), and line spectral frequency (LSF), are introduced. These representations are used in many coding standards for quantization purposes. Interpolation of the LPC is described at the end of the chapter. Note that only scalar quantizers are discussed here, and they are generally the technique of choice for coders standardized prior to 1994. For standards adopted after that year, some form of vector quantization is normally incorporated due to its superior performance. VQ of the LPC is discussed in Chapter 15.

8.1 SPECTRAL DISTORTION

In this section, the commonly accepted objective performance measure in quantization of the LPC, known as spectral distortion (SD), is defined. Based on SD, the requirements for transparent quantization are specified. A spectral sensitivity function is given with the purpose of measuring the magnitude change in spectral distortion due to variation of one of the defining parameter of the PSD.

Defining Spectral Distortion

Given two sets of LPCs $a_{1,i}$ and $a_{2,i}, i = 1, 2, \ldots, M$, we define the polynomials

$$A_1(z) = 1 + \sum_{i=1}^{M} a_{1,i} z^{-i}, \tag{8.1}$$

$$A_2(z) = 1 + \sum_{i=1}^{M} a_{2,i} z^{-i}, \tag{8.2}$$

where M is the prediction order. The above polynomials represent the system functions of two AR process analyzers having different parameters (Chapter 3). Thus, the PSDs of the two AR signals represented above, assuming unit variance input white noise, are given by

$$S_1(e^{j\omega}) = 1/|A_1(e^{j\omega})|^2, \tag{8.3}$$

$$S_2(e^{j\omega}) = 1/|A_2(e^{j\omega})|^2. \tag{8.4}$$

The spectral distortion between S_1 and S_2 is defined by

$$\begin{aligned} SD^2 &= \frac{1}{2\pi}\int_0^{2\pi} [10\log_{10}(S_1(e^{j\omega})) - 10\log_{10}(S_2(e^{j\omega}))]^2 d\omega \\ &= \frac{1}{2\pi}\int_0^{2\pi}\left[10\log_{10}\left(\frac{|A_2(e^{j\omega})|^2}{|A_1(e^{j\omega})|^2}\right)\right]^2 d\omega, \end{aligned} \tag{8.5}$$

where the result is directly expressed in decibels. Equivalently, if f_s is the sampling frequency, we have

$$SD^2 = \frac{1}{f_s}\int_0^{f_s} [10\log_{10}(S_1(e^{j2\pi f})) - 10\log_{10}(S_2(e^{j2\pi f}))]^2 df. \tag{8.6}$$

The integrals in (8.5) and (8.6) can be approximated in practice by sampling the PSDs using N points:

$$SD^2 = \frac{1}{n_1 - n_0}\sum_{n=n_0}^{n_1-1} [10\log_{10}(S_1(e^{j2\pi n/N})) - 10\log_{10}(S_2(e^{j2\pi n/N}))]^2, \tag{8.7}$$

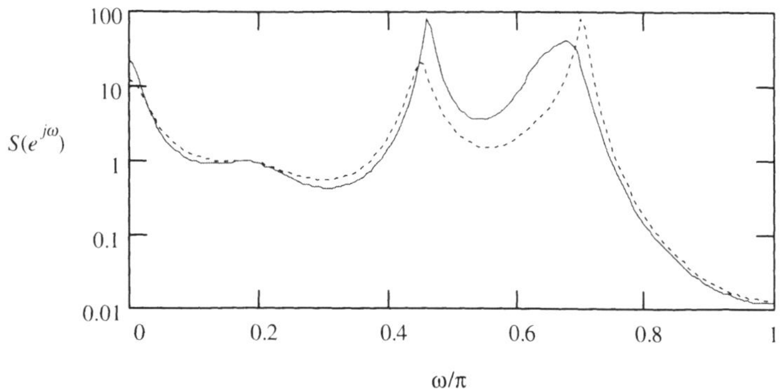

Figure 8.1 Plot of two PSDs corresponding to two different sets of LPCs.

where $0 \le n_0 < n_1 \le N$. Typically, for $f_s = 8$ kHz, $n_0 = 4, n_1 = 100$, and $N = 256$, so that only the spectrum values between 125 Hz and approximately 3.1 kHz are taken into account for the computation of SD. Thus, only the most perceptually sensitive part of the spectrum is considered.

Example 8.1 The following two sets of LPCs are used to illustrate the concept of SD:

$$\begin{array}{lllll} a_{1,1} = -0.6, & a_{1,2} = 1.205, & a_{1,3} = -1.588, & a_{1,4} = 1.153, & a_{1,5} = -1.427, \\ a_{1,6} = 1.018, & a_{1,7} = -0.536, & a_{1,8} = 0.352, & a_{1,9} = -0.314, & a_{1,10} = -0.055, \\ a_{2,1} = -0.56, & a_{2,2} = 1, & a_{2,3} = -1.4, & a_{2,4} = 1.153, & a_{2,5} = -1.427, \\ a_{2,6} = 1, & a_{2,7} = -0.536, & a_{2,8} = 0.352, & a_{2,9} = -0.3, & a_{2,10} = 0. \end{array}$$

The PSDs of the two sets of LPCs are plotted in Figure 8.1, where we can see that even though they have similar shapes, they are not the same. The spectral distortion, according to (8.5), is equal to 3.03 dB (the integral can be solved directly using numerical methods; in the present case, it is solved using Mathcad [MathSoft, 2001]). To evaluate the validity of (8.7), n_0 is set at zero while $n_1 = N$. Figure 8.2 shows the SD results from (8.7) when N changes from 100 to 300. Note how the values of SD from the sum converge toward the exact value of 3.03 dB as the number of frequency sampling points increases.

Requirements for Transparent Quantization of LPC

By "transparent" it means that the LPC quantization does not introduce any perceptible audible distortion in the coded speech; that is, two versions of coded speech—the one obtained by using an unquantized LPC and the other by using a quantized LPC—are indistinguishable through listening.

The average spectral distortion has been used extensively to measure the performance of LPC quantizers. For instance, an average SD of 1 dB has been found as a satisfactory limit to establish transparent quantization. That is, if the average SD is

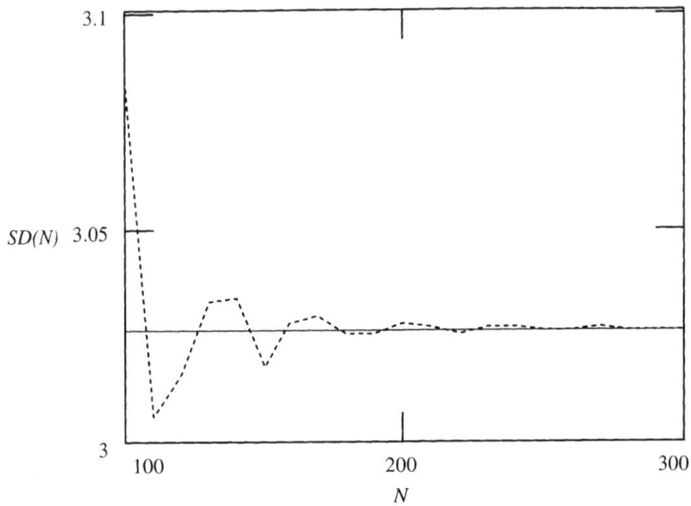

Figure 8.2 Plot of spectral distortion as a function of the number of frequency samples.

less than 1 dB, the quantizer is said to have achieved transparent quantization. In practice, however, it has been observed that too many outlier frames in the speech utterance having large SD can cause audible distortion, even though the average SD is 1 dB. Therefore, it is highly desirable to reduce both the average SD as well as the number of outlier frames. The following conditions are good guidelines as requirements for transparent quantization [Paliwal and Atal, 1993]:

- An average spectral distortion of less than 1 dB.
- Less than 2% outliers having a spectral distortion above 2 dB.
- No outliers with spectral distortion larger than 4 dB.

Note, however, that SD does not account for the frequency-domain or time-domain masking effects of the human auditory system. Therefore, it should be considered as an objective quality, which might not totally correlate with subjective evaluation results.

Spectral Sensitivity Function

Consider the power spectral density denoted by $S(e^{j\omega}, x)$, which is a function of the frequency ω as well as the parameter x. The parameter x controls the PSD; in the present case it can be one of the LPCs or other related variable. To understand the impact of variation of one parameter on spectral distortion, the function

$$\psi(x_o) = \left| \lim_{\Delta x \to 0} \frac{\mathrm{SD}(x_o, x_o + \Delta x)}{\Delta x} \right|, \tag{8.8}$$

known as the spectral sensitivity function, is defined. Spectral distortion (SD) in the above equation is written as

$$SD^2(x_o, x_o + \Delta x) = \frac{1}{2\pi}\int_0^{2\pi} (10\log_{10}(S(e^{j\omega}, x_o)) - 10\log_{10}(S(e^{j\omega}, x_o + \Delta x)))^2 d\omega. \tag{8.9}$$

Thus, the spectral sensitivity function measures the magnitude of change in spectral distortion due to variation of one of the defining parameters of the PSD.

Example 8.2 Spectral sensitivity with respect to the reflection coefficients is illustrated for a typical case. The following set of RCs is used:

$$\begin{array}{lllll} k_1 = 0.639, & k_2 = 0.0667, & k_3 = 0.232, & k_4 = 0.147, & k_5 = 0.294, \\ k_6 = 0.106, & k_7 = -0.123, & k_8 = -0.182, & k_9 = -0.153, & k_{10} = -0.0748. \end{array}$$

Spectral sensitivity with respect to k_1 is considered first. Since the synthesis filter is stable only when the magnitudes of the RCs are less than one, the range considered is $(-1, 1)$. To compute the spectral sensitivity values, k_1 is modified with the rest of the coefficients intact. In the present case, 19 different values of k_1 are utilized, with $k_1 = -0.9, -0.8, \ldots, 0.9$. As shown in (8.8), a small enough step Δk_1 must be selected. We use $\Delta k_1 = k_1/100$, and when $k_1 = 0, \Delta k_1 = 0.001$. To compute spectral sensitivity at $k_1 = -0.9$, for instance, the set of RCs (with $k_1 = -0.9$) is first transformed to LPCs (Chapter 4); then with $k_1 = -0.9 - 0.009 = -0.909$, the new set of RCs is again transformed to LPCs. The two sets of LPCs are then plugged into (8.8) to compute the spectral sensitivity value. This step is repeated for every value of k_1. Figure 8.3 shows the spectral sensitivity results with respect to k_1, and the procedure is repeated for k_5 as well. As we can see, sensitivity is much higher when the magnitude of the RC approaches one. This fact is used for quantizer design later on.

Prototype Spectral Sensitivity Curve with Respect to RC

After experimenting with a large amount of speech data, Viswanathan and Makhoul (1975) found that the spectral sensitivity curves with respect to RC possess a similar shape, where high sensitivity is developed when the magnitudes of the coefficients are close to one. Furthermore, the shapes of the curves are similar regardless of the order of the coefficients. By averaging a large number of sensitivity curves, it was found that the resultant curve can be reasonably approximated with

$$\psi(k) = \frac{1}{c(1 - k^2)}, \tag{8.10}$$

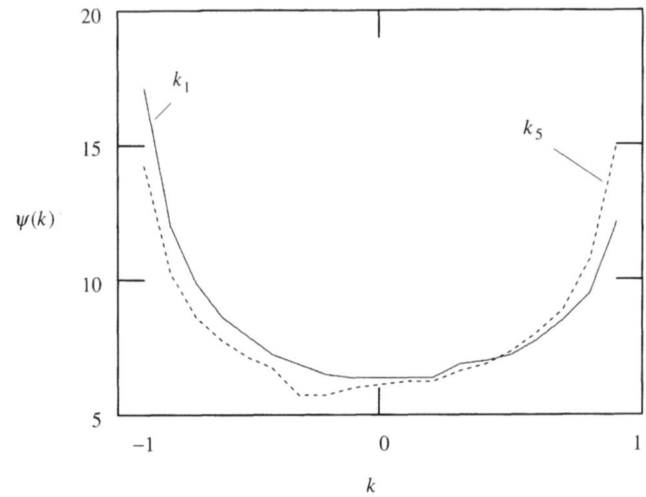

Figure 8.3 Examples of spectral sensitivity curves with respect to reflection coefficients.

where c is a constant. The above equation is known as the prototype spectral sensitivity curve with respect to RC. Figure 8.4 plots the curve when $c = 1$.

8.2 QUANTIZATION BASED ON REFLECTION COEFFICIENT AND LOG AREA RATIO

In scalar quantization of the LPC, each parameter or its alternative form is quantized independently of each other. From the definition of SD, we see that it is a

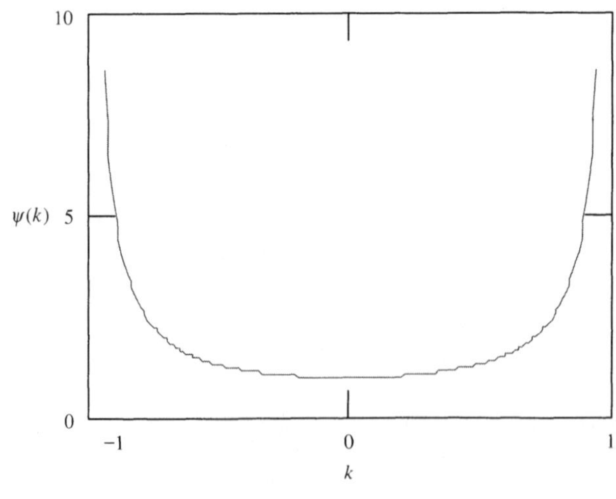

Figure 8.4 Prototype spectral sensitivity curve with respect to reflection coefficients.

function of the entire set of parameters. Thus, scalar quantization applied to the LPC is suboptimal with respect to SD, in the sense that SD is not minimized in the best possible way when each parameter is quantized separately. Due to its simplicity, however, scalar quantization is used in various speech coding standards.

In a scalar quantization scheme, the coefficient to be quantized is compared with a codebook where the quantized values, or codewords, are stored. The codeword closest to the coefficient is selected as the quantized coefficient, with the distance measure given by the squared difference. Thus, as long as the codebook sizes for various coefficients increase, the average spectral distortion will decrease. In practice, by using a set of test data, it is possible to measure the performances of the quantizers at different resolutions and select the ones that provide low enough average SD to meet the requirements of transparent quantization. To design the codebooks, one can rely on the Lloyd algorithm (Chapter 5) and use a set of training data to minimize the sum of squared error. Note, however, that the procedure does not directly minimize the average SD.

The reflection coefficient is a good candidate for quantization, since stability of the synthesis filter can easily be verified from the magnitude. Thus, to design an RC-based LPC scalar quantizer, a set of training coefficients obtained from a speech database is used. The Lloyd algorithm is applied so as to obtain the resultant codebooks.

The problem with the design procedure is that spectral sensitivity functions for the RC tend to have large values when the magnitudes of the coefficients are close to one (Example 8.2). This will lead to problems when magnitudes of the coefficients approach one, making the attempt to satisfy the transparent quantization requirements difficult. What is needed is a quantizer that is much more sensitive (smaller step size) when the magnitudes of the coefficients are near one than when they are near zero. That is, the transfer characteristic of the quantizer is nonuniform. The use of a quantizer having nonuniform characteristic is equivalent to uniform quantization of a transformed coefficient through a nonlinear function; that is,

$$g = f(k), \tag{8.11}$$

where k represents the value of the reflection coefficient, $f(\cdot)$ is a nonlinear function, and g represents the value of the transformed coefficient. Figure 8.5 illustrates the necessary shape of the nonlinear function. Note that uniform quantization intervals on the vertical axis (g) create nonuniform intervals on the horizontal axis (k); with the horizontal intervals shortened as the magnitude of the input coefficient approaches one. Hence, the quantizer is more sensitive for reflection coefficients having magnitudes close to one.

Viswanathan and Makhoul (1975) proposed the function

$$f(k) = \log\left(\frac{1+k}{1-k}\right), \tag{8.12}$$

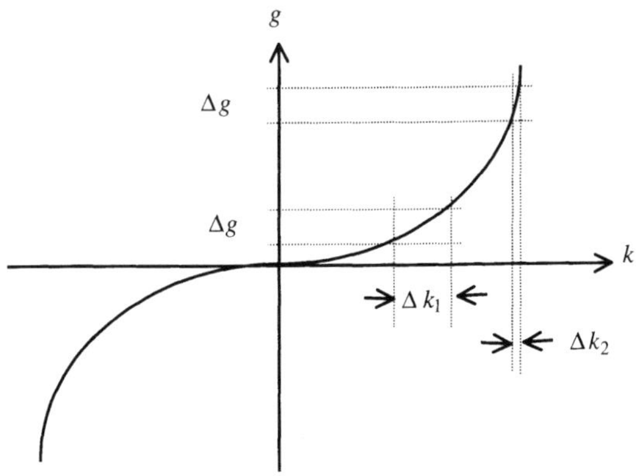

Figure 8.5 Illustration of a nonlinear mapping applied to reflection coefficients.

which is plotted in Figure 8.6. As we can see, the function satisfies the quantization requirements. The RC transformed by (8.12) is known as the log area ratio (LAR). The term of area ratio comes from the association of the RC with the acoustic tube model for speech production, where the areas of the tubes are related to the coefficients themselves. The inverse relation, with $g = f(k)$, is given by

$$k = \frac{10^g - 1}{10^g + 1}. \tag{8.13}$$

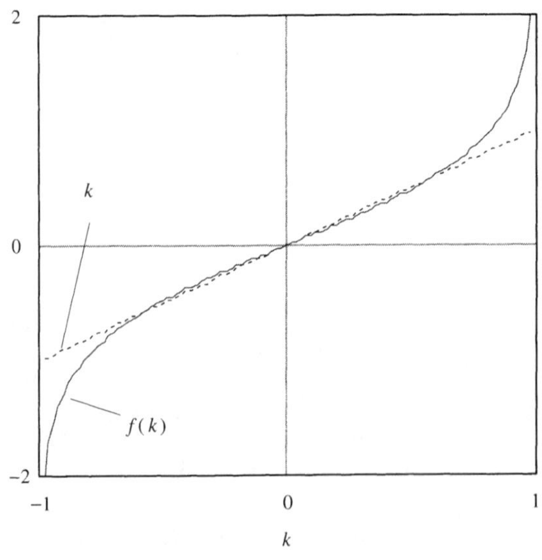

Figure 8.6 Nonlinear function associated with the log area ratio. The identity line is superimposed.

From (8.12) we note that for a stable synthesis filter ($|k| < 1$), the log area ratios take on values in the region $-\infty < g < \infty$, which is essentially unbounded. This wide dynamic range implies a big amount of bits for encoding. In practice, however, the coefficient values seldom surpass certain limits; for instance, $(-2, 2)$ is a practical range over which to limit the LAR.

The plot in Figure 8.6 compares the nonlinear function together with the identity function. Note that for values of k less than 0.7 in magnitude, the LAR curve is almost linear. Therefore, if the magnitude of a given RC has a high probability of being less than 0.7, the quantization performance in the RC domain and LAR domain using a uniform quantizer is similar. In practice, it is found that RCs of order greater than three ($k_1, i > 3$) have in general magnitudes less than 0.7. This fact can be utilized for quantizer design.

Example 8.3 The RCs from Example 4.5 in Chapter 4 are transformed to LARs, with the histograms plotted in Figure 8.7, where g_1, g_2, g_3, g_6, g_7, and g_{10} are displayed. Note that as the order increases, the LARs tend to decrease in magnitude.

Example 8.4 Spectral sensitivity with respect to the LAR is illustrated using the same set of RCs as in Example 8.2. Spectral sensitivity with respect to g_1 is considered first. The range for the LAR is given by $(-2, 2)$. To compute the spectral sensitivity values, g_1 is modified with the rest of the coefficients remaining intact. Nineteen different values of g_1 are utilized, with $g_1 = -1.9, -1.7, \ldots, 1.9$. As shown in (8.8), a small enough step Δg_1 must be selected. We use $\Delta g_1 = g_1/100$. To compute spectral sensitivity at $g_1 = -1.9$, it is first converted to k_1 using (8.13); then the set of RCs is transformed to LPCs. Now with $g_1 = -1.9 - 0.019 = -1.919$, it is again converted to k_1 and the whole set of RCs are transformed to LPCs. The two sets of LPCs are then plugged into (8.8) to compute the spectral sensitivity value. This step is repeated for every value of g_1. Figure 8.8 shows the spectral sensitivity results with respect to g_1, and the procedure is repeated for g_5 as well. Comparing to the spectral sensitivity curves in Figure 8.3, we can see that spectral distortion is far less sensitive with respect to changes in the LAR domain. Thus, high sensitivity for high-magnitude coefficients is eliminated.

Linear Approximation to the LAR Transformation Function

The LAR transformation is relatively complex to implement in practice due to the division as well as the log operations. In practice, a piecewise linear approximation function is often used. See Exercise 8.13 for procedures to find an optimal function. ETSI (1992a) proposed the use of a three-piece approximation function; see Exercise 8.14 for details. Use of the linear approximation significantly reduces the computational load, as well as facilitates fixed-point implementation; a drawback is the introduction of some distortion that is tolerable in many practical situations.

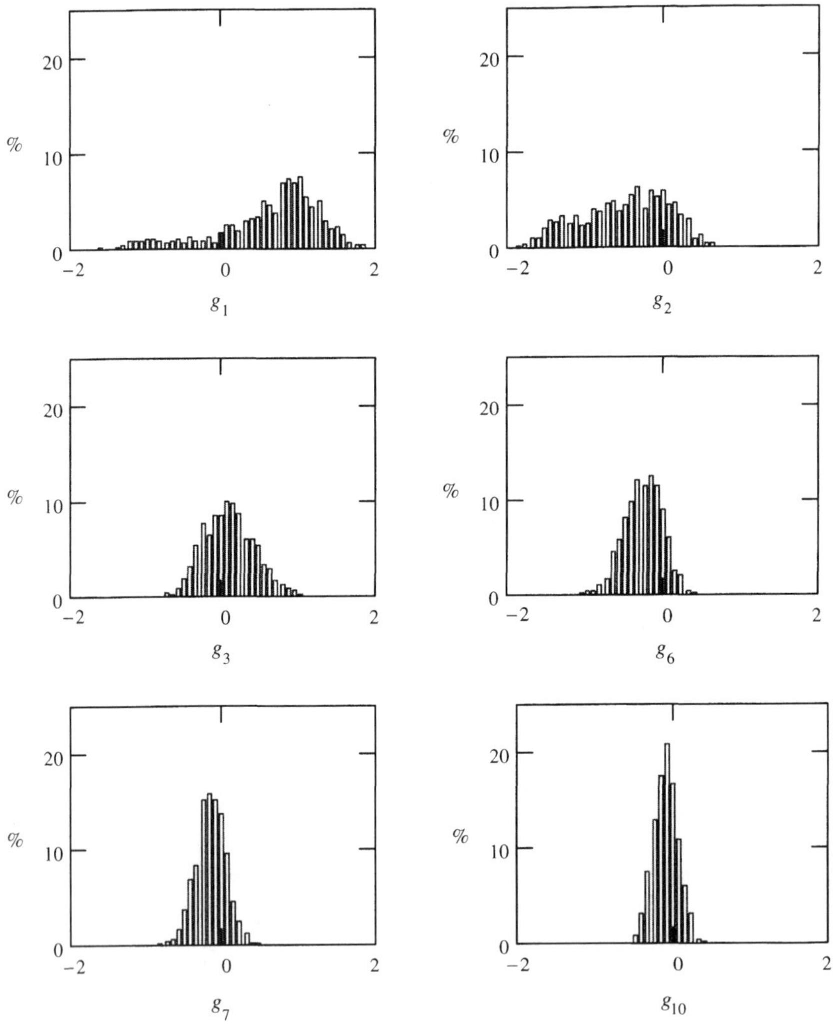

Figure 8.7 Histogram plots of some log area ratios, obtained from 1300 frames of speech material. The vertical axis is the percentage of occurrence while the horizontal axis is the value of the coefficients.

Bit Allocation

In a scalar quantization scheme where each coefficient or parameter is quantized independently, it is not advisable to use an equal number of bits for the quantization of different parameters, since each parameter contributes differently to the average spectral distortion.

During the actual design of scalar quantizer, where a set of training data is available, the following procedure can be used to optimize the bit allocation process.

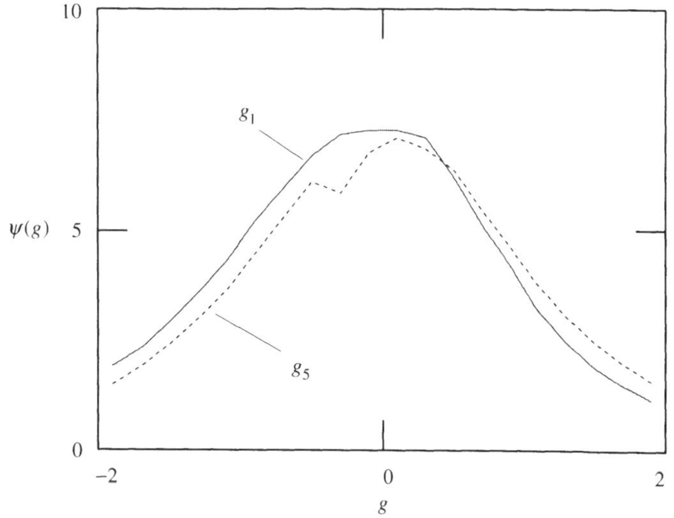

Figure 8.8 Examples of spectral sensitivity curves with respect to log area ratios.

The procedure assigns one bit to a particular quantizer when it offers the minimum average spectral distortion and iterates until the required number of bits is met. The end results depend on the training data set itself; in general, bit allocation for different parameters is nonuniform.

Step 1. Initialization: Set $m = 0$. Start with zero resolution $r_0 = 0$. All quantizer-mapping functions are equal to zero; that is,

$$Q_{m,l}(x) = Q_{0,l}(x) = 0; \quad l = 1, 2, \ldots, M.$$

Step 2. Set $m \leftarrow m + 1$. Increase the number of allocated bits by 1:

$$r_m = r_{m-1} + 1.$$

If r_m exceeds the available number of bits, stop.

Step 3. For $l = 1, 2, \ldots, M$:

- Design the quantizer $Q_{m,l}(x)$ with resolution $r_{m,l} = r_{m-1,l} + 1$.
- Find the average spectral distortion $SD_{m,l}$ using $Q_{m,l}(x)$ and $Q_{m-1,n}(x), n \neq l$.

Step 4. Choose the index l in such a way that $SD_{m,l}$ is minimized. Then the quantizers for the mth iteration are

$$Q_{m,n}(x) = \begin{array}{ll} Q_{m,l}(x), & \\ Q_{m-1,n}(x), & n \neq l. \end{array}$$

Go back to Step 2.

The above procedure is often referred to as *greedy*, where the solution is always selected from the one that looks best at the moment. That is, it makes a locally optimal choice in the hope that it will lead to a globally optimal solution. Greedy algorithms do not always yield optimal solutions; however, due to its simplicity it is applied to solve many practical optimization problems.

TIA IS54 VSELP

This algorithm uses a frame length of 160 samples (20 ms) and a prediction order of 10 to capture the spectral envelope. The ten RCs k_1 to k_{10} are quantized using 6, 5, 5, 4, 4, 3, 3, 3, 3, and 2 bits, respectively, leading to a total of 38 bits/frame. Note that the resolution of the quantizers is reduced for higher-order coefficients since they are less important for the definition of the spectrum. By directly quantizing the RC, the cost associated with the LAR computation is eliminated.

ETSI GSM 6.10 RPE-LTP

The frame length of this coder is equal to 160 samples (20 ms), and a prediction order of 8 is used. The RC is converted to the LAR through a piecewise linear function (Exercise 8.14). Due to their different dynamic ranges and amplitude distributions, different uniform quantizers having different limits are used. Since spectral distortion becomes less dependent on higher-order coefficients, resolutions of the quantizers can gradually decrease as the order increases. The eight LARs g_1 to g_8 are quantized using 6, 6, 5, 5, 4, 4, 3, and 3 bits, respectively, leading to a total of 36 bits/frame.

FS1015 LPC

This coder uses a frame length of 180 samples (22.5 ms) and classifies each frame as voiced or unvoiced according to the periodicity present in the frame. A tenth-order predictor is used for voiced frames, but only a fourth-order predictor is used for unvoiced frames. This is done since, for unvoiced frames, a lower-order predictor is sufficient to describe the spectral envelope.

To quantize the LPC, scalar quantization is used in both the LAR and RC domains. Bit allocation is summarized in Table 8.1. As we can see, only the first two parameters are quantized in the LAR domain. This is because, statistically, only the first two RCs have a significant probability of having magnitudes close to one. In fact, as illustrated in the histogram plots (Figure 8.7), the magnitude tends to diminish as the order of the coefficient increases. Also, when the magnitude of the RC is less than 0.7, performances of quantization in the RC and LAR domains are similar, due to the fact that the transformation function associated with the LAR is relatively linear under that condition. Limiting LAR quantization to the first two coefficients has the advantage of computational saving. Uniform quantization is used for all quantizers. A maximum of 41 bits/frame is used for the transmission of the LPC information.

TABLE 8.1 Bit Allocation for LPC Quantization of the FS1015 Coder[a]

Parameter	Resolution: Voiced	Resolution: Unvoiced
g_1 (LAR)	5	5
g_2 (LAR)	5	5
k_3 (RC)	5	5
k_4 (RC)	5	5
k_5 (RC)	4	—
k_6 (RC)	4	—
k_7 (RC)	4	—
k_8 (RC)	4	—
k_9 (RC)	3	—
k_{10} (RC)	2	—

[a] Data from Tremain [1982], Figure 8.

8.3 LINE SPECTRAL FREQUENCY

Line spectral frequency (LSF) was first introduced by Itakura (1975) as an alternative representation of LPC. Due to many desirable properties, the LSF has received widespread acceptance in speech coding applications. In this section, the origin of LSF is explained, a procedure to convert LPC to LSF is given, and properties of this alternative representation are described.

Definition of Line Spectral Frequency

We are given the prediction-error filter with system function

$$A(z) = 1 + a_1 z^{-1} + \cdots + a_M z^{-M} = \prod_{i=1}^{M}(1 - z_i z^{-1}), \tag{8.14}$$

where z_i denotes the zeros or roots of $A(z)$. Since the LPCs are real, the z_i either are real or form complex conjugate pairs. Note that (8.14) is a polynomial in z^{-1} of order M. We can form two polynomials according to

$$P(z) = A(z)\left(1 + z^{-(M+1)}\frac{A(z^{-1})}{A(z)}\right) = A(z)(1 + G(z)), \tag{8.15}$$

$$Q(z) = A(z)\left(1 - z^{-(M+1)}\frac{A(z^{-1})}{A(z)}\right) = A(z)(1 - G(z)) \tag{8.16}$$

with

$$G(z) = z^{-(M+1)} \frac{A(z^{-1})}{A(z)}. \tag{8.17}$$

Then

$$A(z) = (P(z) + Q(z))/2. \tag{8.18}$$

The LSFs are defined as those values of frequency ω such that

$$\{\omega | P(e^{j\omega}) = 0 \quad \text{or} \quad Q(e^{j\omega}) = 0; \quad 0 < \omega < \pi\}. \tag{8.19}$$

That is, they are the frequency values associated to the unit-magnitude zeros of $P(z)$ or $Q(z)$. Since we are interested only in polynomials with real coefficients, the zeros occur in complex conjugate pairs. Therefore, only the positive frequency values need be considered, with the corresponding negative frequency values obtained through sign inversions.

Line spectral frequency is also known as line spectral pair (LSP) in the literature. However, there has been some confusion on the naming convention; for instance, some sources refer to $\cos(\omega)$ as the LSP and ω as the LSF. In fact, many authors disagree on this matter; in Soong and Juang [1984], ω is called the LSP; while in Kabal and Ramachandran [1986] it is called the LSF. In this book we assume that the two are referring to the same parameter ω, with LSF the preferred term since the underlying quantities are frequency values.

Symmetric and Antisymmetric Properties

The polynomials $P(z)$ and $Q(z)$ can be written as

$$P(z) = A(z) + z^{-(M+1)} A(z^{-1}), \tag{8.20}$$

$$Q(z) = A(z) - z^{-(M+1)} A(z^{-1}), \tag{8.21}$$

or

$$P(z) = 1 + (a_1 + a_M)z^{-1} + (a_2 + a_{M-1})z^{-2} + \cdots + (a_1 + a_M)z^{-M} + z^{-(M+1)}, \tag{8.22}$$

$$Q(z) = 1 + (a_1 - a_M)z^{-1} + (a_2 - a_{M-1})z^{-2} + \cdots - (a_1 - a_M)z^{-M} - z^{-(M+1)}, \tag{8.23}$$

denoting the polynomials by

$$P(z) = \sum_{i=0}^{M+1} p_i z^{-i}, \tag{8.24}$$

$$Q(z) = \sum_{i=0}^{M+1} q_i z^{-i}. \tag{8.25}$$

Comparing (8.22), (8.23) with (8.24), (8.25) we conclude that the coefficients p_i and q_i are given by

$$p_0 = p_{M+1} = 1, \tag{8.26a}$$

$$p_i = p_{M-i+1} = a_i + a_{M-i+1}, \tag{8.26b}$$

$$q_0 = -q_{M+1} = 1, \tag{8.27a}$$

$$q_i = -q_{M-i+1} = a_i - a_{M-i+1}, \tag{8.27b}$$

with $i = 1, \ldots, M$. Due to this structure in the coefficients, $P(z)$ is known as a symmetric polynomial while $Q(z)$ is known as an antisymmetric polynomial.

For M odd, we have

$$p_{(M+1)/2} = 2a_{(M+1)/2}, \tag{8.28}$$

$$q_{(M+1)/2} = 0. \tag{8.29}$$

Zeros at $z = \pm 1$

The zeros of $P(z)$ and $Q(z)$ at $z = \pm 1$ are found as follows:

For M even: $P(z)$ has zero at $z = -1, Q(z)$ has zero at $z = 1$.
For M odd : $Q(z)$ has zeros at $z = \pm 1$.

See Exercise 8.2 for a proof of the above claims.

Order Reduction

Since we are not interested in the zeros at $z = \pm 1 (\omega = 0$ or $\omega = \pi)$, the order of the polynomials $P(z)$ and $Q(z)$ can be reduced by removing these zeros. This can be done with polynomial divisions as follows:

$$P'(z) = \frac{P(z)}{1 + z^{-1}}, \quad Q'(z) = \frac{Q(z)}{1 - z^{-1}}; \quad \text{for } M \text{ even}, \tag{8.30a}$$

$$P'(z) = P(z), \qquad Q'(z) = \frac{Q(z)}{1 - z^{-2}}; \quad \text{for } M \text{ odd}. \tag{8.30b}$$

For M even, $P'(z)$ is an Mth-order polynomial with the following form:

$$P'(z) = \sum_{i=0}^{M} p'_i z^{-i}. \tag{8.31}$$

Performing the division described in (8.30a), it is possible to show that the coefficients p'_i are given by (Exercise 8.3)

$$\begin{aligned} p'_0 &= 1, \\ p'_1 &= p_1 - 1 = p_1 - p'_0, \\ p'_2 &= p_2 - p_1 + 1 = p_2 - p'_1, \\ &\vdots \\ p'_{M-1} &= p_{M-1} - p'_{M-2} = p_1 - 1, \\ p'_M &= p_M - p'_{M-1} = 1, \end{aligned} \tag{8.32}$$

where relation (8.26) is used. Thus, we have

$$p'_0 = 1, \tag{8.33a}$$

$$p'_i = p_i - p'_{i-1}, \tag{8.33b}$$

for $i = 1, \ldots, M$. $Q'(z)$ is another Mth-order polynomial with the following form:

$$Q'(z) = \sum_{i=0}^{M} q'_i z^{-i}. \tag{8.34}$$

Performing the division in (8.30a) shows that

$$\begin{aligned} q'_0 &= 1, \\ q'_1 &= q_1 + 1 = q_1 + q'_0, \\ q'_2 &= q_2 + q_1 + 1 = q_2 + q'_1, \\ &\vdots \\ q'_{M-1} &= q_{M-1} + q'_{M-2} = q_1 + 1, \\ q'_M &= q_M + q'_{M-1} = 1, \end{aligned} \tag{8.35}$$

where relation (8.27) is used. Hence,

$$q'_0 = 1, \tag{8.36a}$$

$$q'_i = q_i + q'_{i-1}, \tag{8.36b}$$

for $i = 1, \ldots, M$.

For M odd, $P'(z)$ is equal to $P(z)$, which is a $(M+1)$st-order polynomial:

$$P'(z) = \sum_{i=0}^{M+1} p'_i z^{-i} = \sum_{i=0}^{M+1} p_i z^{-i}. \tag{8.37}$$

$Q'(z)$ is an $(M-1)$st-order polynomial written as

$$Q'(z) = \sum_{i=0}^{M-1} q'_i z^{-i}. \tag{8.38}$$

Performing the division described in (8.30b) gives

$$\begin{aligned} q'_0 &= 1, \\ q'_1 &= q_1, \\ q'_2 &= q_2 + 1 = q_2 + q'_0, \\ &\vdots \\ q'_{M-2} &= q_{M-2} + q'_{M-4} = q_1, \\ q'_{M-1} &= q_{M-1} + q'_{M-3} = 1. \end{aligned} \tag{8.39}$$

Thus,

$$q'_0 = 1, \tag{8.40a}$$

$$q'_1 = q_1, \tag{8.40b}$$

$$q'_i = q_i + q'_{i-2}, \tag{8.40c}$$

for $i = 2, \ldots, M-1$. From these results it follows that both $Q'(z)$ and $P'(z)$ are symmetric polynomials of even order. Let the order of the polynomials $P'(z)$ and $Q'(z)$ be $2M_1$ and $2M_2$, respectively. Then

$$M_1 = M/2, \qquad M_2 = M/2; \qquad M \text{ even}, \tag{8.41a}$$

$$M_1 = (M+1)/2, \quad M_2 = (M-1)/2; \quad M \text{ odd}. \tag{8.41b}$$

Hence, $P'(z)$ contributes M_1 pairs of conjugate zeros while $Q'(z)$ contributes M_2 pairs of conjugate zeros. Together this gives a total of $M_1 + M_2 = M$ pairs of conjugate zeros on the unit circle.

Evaluating the polynomials at $z = e^{j\omega}$ gives

$$
\begin{aligned}
P'(e^{j\omega}) &= \sum_{i=0}^{2M_1} p'_i e^{-ij\omega} \\
&= e^{-j\omega M_1}[p'_0 e^{j\omega M_1} + p'_1 e^{j\omega(M_1-1)} + p'_2 e^{j\omega(M_1-2)} + \cdots + p'_{2M_1} e^{-j\omega M_1}] \\
&= e^{-j\omega M_1}[2p'_0\cos(M_1\omega) + 2p'_1\cos((M_1-1)\omega) + \cdots + 2p'_{M_1-1}\cos\omega + p'_{M_1}],
\end{aligned} \tag{8.42}
$$

$$
\begin{aligned}
Q'(e^{j\omega}) &= \sum_{i=0}^{2M_2} q'_i e^{-ij\omega} \\
&= e^{-j\omega M_2}[q'_0 e^{j\omega M_2} + q'_1 e^{j\omega(M_2-1)} + q'_2 e^{j\omega(M_2-2)} + \cdots + q'_{2M_2} e^{-j\omega M_2}] \\
&= e^{-j\omega M_2}[2q'_0\cos(M_2\omega) + 2q'_1\cos((M_2-1)\omega) + \cdots + 2q'_{M_2-1}\cos\omega + q'_{M_2}].
\end{aligned} \tag{8.43}
$$

The values of LSFs are found by locating the roots of (8.42) and (8.43), that is, values of ω with $0 < \omega < \pi$ such that $P'(e^{j\omega}) = 0$ or $Q'(e^{j\omega}) = 0$. In order to find the LSF, we only have to consider the following two functions of ω:

$$P_o(\omega) = 2\cos(M_1\omega) + 2p'_1\cos((M_1-1)\omega) + \cdots + 2p'_{M_1-1}\cos\omega + p'_{M_1}, \tag{8.44}$$

$$Q_o(\omega) = 2\cos(M_2\omega) + 2q'_1\cos((M_2-1)\omega) + \cdots + 2q'_{M_2-1}\cos\omega + q'_{M_2}. \tag{8.45}$$

Example 8.5: Solving the LSF when $M = 10$ For $M = 10 : M_1 = M_2 = 5$. The function (8.44) reduces to

$$
\begin{aligned}
P_o(\omega) &= 32\cos^5\omega + 16p'_1\cos^4\omega \\
&\quad + (8p'_2 - 40)\cos^3\omega + (4p'_3 - 16p'_1)\cos^2\omega \\
&\quad + (2p'_4 - 6p'_2 + 10)\cos\omega + p'_5 - 2p'_3 + 2p'_1.
\end{aligned} \tag{8.46}
$$

The corresponding expression for $Q_o(\omega)$ can be obtained by replacing "p" with "q" in the above equation. Note that elementary trigonometric relations are used to derive (8.46) (Exercise 8.4). Since the coefficients p'_i and q'_i are found directly from the LPC, $P_o(\omega)$ and $Q_o(\omega)$ can be solved using a root-finding algorithm, leading to the LSFs.

Summary of the LPC-to-LSF Conversion Procedure

Given the set of LPCs $a_i, i = 1, \ldots, M$, the objective is to convert them to the alternative LSF representation, denoted by $\omega_i, i = 1, \ldots, M$, with $0 < \omega_i < \pi$. The procedure is summarized below.

- Find the polynomial coefficients.

For M even: $M_1 = M_2 = M/2$.
For M odd: $M_1 = (M+1)/2, \quad M_2 = (M-1)/2$.

$$\begin{aligned}
p_i &= a_i + a_{M-i+1}; \quad i = 1, \ldots, M_1, \\
q_i &= a_i - a_{M-i+1}; \quad i = 1, \ldots, M_2, \\
p'_0 &= 1, \\
p'_i &= p_i - p'_{i-1}; \quad i = 1, \ldots, M_1.
\end{aligned}$$

For M even:

$$\begin{aligned}
q'_0 &= 1, \\
q'_i &= q_i + q'_{i-1}; \quad i = 1, \ldots, M_2.
\end{aligned}$$

For M odd:

$$\begin{aligned}
q'_0 &= 1, \qquad q'_1 = q_1, \\
q'_i &= q_i + q'_{i-2}; \quad i = 2, \ldots, M_2.
\end{aligned}$$

- Construct the functions $P_o(\omega)$ and $Q_o(\omega)$ ((8.44) and (8.45)) and solve for roots, that is, the frequency values when $P_o(\omega) = 0$ or $Q_o(\omega) = 0$, for $0 < \omega < \pi$. These frequency values are the LSFs.

Note that the functions $P_o(\omega)$ and $Q_o(\omega)$ depend on the order M and must be found for each order of interest. For the commonly used value of $M = 10$, for instance, the functions are derived in Example 8.5. For practical implementation, the cosine calculation is relatively expensive to deploy. Thus, it is convenient to make the substitution $x = \cos(\omega)$ so that during root solving, the root values in x are found first. Denoting the roots as x_i, the LSFs are given by

$$\omega_i = \cos^{-1}(x_i). \tag{8.47}$$

The two frequency functions for $M = 10$ are rewritten below using the substitution $x = \cos(\omega)$

$$\begin{aligned}
P_o(x) = {} & 32x^5 + 16p'_1x^4 + 8(p'_2 - 5)x^3 + 4(p'_3 - 4p'_1)x^2 \\
& + 2(p'_4 - 3p'_2 + 5)x + p'_5 - 2p'_3 + 2p'_1,
\end{aligned} \tag{8.48}$$

$$\begin{aligned}
Q_o(x) = {} & 32x^5 + 16q'_1x^4 + 8(q'_2 - 5)x^3 + 4(q'_3 - 4q'_1)x^2 \\
& + 2(q'_4 - 3q'_2 + 5)x + q'_5 - 2q'_3 + 2q'_1.
\end{aligned} \tag{8.49}$$

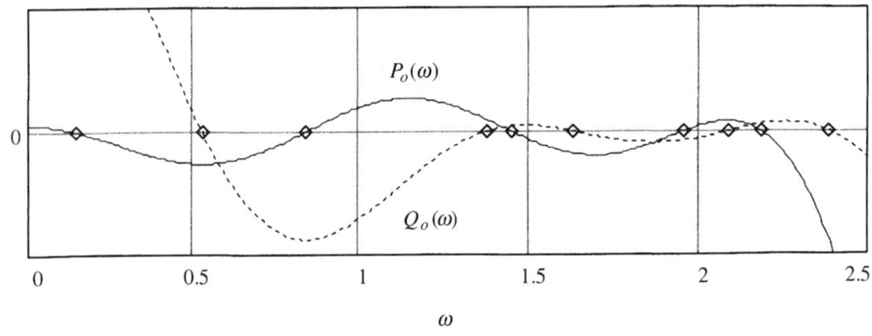

Figure 8.9 Plot of some example frequency functions. LSF values are marked on the horizontal axis.

Example 8.6 Consider the first set of LPCs in Example 8.1. Following the LPC-to-LSF conversion procedure results in the coefficients:

$$p'_1 = -1.655,\quad p'_2 = 2.546,\quad p'_3 = -3.782,\quad p'_4 = 4.399,\quad p'_5 = -4.808,$$
$$q'_1 = 0.455,\quad q'_2 = 1.974,\quad q'_3 = 0.034,\quad q'_4 = 1.723,\quad q'_5 = -0.722.$$

The resultant frequency functions are

$$P_o(\omega) = 32\cos^5\omega - 26.5\cos^4\omega - 19.6\cos^3\omega + 11.4\cos^2\omega + 3.52\cos\omega - 0.554,$$
$$Q_o(\omega) = 32\cos^5\omega + 7.28\cos^4\omega - 24.2\cos^3\omega - 7.14\cos^2\omega + 1.60\cos\omega + 0.12.$$

These functions are plotted in Figure 8.9, where the LSF values are found from the intersections with the frequency axis. Using the mapping $x = \cos(\omega)$, the equivalent functions are plotted in Figure 8.10, with $-1 < x < 1$. The resultant LSFs are found to be

$$\omega_1 = 0.143,\quad \omega_2 = 0.529,\quad \omega_3 = 0.840,\quad \omega_4 = 1.374,\quad \omega_5 = 1.449,$$
$$\omega_6 = 1.632,\quad \omega_7 = 1.956,\quad \omega_8 = 2.089,\quad \omega_9 = 2.183,\quad \omega_{10} = 2.388.$$

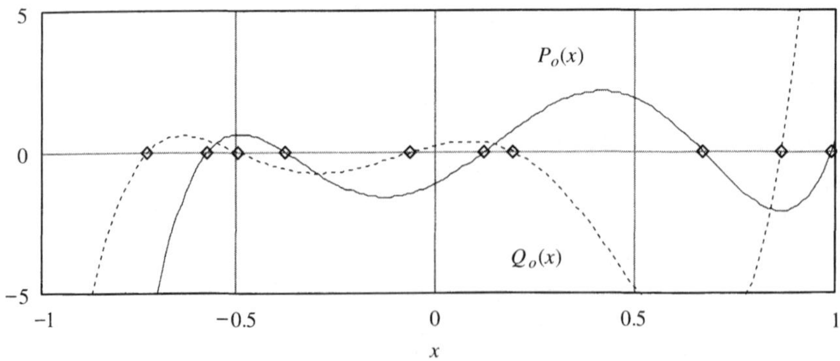

Figure 8.10 Plot of some example frequency functions with $x = \cos(\omega)$.

Note that $\omega_1, \omega_3, \omega_5, \omega_7$, and ω_9 are from $P_o(\omega)$, while the even-indexed parameters are from $Q_o(\omega)$. This example illustrates an important property of the LSF: namely, the zeros of $P_o(\omega)$ and $Q_o(\omega)$ are interlaced with each other, in the sense that

$$\omega_1 > \omega_2 > \omega_3 > \omega_4 > \cdots > \omega_{10}$$

and this property is satisfied as long as the LPCs are from a minimum-phase system. A formal statement about this property is given at the end of the section.

Root Finding Routine

The method included here for root location is proposed in Soong and Juang [1984], and a similar approach is described in Kabal and Ramachandran [1986], where the interlacing property is taken into account for efficient computation. The roots of $P_o(x)$ and $Q_o(x)$ can be found by searching incrementally for intervals in which the sign changes. The search proceeds backward from $x = 1$ since $P_o(x)$ has the first root close to $x = 1$. The location of the root in an interval containing a sign change is refined by successive bisection of the root interval. Given the interlacing property of the roots, the search for a root of $Q_o(x)$ starts from the position of the root of $P_o(x)$ just found.

The initial evaluation interval δ must be sufficiently small so that two or more roots of the same function do not occur in the same interval. Let the roots be

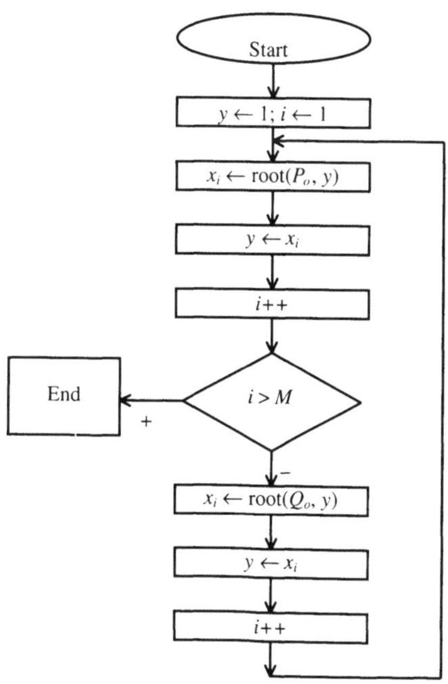

Figure 8.11 Flowchart of the main routine for root finding.

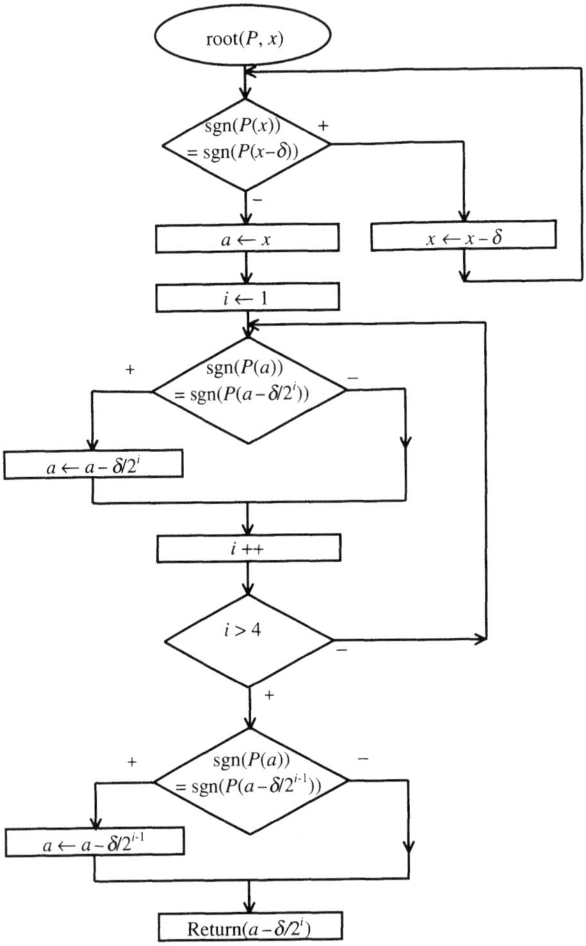

Figure 8.12 Flowchart of the root function. The function sgn(x) returns the sign of the variable x.

denoted by $\{x_i\}$ for $i = 1, 2, \ldots, M$, and let them be ordered such that $x_i > x_{i-1}$. The roots of $P_o(x)$ (x_i with i odd) interlace with the roots of $Q_o(x)$ (x_i with i even). Then the initial evaluation interval must satisfy

$$\delta < \min_i (x_i - x_{i-2})$$

In practice, $\delta = 0.02$ is sufficiently small to avoid missing sign changes.

The main routine for root finding is shown in Figure 8.11. The routine makes use of the function root(P, y), which solves for the root of the function P with a starting value of y. Since for M odd, $P_o(x)$ has one more root than $Q_o(x)$, the maximum count is checked after finding the root of $P_o(x)$. Figure 8.12 shows the flowchart of the root function.

As we can see, the root function bisects a given interval sequentially to search for a sign change. In Figure 8.12 the shortest interval applied is $\delta/16$; by increasing the number of iterations through the loop, it is possible to augment the precision to a higher level. The function returns the linearly interpolated value of the final interval. After all the roots are found, the LSFs are given by (8.47).

LSF-to-LPC Conversion

Given the set of LSFs $\omega_i, i = 1, \ldots, M$, ordered in the form $0 < \omega_1 < \omega_2 < \cdots < \omega_M < \pi$, the objective is to find the set of LPCs $a_i, i = 1, \ldots, M$. The method explained for LPC-to-LSF conversion can be applied in reverse to accomplish the task. First, the LSFs are transformed according to

$$x_i = \cos \omega_i; \quad i = 1, \ldots, M,$$

so that

$$x_1 > x_2 > \cdots > x_M.$$

For i odd, the x_i are the zeros of $P_o(x)$ (8.48), and for i even, the x_i are the zeros of $Q_o(x)$ (8.49). Substituting the values of x_i into the respective function of x, one can solve the coefficients of $P_o(x)$ and $Q_o(x)$. For the case of $M = 10$, the results are

$$p'_1 = -2(x_1 + x_3 + x_5 + x_7 + x_9), \tag{8.50}$$

$$p'_2 = 4(x_1x_3 + x_1x_5 + x_3x_5 + x_1x_7 + x_3x_7 + x_5x_7 + x_1x_9 + x_3x_9 + x_5x_9 + x_7x_9) + 5, \tag{8.51}$$

$$p'_3 = -8(x_1x_3x_5 + x_1x_3x_7 + x_1x_5x_7 + x_3x_5x_7 + x_1x_3x_9 + x_1x_5x_9 + x_3x_5x_9 + x_1x_7x_9 + x_3x_7x_9 + x_5x_7x_9) + 4p'_1, \tag{8.52}$$

$$p'_4 = 16(x_1x_3x_5x_7 + x_1x_3x_5x_9 + x_1x_3x_7x_9 + x_1x_5x_7x_9 + x_3x_5x_7x_9) + 3p'_2 - 5, \tag{8.53}$$

$$p'_5 = -32x_1x_3x_5x_7x_9 - 2p'_1 + 2p'_3, \tag{8.54}$$

$$q'_1 = -2(x_2 + x_3 + x_6 + x_8 + x_{10}), \tag{8.55}$$

$$q'_2 = 4(x_2x_{10} + x_4x_{10} + x_2x_4 + x_6x_{10} + x_2x_6 + x_4x_6 + x_8x_{10} + x_2x_8 + x_4x_8 + x_6x_8) + 5, \tag{8.56}$$

$$q'_3 = -8(x_2x_4x_{10} + x_2x_6x_{10} + x_4x_6x_{10} + x_2x_4x_6 + x_2x_8x_{10} + x_4x_8x_{10} + x_2x_4x_8 + x_6x_8x_{10} + x_2x_6x_8 + x_4x_6x_8) + 4q'_1, \tag{8.57}$$

$$q'_4 = 16(x_2x_4x_6x_{10} + x_2x_4x_8x_{10} + x_2x_6x_8x_{10} + x_4x_6x_8x_{10} + x_2x_4x_6x_8) + 3q'_2 - 5, \tag{8.58}$$

$$q'_5 = -32x_2x_4x_6x_8x_{10} - 2q'_1 + 2q'_3. \tag{8.59}$$

Then, from (8.33),

$$p'_0 = 1, \tag{8.60a}$$

$$p_k = p'_k + p'_{k-1}, \tag{8.60b}$$

for $k = 1, \ldots, M_1$.

For M even using (8.36),

$$q'_0 = 1, \tag{8.61a}$$

$$q_k = q'_k - q'_{k-1} \tag{8.61b}$$

for $k = 1, \ldots, M_2$.

For M odd using (8.40),

$$q'_0 = 1, \tag{8.62a}$$

$$q_1 = q'_1, \tag{8.62b}$$

$$q_k = q'_k - q'_{k-2}, \tag{8.62c}$$

for $k = 2, \ldots, M_2$.

Once the polynomials' coefficients are found, the LPC can be solved from (8.26) and (8.27):

$$a_i = \frac{p_i + q_i}{2}, \tag{8.63a}$$

$$a_{M-i+1} = \frac{p_i - q_i}{2} \tag{8.63b}$$

for M even and $i = 1, \ldots, M_1$. For M odd, (8.63) applies with no modification for $i = 1, \ldots, M_2$, and

$$a_{M_1} = p_{M_1}/2. \tag{8.64}$$

Some Properties of the LSF

As defined before, the LSFs are those frequency values inside the interval $(0, \pi)$, where $P(e^{j\omega}) = 0$ or $Q(e^{j\omega}) = 0$, or equivalently $P_o(\omega) = 0$ or $Q_o(\omega) = 0$. Two important properties of the LSF are given below.

PROPERTY 1: *If $A(z)$ is minimum-phase, then all zeros of $P(z)$ and $Q(z)$ are on the unit circle.*

This property guarantees the existence of the LSF when the system $A(z)$ (8.14) is minimum-phase.

PROPERTY 2: *If $A(z)$ is minimum-phase, then the zeros of $P(z)$ and $Q(z)$ are interlaced with each other.*

This property allows the verification of the minimum-phase status, and hence the stability of the underlying synthesis filter. Therefore, if the LSFs are modified by some operator—like quantization—stability of the resultant synthesis filter is guaranteed if the interlacing property is preserved. See Appendix B for proof of the above properties.

Another benefit of using the LSF is that the PSD at a particular frequency value tends to depend on the close-by LSF. In other words, a LSF of a certain frequency value affects mainly the PSD at the same frequency value. This is sometimes known as the localization property, where modifications to the LSF have a local effect on the PSD. This is in contrast to other parameters, like the LPC or RC, where changes to one particular parameter affect the whole spectrum. The localization property is highly desirable in quantization since distortion at a certain frequency range can be predicted from the values of the LSFs; thus, more bits can be allocated to those LSF values that tend to be more sensitive to a human listener, for example, between 200 and 2000 Hz. As we will study later, some quantization schemes take advantage of the localization property and put more emphasis on the sensitive frequency regions. The localization property is also due to the simple fact that the LSF parameters are themselves frequency values directly linked to the signal's frequency description.

Example 8.7 The same set of LPCs as in Example 8.6 is used, where a_{10} is changed from -0.055 to -0.4. This modification causes the three zeros of $A(z)$ to have magnitude greater than one and hence are outside the unit circle. The functions $P_o(x)$ and $Q_o(x)$ are plotted in Figure 8.13. We can see that only nine zero crossings are present; therefore, some zeros of $P(z)$ and $Q(z)$ are not located on the unit circle (Property 1 is not satisfied). On the other hand, the interlacing property is also violated.

Example 8.8 The localization property of the LSF is illustrated in this example. The same set of LSFs as in Example 8.6 is used. First, ω_4 is modified by 1% with the rest of the parameters intact; the squared magnitude difference between the original transfer function and the modified transfer function is calculated, which is given by

$$(|H(e^{j\omega})| - |H'(e^{j\omega})|)^2, \tag{8.65}$$

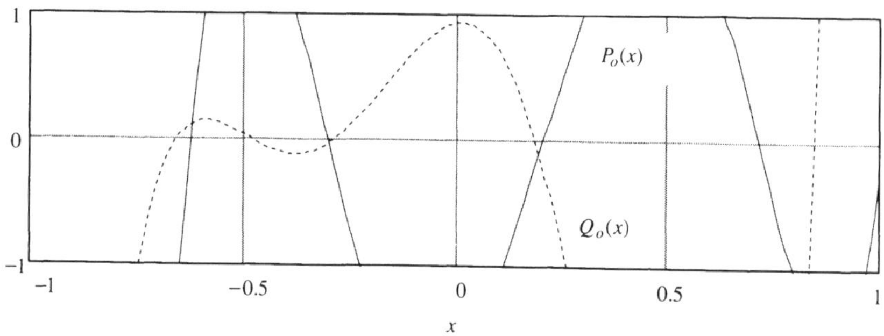

Figure 8.13 Plot of frequency functions for a non-minimum-phase system.

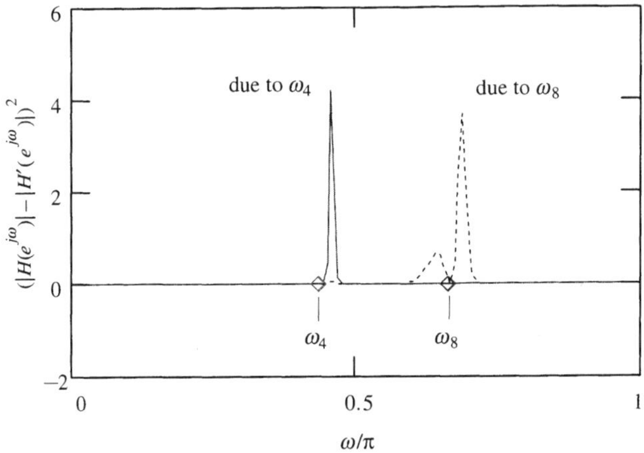

Figure 8.14 Squared magnitude difference between the original transfer function and a modified transfer function, with the modification due to a 1% change in one of the LSFs.

where H is defined using the original set of parameters with H' is defined using the new set of parameters. Plotting the above equation reveals the spectrum difference between the two transfer functions. Figure 8.14 shows the results, where the experiment is repeated for ω_8. Note that the highest spectrum distortion incurred by modifying a certain LSF happens in a frequency range close to the value of the LSF itself. Thus, changes in the LSF have a local effect on the spectrum.

8.4 QUANTIZATION BASED ON LINE SPECTRAL FREQUENCY

Line spectral frequency, as presented in the last section, possesses several desirable features that make it attractive as an alternative LPC representation. First, the minimum-phase property of the associated synthesis filter is preserved after quantization, as long as the interlacing condition is satisfied. On the other hand, the values of the LSFs directly control the property of the signal in the frequency domain, and changes of one parameter have a local effect on the spectrum. Also, the LSFs are bounded: they are located inside the $(0, \pi)$ interval and hence are highly suitable for fixed-point implementation. By using the LSF, stability of the synthesis filter can be restrained easily and the amount of distortion in a different frequency zone can be regulated. Use of the LSF has also been found to be advantageous in interpolation, a topic discussed at the end of the chapter. In this section, the method adopted in the FS1016 CELP coder (Chapter 12) is explained, which consists of scalar quantization of the LSF parameters.

As described in the previous section, the LSF is first found in the cosine domain $x = \cos(\omega), x \in [-1, 1]$ and later is mapped to the frequency domain $\omega = \cos^{-1}(x)$, $\omega \in [0, \pi]$. For the purpose of quantization, the parameters are sometimes directly quantized in the cosine domain, since the mapping using $\cos^{-1}$ represents extra computation, and the interlacing property can be verified in both domains.

Example 8.9 The LPCs from Example 4.5 in Chapter 4 are transformed to LSFs, with the histograms plotted in Figure 8.15. Note how the individual parameter has its own range within the interval (0, π).

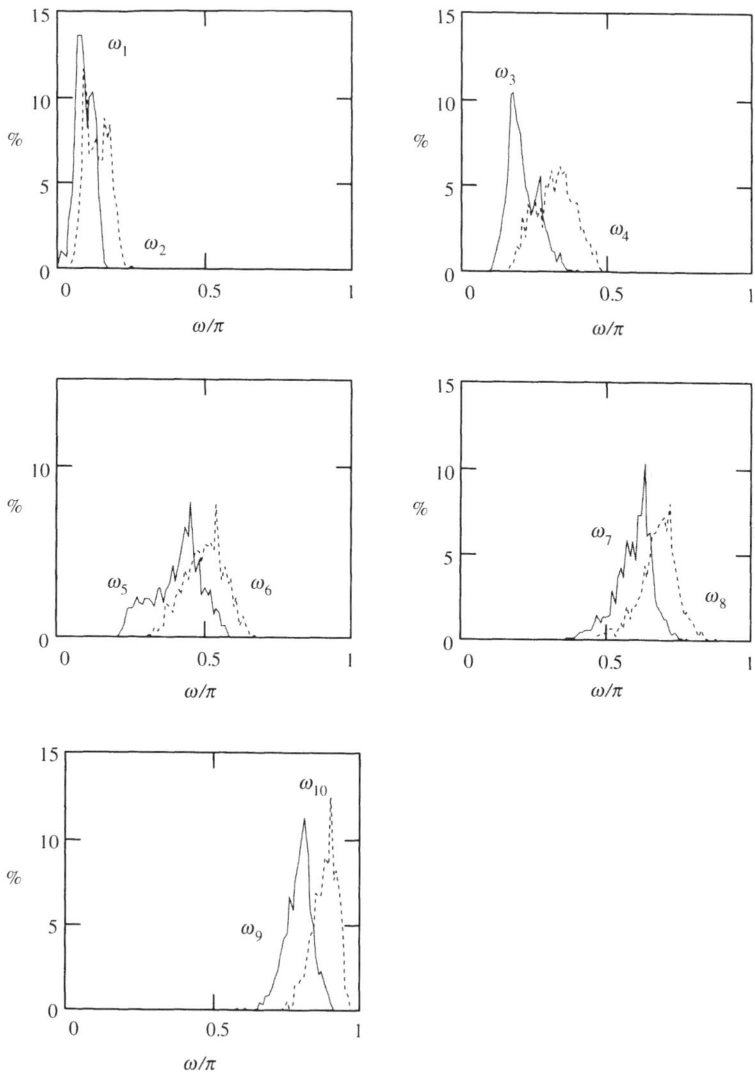

Figure 8.15 Histogram plots of some line spectral frequencies, obtained from 1300 frames of speech material. The vertical axis is the percentage of occurrence while the horizontal axis is the value of the parameters.

FS1016 CELP

This algorithm uses a frame length of 240 samples (30 ms). A tenth-order LP analysis is performed. The LPCs are quantized as LSFs, where ten different nonuniform quantizers are used. The codewords are summarized in Figure 8.16. Note that more bits are allocated for the second to fifth parameters, due to the fact

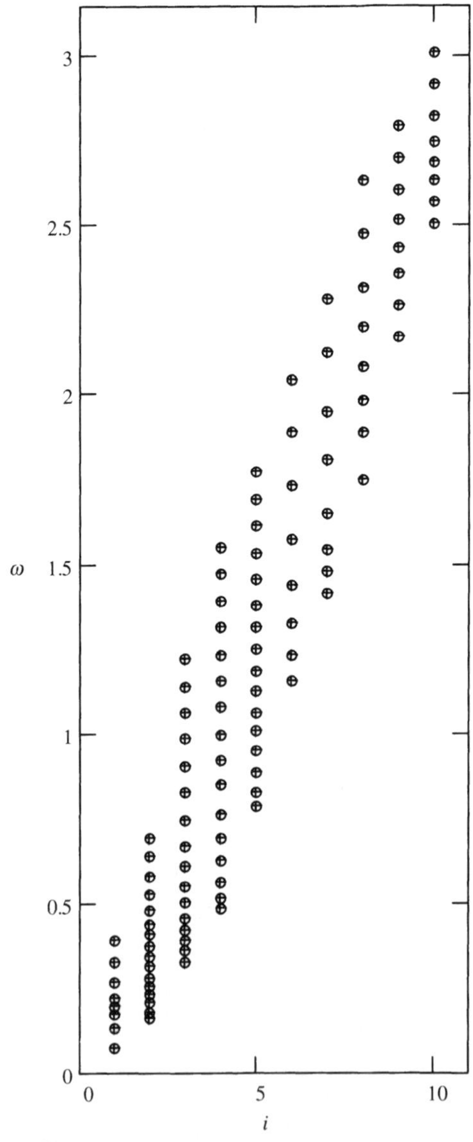

Figure 8.16 Plot of LSF quantization codewords for the FS1016 coder. Data from National Communications System [1992], Table 8.

that the frequency range covered by these four parameters is the most sensitive for a human listener. The ten LSFs ω_1 to ω_{10} are quantized using 3, 4, 4, 4, 4, 3, 3, 3, 3, and 3 bits, respectively, leading to a total of 34 bits/frame.

Conversion of LPC to Quantized LSF

For the LPC quantizer of the FS1016 CELP coder, we denote the codewords by $f_{i,j}$, where $i = 1$ to 10 indicates the quantizer number; and $j = 1$ to α_i indicates the codeword number, with $\alpha_i = 8$ for $i = 1$ and $i = 6$ to 10, and $\alpha_i = 16$ for $i =$ 2 to 5. These codewords are first converted to

$$x_{i,j} = \cos\left(\frac{2\pi f_{i,j}}{f_s}\right). \quad (8.66)$$

The root function, as described in Section 8.3 for the conversion of the LPC to the LSF, can be simplified due to the fact that the final LSF values are fixed and known: they are already contained in Figure 8.16. Thus, availability of the table implies saving in computation during LPC-to-LSF conversion. The flowchart of the new root routine is shown in Figure 8.17; in this case, however, the index of the quantizer i is passed to the routine as an input parameter.

In this routine, the function type (P or Q), initial search value x, and the index of the quantizer i are input parameters. In the first loop, the codeword that is just smaller than the initial search value x is determined. Then in the next loop, a sign change is detected by going through the remaining codewords one at a time. The routine is

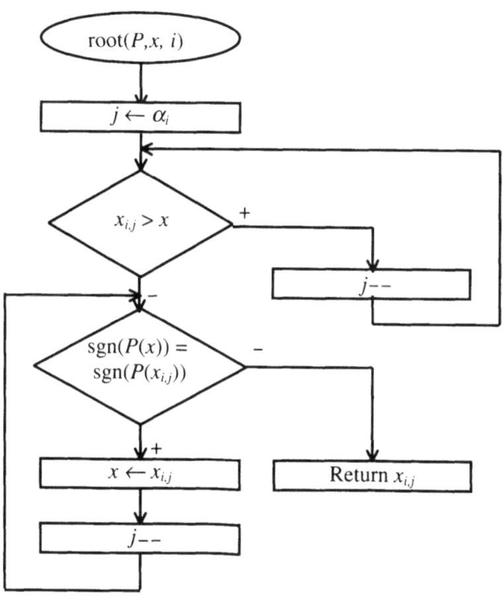

Figure 8.17 Flowchart of the root function, applicable when the quantized LSF values are known.

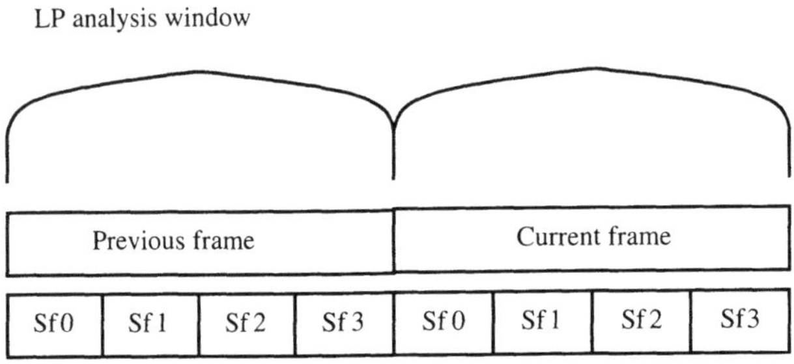

Figure 8.18 Positions of frame, subframe, and LP analysis window for the TIA IS54 VSELP coder.

obviously much simpler than the one presented in Figure 8.12, with much lower complexity.

8.5 INTERPOLATION OF LPC

It is common for LP-based speech coding schemes to perform a LP analysis procedure for each frame and quantize the resultant LPC as information on the frame. Since the length of a frame is relatively long (20 to 30 ms), changes to the LPC for adjacent frames can introduce undesired transients in the synthesized speech signal. To alleviate the problem, the LPC parameters are often interpolated so as to smooth out the transition. Interpolation is frequently done within the frame/subframe context (Chapter 4), where the LPCs of each subframe are obtained through interpolation from two adjacent frames. Some interpolation schemes adopted by various standards are described in this section.

TIA IS54 VSELP

This coder has a frame length of 160 samples, divided into four subframes having 40 samples each (5 ms). LP analysis is performed using a Hamming window (Figure 8.18). The RCs are quantized, with the quantized coefficients converted to LPCs. Combining with the LPCs of the previous frame, interpolation is done using the weights shown in Figure 8.19. As an example, consider Subframe 0 of

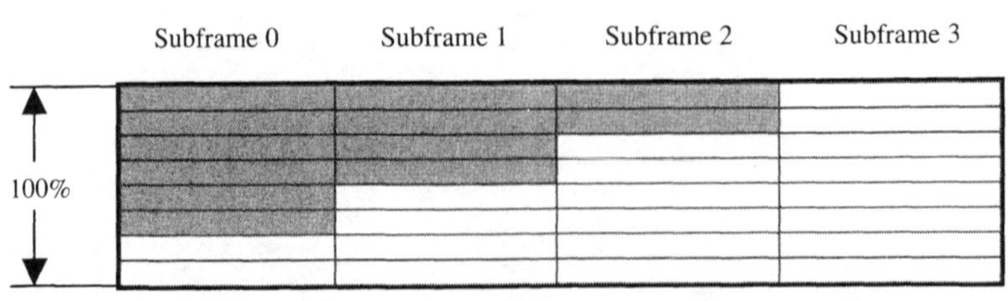

Figure 8.19 Interpolation weights for the TIA IS54 VSELP coder: previous LPC in grey and current LPC in white. Data from Macres [1994].

the current frame. Its LPCs are obtained by the sum of two products: previous LPCs by 3/4, and current LPCs by 1/4. The weights are adjusted accordingly for other subframes depending on their relative positions with respect to the frame.

Interpolating directly using the LPC can result in an unstable filter; therefore, the resulting coefficients must be checked for stability. For a stability check, the interpolated LPCs of Subframes 0, 1, and 2 are converted to RCs (Subframe 3 need not be considered, why?); if any of the resulting RCs have magnitudes greater than 1, the associated filter is unstable. To remedy the situation, the LPCs for the subframe are replaced by the original, noninterpolated LPCs. The replacement rules are specified as follows:

- Subframe 0: Use the previous LPC.
- Subframe 1: If the energy of the previous frame is greater than or equal to that of the current frame, use the previous LPC; otherwise use the current LPC.
- Subframe 2: Use the current LPC.

ETSI GSM 6.10 RPE-LTP

This coder has the same frame and subframe lengths as the TIA VSELP coder. Positions of the frame, subframe, and LP analysis window are also identical (Figure 8.18). The LPCs corresponding to an eighth-order predictor are transformed to the LAR. Interpolation is done with the quantized LAR, with the same weights as shown in Figure 8.19.

FS1016 CELP

The FS1016 coder has a frame length of 240 samples, divided into four subframes having 60 samples each (7.5 ms). LP analysis is performed once per frame using a Hamming window, where the window is centered at the end of the frame (Figure 8.20). The LPCs are transformed to quantized LSFs and are linearly interpolated to form an intermediate set for each of the four subframes. The interpolation is done from two sets of LSFs. For the case of the "current frame" in Figure 8.20, the first set of LSFs is obtained from the analysis window centered at the end of the previous frame, or the beginning of the current frame. The second set of LSFs is obtained

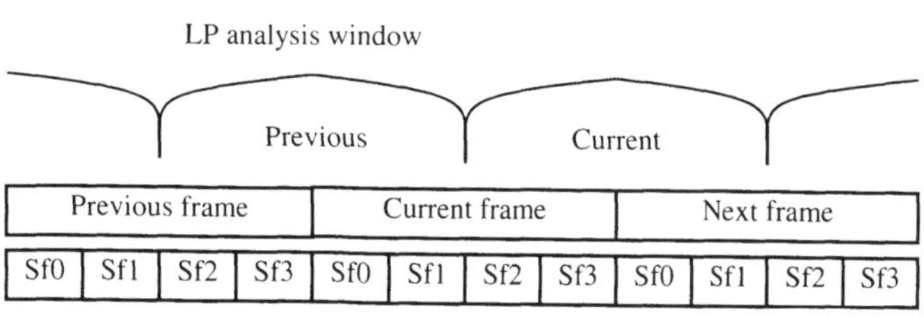

Figure 8.20 Positions of frame, subframe, and LP analysis window for the FS1016 CELP coder.

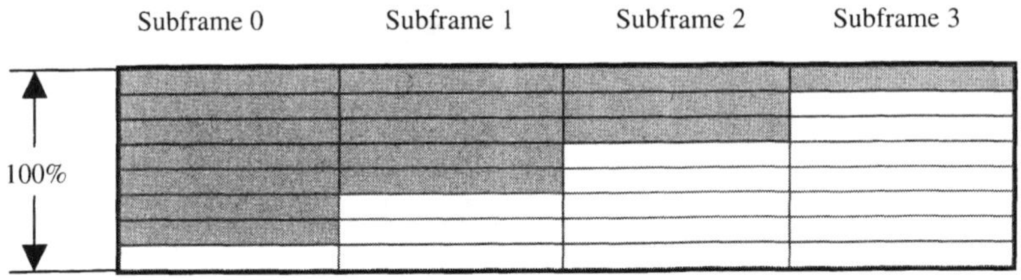

Figure 8.21 Interpolation weights for the FS1016 CELP coder: previous LSF in grey and current LSF in white. Data from National Communications System [1992].

from the analysis window centered at the end of the current frame, or the beginning of the next frame. The interpolation weights are summarized in Figure 8.21.

Note that the positions of the frame and LP analysis window introduce an extra half-frame delay, since the coder must wait for an addition of one-half frame in order to begin LP analysis to find out the LPCs of each subframe in the current frame.

Discussion

In general, it is preferable not to use the LPCs directly for interpolation, since the stability of the resultant synthesis filter cannot be ensured. Other LPC representations having a one-to-one correspondence to the LPC can be used for interpolation, including the RC, LAR, and LSF parameters. The advantage of these parameters is the fact that stability is maintained after interpolation. Though each of these representations provides equivalent information, their performances under interpolation differ. Experimental results from Paliwal and Kleijn [1995] show that the LSF provides the best interpolation performance, leading to the lowest average SD. In addition, it always results in a stable synthesis filter, if the original filters are stable (Exercise 8.9). For the LSF, interpolation in the cosine domain and frequency domain is essentially equivalent in practice, with negligible difference in performance. Due to the many advantages of the LSF, it is the dominant LPC representation for modern speech coders.

8.6 SUMMARY AND REFERENCES

Several alternative LPC representations are introduced in this chapter, with their performances in quantization discussed. Spectral distortion is defined as an objective measure of quantization performance, a commonly used reference for comparison among different schemes. Most of the earlier speech coders use scalar quantization; for higher performance, however, VQ is becoming the method of choice.

To conclude this chapter, some experimental data from Paliwal and Kleijn [1995] are presented to illustrate the performance of various quantization schemes. The experiment is performed in the following way: 1200 s of speech is used for training while 160 s of speech is used for testing; the signal was collected from

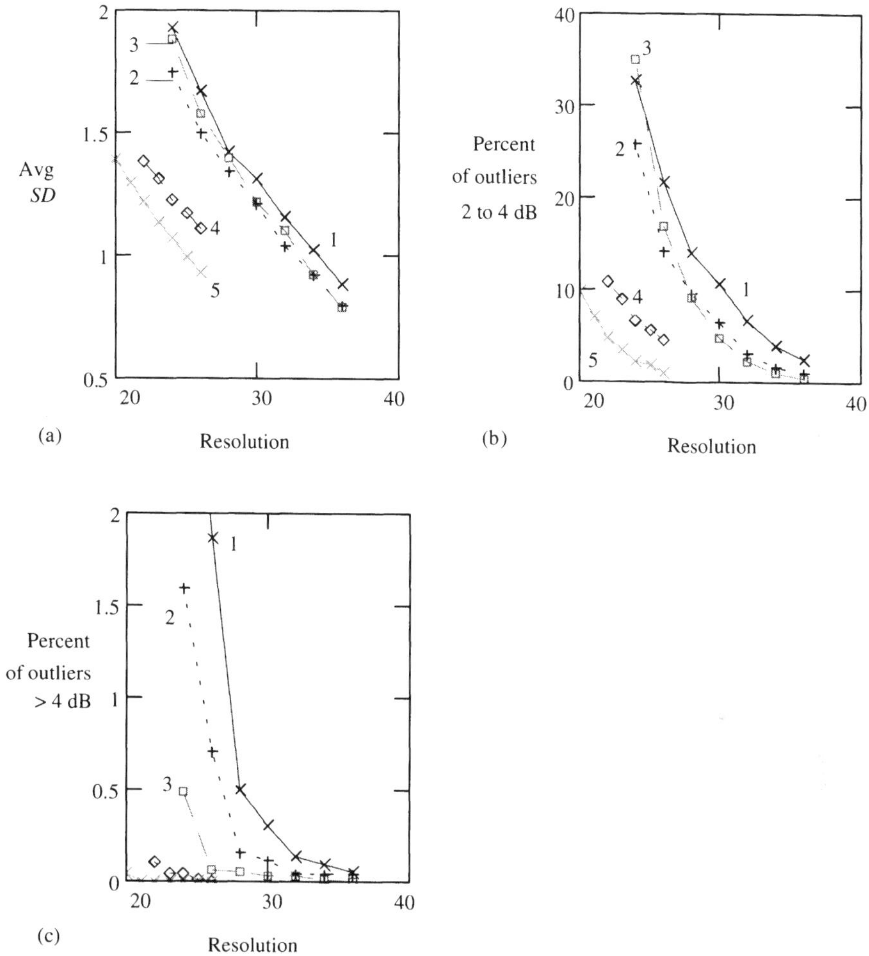

Figure 8.22 Plots of quantization performance for five different schemes: (a) average SD, (b) percentage of outliers having SD between 2 and 4 dB, and (c) percentage of outliers having SD $>$ 4 dB. Data from Paliwal and Kleijn [1995], Tables 2, 3, 5, 8, and 9.

FM radio programs. Prediction order is equal to ten with a 20-ms analysis window. The results are plotted in Figure 8.22, where the following five schemes are included:

1. Scalar quantization of the RC.
2. Scalar quantization of the LAR.
3. Scalar quantization of the LSF.
4. MSVQ of the LSF using squared Euclidean distance.
5. MSVQ of the LSF using weighted squared Euclidean distance.

All scalar quantizers are designed using the Lloyd algorithm and therefore are nonuniform. The bit allocation procedure, as explained in Section 8.2 is applied. In the multistage vector quantization (MSVQ) schemes, two stages are used; see Chapter 15 for details on MSVQ design. From Figure 8.22 we can see that the average SD result for scalar quantization of the RC is the highest, while Schemes 2 and 3 are close to each other. In MSVQ of the LSF, use of weighting in the distance calculation roughly saves 2 bits/frame. Compared to scalar quantization, the saving by using MSVQ is more than 10 bits/frame. Plots of the percentage of outliers also show the superiority of MSVQ. VQ is the dominant LPC quantization technique for most modern coders; see Chapter 15 for details of this topic.

EXERCISES

8.1 Inverse sine is a proposed nonlinear transformation method to desensitize the reflection coefficient in a similar way as for log area ratio. The transformation function is given by

$$g = f(k) = \sin^{-1}(k).$$

(a) Plot the above function and compare it with the LAR function.

(b) Discuss any possible advantage/disadvantage of using the above function.

(c) Repeat Example 8.4 using inverse sine. Is there any major difference with respect to the LAR?

8.2 Given the polynomials $P(z)$ and $Q(z)$ ((8.22) and (8.23)), show the following:

(a) For M even: $P(z)$ has zero at $z = -1, Q(z)$ has zero at $z = 1$.

(b) For M odd: $Q(z)$ has zeros at $z = \pm 1$.

Hint: Use the expressions of $P(z)$ and $Q(z)$ directly together with symmetric and antisymmetric properties of the coefficients.

8.3 Perform the polynomial divisions (8.30) to find the relationships for p_i' and q_i' ((8.32), (8.35), and (8.39)).

8.4 Using the trigonometric relations

$$\cos 2x = 2\cos^2 x - 1,$$

$$\cos 3x = 4\cos^3 x - 3\cos x,$$

$$\cos 4x = 8\cos^4 x - 8\cos^2 x + 1,$$

$$\cos 5x = 16\cos^5 x - 20\cos^3 x + 5\cos x.$$

Derive the expression for $P_o(\omega)$ (8.46) from (8.44).

8.5 Using a procedure similar to the one used to derive (8.46), find the expressions of $P_o(\omega)$ for $M = 8$ and $M = 9$.

8.6 The root function as shown in Figure 8.12 will work only if the input LPCs correspond to a stable synthesis filter. In other words, the interlacing property of the LSF is satisfied. Modify the flowchart in such a way that an error status is returned when the interlacing property is false.

8.7 Derive the LSF-to-LPC conversion procedure for $M = 8$ and $M = 9$.

8.8 Given the reflection coefficients $|k_1| < 1$ and $|k_2| < 1$, show that

$$|\alpha k_1 + (1 - \alpha)k_2| < 1,$$

where $0 \leq \alpha \leq 1$ is a constant. This result indicates that interpolating the RCs from two stable synthesis filters generates a stable synthesis filter, since the magnitudes of the resultant RCs are less than one.

8.9 Given the LSF parameters $\omega_1 > \omega_2$ and $\omega_a > \omega_b$, show that

$$\alpha\omega_1 + (1 - \alpha)\omega_a > \alpha\omega_2 + (1 - \alpha)\omega_b,$$

where $0 \leq \alpha \leq 1$ is a constant. The result indicates that interpolating the LSFs from two stable synthesis filters generates a stable synthesis filter, since the interlacing property is preserved.

8.10 Soong and Juang (1990) proposed the use of LSF differences for quantization, defined by

$$\Delta\omega_i = \omega_{i+1} - \omega_i; \quad i = 1 \text{ to } 10,$$

with $\omega_{11} = \pi$. The advantage of using this scheme instead of the LSF is that the variability of the parameters is reduced; thus, the quantizers can be deployed more efficiently. Using a set of LSF data from a speech source, find the corresponding LSF differences and plot the histograms; compare to the histograms of the original LSF.

8.11 The zeros or roots of the polynomial (8.14) can be considered as an alternative LPC representation employable for quantization purposes, since there is a one-to-one mapping between the LPC and the zeros. Elaborate a list of advantages/disadvantages of using this approach. Consider the aspects of computational complexity, filter stability, and interpolation.

8.12 Draw the flowchart of the procedure required for LPC interpolation for the TIA IS54 VSELP coder. Inputs to the algorithm are the sets of LPCs (previous and current) with outputs being the four sets of LPCs corresponding to the four subframes.

8.13 Consider the linear approximation to the LAR transformation function using two slopes; that is, the approximation is of form

$$f_o(k) = \begin{cases} \alpha_1 k, & |k| < k_1 \\ \text{sgn}(k)(\alpha_2|k| - \alpha_3), & k_1 \leq |k| < 1. \end{cases}$$

The constants $\alpha_1, \alpha_2, \alpha_3$, and k_1 must be found so as to minimize the approximation error

$$J = \int_0^1 (f(k) - f_o(k))^2 dk.$$

Assuming that k_1 is known:

(a) Prove that

$$\alpha_1 = \frac{2}{k_1^2} \int_0^{k_1} f(k)\, dk$$

(b) Under the constraint that $f_o(k)$ is continuous at $k = k_1$, show that

$$\alpha_3 = k_1(\alpha_2 - \alpha_1).$$

(c) Prove that

$$\alpha_2 = \frac{\int_{k_1}^{1} (f(k) - \alpha_1 k_1)(k - k_1)\, dk}{\frac{1}{3}k_1^3 - k_1^2 + k_1 - \frac{1}{3}}$$

(c) The above relations can be solved repeatedly for a range of k_1 values, with the approximation error found for each k_1, and the optimal set of constants determined. What is the optimal value of k_1? The integrals can be solved numerically with math software such as Mathcad [MathSoft, 2001].

8.14 ETSI (1992a) proposed the use of a three-slope linear approximation function for the LAR transformation, with the input interval $k \in [-1, 1]$ partitioned into $|k| < 0.675, 0.675 \leq |k| < 0.950$, and $0.950 \leq |k| \leq 1$. Extend the two-slope methodology described in Exercise 8.13 to the three-slope case and find all parameters for the linear approximation problem.

8.15 In some applications the RC-to-LAR transformation function is defined with the natural logarithm instead of the base-10 logarithm. Show that this alternative function is related to the original by a scaling constant. What is the value of the scaling constant?

CHAPTER 9

LINEAR PREDICTION CODING

Based on a highly simplified model for speech production, the linear prediction coding (LPC) algorithm is one of the earliest standardized coders that works at low bit-rate. At 2.4 kbps, the FS1015 LPC coder [Tremain, 1982] is a breakthrough in speech coding development; even though the quality of the decoded speech is low, it is quite intelligible. The name "linear prediction coding" has since been linked to any algorithm utilizing the LPC model of speech production, with the FS1015 standard being the most outstanding representative.

Originally developed for military applications of secure communication, the FS1015 coder is characterized by its synthetic output speech that often requires trained operators for reliable usage. Although most modern LP-based speech coders achieve higher performance at similar bit-rate, their operating principles are derived from LPC, with modifications to improve quality and coding effectiveness. Thus, a good understanding of LPC helps in the comprehension of more complex coders covered in subsequent chapters.

In this chapter, the speech production model that the LPC coder relies on is described. Detail structures of the algorithm are given, which essentially incorporates the speech production model at its core. One important component of the LPC coder is the voicing detector; its design is thoroughly explained. The main features of the FS1015 coder are highlighted for reference purposes. At the end of the chapter, the shortcomings of the LPC coder are depicted; these flaws are targets for correction by the next generation of speech coders.

9.1 SPEECH PRODUCTION MODEL

Linear prediction coding relies on a highly simplified model for speech production, with the block diagram shown in Figure 9.1. The model is inspired by observations of the basic properties of speech signals and represents an attempt to mimic the human speech production mechanism. The combined spectral contributions of the glottal flow, the vocal tract, and the radiation of the lips are represented by the synthesis filter. The driving input of the filter or excitation signal is modeled as either an impulse train (voiced speech) or random noise (unvoiced speech). Therefore, depending on the voiced or unvoiced state of the signal, the switch is set to the proper location so that the appropriate input is selected. Energy level of the output is controlled by the gain parameter.

How does the model fit into the context of speech coding? Consider speech samples separated into nonoverlapping frames. For a short enough length of the frame, properties of the signal essentially remain constant. In each frame, parameters of the model are estimated from the speech samples; in the present case, these parameters are as follows:

- Voicing: whether the frame is voiced or unvoiced.
- Gain: mainly related to the energy level of the frame.
- Filter coefficients: specify the response of the synthesis filter.
- Pitch period: in the case of voiced frames, time length between consecutive excitation impulses.

The parameter estimation process is repeated for each frame, with the results representing information on the frame. Thus, instead of transmitting the PCM samples, parameters of the model are sent. By carefully allocating bits for each parameter so as to minimize distortion, an impressive compression ratio can be achieved. For instance, the bit-rate of 2.4kbps for the FS1015 coder is 53.3 times lower than the corresponding bit-rate for 16-bit PCM. The price to pay is an

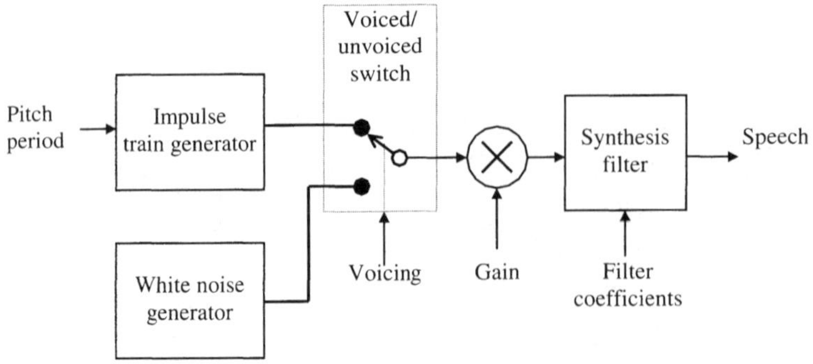

Figure 9.1 The LPC model of speech production.

irreversible loss of quality. However, high-quality reproduction is not required in many practical applications.

Estimating the parameters is the responsibility of the encoder. The decoder takes the estimated parameters and uses the speech production model to synthesize speech. The approach of utilizing noise to generate the output signal is somehow mystifying. How can the scheme work if the output waveform is completely different from the original? In fact, output waveforms using the same set of parameters and filter initial conditions are different since the white noise generator is random. The point to note is that the power spectral density of the original speech is captured by the synthesis filter; therefore, the PSD of the synthetic speech is close to the original due to the flat spectrum of input excitation. The approach throws away all phase information of the original waveform, preserving only the magnitude of the frequency spectrum. The synthetic waveform sounds like the original because, for a human listener, phase has a relatively lower rank than magnitude information. This phenomenon is the reason why signal-to-noise ratio is a poor, and sometimes, senseless measure of speech quality.

As shown in Chapter 4, linear prediction is a practical method of spectrum estimation, where the PSD can be captured using a few coefficients. These coefficients or linear prediction coefficients (LPCs, not to be confused with the coding acronym!) can be used to construct the synthesis filter. The synthesis filter shapes the flat spectrum of the noise input so that the output imitates the envelope of the original spectrum. It is important to note that this is true only for noise excitation in the unvoiced case; for the voiced case, however, the input is an impulse train—a sequence of regularly spaced impulses—violating therefore the basic assumptions of the AR model (Chapter 3). Recall that in an AR model, the excitation signal has a flat spectrum, which is satisfied by white noise or a single impulse. For a train of impulses, the corresponding spectrum is roughly flat only when the distance between impulses is sufficiently high. This violation of the AR model for voiced signal is one of the fundamental limitations of the LPC model for speech production and is analyzed in more detail at the end of the chapter. The impulse train for excitation is given by

$$\sum_{i=-\infty}^{\infty} \delta[n - iT] \tag{9.1}$$

with

$$\delta[n] = \begin{cases} 1, & \text{if } n = 0, \\ 0, & \text{otherwise,} \end{cases} \tag{9.2}$$

and T a positive constant being the period. The use of a periodic impulse train is to create periodicity in the output waveform, so that the resulting signal possesses a PSD that resembles voiced signals.

Since the coefficients of the synthesis filter must be quantized and transmitted, only a few of them are calculated so as to maintain low bit-rate. As indicated in

Chapter 4, a prediction order of ten is in general enough to capture the spectrum envelope. This prediction order is adequate for unvoiced frames; for voiced frames, however, a much higher order is required due to correlation of distant samples. The LPC coder solves this by using an impulse train input: if the period of the input excitation matches the original pitch, periodicity is introduced to the synthetic speech with a PSD that is similar to the original. In this manner, high prediction order is avoided, thus achieving the low bit-rate objective.

Besides implementing the model as a discrete-time system, it is interesting to mention that mechanical and electrical means were used for its realization. See Deller et al. [1993] for an engaging discussion of speech modeling efforts starting in the 1930s.

Example 9.1 Validity of the LPC model for speech production is illustrated here using some typical speech waveforms. Figure 9.2 shows an unvoiced frame having 180 samples (in accordance with the FS1015 coder). The original samples are LP analyzed, with the resultant LPC used for speech synthesis based on the described model. Since the frame is unvoiced, white noise with uniform distribution is used as excitation. The generated white noise has unit variance, with the gain estimated using a method described in the next section. Gain estimation sets the energy levels to be the same for the two frames. As we can see, in the time domain the two waveforms are completely different. They sound similar because the power spectral densities have similar shapes, which are plotted in Figure 9.3.

Waveforms for a voiced frame are shown in Figure 9.4. The pitch period is roughly equal to 75. Synthetic speech is generated using a train of impulses with unit amplitude. This unit-amplitude impulse train is scaled by a gain term (see the

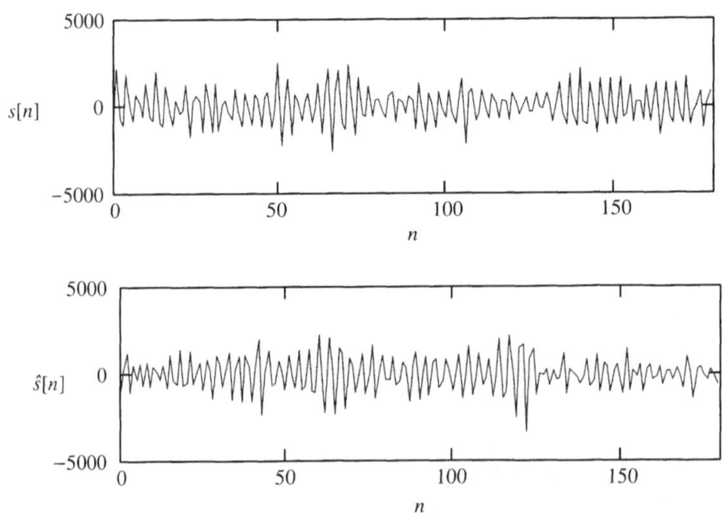

Figure 9.2 Plots of unvoiced frames. *Top*: Original; *bottom*: synthetic.

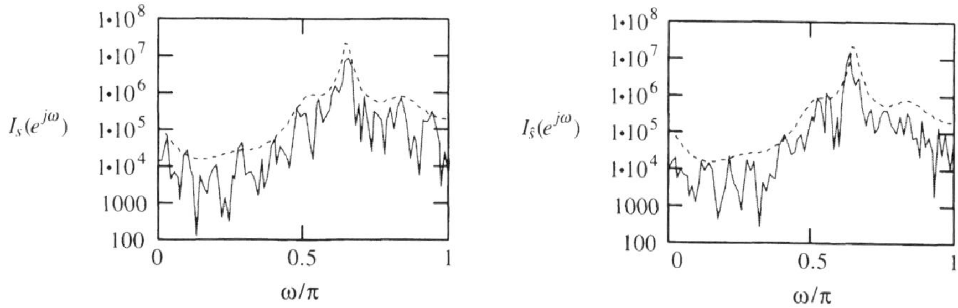

Figure 9.3 Plots of periodograms for an unvoiced frame. *Left*: Original; *right*: synthetic. The PSD using the estimated LPC is superimposed (dotted line).

next section for gain computation) so that the energy level of the synthetic speech matches the original. In the frequency domain we can see that the periodograms of the two frames have similar appearance (Figure 9.5). Presence of harmonic components due to periodicity of the frame is also evident from the regularly separated peaks in the periodograms. For the original signal, however, the structure of the harmonics looks more irregular and randomized, while for the synthetic case, a more regular structure appears. The harmonic structure for the synthetic voiced frame is

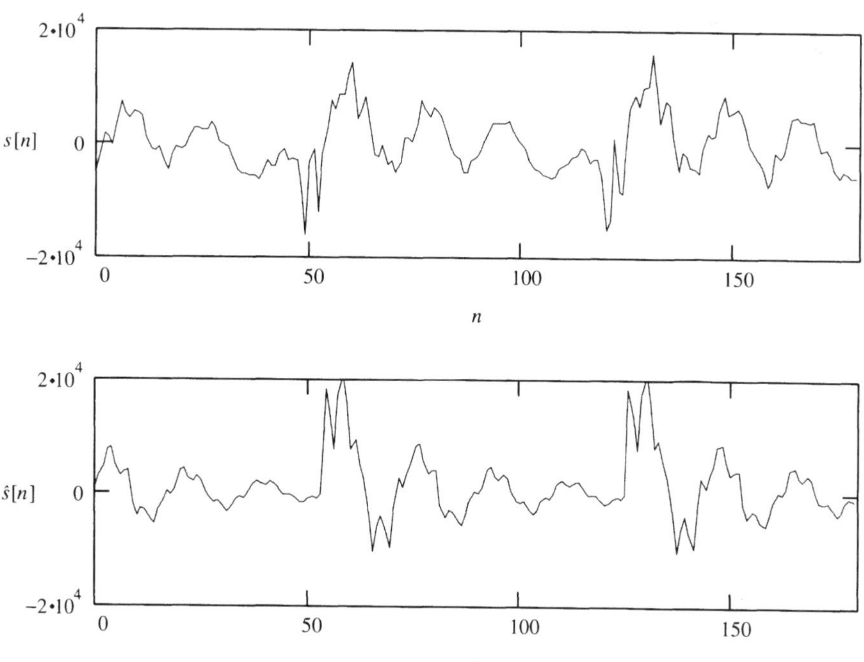

Figure 9.4 Voiced frames plots. *Top*: Original; *bottom*: synthetic.

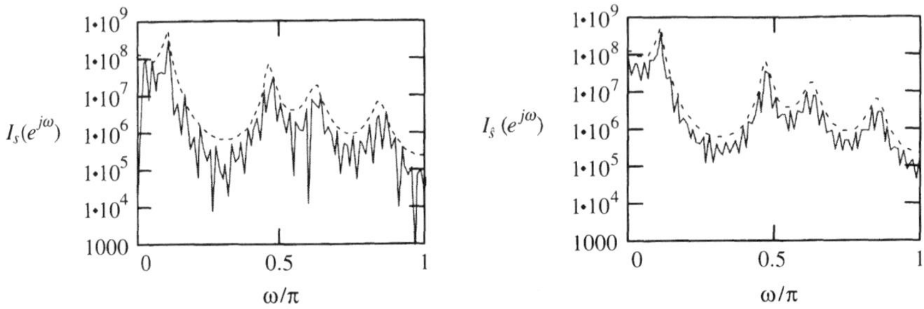

Figure 9.5 Plots of periodograms for a voiced frame. *Left*: Original; *right*: synthetic. The PSD using the estimated LPC is superimposed (dotted line).

created mainly by the input excitation, which is a train of impulses with separation given by the pitch period.

9.2 STRUCTURE OF THE ALGORITHM

In this section, the structure of a speech coding algorithm that utilizes the LPC model of speech production is presented. The structure closely resembles the FS1015 coder but is not identical.

Encoder

Figure 9.6 shows the block diagram of the encoder. The input speech is first segmented into nonoverlapping frames. A pre-emphasis filter is used to adjust the spectrum of the input signal; its purpose was already explained in Chapter 4.

The voicing detector, discussed in the next section, classifies the current frame as voiced or unvoiced and outputs one bit indicating the voicing state.

The pre-emphasized signal is used for LP analysis, where ten LPCs are derived. These coefficients are quantized (Chapter 8) with the indices transmitted as information of the frame. The quantized LPCs are used to build the prediction-error filter, which filters the pre-emphasized speech to obtain the prediction-error signal at its output (internal prediction, Chapter 4).

Pitch period is estimated from the prediction-error signal if and only if the frame is voiced. Methods already described in Chapter 2 can be used for the task. By using the prediction-error signal as input to the pitch period estimation algorithm, a more accurate estimate can be obtained since the formant structure (spectrum envelope) due to the vocal tract is removed.

Example 9.2 Pitch period estimation using the magnitude difference function over the prediction-error signal is illustrated where the same frame as in

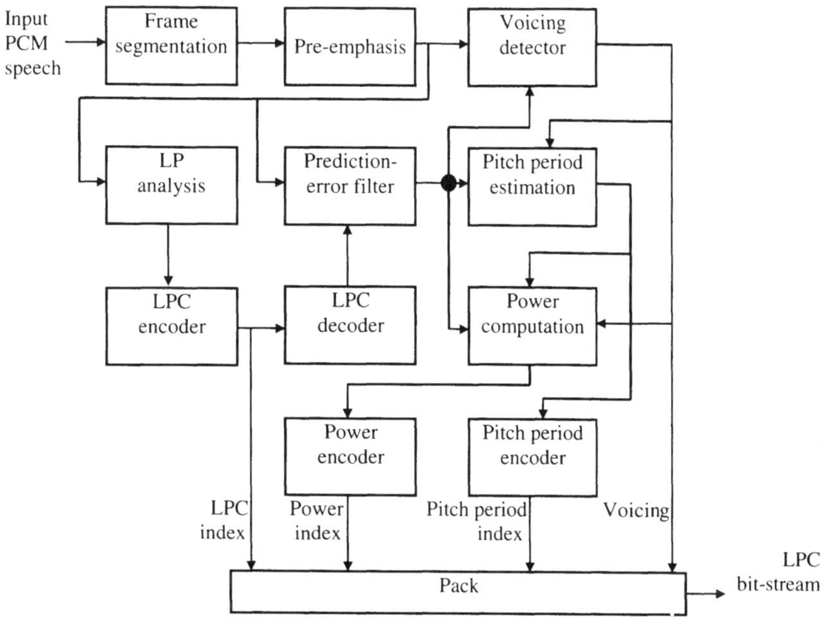

Figure 9.6 Block diagram of the LPC encoder.

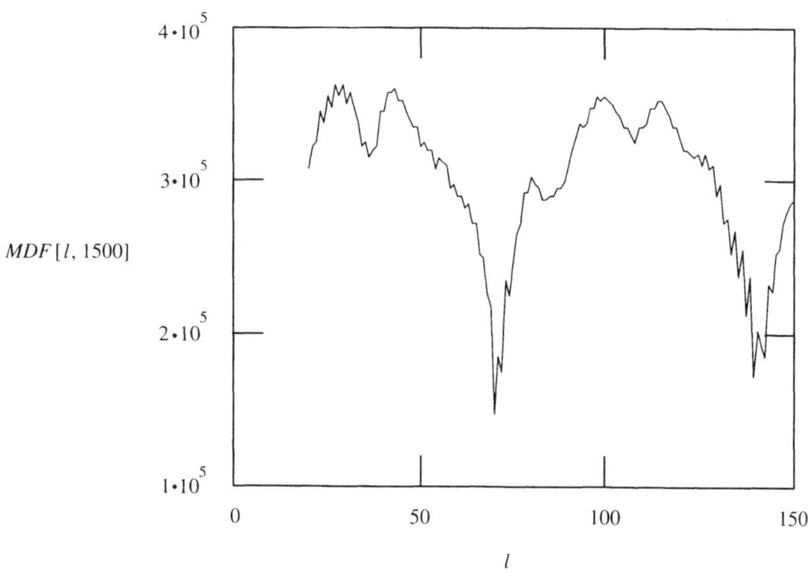

Figure 9.7 Magnitude difference values obtained from a voiced frame. Prediction error is used for its computation.

Chapter 2 is considered. Here, however, the prediction error is first calculated by LP analyzing the frame with a prediction order of ten. The frame is then filtered with the resultant prediction-error filter. Figure 9.7 shows the plot of the magnitude difference function when the lag is changed from 20 to 150. Comparing with the result in Chapter 2, the curve is less oscillatory, and transition toward the minimum is more abrupt, thus leading to a more accurate pitch period estimate.

Power Calculation

Power of the prediction-error sequence is calculated next, which is different for voiced and unvoiced frames. Denoting the prediction-error sequence as $e[n], n \in [0, N-1]$, with N being the length of the frame, we have for the unvoiced case

$$p = \frac{1}{N}\sum_{n=0}^{N-1} e^2[n]. \tag{9.3}$$

For the voiced case, power is calculated using an integer number of pitch periods:

$$p = \frac{1}{\lfloor N/T \rfloor T}\sum_{n=0}^{\lfloor N/T \rfloor T-1} e^2[n] \tag{9.4}$$

with $\lfloor \cdot \rfloor$ the floor function (returns the greatest integer less than or equal to the operand). It is assumed that $N > T$, and hence use of the floor function ensures that the summation is always performed within the frame's boundaries. The use of an integer number of pitch periods to compute the power is largely motivated by the objective of synchronizing with the pitch period. The limits in the summation are selected so as to facilitate gain computation on the decoder side.

The voicing bit, pitch period index, power index, and LPC index are packed together to form the bit-stream of the LPC coder.

Decoder

Figure 9.8 shows the block diagram of the decoder and is essentially the LPC model of speech production with parameters controlled by the bit-stream. It is assumed that the output of the impulse train generator is comprised of a series of unit-amplitude impulses, while the white noise generator has unit-variance output.

Gain computation is performed as follows. For the unvoiced case, the power of the synthesis filter's input must be the same as the prediction error on the encoder side. Denoting the gain by g, we have

$$g = \sqrt{p}, \tag{9.5}$$

since the white noise generator has unit-variance output.

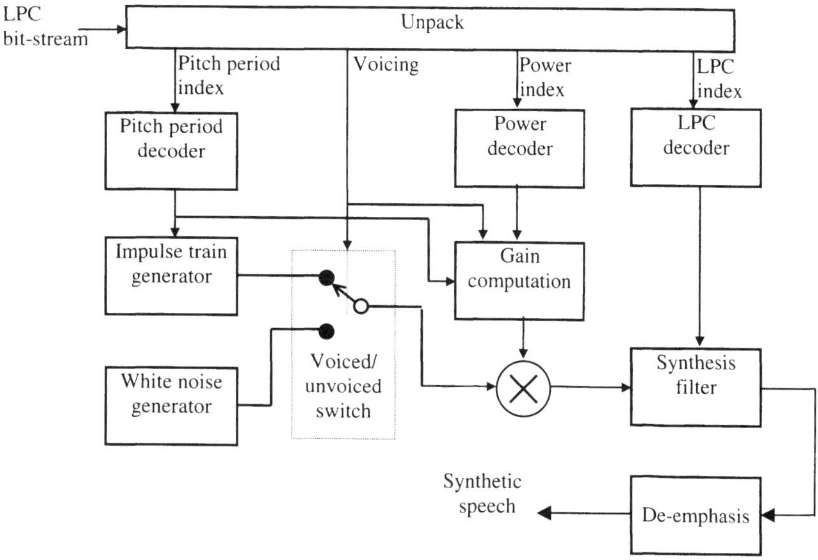

Figure 9.8 Block diagram of the LPC decoder.

For the voiced case, the power of the impulse train having an amplitude of g and a period of T, measured over an interval of length $\lfloor N/T \rfloor T$, must equal p. Carrying out the operation yields

$$g = \sqrt{Tp}. \tag{9.6}$$

Finally, the output of the synthesis filter is de-emphasized to yield the synthetic speech.

9.3 VOICING DETECTOR

The purpose of the voicing detector is to classify a given frame as voiced or unvoiced. In many instances, voiced/unvoiced classification can easily be accomplished by observing the waveform: a frame with clear periodicity is designated as voiced, and a frame with noise-like appearance is labeled as unvoiced. In other instances, however, the boundary between voiced and unvoiced is unclear; this happens for transition frames, where the signal goes from voiced to unvoiced or vice versa. The necessity to perform a strict voiced/unvoiced classification is indeed one of the fundamental limitations of the LPC model.

In this section we discuss some measurements that a voicing detector relies on to accomplish its task. For reliable operation, the detector must take into account as many parameters as possible so as to achieve a high degree of robustness. These parameters are input to a linear classifier having binary output. The voicing detector is one of the most critical components of the LPC coder, since misclassification of voicing states can have disastrous consequences on the quality of the synthetic speech.

Energy

This is the most obvious and simple indicator of "voicedness." Typically, voiced sounds are several order of magnitude higher in energy than unvoiced signals. For the frame (of length N) ending at instant m, the energy is given by

$$E[m] = \sum_{n=m-N+1}^{m} s^2[n]. \tag{9.7}$$

For simplicity, the magnitude sum function defined by

$$MSF[m] = \sum_{n=m-N+1}^{m} |s[n]| \tag{9.8}$$

serves a similar purpose.

Since voiced speech has energy concentrated in the low-frequency region, due to the relatively low value of the pitch frequency, better discrimination can be obtained by lowpass filtering the speech signal prior to energy calculation. That is, only energy of low-frequency components is taken into account. A bandwidth of 800 Hz is adequate for the purpose since the highest pitch frequency is around 500 Hz.

Zero Crossing Rate

The zero crossing rate of the frame ending at time instant m is defined by

$$ZC[m] = \frac{1}{2} \sum_{n=m-N+1}^{m} |\text{sgn}(s[n]) - \text{sgn}(s[n-1])|, \tag{9.9}$$

with $\text{sgn}(\cdot)$ the sign function returning ± 1 depending on the sign of the operand. Equation (9.9) computes the zero crossing rate by checking the samples in pairs to determine where the zero crossings occur. Note that a zero crossing is said to occur if successive samples have different signs.

For voiced speech, the zero crossing rate is relatively low due to the presence of the pitch frequency component (of low-frequency nature), whereas for unvoiced speech, the zero crossing rate is high due to the noise-like appearance of the signal with a large portion of energy located in the high-frequency region.

Prediction Gain

Prediction gain was already defined in Chapter 4 as the ratio between the energy of the signal and the energy of the prediction error

$$PG[m] = 10\log_{10}\left(\frac{\sum_{n=m-N+1}^{m} s^2[n]}{\sum_{n=m-N+1}^{m} e^2[n]}\right). \tag{9.10}$$

Voiced frames on average achieve 3 dB or more in prediction gain than unvoiced frames, mainly due to the fact that periodicity implies higher correlation among samples, and thus more predictability. Unvoiced frames, on the other hand, are more random and therefore less predictable. For very low-amplitude frames, prediction gain is normally not calculated to avoid numerical problems; in this case, the frame can be assigned as unvoiced just by verifying the energy level.

Example 9.3 Figure 9.9 shows an example of a speech waveform together with its magnitude sum function, zero crossing, and prediction gain; all these parameters are calculated using a frame length of 180 samples. The parameters are assumed to be constant within the frame, and that is the reason why plots of the parameters take on a step appearance.

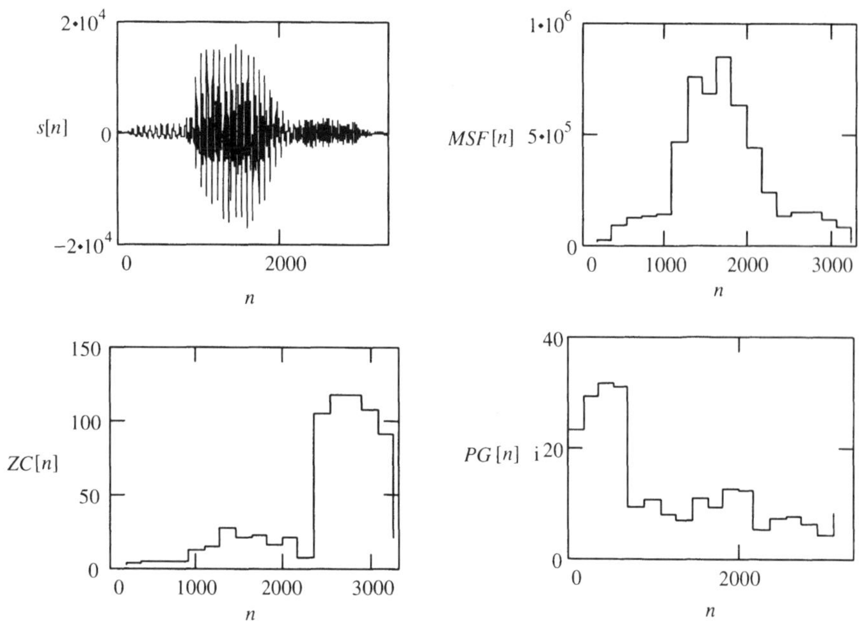

Figure 9.9 *Top left*: A speech waveform. *Top right*: Magnitude sum function. *Bottom left*: Zero crossing rate. *Bottom right*: Prediction gain.

Roughly speaking, the signal is voiced for $n < 2200$, and unvoiced beyond that limit; for $n < 1000$, the signal has low amplitude but is periodic. The calculated parameters reflect this property of the signal. For $n < 1000$, the magnitude sum and zero crossing rate are low, with high prediction gain, typical of a low-amplitude voiced frame. For $1000 < n < 2200$, the energy is high, and the zero crossing rate is low with medium prediction gain, common for most voiced frames. For $n > 2200$, energy and prediction gain are low with high zero crossing rate, typical characteristics of unvoiced frames. Thus, by deciphering the information of these parameters, one can make the proper voicing decision.

Voicing Detector Design

A voicing detector can rely on the parameters discussed so far (energy, zero crossing rate, etc.) to make the proper decision. A simple detector can be implemented by using just one parameter as input. For instance, the zero crossing rate can be used for voicing detection in the following manner: if the rate is lower than a certain threshold, the frame is declared voiced; otherwise, it is unvoiced. The design problem is therefore to find the proper threshold so that a voicing decision can be accomplished reliably. By analyzing a large amount of speech signals, it is possible to come up with a reasonable value of a decision threshold so as to minimize the total classification error.

Relying on just one parameter, however, limits the robustness of the system. For the voicing detector using the zero crossing rate alone, noise contamination can increase the rate in such a way that voiced frames are classified as unvoiced frames. Thus, using more parameters of the frame is necessary to improve the reliability in voicing detection.

Consider the parameters of the frame grouped into a vector of form

$$\mathbf{x}[m] = [x_1[m] \quad x_2[m] \quad \cdots \quad x_M[m]]^T, \tag{9.11}$$

corresponding to the frame ending at time instant m. Elements of the vector are parameters of the frame, which could be any of the quantities described so far. The voicing detector utilizes the elements of the vector to make the proper decision. This situation corresponds to a *pattern classification* problem, where the pattern vector is classified in two different ways: voiced and unvoiced.

A pattern classifier based on a linear combiner is a simple and effective method to accomplish the voicing detection task. In this approach, the frame ending at time m is classified according to

$$w_0 + \sum_{k=1}^{M} x_k[m] w_k > 0. \tag{9.12}$$

That is, if the quantity is greater than zero, the frame is declared voiced; otherwise it is unvoiced. The problem of voicing detector design reduces to finding the proper set of weights w_k, $k = 0, \ldots, M$. The design procedure for pattern classifier based on a linear combiner is covered in Appendix D. Here, the way to obtain the training data or exemplars is described.

By analyzing a large amount of speech data on a frame-by-frame basis, the parameters of interest are extracted and stored. Each frame is classified as voiced or unvoiced by a human observer; that is, by visual evaluation of the waveform, the human operator decides about the voicing status. The decisions of the human operator constitute the desired response on which the pattern classifier is designed. Hence, the extracted parameters (pattern vectors) together with voicing decision (desired response) form the training exemplars. These exemplars are used to design the pattern classifier, which basically consists of finding the set of weights in (9.12). Readers are referred to Appendix D for details related to pattern classifier design.

9.4 THE FS1015 LPC CODER

Highlights are provided for the FS1015 LPC coder, with some of the most important aspects explained. Readers must be aware that in order to implement a bit-stream compatible version of the coder, consultation with official documentation is mandatory.

Speech Input and Pre-emphasis

The 8-kHz sampled speech is quantized to uniform PCM with 12 bits/sample. It is then pre-emphasized with the first-order filter having system function $(1 - 0.9375z^{-1})$.

LP Analysis

A covariance method with stabilization is the official approach used by the FS1015 for LP analysis. However, most modern coders are in favor of the autocorrelation method, explained in Chapter 4, due to its computational efficiency and ease of stability check.

The FS1015 coder incorporates a *pitch synchronous* approach during LP analysis. In this method, 130 samples are windowed from the current frame, with the position of the window adjusted with respect to a certain reference point of the waveform, like the beginning of a pitch cycle. By doing so, smoothness is added to the synthesized speech, especially during voiced frame synthesis, since variations of the LPC for consecutive voiced frames are diminished. See Exercise 9.5 for additional details.

Quantization of the LPC is covered in Chapter 8.

Pitch Period Estimation

For pitch period estimation, 60 values of pitch period are considered, given by $T = 20, 21, \ldots, 39, 40, 42, \ldots, 78, 80, 84, \ldots, 156$. These values of pitch period correspond to the frequency range between 50 and 400 Hz. The magnitude difference function is applied to a pseudo-prediction-error signal obtained by lowpass filtering the speech signal and inverse filtering by a second-order filter.

Voicing Detector

A linear pattern classifier (Section 9.3) is used to perform the voicing determination task. The pattern vectors are comprised of the following parameters:

- Low-band energy (Section 9.3).
- Peak to bottom ratio of the magnitude difference function. This parameter indicates the amount of periodicity of the frame. For voiced frames the ratio is in general much higher than for unvoiced frames.
- Zero crossing rate (Section 9.3).

The situation where a single voiced frame is located between unvoiced frames can cause annoying artifacts in the synthetic speech. To prevent this from happening, voicing decisions of neighboring frames are taken into consideration in the final decision.

Bit Allocation

Table 9.1 summarizes the bit allocation scheme for the FS1015 coder. The 60 pitch period values are jointly encoded with the voicing state using 7 bits. Power is encoded using 5 bits. Details of LPC encoding are presented in Chapter 8. Synchronization is an alternating one/zero pattern. Error protection is provided for unvoiced frames only, where 21 bits are transmitted. The scheme utilizes a total of 54 bits per frame; for a frame length of 22.5 ms, the bit-rate of 2400 bits per second (bps) results.

TABLE 9.1 Bit Allocation for the FS1015 Coder[a]

	Resolution	
Parameter	Voiced	Unvoiced
Pitch period / voicing	7	7
Power	5	5
LPC	41	20
Synchronization	1	1
Error protection	—	21
Total	**54**	**54**

[a] Data from Tremain [1982], Figure 8.

9.5 LIMITATIONS OF THE LPC MODEL

The overly simplistic model that the LPC coder relies on has relatively low computational cost and makes the low bit-rate speech coder a practical reality. The uncomplicated model, however, is also highly inaccurate in various circumstances, creating annoying artifacts in the synthetic speech. In this section, fundamental limitations of the LPC model are analyzed. These limitations are targets for improvement by the next generation of speech coders.

LIMITATION 1: *In many instances, a speech frame cannot be classified as strictly voiced or strictly unvoiced.*

Indeed, there are transition frames (voiced to unvoiced and unvoiced to voiced) that the LPC model fails to correctly sort. This inaccuracy of the model generates annoying artifacts such as buzzes and tonal noises.

LIMITATION 2: *The use of strictly random noise or a strictly periodic impulse train as excitation does not match practical observations using real speech signals.*

The excitation signal can be observed in practice as prediction error and is obtained by filtering the speech signal using the prediction-error filter. In general, the excitation for unvoiced frames can be reasonably approximated with white noise. For voiced frames, however, the excitation signal is a combination of a quasiperiodic component with noise. Thus, the use of an impulse train is a coarse approximation that degrades the naturalness of synthetic speech. In addition, the quasiperiodic excitation component often displays a changing period, and the shape of the pulse is not exactly an impulse. For the FS1015 coder, the excitation pulses are obtained by exciting an allpass filter using an impulse train.

The above discussion is applicable for a typical prediction order of ten, such as the case of the FS1015 coder.

LIMITATION 3: *No phase information of the original signal is preserved.*

As we have seen, no phase information is captured by the LPC model: neither voiced nor unvoiced frames have explicit parameters containing clues about the phase. The synthetic speech sounds like the original because the magnitude spectrum, or power spectral density, is similar to the original signal. Even though a human listener is relatively insensitive to the phase, retaining some phase information adds naturalness to the synthetic speech, leading to an improvement in quality.

It is important to note that during speech synthesis of the LPC decoder, phase information for unvoiced frames can generally be ignored, since noise perception is practically phaseless. Not quite for voiced frames where a nonsmooth pitch period contour is perceived as being unnatural and distorted. Thus, voiced frames synthesis requires the maintenance of a (ideally) continuous pitch contour so that transitions (normally of differing pitch periods) become as transparent as possible. This can be achieved by controlling the timing of the excitation pules (Exercise 9.4) while decoding. Therefore for voiced frames, even though the original phase is lost, the relative phase or position among the excitation pulses must be re-created in such a way that transition is smooth between frames with the smallest amount of artifacts introduced.

LIMITATION **4:** *The approach used to synthesize voiced frames, where an impulse train is used as excitation to a synthesis filter with coefficients obtained by LP analysis, is a violation of the foundation of AR modeling.*

The violation introduces spectral distortion into the synthetic speech, which becomes more and more severe as the pitch period decreases. This is the reason why the LPC coder does not work well for low-pitch-period or high-pitch-frequency talkers, like women and children. For the typical male, however, spectral distortion is moderate, and reasonable quality can be obtained with the LPC coder.

Recall from Chapter 3 that the AR model is based on flat-spectrum excitation. The flat-spectrum requirement is satisfied by white noise, but a single impulse also meets this constraint since the spectrum is constant for this latter case. For a periodic train of impulses, the spectrum is not flat in general; thus, the basic assumption of the AR model is not fulfilled. In fact, the spectrum becomes more oscillatory with decreasing period. Figure 9.10 shows the periodograms of some impulse trains, where we can see that oscillations exist. The amplitude of the oscillation increases when the period decreases.

Example 9.4 An experiment is used to illustrate the behavior of LP analysis when the excitation to a synthesis filter is an impulse train. The setup is shown in Figure 9.11. A set of reference LPCs is selected and corresponds to a stable synthesis filter. The synthesis filter is excited with its output used for LP analysis, which generates a set of estimated LPCs. Using enough samples and the appropriate

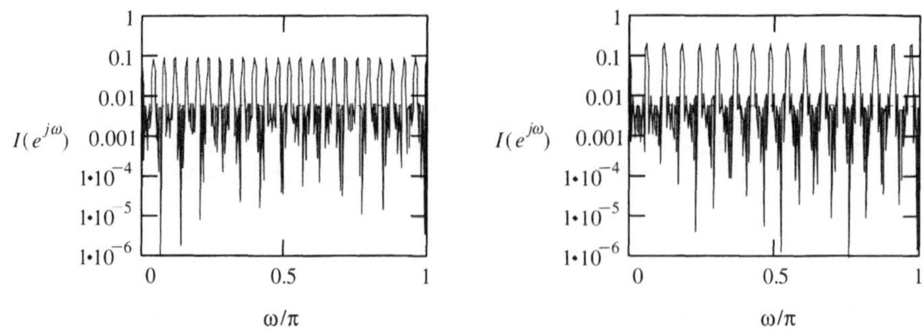

Figure 9.10 Plots of periodograms for impulse trains with a period of 50 (*left*) and a period of 33 (*right*). The impulses have unit amplitude and a Hamming window of length equal to 180 samples is used.

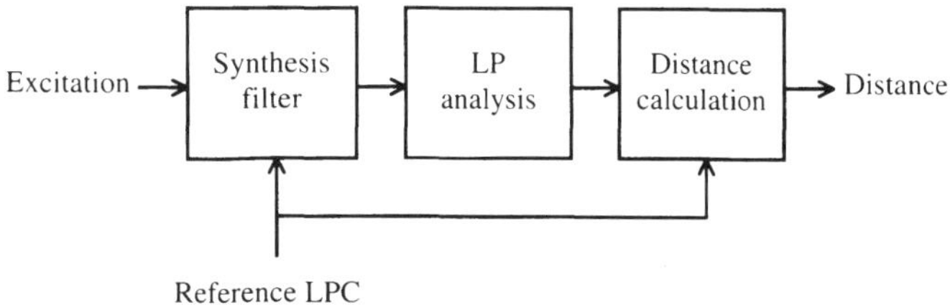

Figure 9.11 Setup of an experiment in LP analysis.

excitation signal (white noise or a single impulse), the estimated LPCs found from LP analysis closely match the reference LPC (system identification, Chapter 4). The degree of closeness can be found by comparing the estimated and reference coefficients. Figure 9.12 shows the filter's output when the input is an impulse train; the synthesis filter has a prediction order of ten. As we can see, the filter's output consists of a periodic signal, formed by the superposition of impulse responses of the filter. Since the filter is stable, its impulse response decays with time. For a long enough period, interference between consecutive impulse responses is negligible. For a short period, however, the impulse responses overlap.

Figure 9.13 shows the experimental results, where the distance measure between the estimated and reference LPC is plotted as a function of the period. For simplicity, the sum of squared error between the two sets of coefficients is used as the distance measure. As we can see, the estimation is quite good for high values of period, say, $T > 30$, with low distance between coefficients. As the period decreases, estimation error increases rapidly. This is because the impulse responses overlap, thus distorting the autocorrelation values on which the LP analysis procedure relies. For high period values, the impulse responses are well separated, and the autocorrelation values are close to that obtained with single impulse excitation.

We conclude from the experiment that the LP analysis procedure, applied to a synthesis filter excited by a periodic impulse train, will not perform well—as a system identification tool—when the period of the impulse train decreases past a certain threshold. The threshold depends on the decay rate of the filter's impulse

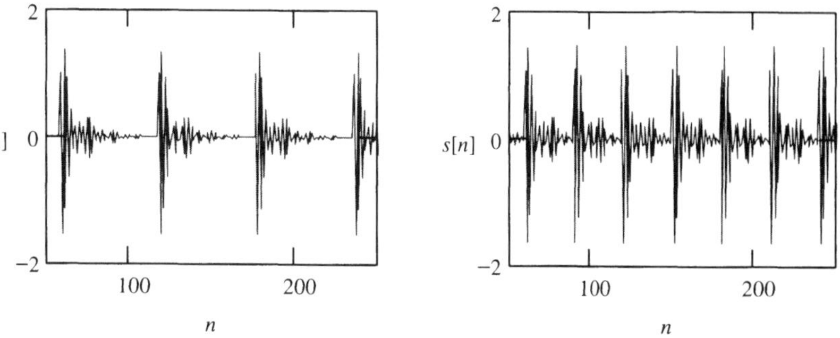

Figure 9.12 Output of a synthesis filter excited by an impulse train. *Left*: Period equal to 59. *Right*: Period equal to 30.

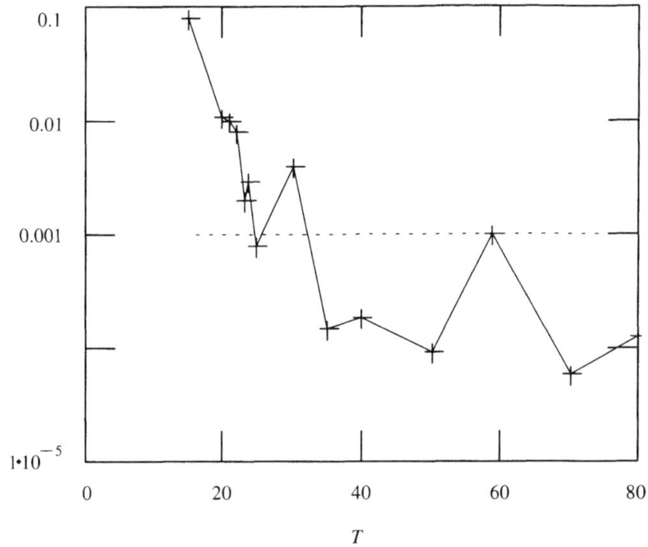

Figure 9.13 Sum of squared error (*SSE*) between the estimated and reference LPC as a function of the period of the excitation impulse train in a single experiment. The dotted line represents the *SSE* obtained from a single impulse input.

response. It is important to emphasize the fact that failure of LP analysis is due to violation of the assumptions for the AR model, since the excitation signal is invalid.

9.6 SUMMARY AND REFERENCES

Principles of linear prediction coding are presented in this chapter beginning with the speech production model, followed by structure of the algorithm, and finishing with the main limitations. The LPC coder provides intelligible reproduction at low bit-rate. However, the use of only two kinds of excitation signal gives an artificial quality to the synthetic speech. This approach also suffers when used in noisy environments since the encoder can be confused by the background noise, declaring a frame as unvoiced even though it is voiced.

The LPC coder pertains to the class of parametric coders (Chapter 1) where the synthetic speech does not assume the shape of the original signal in the time domain. For these types of coders, the speech signal is characterized in terms of a set of model parameters, where the quality of the synthetic speech depends largely on the accuracy of the model. Also, measurement of SNR is meaningless, having poor correlation with subjective quality. The LPC coder can also be classified as a source-controlled multimode coder, with the coding technique adapted to the local properties of the signal. That is, depending on whether the signal frame is voiced or unvoiced, different methodologies are applied.

Due to the poor quality of the LPC coder, it is no longer active for communication purposes. However, there are still applications where the LPC principle is appropriate, such as low-quality reproduction and speech synthesis. As we will see in subsequent chapters, the principles of the LPC coder can be refined so as to create coders with higher performance. The code-excited linear prediction (CELP) coder, for instance, utilizes a long-term and short-term synthesis strategy to avoid voiced/unvoiced decision, and phase information is partially retained (Chapter 11). On the other hand, the mixed excitation linear prediction (MELP) coder employs a more sophisticated excitation signal to improve naturalness of the synthetic speech (Chapter 17).

Major contributions of the LPC coder can be summarized as follows:

- Demonstration of low bit-rate speech coding based on linear prediction.
- Use of a simple speech production model to achieve high coding efficiency.
- Provision of a reference framework for next generation speech coders.

See Deller et al. [1993] for early development of speech production modeling. Incomplete information on the FS1015 coder can be found in Tremain [1982]. A more robust voicing detector is presented in Campbell and Tremain [1986] and is incorporated in an enhanced version of the coder known as LPC-10E. See Exercise 9.4 for guidelines on coder implementation.

EXERCISES

9.1 Consider the unit-amplitude impulse train

$$x[n] = \sum_{i=-\infty}^{\infty} \delta[n - iT],$$

with $T > 0$ the period. Prove that the PSD is given by

$$S_x(e^{j\omega}) = \frac{2\pi}{T^2} \sum_{k=-\infty}^{\infty} \delta\left(\omega - \frac{2\pi k}{T}\right).$$

Hence, the PSD of an impulse train is another impulse train in the frequency domain with amplitude $2\pi/T^2$ and period $2\pi/T$.

9.2 Using some speech signal, calculate the magnitude difference function on a frame-by-frame basis. Use a frame length of 180 samples and a lag in the range of 20 to 150 (Chapter 2). Determine the maximum and minimum in the resultant functions and compute their ratios. What conclusion can be obtained regarding the magnitude of the ratio and voicing state?

9.3 The first reflection coefficient is included in the enhanced version of the FS1015 coder for voicing detection. The parameter reflects the spectral tilt of the speech waveform. Voiced frames typically have a significant tilt of decreasing magnitude with increasing frequency. Using some speech signal, obtain the first reflection coefficient by LP analyzing on a frame-by-frame basis. What conclusion can be obtained for the first reflection coefficient and voicing state?

9.4 Below are the steps to follow for the implementation of the LPC coder. First, the encoder is constructed without the quantizers (parameter encoders); then, the decoder is constructed without the parameter decoders. Functionality of the encoder/decoder constructed in this way is tested; if the system is working properly, the parameter encoders and decoders are incorporated to evaluate the final system.

1. *Encoder.*
   - Pre-emphasis filter: This is a first-order difference equation (Chapter 4).
   - Voicing detector: Before exploring a multiparameter detector, a simple detector based on just one parameter, like the zero crossing rate, can easily be implemented. In this case, find out the threshold value that seems to do a reasonably good job in voiced/unvoiced classification by using some speech material.
   - LP analysis: See Chapter 4 for details regarding autocorrelation estimation and algorithms to solve the normal equation. Prediction order is equal to ten.
   - Prediction-error filter and power computation: The filter is a difference equation with coefficients given by LP analysis. Power computation makes use of the prediction error.
   - Pitch period estimation: The magnitude difference function (Chapter 2) can be used.
   - Integration: The blocks from previous steps are joined together following the diagram of Figure 9.6. However, no quantization or encoding of an individual parameter is considered yet. The decoder is first built to perform a system test.
2. *Decoder.*
   - Impulse train generator: Given a period value, a train of impulses is generated, with each impulse having unit amplitude. The instant when the first impulse appears can be adjusted for consecutive voiced frames so that variation of the pitch period across frame boundaries is minimized.
   - White noise generator: This is essentially a random number generator. Uniform distribution is appropriate. Most programming languages have a built-in random number generator. Many issues exist in the design of a random number generator; as a reference, see Banks and Carson [1984]. To reduce computational load, an array of random numbers can be used; that is, a long sequence of random numbers is stored in memory as an array. To generate the sequence of white noise, one single random number is created and is used to index the first element of the sequence from the array. That is, instead of generating N numbers, only one number is generated per frame. This method was found to work well in practice.
   - Synthesis filter: This is a tenth-order difference equation.

- De-emphasis filter: This is a first-order difference equation.
- Integration: From the diagram in Figure 9.8, join the different blocks to form the decoder.

3. *Test of encoder–decoder operation.*
   - Using the system developed so far, speech is input to the encoder with the resultant parameters processed by the decoder. Verify the intelligibility of the synthetic speech. Note that this step is done without any quantization of individual parameters. If the quality of synthetic speech is as expected, proceed to the next steps. Otherwise, review the different blocks to discover the problem.
4. *Incorporation of parameter-quantizers.*
   - LPC: See Chapter 8 for options and ideas. You can opt for the same technique as used by the FS1015 or methods adopted by other standards, as well as creating your own scheme.
   - Power: Uniform quantization can be used for a first test. However, nonuniform quantization will definitely provide higher quality. This can be designed using the Lloyd algorithm with training data collected from real speech signals (Chapter 5).
   - Pitch period: Many options are possible. For instance, with 7-bit encoding, 128 values of pitch period are covered; the interval of [20, 147] is reasonably good in practice. A scheme similar to FS1015 can also be applied.
   - Integration (encoder): Encoders for the various parameters can be integrated into the encoder according to Figure 9.6; outputs are packed together as the LPC bit-stream.
   - Integration (decoder): Decoders for the various parameters are integrated to the decoder according to Figure 9.8. The LPC bit-stream is unpacked with the indices directed to the right places.
5. *Overall test and system improvement.*
   - The system is tested again for correct operation and quality. For quality improvement, more bits can be allocated to a particular quantizer, at the expense of a higher bit-rate. After the basic system is working properly, the different blocks can be modified individually. For instance, a more sophisticated voicing detector using multiple parameters can be developed or a better LPC quantization scheme can be utilized.

9.5 A 130-sample block is selected for pitch-synchronous LP analysis within each 180-sample voiced frame, specified with $n \in [0, 179]$. The 130-sample block is indicated with a starting position $n_o \in [-20, 50]$; that is, the interval on which LP analysis is performed is given by $n \in [n_o, n_o + 129]$. The starting position is generally found in such a way that consecutive frames rely on approximately the same cycle of the waveform to perform LP analysis. In this way, synchronization is maintained. Design an algorithm to determine n_o based on peak location. Propose alternative realizations to boost robustness.

9.6 To improve robustness in voicing detection, a smoothing stage is often deployed so as to process the voicing state data with the objective of eliminating spurious parameters that might compromise the quality of synthetic speech. The smoothing mechanism operates by examining the past and future to come up with the final voicing decisions. Propose some schemes to achieve the objective of smoothing. What is the disadvantage, if any, for the incorporation of such a mechanism?

9.7 *Time-scale modification* consists of expanding or compressing the time length of a signal, with minimum effect on the perception of its frequency content. A time-scale-expanded signal takes longer to play while a time-scale-compressed signal takes lesstime to play. In both cases the signals have frequency contents similar to the original and thus are perceived as originating from the same speaker. One application of the technique is in speech synthesis where the duration of the signal or "speed" can be modified at will. Do you think that the basic LPC decoding scheme supports time-scale modification? That is, what happens when the length of a frame is modified during decoding? How would you improve the quality of the synthetic speech under variable frame-length decoding?

9.8 Many speech coding algorithms rely on interpolation to ensure a smooth transition between frames during decoding. That is, the actual parameters used for decoding are obtained through interpolation of two sets of parameters: past and present frames. For the case of the LPC coder, the parameters are pitch period, voicing, power, and LPC. Which of these parameters are not suitable for interpolation? Propose some interpolation schemes that would work reasonably well for signal decoding.

CHAPTER 10

REGULAR-PULSE EXCITATION CODERS

The idea of DPCM, as presented in Chapter 6, is based on the quantization of prediction error, instead of quantizing the signal itself. DPCM yields far better quality than plain PCM at the same bit-rate since the dynamic range of prediction error, for a well-designed predictor, is often much lower than that of the original signal. Reduced dynamic range lowers the quantization error since step size of the quantizer can be decreased. It is possible to apply the same principle to speech; however, due to its nonstationary nature, adaptation is required. ADPCM has achieved various levels of success at bit-rates greater than or equal to 16 kbps and is the core technology of several speech coding standards, like the ITU-T G.726 [ITU, 1990].

Within the context of LP, prediction error is obtained by filtering the speech signal using the prediction-error filter. The original signal can be reconstructed by passing the prediction-error sequence through the synthesis filter, with the synthesis filter and prediction-error filter the inverse of each other. Thus, using LP (with the coefficients updated on a frame-by-frame basis), an ADPCM system can be designed by quantizing the samples of the prediction error.

Atal and Remde (1982) observed that it was not necessary to encode all prediction-error samples in order to achieve good quality for the reconstructed speech. In fact, many prediction-error samples have relatively low amplitudes. Typically, by preserving only 10% of the samples (i.e, the other 90% are set to zero), different kinds of speech sounds can be generated with little perceptual distortion. The resultant approach is called the multipulse excitation model since the excitation sequence is formed by scattered pulses, with most of the samples set to zero. The biggest challenge for the model to become practical is the development of an efficient procedure for determining the pulses so as to minimize distortion of the synthetic speech. The genesis of the multipulse model was also aroused by the LPC coder, where utilization of only two classes of excitation is to be avoided for quality improvement (Chapter 9).

In this chapter, the multipulse excitation model is presented, followed by the most successful coder in the multipulse class, namely, the regular-pulse-excited–long-term prediction (RPE-LTP) coder. The principle of the coder is utilized by the GSM 6.10 standard, adopted by the ETSI for digital mobile radio applications [Vary et al., 1988; ETSI, 1992a].

10.1 MULTIPULSE EXCITATION MODEL

Consider the following speech coding method. The speech signal is segmented into frames, with each frame LP analyzed, and the resultant LPC used to construct the prediction-error filter, where samples of prediction error are derived. These samples are quantized and then transmitted to the decoder together with the LPC. On the decoder side, by filtering the quantized prediction error with the synthesis filter constructed with the received LPC, a waveform that is closed to the original speech is generated.

The mentioned scheme is essentially an ADPCM system, where the predictor is made adaptive. Even though the method is practical, it is highly redundant in the sense that unnecessary information is being sent as part of the bit-stream, needlessly increasing the resultant bit-rate. The main source of redundancy is from the prediction-error signal itself; since many samples are low amplitude in nature, their elimination from the bit-stream causes diminutive perceptual distortion.

The multipulse excitation model is based on the observation that only a small percentage of nonzero samples are necessary to excite the synthesis filter so as to produce good quality speech. The nonzero samples are encoded by their amplitudes and positions. The problem is therefore the finding of a set of pulses that minimizes distortion. The number of pulses per frame is directly related to bit-rate and quality and must be selected to satisfy the constraints of an application. Many methods have been proposed to solve the problem of multipulse excitation; in general, they can be classified as *open loop* and *closed loop*, explained next.

Open-Loop Method

Figure 10.1 shows the block diagram of a coding system based on the multipulse model, with the pulses selected via an open-loop approach. The input speech is LP analyzed to obtain the LPCs, which are used to construct the prediction-error filter. The resultant prediction-error sequence is processed by the pulse selection block, where a certain criterion is applied to eliminate the majority of samples. The selected samples are encoded by their amplitudes and positions. The decoder recovers the information, regenerates the excitation sequence, and passes it through the synthesis filter to obtain the synthetic speech.

Many criteria can be used in the pulse selection block. A simple scheme can be based on the magnitude of the pulses, where the samples are first sorted

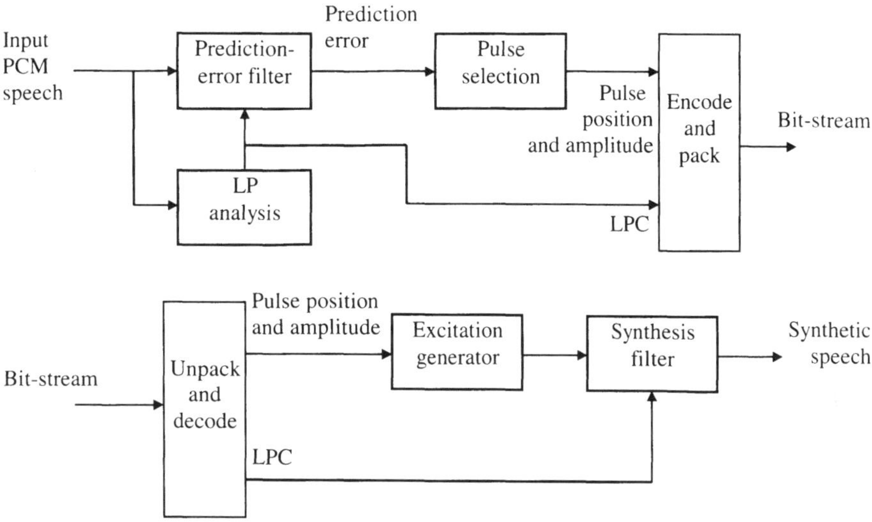

Figure 10.1 Encoder (*top*) and decoder (*bottom*) of an open-loop multipulse coder.

accordingly, and only a fixed number of the highest magnitude samples are retained. This approach, however, requires the coder to take note of the amplitude as well as position of every selected pulse. One popular method is the regular-pulse excitation scheme, illustrated in Figure 10.2. In this approach, the prediction-error

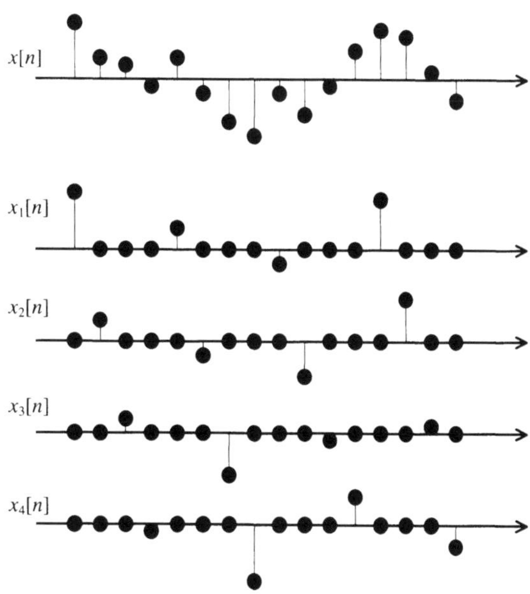

Figure 10.2 Illustration of regular-pulse excitation, using a down-sampling factor of 4. The original prediction-error sequence $x[n]$ is down-sampled to four different sequences.

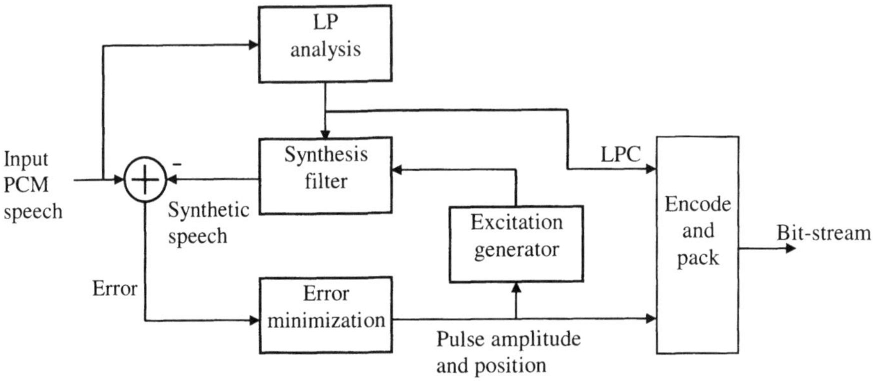

Figure 10.3 Encoder of a closed-loop multipulse coder.

sequence is down-sampled to a number of different sequences, with the highest energy sequence selected. In this way, only the amplitude of the pulses and the sequence number need to be encoded, thus reducing the number of bits needed to carry information.

Closed-Loop Method

Figure 10.3 shows the encoder of a closed-loop system. In this approach, the excitation sequence is selected through a feedback loop, where the difference between input speech and synthetic speech is minimized. Note that speech synthesis is actually performed during encoding; hence, the method is also known as *analysis-by-synthesis*, due to the fact that the signal is analyzed by synthesizing a replica for comparison. This is an important concept that will be revisited in subsequent chapters.

Within the loop, parameters of the signal are determined so as to minimize error. In the present case, the parameters are the amplitudes and positions of the pulses that constitute the excitation sequence.

Comparison

Since the closed-loop method optimizes the excitation sequence, lower error (with respect to the open-loop method) is normally achievable, at the expense of higher computational load. The open-loop method has the advantage of being relatively simple; and for the case of regular-pulse excitation, it is highly efficient for encoding as well, since only one position is required per group of samples.

In 1987 [Vary et al., 1988], the Groupe Special Mobile (GSM) of the Conference of European Posts and Telephones (CEPT) was searching for an appropriate coder for use by the pan-European digital mobile radio system. Initially, more than 20 different proposals were under consideration. Among them was the regular-pulse excitation coder and other coders of the multipulse family. The effort led to the standardization of the GSM 6.10 regular-pulse-excited–long-term prediction

(RPE-LTP) coder in 1988 [ETSI, 1992a]. The success was due to a combination of advantages, like low computational cost, high-quality reproduction, robustness against channel errors, and coding efficiency. In addition to digital cellular telephony, this coder has since been used for many additional applications, such as messaging, because of its low complexity.

The rest of the chapter is dedicated exclusively to the study of the RPE-LTP coder.

10.2 REGULAR-PULSE-EXCITED–LONG-TERM PREDICTION

Structural details of the GSM 6.10 RPE-LTP coder are presented in this section. The coder is essentially an ADPCM system, where a predictor is computed from the signal, with the prediction error found, and is subsequently quantized using an adaptive scheme. The predictor is implemented as a cascade connection of short-term and long-term predictors; inclusion of the long-term predictor, as shown in Chapter 4, greatly increases the average prediction gain, thus elevating the overall performance. Readers must be aware that in order to implement a bit-stream compatible version of the coder, consultation with official documentation is mandatory.

Encoder

Figure 10.4 shows the block diagram of the encoder, where parameters of each frame/subframe are extracted and packed to form the bit-stream. Like other coders, the encoder divides the input speech samples into frames for further processing. In the present case, each frame has a length of 160 samples (20 ms) and is subdivided into four subframes of 40 samples each. Functions of each block in the encoder structure are explained as follows.

Preprocessing

The filter with system function $(1 - z^{-1})/(1 - 0.999z^{-1})$ is used to highpass filter the input speech so as to eliminate any zero-frequency (DC) component. The resultant signal is pre-emphasized (Chapter 4) using a filter with system function $1 - 0.86z^{-1}$.

LP Analysis

LP analysis is performed on every 160-sample frame. A prediction order of eight is employed. Nine autocorrelation values are calculated from the frame using a rectangular window. The autocorrelation values are solved to obtain eight reflection coefficients.

LPC Quantization and Interpolation

The RCs are transformed to LARs through a piecewise linear function and are then scalar quantized using 36 bits. An index of the quantized LPCs is transmitted as

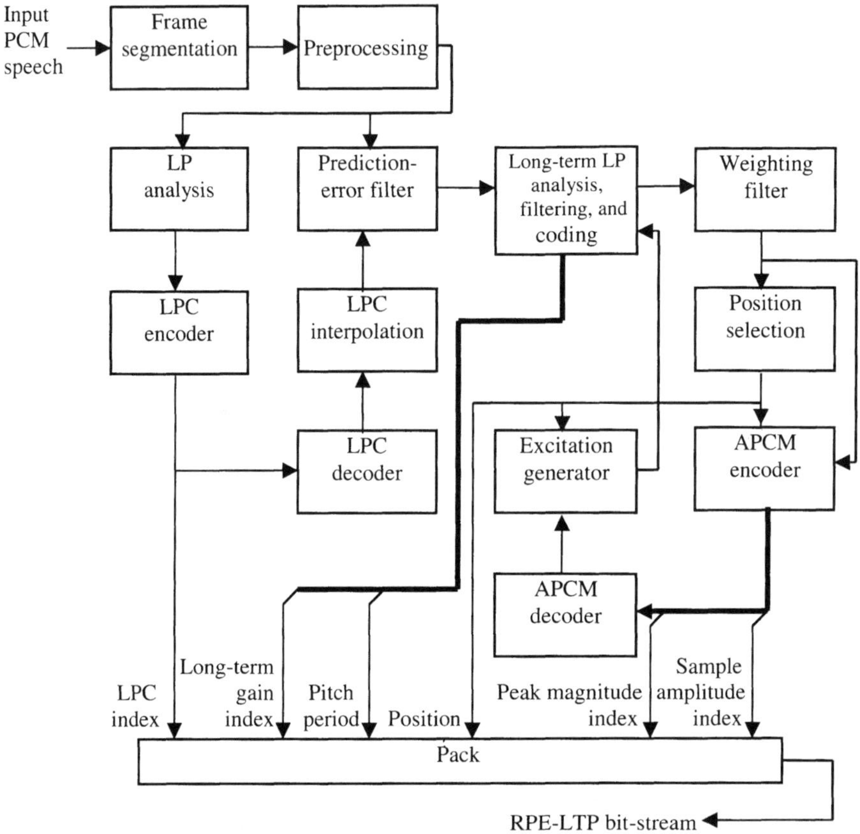

Figure 10.4 Block diagram of the RPE-LTP encoder.

part of the bit-stream. The quantized LARs are interpolated to four different sets to be used by the four 40-sample subframes (Chapter 8).

Prediction-Error Filter

The interpolated LARs are transformed back to RCs, which are used to form the prediction-error filter. This filter is implemented in lattice form (Chapter 2). The filter produces the (short-term) prediction-error signal (internal prediction, Chapter 4) at its output.

Long-Term LP Analysis, Filtering, and Coding

The "long-term LP analysis, filtering, and coding" block of Figure 10.4 is redrawn in Figure 10.5, where all details are shown. Note that the operations described below are repeated once every subframe, or every 40 samples. The short-term

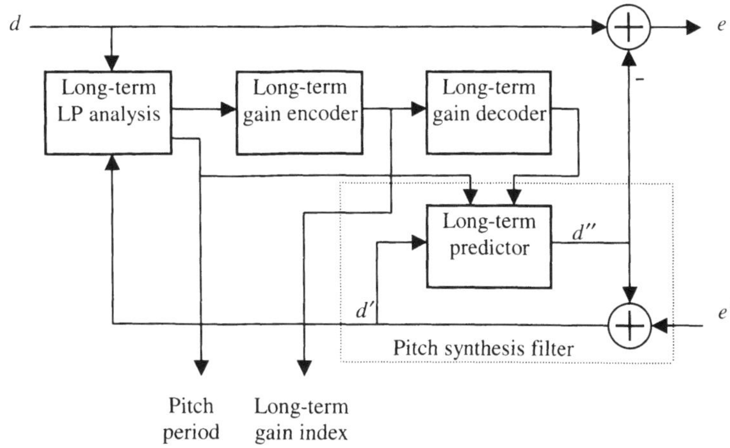

Figure 10.5 Implementation of the long-term LP analysis, filtering, and coding block.

prediction error d from the (short-term) prediction-error filter together with the reconstructed prediction error d' are used for long-term LP analysis. First, the cross-correlation function $R[T]$ is calculated according to

$$R[T] = \sum_{n} d[n]d'[n - T] \tag{10.1}$$

for $T = 40$ to 120; that is, a total of 81 pitch period values are utilized, encodable with 7 bits. In (10.1), the range of n corresponds to the subframe under consideration and takes into account 40 samples. Thus, the range of T is selected in such a way that the delayed samples (from d') used in the cross-correlation calculation are outside the current subframe. Pitch period of the subframe is defined as the one that maximizes (10.1). Once the pitch period of the subframe is found, the long-term gain is calculated as

$$b = -\frac{R[T]}{\sum_{n} (d'[n - T])^2}. \tag{10.2}$$

Note the differences between the long-term LP analysis procedure of the present approach and that of Chapter 4. First, the reconstructed prediction error d' is involved with the correlation calculation; since d' is the actual sequence employed for speech synthesis, its usage yields higher performance due to the fact that both encoder and decoder can operate using the exact same signals. Second, instead of using the full expression to evaluate the sum of squared prediction error, a relatively simple cross-correlation expression is deployed. The use of (10.2) is suboptimal in the sense that the overall prediction error may not be minimized within the range of interest. Also, only negative long-term gain is considered in the present case since the peak of $R[T]$ is in general positive; restricting the sign of the gain limits the maximum achievable prediction gain. The present approach, however, has a much-reduced computational cost.

Since the pitch period values are between 40 and 120, they are transmitted directly as binary values using 7 bits. Long-term gain is quantized with 2 bits resulting in four quantized values; these are $\{-0.1, -0.35, -0.65, -1\}$ assigned to the intervals $b \in [-0.2, \infty)$, $[-0.5, -0.2)$, $[-0.8, -0.5)$, and $(-\infty, -0.8)$, respectively.

The pitch synthesis filter takes the reconstructed overall prediction error e' to generate the reconstructed short-term prediction error d'

$$d'[n] = e'[n] + d''[n], \tag{10.3}$$

where d'' is the prediction given by

$$d''[n] = -\hat{b}d'[n - T]. \tag{10.4}$$

Note that the quantized long-term gain is used to calculate the prediction. The overall prediction error is then

$$e[n] = d[n] - d''[n]. \tag{10.5}$$

It is interesting to note how the relatively simple long-term processing scheme described in Chapter 4 is modified to accommodate all the quantization blocks and quantized signals in the RPE-LTP coder. The changes are necessary for improved accuracy and reduced computational cost.

Weighting Filter

The overall prediction-error sequence of the current subframe is filtered by an 11-tap FIR filter. Plots of impulse response and frequency response are shown in Figure 10.6. As we can see, the filter is lowpass in nature. Its incorporation

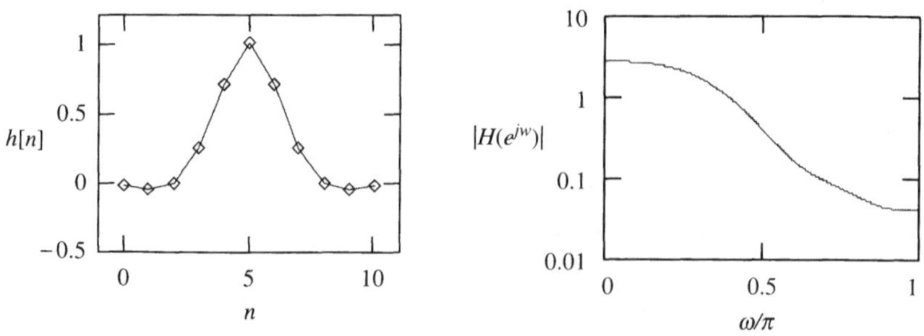

Figure 10.6 Impulse response (*left*) and frequency response (magnitude, *right*) of the weighting filter. Data from ETSI [1992a], Table 4.4.

smoothes out variations between samples, suppresses high-frequency noise, and makes transitions between subframes smoother, thus improving the subjective quality of the synthetic speech.

Convolution between the 40-sample sequence $e[n]$ and the 11-sample sequence $h[n]$ (impulse response) produces $40 + 11 - 1 = 50$ samples. In the present case, however, only the 40 middle samples of the convolution are calculated. For the 40-sample error sequence $e[n]$, $n = 0, \ldots, 39$, the convolution operation is described by

$$x[n] = \sum_{k=0}^{10} h[k]e[n + 5 - k] \tag{10.6}$$

for $n = 0, \ldots, 39$. The operation is done under the assumption that the current subframe is isolated through a rectangular window; that is, $e[n + 5 - k] = 0$ for $n + 5 - k < 0$ and $n + 5 - k > 39$.

Position Selection

The filtered prediction-error sequence is down-sampled by a ratio of 3, resulting in four interleaved sequences with regularly spaced pulses. These are defined with

$$x_m[n] = x[m + 3n]; \qquad m = 0, 1, 2, 3; \quad n = 0, 1, \ldots, 12. \tag{10.7}$$

The above equation defines four sampling grids illustrated in Figure 10.7. The best candidate sequence $x_m[n]$ is selected by verifying the energy

$$E_m = \sum_{n=0}^{12} x_m^2[n]; \qquad m = 0, 1, 2, 3, \tag{10.8}$$

with the one producing the highest energy selected. The resultant grid position m is coded with 2 bits.

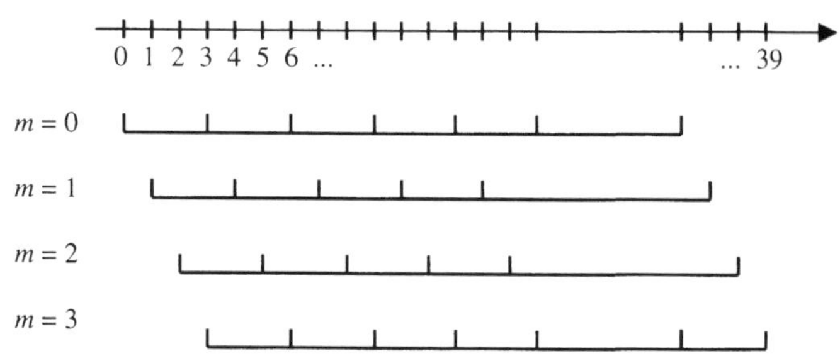

Figure 10.7 Sampling grids used in position selection.

APCM

The selected subsequence $x_m[n]$ is quantized using a forward gain-adaptive quantizer (Chapter 6). For each 13-sample sequence, the peak magnitude is determined with

$$g = \max_n |x_m[n]|, \tag{10.9}$$

which is quantized using a nonlinear (logarithmic) quantizer having 6 bits. The quantized peak magnitude is used to normalize the samples

$$x'_m[n] = x_m[n]/\hat{g} \tag{10.10}$$

and are quantized using a 3-bit uniform quantizer.

Excitation Generator

The quantized samples of the normalized sequence are denormalized with the quantized peak magnitude. The resultant sequence is up-sampled by a ratio of 3 by inserting zero samples according to the grid position. The final signal is denoted by e', as shown in Figure 10.5.

Decoder

Figure 10.8 shows the block diagram of the RPE-LTP decoder. Its operation is quite obvious after studying the encoder. It is left as an exercise for readers to figure out the equations and operational details.

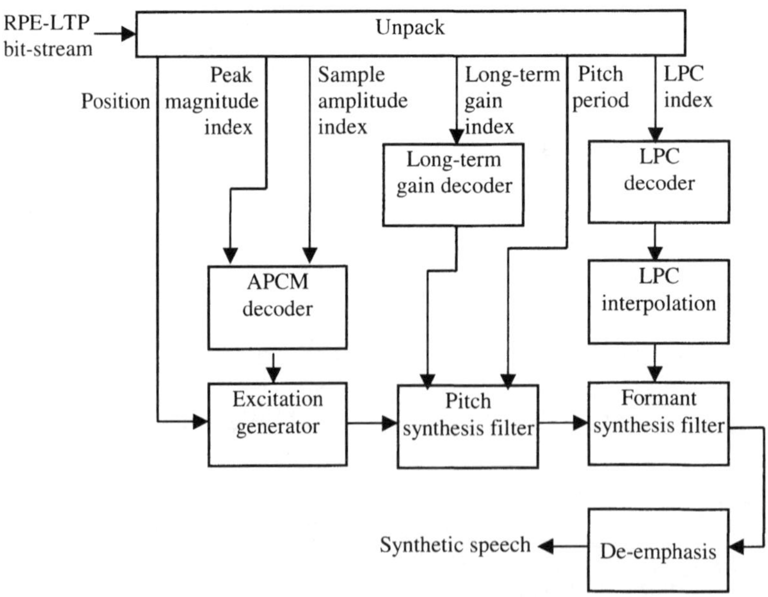

Figure 10.8 Block diagram of the RPE-LTP decoder.

TABLE 10.1 Bit Allocation for the GSM 6.10 RPE-LTP Coder[a]

Parameter	Number per Frame	Resolution	Total Bits per Frame
LPC	8	6, 6, 5, 5, 4, 4, 3, 3	36
Pitch period	4	7	28
Long-term gain	4	2	8
Position	4	2	8
Peak magnitude	4	6	24
Sample amplitude	$4 \cdot 13$	3	156
Total			**260**

[a]Data from ETSI [1992a], Table 1.1a.

Bit Allocation

Table 10.1 summarizes the bit allocation scheme for the GSM 6.10 RPE-LTP coder. Details of LPC encoding are presented in Chapter 8. The scheme utilizes a total of 260 bits/frame, resulting in a bit-rate of 13,000 bits per second (frame length equal to 20 ms).

10.3 SUMMARY AND REFERENCES

Concepts of the multipulse excitation model are presented in this chapter, with emphasis on the RPE-LTP coder, which is by far the most triumphant coder of the family. This coder is designated as GSM "full rate" and is essentially an ADPCM system. It can be classified as belonging to the hybrid type since it is based on a parametric model, with an attempt to match the waveform.

Generally, full potential of a multipulse coder is achieved only when all combinations of pulse amplitudes and positions are taken into account. This approach, however, is computationally too expensive to implement in practice. Several suboptimal methods have been proposed but failed to receive widespread acceptance. This is partly because of the relatively high computational burden, as well as coding inefficiency, since the amplitude and position of each pulse must be transmitted. See Singhal and Atal [1989] for some studies on the topic. The idea of encoding the prediction error was evaluated in many studies in the past, like Un and Magill [1975]. The main ideas of regular-pulse excitation were published in Kroon et al. [1986].

ETSI published a series of recommendations, known as the GSM recommendations, in 1992. They set the standards in many aspects of telecommunication systems, including speech processing and coding, correction for channel errors, voice activity detection, and others. Readers are referred to the bibliography at the end of this book for reference on some of the documents.

In general, the RPE-LTP coder provides good quality and is quite robust under various channel errors and noise conditions. Due to its open-loop operational nature, however, full potential has not been achieved. As we can see, a large amount of bits are allocated to the excitation signal, which is quite inefficient due to the scalar quantization technique involved. As we will see in subsequent chapters, the CELP type of algorithm improves on this by utilizing vector quantization schemes. In the early 1990s, the RPE-LTP coder was designated for replacement, and in 1996, the GSM EFR (Enhanced Full Rate) coder was standardized. This latter coder operates at 12.2 kbps and is based on the ACELP algorithm (Chapter 16).

EXERCISES

10.1 Plot the magnitude response of the preprocessing filter with system function $(1 - z^{-1})/(1 - 0.999z^{-1})$. From the response, what purpose does the filter serve?

10.2 During position selection in RPE-LTP encoding, energy values of the four subsampled sequences must be calculated. However, two of the sequences share most of the samples. Develop an efficient energy calculation procedure based on this fact.

10.3 Modify the basic open-loop RPE-LTP scheme to develop a closed-loop coder. In this case, all four grid positions are evaluated by synthesizing the corresponding speech sequence and comparing to the original, with the grid position resulting in the lowest error energy selected. Draw the block diagram of the encoder and specify the relevant procedures and equations. What does the decoder look like?

10.4 Suggestions for the implementation of an RPE-LTP coder: First, the encoder is constructed without the quantizers (parameter encoders). Then the decoder is constructed without the parameter decoders. Functionality of the encoder/decoder constructed in this way is tested; if the system is working properly, the parameter encoders and decoders are incorporated to evaluate the final system.

1. *Encoder.*
    - Preprocessing: Two filters are involved in this stage, the DC-cancellation filter and the pre-emphasis filter.
    - LP analysis: See Chapter 4 for details regarding autocorrelation estimation and algorithms to solve the normal equation. Prediction order is equal to eight.
    - Prediction-error filter: For the first trial, the filter can be implemented using the coefficients obtained from the last step with neither quantization nor interpolation.
    - Long-term LP analysis, filtering, and coding: This involves the cross-correlation calculation, finding of the peak and hence the pitch period, followed by a long-term gain calculation. Three more signals must be calculated: reconstructed short-term prediction error, predicted short-term prediction error, and overall prediction error. The procedure is repeated for every subframe.

- Weighting error: This is not important in the first try and can be omitted completely.
- Position selection: This involves down-sampling and energy calculation. The sequence with the highest energy is selected.
- Excitation generator: The selected down-sampled sequence is up-sampled and zero-padded at the right position, resulting in the reconstructed overall prediction-error signal.
- Integration: The blocks from previous steps are joined together following the diagram of Figure 10.4. However, no quantization or encoding of individual parameters is considered yet. The decoder is first built to perform a system test.

2. *Decoder.*
   - Excitation generator: Same functionality as for the encoder.
   - Pitch synthesis filter: This filter involves long-term gain and pitch period delay. See Chapter 4 for details.
   - Formant synthesis filter: This filter requires the interpolated RC and is implemented in lattice form.
   - De-emphasis filter: This is a first-order difference equation.
   - Integration: From the diagram in Figure 10.8, join the different blocks to form the decoder.
3. *Test of encoder–decoder operation.*
   - Using the system developed so far, speech is input to the encoder with the resultant parameters processed by the decoder. Verify the intelligibility of the synthetic speech. Note that this step is done without any quantization of individual parameters. If the quality of synthetic speech is as expected, proceed to the next steps. Otherwise, review the different blocks to discover the problem.
4. *Incorporation of parameter quantizers.*
   - LPC: See Chapter 8 for the approach utilized by the GSM standard. Note, however, that other methods are also applicable. Transform the resultant RC to the LAR; and apply interpolation according to the rules. The resultant coefficients are used to implement the prediction-error filter in lattice form (Chapter 2).
   - Long-term gain: A 2-bit quantizer is specified in the GSM standard.
   - Peak magnitude and sample amplitude: Uniform quantization can be used for a first test. However, nonuniform quantization will definitely provide higher quality. This can be designed using the Lloyd algorithm with training data collected from real speech signals (Chapter 5).
   - Integration (encoder): Encoders and decoders for the various parameters can be integrated into the encoder according to Figures 10.4 and 10.5; outputs are packed together as the LPC bit-stream.

- Integration (decoder): Decoders for the various parameters are integrated to the decoder according to Figure 10.8. The RPE-LTP bit-stream is unpacked with the indices directed to the right places.

5. *Overall test and system improvement.*
   - The system is tested again for correct operation and quality. Using a bit allocation scheme similar to the GSM specification, the system should be able to generate good quality synthetic speech. Parameters of various blocks can be modified according to needs.

10.5 The amplitude optimization procedure proposed by Singhal and Atal [1989] can be applied to the RPE-LTP coder in the following way. After determining the grid position ($m = 0, 1, 2$, or 3), the amplitudes of the pulses are found so that the sum of squared error between the synthetic speech and input speech is minimized. This problem is solved by differentiating the sum of squared error with respect to the pulse amplitudes and manipulating the resultant linear equations. Derive the corresponding equations to implement the technique. What changes must be introduced to the encoder?

10.6 The GSM 6.10 standard is modified in such a way that a down-sampling factor of 2 is applied, with only two possible grid positions. Thus, the two interleaved sequences are defined with

$$x_m[n] = x[m + 2n]; \quad m = 0, 1; \; n = 0, 1, \ldots, 19,$$

with the assumption that the rest of the coder is not modified. What is the bit-rate of this modified RPE-LTP coder? Repeat for the case of a down-sampling factor of 4, with the four interleaved sequences being

$$x_m[n] = x[m + 4n]; \quad m = 0, 1, 2, 3; \; n = 0, 1, \ldots, 9.$$

CHAPTER 11

CODE-EXCITED LINEAR PREDICTION

The idea of code-excited linear prediction (CELP) was born as another attempt to improve on the existent LPC coder (Chapter 9). The acronym of CELP* itself was first created by Schroeder and Atal in their 1985 paper. Among the core ideas are the utilization of long-term and short-term linear prediction models for speech synthesis (Chapter 4), thus avoiding the strict voiced/unvoiced classification of LPC, and the incorporation of an excitation codebook, which is searched during encoding to locate the best excitation sequence. The name of "code-excited" comes from the excitation codebook, containing the "code" to "excite" the synthesis filters. The high complexity of CELP coders was originally thought to be an impractical proposition; throughout the years, however, researchers have found many ways to speed up the encoding process, making CELP a practical reality.

CELP is among one of the most influential ideas in speech coding. Its principles lay the foundation for many standardized coders. In this chapter, generic techniques of CELP are given; the architecture, functionality, and constraint are described without referring to any particular standard. Thorough understanding of the material in this chapter is crucial for later chapters, where details of several CELP-based standards are described. The chapter begins with the speech production model, followed by an explanation of the principle of analysis-by-synthesis; operations of the encoder and decoder are described. Two important components—excitation codebook search and the postfilter—are analyzed in different sections.

*It is pronounced as either "KELP" or "SELP." The recommendation is to be flexible regarding the sound and to be ready to adapt to the other party's preference to show civility and politeness. When in doubt, spell out the acronym letter-by-letter or use the entire name.

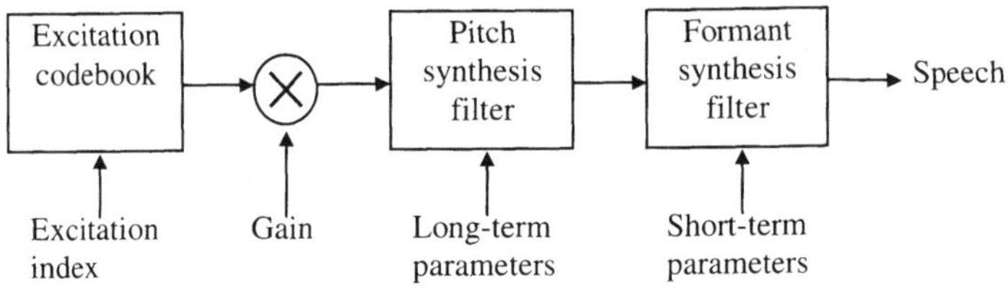

Figure 11.1 The CELP model of speech production.

11.1 THE CELP SPEECH PRODUCTION MODEL

The CELP coder relies on the long-term and short-term linear prediction models, which are thoroughly explained in Chapter 4. Figure 11.1 shows the block diagram of the speech production model, where an excitation sequence is extracted from the codebook through an index. The extracted excitation is scaled to the appropriate level and filtered by the cascade connection of pitch synthesis filter and formant synthesis filter to yield the synthetic speech. The pitch synthesis filter creates periodicity in the signal associated with the fundamental pitch frequency, and the formant synthesis filter generates the spectral envelope.

What type of signal is contained in the excitation codebook? The codebook can be fixed or adaptive and can contain deterministic pulses or random noise. For simplicity, we assume for now that the codebook is fixed and contains white noise samples. Thus, the excitation index selects a white noise sequence from the codebook as input to the cascade connection of the synthesis filters. The speech production model that the CELP coder relies on consists simply of a white noise source exciting the synthesis filters. What are the advantages of this model? The following can be pointed out when compared with other coders.

- A strict voiced/unvoiced classification, as for the LPC coder, is eliminated.

As explained in Chapter 9, rigid classification of a speech frame is one of the main limitations of the LPC coder and is responsible for major artifacts in the synthetic speech. Use of the two synthesis filters in cascade allows an efficient and accurate modeling of transition frames with smoothness and continuity, producing a much more naturally sounding synthetic speech.

- Partial phase information of the original signal is preserved.

As explained in Chapter 9, the LPC model does not retain any phase information of the original signal. The CELP model captures some phase information through a closed-loop analysis-by-synthesis method (described later). In this approach, the best excitation sequence from the codebook is selected, where "best" means the particular sequence is capable of generating the synthetic speech that is as close as possible to the original, where "closeness" is often measured using time-domain techniques, such as signal-to-noise ratio. By doing so, the synthetic speech is not only matched in the magnitude spectrum domain (captured by the linear prediction coefficients) but also in the time domain, where a difference in phase plays an important role. Even though a human listener is relatively insensitive to the phase, retaining some phase information adds naturalness to the synthetic speech, leading to an improvement in quality.

Note the resemblance between the CELP model and the RPE-LTP coder described in Chapter 10, where both rely on the cascade connection of synthesis filters. The RPE-LTP coder, however, encodes samples of the prediction error through an adaptive scalar quantization scheme, which are then decoded and used as excitation to the synthesis filters. Even though the approach is quite effective for the synthesis of high-quality speech, it is not efficient in the sense that a relatively large number of bits must be allocated to accomplish the task. This is the reason why the approach works well only at a relatively high bit-rate of 13 kbps.

The CELP coder works effectively in the mid bit-rate region and is achieved by encoding the excitation sequence through a codebook, which in a sense is a vector quantization approach, where the whole sequence is encoded using a single index. CELP pertains to the hybrid-type coders, where an underlying model is utilized and, at the same time, attempts are made to approximate the original waveform.

11.2 THE PRINCIPLE OF ANALYSIS-BY-SYNTHESIS

In a speech coder, the speech signal is represented by a combination of parameters: gain, filter coefficients, voicing strengths, and so on. In an *open-loop* system, the parameters are extracted from the input signal, which are quantized and later used for synthesis. This is the principle in several coders studied in separate chapters, like LPC (Chapter 9), RPE-LTP (Chapter 10), and MELP (Chapter 17).

A more effective method is to use the parameters to synthesize the signal during encoding and fine-tune them so as to generate the most accurate reconstruction. Conceptually, this is a *closed-loop* optimization procedure, where the goal is to choose the best parameters so as to match as much as possible the synthetic speech with the original speech. Figure 11.2 shows the block diagram of an encoder with

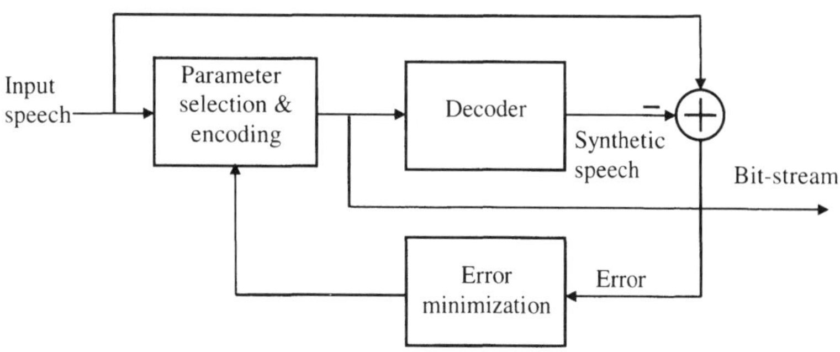

Figure 11.2 Block diagram showing an encoder based on the analysis-by-synthesis principle.

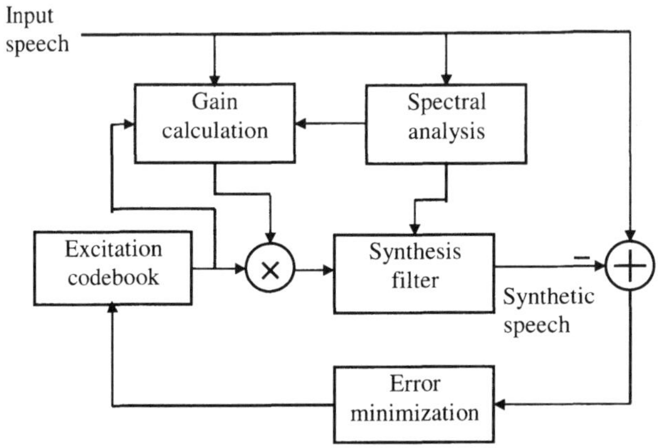

Figure 11.3 Block diagram showing the key components of a CELP encoder.

the closed-loop approach. Since the signal is synthesized during encoding for analysis purposes, the principle is known as analysis-by-synthesis (abbreviated as AbS in some of the literature).

Theoretically, all parameters of the speech coder can be optimized jointly so as to yield the best result. This approach, however, is too complex due to the computation involved. In practice, only a subset of parameters is selected for closed-loop optimization, while the rest are determined through an open-loop approach. The CELP coder is based on the analysis-by-synthesis principle, where the excitation sequences contained in a codebook are selected according to a closed-loop method. Other parameters, such as the filter coefficients, are determined in an open-loop fashion. Figure 11.3 illustrates a simplified block diagram of the CELP encoder. A commonly used error criterion, such as the sum of squared error, can be applied to select the final excitation sequence; hence, waveform matching in the time domain is performed, leading to a partial preservation of phase information.

Since the model requires frequent updating of the parameters to yield a good match to the original signal, the analysis procedure of the system is carried out in blocks. That is, the input speech is partitioned into suitable blocks of samples to determine the time-varying filter parameters; with these parameters fixed, the excitation is optimized. The excitation sequence and other parameters are then encoded to form the compressed bit-stream. The length of the analysis block or frame affects the bit-rate of the coding scheme.

11.3 ENCODING AND DECODING

CELP is an analysis-by-synthesis method, where the excitation signal is selected by a closed-loop search procedure and applied to the synthesis filters. The synthesized waveform is compared to the original speech segment, the distortion is measured, and the process is repeated for all excitation codevectors stored in a codebook. The

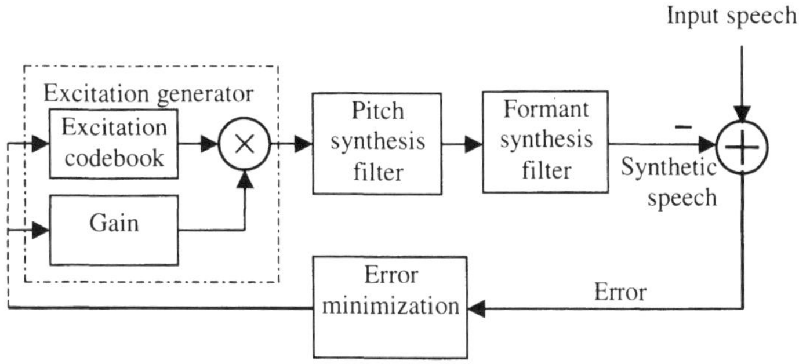

Figure 11.4 Analysis-by-synthesis loop of a CELP encoder.

index of the "best" excitation sequence is transmitted to the decoder, which retrieves the excitation codevectors from a codebook identical to that at the encoder. In this section, generic structures of the CELP encoder and decoder are described.

Perceptual Weighting

The analysis-by-synthesis loop that the CELP encoder is based on is shown in Figure 11.4, where the system function of the formant synthesis filter is (Chapter 4)

$$H_f(z) = \frac{1}{A(z)} = \frac{1}{1 + \sum_{i=1}^{M} a_i z^{-i}}, \tag{11.1}$$

with $A(z)$ denoting the system function of the formant analysis filter. Assume that the size of the excitation codebook is L; that is, it contains a total of L excitation codevectors. The encoder will pass through the loop L times for each short segment of input speech; a mean-squared error value is calculated after each pass. The excitation codevector providing the lowest error is selected at the end. Dimension of the codevector depends on the length of the speech segment under consideration; for most CELP coders, the length is equal to that of the subframe (Chapter 4). For the FS1016 coder (next chapter), for instance, the subframe consists of 60 samples (7.5 ms), with a codebook size of $L = 512$. Thus, the excitation codebook can be considered as a matrix of dimension 512×60.

The masking phenomenon in the human auditory system can be explored to yield a more suitable error measure. It is well known that we can tolerate much more noise in frequency regions where the speech has significant energy. Therefore, the noise components in these regions can have higher energy relative to the components in low-energy regions without increasing the perceptual distortion. In other words, frequency regions where the speech signal has high energy can cover or "mask" a proportionately higher amount of noise, whereas in low-energy regions, the amount of tolerable noise is proportionately lower. Therefore, in a typical speech spectrum, the amount of noise at the peaks can be higher than the amount of noise at the valleys. A simple way of controlling the noise spectrum is by filtering

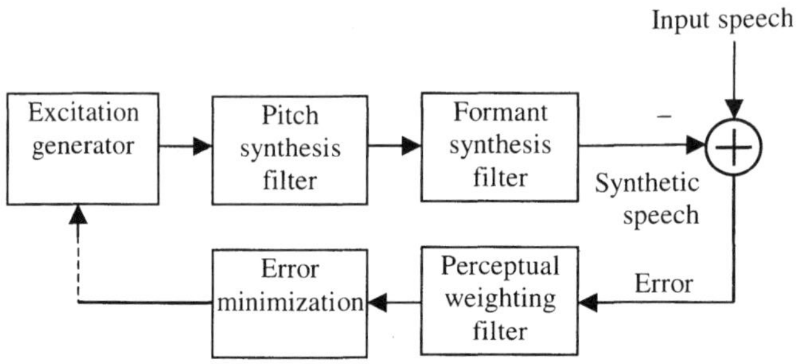

Figure 11.5 Analysis-by-synthesis loop of a CELP encoder with perceptual weighting.

the error signal through a weighting filter before minimization. One efficient way to implement the weighting filter is by using the system function

$$W(z) = \frac{A(z)}{A(z/\gamma)} = \frac{1 + \sum_{i=1}^{M} a_i z^{-i}}{1 + \sum_{i=1}^{M} a_i \gamma^i z^{-i}}, \tag{11.2}$$

with γ being a constant in the interval [0, 1]. Position of this filter within the loop is indicated in Figure 11.5. The constant γ determines the degree to which the error is de-emphasized in any frequency region. The filter amplifies the error signal spectrum in nonformant regions of the speech spectrum, while attenuating the error signal spectrum in formant regions. Hence, an error signal whose spectral energy is concentrated in formant regions of the input spectrum is considered better than one whose spectral energy is not located under formants.

From (11.2) we can see that as $\gamma \to 1$, $W(z) \to 1$, and hence no modification of the error spectrum is performed. On the other hand, if $\gamma \to 0$, $W(z) \to A(z)$, which is the formant analysis filter. The constant γ introduces a broadening effect (bandwidth expansion) to the error weighting filter. The most suitable value of γ is selected subjectively by listening tests, and for 8-kHz sampling, γ is usually between 0.8 and 0.9.

Example 11.1 Consider the following LPCs:

$$a_1 = -1.286, \quad a_2 = 1.138, \quad a_3 = -1.047, \quad a_4 = 0.691, \quad a_5 = -0.304,$$
$$a_6 = 0.373, \quad a_7 = -0.071, \quad a_8 = 0.012, \quad a_9 = 0.048, \quad a_{10} = 0.064.$$

Magnitudes of the transfer functions for the formant synthesis filter and the formant analysis filter are plotted in Figure 11.6, with the characteristics of the corresponding perceptual weighting filter shown in Figure 11.7 for various values of γ. Note that the weighting filter attenuates the error energy around the peaks of the spectrum envelope in the formant synthesis filter. Thus, by using this filter to process the error signal, the error energy corresponding to the valleys of the input spectrum

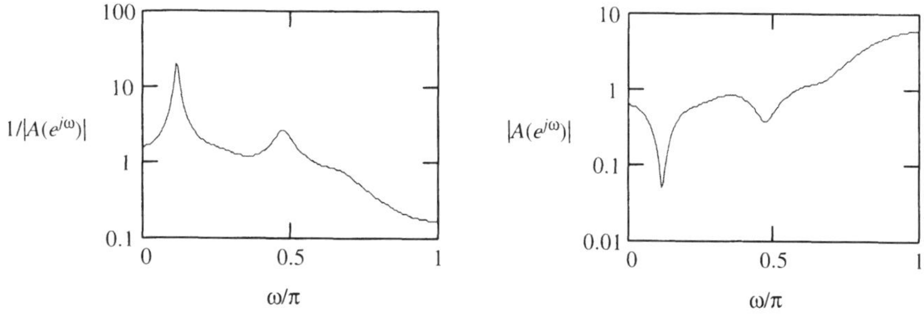

Figure 11.6 Magnitude plots of an example formant synthesis filter (*left*) and associated formant analysis filter (*right*).

receives higher weighting. This is important from a perceptual point of view, since error in the spectrum peaks is easily masked due to the high signal energy at these points; while at the spectrum valleys, error becomes more noticeable due to the low signal energy in such regions.

Complexity Issues

Due to the linearity of the weighting filter, it is possible to translate it so as to obtain the equivalent systems shown in Figure 11.8. Since the formant synthesis filter and the weighting filter are in cascade, they can be merged to form the *modified formant synthesis filter* with system function

$$H_f(z/\gamma) = \frac{1}{A(z/\gamma)} = \frac{1}{1 + \sum_{i=1}^{M} a_i \gamma^i z^{-i}}. \tag{11.3}$$

The system of Figure 11.8 is computationally more efficient than the equivalent system of Figure 11.5 and is due to the fact that computation within the loop is reduced. In Figure 11.5, each excitation codevector is filtered by two synthesis

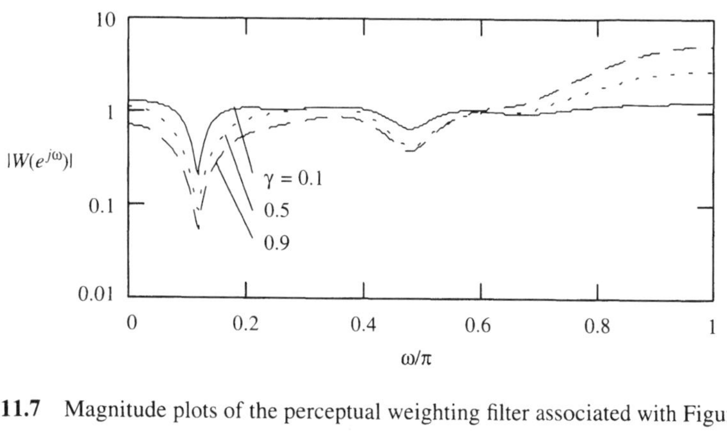

Figure 11.7 Magnitude plots of the perceptual weighting filter associated with Figure 11.6.

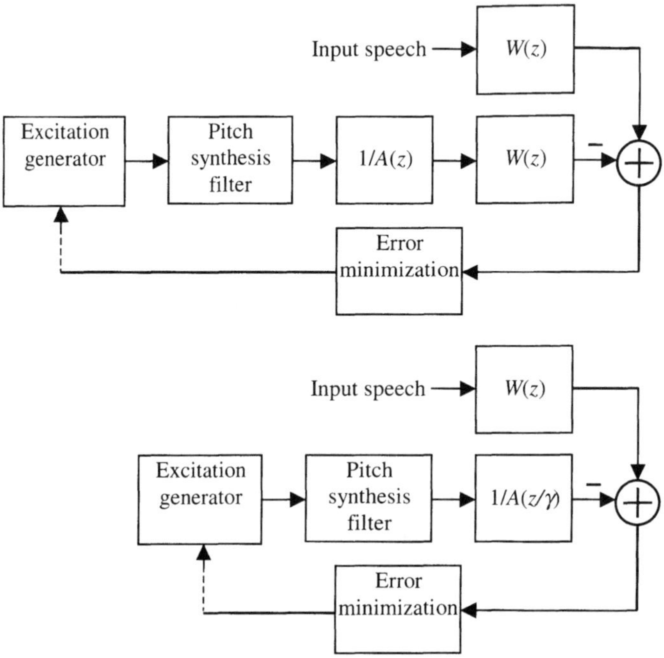

Figure 11.8 Equivalent realizations of the system in Figure 11.5. *Top*: By translating the perceptual weighting filter. *Bottom*: After merging the formant synthesis filter with the perceptual weighting filter.

filters, with the error processed by the weighting filter. In Figure 11.8, however, there is no need to filter the difference signal. And since typically there are a relatively high number of excitation codevectors, the computation saving is quite significant (Exercise 11.1).

Encoder Operation

A block diagram of a generic CELP encoder is shown in Figure 11.9. This encoder is highly simplistic and serves only as an illustration. Subsequent chapters contain the details of operation of different standard CELP coders. The encoder works as follows:

- Input speech signal is segmented into frames and subframes. As explained in Chapter 4, the scheme of four subframes in one frame is a popular choice. Length of the frame is usually around 20 to 30 ms, while for the subframe it is in the range of 5 to 7.5 ms.
- Short-term LP analysis is performed on each frame to yield the LPC. Afterward, long-term LP analysis is applied to each subframe (Chapter 4). Input to short-term LP analysis is normally the original speech, or pre-emphasized speech; input to long-term LP analysis is often the (short-term)

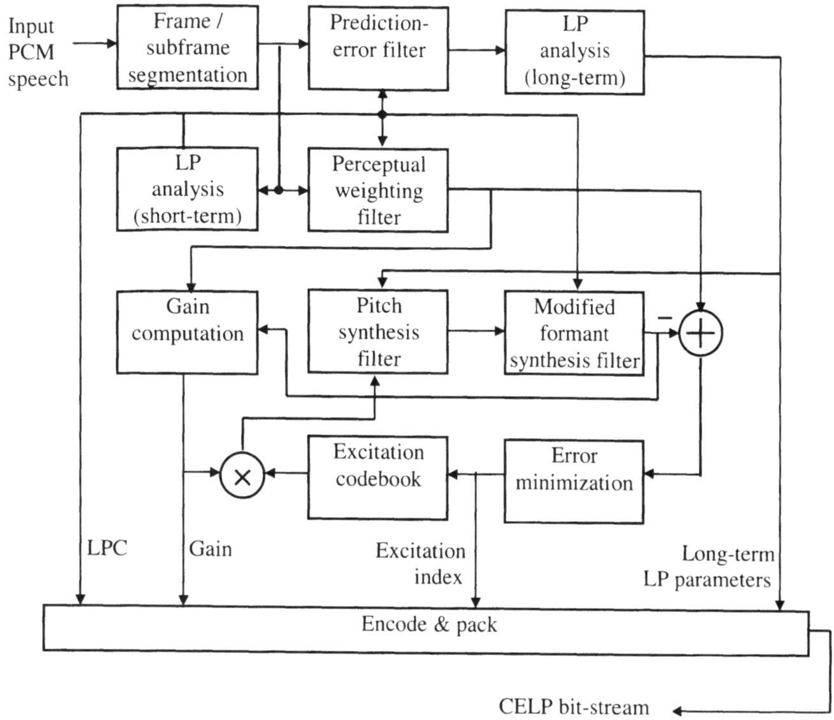

Figure 11.9 Block diagram of a generic CELP encoder.

prediction error. Coefficients of the perceptual weighting filter, pitch synthesis filter, and modified formant synthesis filter are known after this step.

- The excitation sequence can now be determined. The length of each excitation codevector is equal to that of the subframe; thus, an excitation codebook search is performed once every subframe. The search procedure begins with the generation of an ensemble of filtered excitation sequences with the corresponding gains; mean-squared error (or sum of squared error) is computed for each sequence, and the codevector and gain associated with the lowest error are selected.
- The index of excitation codebook, gain, long-term LP parameters, and LPC are encoded, packed, and transmitted as the CELP bit-stream.

Decoder Operation

A block diagram of the CELP decoder is shown in Figure 11.10. It basically unpacks and decodes various parameters from the bit-stream, which are directed to the corresponding block so as to synthesize the speech. A postfilter is added at the end to enhance the quality of the resultant signal; structure of this filter is described in Section 11.5.

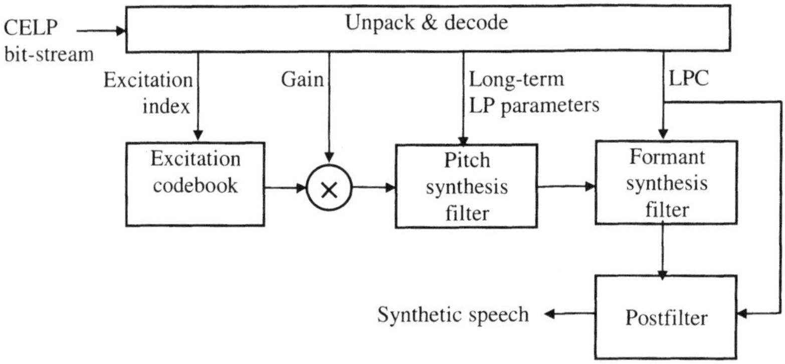

Figure 11.10 Block diagram of a generic CELP decoder.

11.4 EXCITATION CODEBOOK SEARCH

Excitation codebook search is the most computationally intensive part of CELP coding. Throughout the years, many ideas have been proposed and tested surrounding the topic, with the sole purpose of accelerating the search process without compromising significantly the output quality. In this section, the precise mathematical framework is provided to describe the search procedure.

Preliminaries

A step-by-step procedure for excitation codebook search, described superficially in the last section, is explained in more detail below.

1. Filter the input speech subframe with the perceptual weighting filter.
2. For each codevector in the excitation codebook:
   - Calculate the optimal gain (described later) and scale the codevector using the value found.
   - Filter the scaled excitation codevector with the pitch synthesis filter.
   - Filter the pitch synthesis filter's output with the modified formant synthesis filter.
   - Subtract the perceptually filtered input speech from the modified formant synthesis filter's output; the result represents an error sequence.
   - Calculate the energy of the error sequence.
3. The index of the excitation codevector associated with the lowest error energy is retained as information on the input subframe.

The above procedure is repeated for every input subframe. It is possible to improve computational efficiency by exploring the redundancies in the search loop. In particular, by decomposing the filters' responses in zero-state and

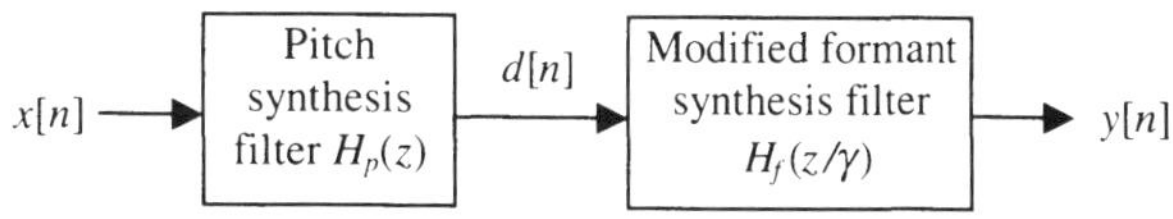

Figure 11.11 Cascade connection of pitch synthesis filter and modified formant synthesis filter.

zero-input (Chapter 2), substantial computational savings are obtained. Figure 11.11 shows the system under consideration, where a one-tap pitch synthesis filter is connected in cascade with the modified formant synthesis filter. The difference equations of the system are

$$y[n] = d[n] - \sum_{i=1}^{M} a_i r^i y[n-i], \tag{11.4}$$

$$d[n] = x[n] - bd[n-T], \tag{11.5}$$

where

$M =$ Prediction order (short-term),
$a_i =$ LPC (short-term),
$b =$ Long-term gain,
$T =$ Pitch period.

Equations (11.4) and (11.5) are used to compute the filters' outputs.

State-Save Method

This and the next method are already described in Chapter 2; however, the discussion is only for the case of a single-stage filter. Here, the difference equations (11.4) and (11.5) are applied to each of the finite-length input subframes $x_r[n]$, $n = 0$ to $N-1$ in the following way ($r > 0$):

$$d_r[n] = d_{r-1}[n+N], \qquad -T \le n \le -1; \tag{11.6}$$

$$d_r[n] = x_r[n] - bd_r[n-T], \qquad 0 \le n \le N-1; \tag{11.7}$$

$$y_r[n] = y_{r-1}[n+N], \qquad -M \le n \le -1; \tag{11.8}$$

$$y_r[n] = d_r[n] - \sum_{i=1}^{M} a_i \gamma^i y_r[n-i], \qquad 0 \le n \le N-1. \tag{11.9}$$

Note that the subframe timing notation, as explained in Chapter 2, is employed for signal description.

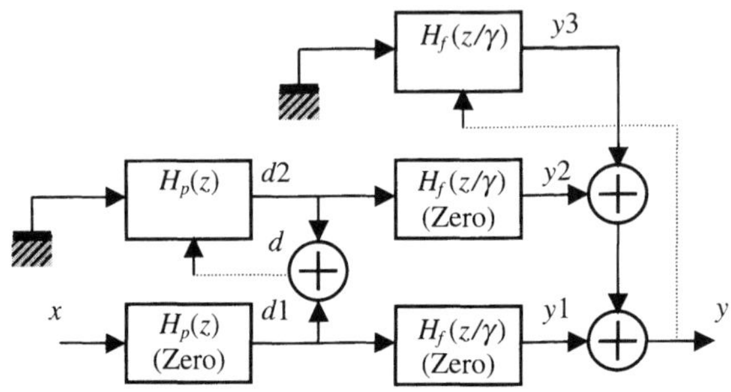

Figure 11.12 Block diagram showing the signals involved in the zero-input zero-state method for the computation of output signal from a cascade connection of pitch synthesis filter and modified formant synthesis filter.

Zero-Input Zero-State Method

This method computes the total response by calculating separately the zero-input response and zero-state response. These two responses are added to form the overall response. The signals involved are depicted in Figure 11.12. Details for each signals are explained below.

- Zero-state response of pitch synthesis filter ($d1$):

$$d1_r[n] = 0, \qquad -T \le n \le -1; \tag{11.10}$$

$$d1_r[n] = x_r[n] - b\,d1_r[n-T], \qquad 0 \le n \le N-1. \tag{11.11}$$

  In (11.11), the beginning T samples can be computed without the product due to the zero-state condition. Hence, we can rewrite the equation as

$$d1_r[n] = x_r[n], \qquad 0 \le n \le T-1; \tag{11.12}$$

$$d1_r[n] = x_r[n] - b\,d1_r[n-T], \qquad T \le n \le N-1. \tag{11.13}$$

  We assume here that $T < N$. If $T \ge N$, then the zero-state response is equal to the input signal.

- Zero-input response of pitch synthesis filter ($d2$):

$$d2_r[n] = d_{r-1}[n+N], \qquad -T \le n \le -1; \tag{11.14}$$

$$d2_r[n] = -b\,d2_r[n-T], \qquad 0 \le n \le N-1. \tag{11.15}$$

- Total response of pitch synthesis filter (d):

$$d_r[n] = d1_r[n] + d2_r[n], \qquad 0 \le n \le N-1. \tag{11.16}$$

- Total zero-state response: zero-state response of the pitch synthesis filter, filtered by the formant synthesis filter in zero-state ($y1$):

$$y1_r[n] = 0, \qquad -M \le n \le -1; \tag{11.17}$$

$$y1_r[n] = d1_r[n] - \sum_{i=1}^{M} a_i \gamma^i y1_r[n-i], \qquad 0 \le n \le N-1. \tag{11.18}$$

- Zero-input response of pitch synthesis filter, filtered by the formant synthesis filter in zero-state ($y2$):

$$y2_r[n] = 0, \qquad -M \le n \le -1; \tag{11.19}$$

$$y2_r[n] = d2_r[n] - \sum_{i=1}^{M} a_i \gamma^i y2_r[n-i], \qquad 0 \le n \le N-1. \tag{11.20}$$

- Zero-input response of formant synthesis filter ($y3$):

$$y3_r[n] = y_{r-1}[n+N], \qquad -M \le n \le -1; \tag{11.21}$$

$$y3_r[n] = -\sum_{i=1}^{M} a_i \gamma^i y3_r[n-i], \qquad 0 \le n \le N-1. \tag{11.22}$$

- Total response (formant synthesis filter):

$$y_r[n] = y1_r[n] + y2_r[n] + y3_r[n], \quad 0 \le n \le N-1. \tag{11.23}$$

See Exercise 11.2 for a summary of the amount of computation required for each method.

Computational Cost

Consider the problem of calculating the L total responses using the two described methods, with L being the total number of codevectors, or size of the excitation codebook. Further assume that subframe length (N), prediction order (M), and pitch period (T) are known. For the state-save method,

$$\#\text{sums} = \#\text{products} = (M+1)N \cdot L; \tag{11.24}$$

while for the zero-input zero-state method,

$$\#\text{sums} = (N(M+2) - T)L + (2M+1)N, \tag{11.25}$$

$$\#\text{products} = (N(M+1) - T)L + (2M+1)N. \tag{11.26}$$

The above equations are obtained from the results of Exercise 11.2, and by realizing that for the latter procedure, only $d1$, $y1$, and y need be computed L times.

TABLE 11.1 Computational Cost (per Codebook Search) with $M = 10$, $N = 60$, and $L = 512$

	State-Save	Zero-Input Zero-State,[a] $T = 50$	Zero-Input Zero-State,[a] $T = 80$
#Sums	337920	344300 (+1.9%)	339180 (+0.4%)
#Products	337920	313580 (−7.2%)	308460 (−8.7%)

[a] Percentage change with respect to the state-save method is shown for the zero-input zero-state method.

The rest of the signals ($d2$, d, $y2$, and $y3$) are calculated just once per excitation codebook search, since they are associated with the zero-input responses. Note also that for y, half of the sums are done outside the loop ($y2 + y3$), while the other half are done inside the loop ($y1 + (y2 + y3)$). Equations (11.25) and (11.26) are valid only when $T \leq N$. For $T > N$, we have

$$\#\text{sums} = (M + 1)N \cdot L + (2M + 1)N, \tag{11.27}$$

$$\#\text{products} = M \cdot N \cdot L + (2M + 1)N, \tag{11.28}$$

since in this case, $d1$ needs no operation at all ($d1 = x$, see (11.12)).

For fixed M and N, (11.24), (11.25), (11.26), (11.27), and (11.28) are linear functions of the excitation codebook size L and represent the number of operations required to exhaustively search the entire excitation codebook once. For very low values of L (small codebook), the state-save method is less costly to implement; as the codebook size increases, the zero-input zero-state method becomes more and more attractive. The reduced computational load is associated with the zero-state response of the pitch synthesis filter ($d1$, Figure 11.12), where the first T samples need no operation at all (see (11.12)). Less computation is required for higher values of T; however, if $T > N$, no further reduction is obtainable ((11.27) and (11.28)).

Table 11.1 shows some typical numbers for the two methods, assuming a 9-bit excitation codebook. As we can see, the number of sums for the state-save method requires slightly fewer additions but more products. Since traditionally multiplication is a huge speed bottleneck for most processors, the saving in the number of multiplications is very attractive, and the zero-input zero-state method has since been widely adopted for excitation codebook search in CELP. Exercise 11.3 contains an alternative zero-input zero-state method with even less computation.

Error Calculation and Optimal Scaling

So far we have seen the procedure to compute an ensemble of output sequences from the cascade connection of two filters. In order to complete the excitation codebook search process, it is necessary to take into account the input speech signal so as to obtain an error sequence, based on which the excitation codevector is selected.

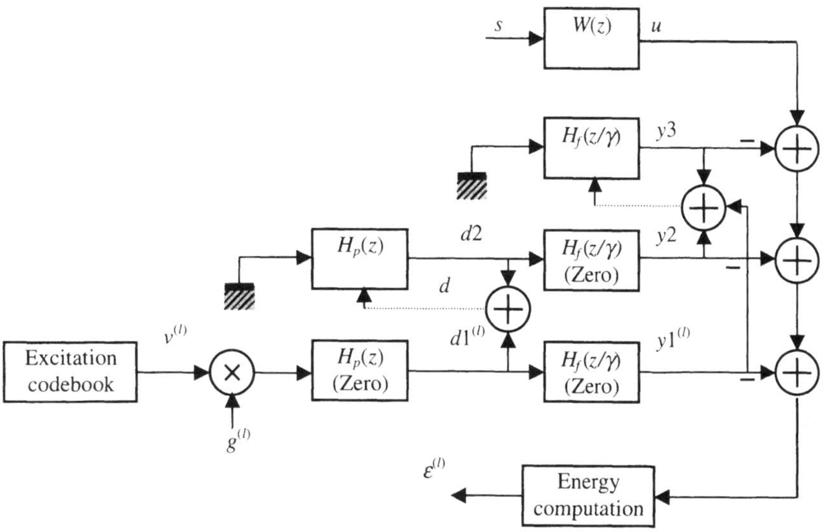

Figure 11.13 Signals involved in the excitation codebook search.

The target system is shown in Figure 11.13 and is obtained by inserting the scheme of Figure 11.12 into Figure 11.8.

The excitation codebook is fixed and consists of L codevectors of dimension equal to N, denoted by

$$v^{(l)}[n]; \quad l = 0, \ldots, L-1; \quad n = 0, \ldots, N-1.$$

These vectors are scaled by the gain $g^{(l)}$, computed individually for each l. The scaled codevectors are utilized as inputs to the filters.

As explained in previous sections, the input speech subframe $s[n]$ is filtered by the perceptual weighting filter $W(z)$, the output is subtracted from the zero-input responses $y3$ and $y2$, and finally combined with the zero-state response $y1^{(l)}$, resulting in an error sequence with energy

$$\varepsilon^{(l)} = \sum_{n=0}^{N-1} \left(u[n] - y1^{(l)}[n] - y2[n] - y3[n]\right)^2. \tag{11.29}$$

The index l resulting in the lowest $\varepsilon^{(l)}$ is transmitted as information on the subframe. To complete the search procedure, we need to find the gain $g^{(l)}$. Let

$$y1_o^{(l)}[n] = y1^{(l)}[n]/g^{(l)}; \quad n = 0, \ldots, N-1 \tag{11.30}$$

the zero-state response to the lth excitation codevector with no scaling (unity gain), and

$$u_o[n] = u[n] - y2[n] - y3[n]. \tag{11.31}$$

Using the new terminology in (11.29), we have

$$\varepsilon^{(l)} = \sum_{n=0}^{N-1} \left(u_o[n] - g^{(l)} y1_o^{(l)}[n]\right)^2. \tag{11.32}$$

The gain $g^{(l)}$ is calculated so as to minimize $\varepsilon^{(l)}$. Standard optimization procedure calls for differentiating (11.32) and equating the result to zero, leading to the optimal scaling factor of

$$g^{(l)} = \frac{\sum_{n=0}^{N-1} u_o[n] y1_o^{(l)}[n]}{\sum_{n=0}^{N-1} \left(y1_o^{(l)}[n]\right)^2}. \tag{11.33}$$

It is left to the reader as an exercise to prove the following relations:

$$\varepsilon^{(l)} = \left(\sum_{n=0}^{N-1} (u_o[n])^2\right) - P^{(l)}, \tag{11.34}$$

where

$$P^{(l)} = \frac{\left(\sum_{n=0}^{N-1} u_o[n] y1_o^{(l)}[n]\right)^2}{\sum_{n=0}^{N-1} \left(y1_o^{(l)}[n]\right)^2}. \tag{11.35}$$

Note that minimization of $\varepsilon^{(l)}$ is equivalent to the maximization of (11.35) (Why?). In practice, it is preferable to use (11.35) to find the optimal index, since the final result can be located with less computation. Figure 11.14 shows the signal flow graph of the excitation codebook search procedure, where the signal calculation steps are put into order. Two parameters are returned upon its completion: the index *lopt* for the excitation codevector and the associated gain $g^{(lopt)}$.

Calculating the Zero-State Response Using the Convolution Sum

From Chapter 2 the zero-state response $y1^{(l)}[n]$ can be calculated with the convolution sum

$$y1^{(l)}[n] = g^{(l)} \sum_{k=0}^{N-1} h[n-k] v^{(l)}[k]; \quad 0 \le n \le N-1, \tag{11.36}$$

with $h[n]$ being the impulse response of the cascade connection of filters (pitch synthesis and formant synthesis). Consider the vectors

$$\mathbf{y1}^{(l)} = \left[y1^{(l)}[0] \quad y1^{(l)}[1] \quad \cdots \quad y1^{(l)}[N-1]\right]^T$$

and

$$\mathbf{v}^{(l)} = \left[v^{(l)}[0] \quad v^{(l)}[1] \quad \cdots \quad v^{(l)}[N-1]\right]^T.$$

Equation (11.36) can be written in matrix form:

$$\mathbf{y}1^{(l)} = g^{(l)}\mathbf{H}\mathbf{v}^{(l)} \tag{11.37}$$

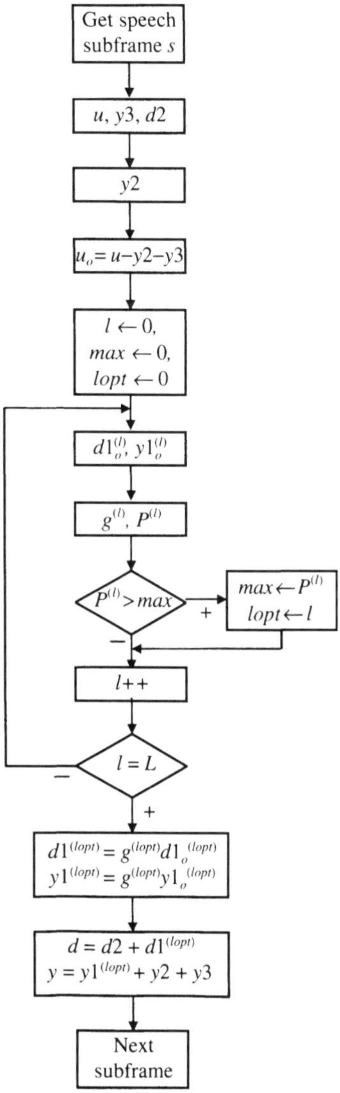

Figure 11.14 Signal flow graph illustrating the excitation codebook search procedure.

with **H** being the $N \times N$ impulse response matrix. Similarly, the error expression (11.29) can be written as

$$\varepsilon^{(l)} = \left\|\mathbf{u} - \mathbf{y}1^{(l)} - \mathbf{y}2 - \mathbf{y}3\right\|^2 = \left\|\mathbf{u}_o - g^{(l)}\mathbf{H}\mathbf{v}^{(l)}\right\|^2 \quad (11.38)$$

with $\|\mathbf{x}\|$ denoting the Euclidean norm* of the vector $\mathbf{x}$. The optimal gain $g^{(l)}$ is the one that minimizes (11.38) and can be found with

$$\begin{aligned}\frac{\partial\varepsilon^{(l)}}{\partial g^{(l)}} &= \frac{\partial}{\partial g^{(l)}}\left[\left(\mathbf{u}_o - g^{(l)}\mathbf{H}\mathbf{v}^{(l)}\right)^T\left(\mathbf{u}_o - g^{(l)}\mathbf{H}\mathbf{v}^{(l)}\right)\right] \\ &= -2\mathbf{u}_o^T\mathbf{H}\mathbf{v}^{(l)} + 2g^{(l)}\left\|\mathbf{H}\mathbf{v}^{(l)}\right\|^2 = 0,\end{aligned} \quad (11.39)$$

leading to

$$g^{(l)} = \frac{\mathbf{u}_o^T\mathbf{H}\mathbf{v}^{(l)}}{\left\|\mathbf{H}\mathbf{v}^{(l)}\right\|^2}, \quad (11.40)$$

and represents the optimal gain. Expanding (11.38) gives

$$\varepsilon^{(l)} = \|\mathbf{u}_o\|^2\left[1 - \frac{g^{(l)}}{\|\mathbf{u}_o\|^2}\left(2\mathbf{u}_o^T\mathbf{H}\mathbf{v}^{(l)} - g^{(l)}\left\|\mathbf{H}\mathbf{v}^{(l)}\right\|^2\right)\right]. \quad (11.41)$$

Substituting the expression for optimal gain ((11.40) into (11.41)) leads to

$$\varepsilon^{(l)} = \|\mathbf{u}_o\|^2\left[1 - \frac{\left(\mathbf{u}_o^T\mathbf{H}\mathbf{v}^{(l)}\right)^2}{\|\mathbf{u}_o\|^2\left\|\mathbf{H}\mathbf{v}^{(l)}\right\|^2}\right] = \|\mathbf{u}_o\|^2 - P^{(l)}, \quad (11.42)$$

where

$$P^{(l)} = \frac{\left(\mathbf{u}_o^T\mathbf{H}\mathbf{v}^{(l)}\right)^2}{\left\|\mathbf{H}\mathbf{v}^{(l)}\right\|^2}. \quad (11.43)$$

The convolution sum represents an alternative way to find the zero-state response. In Chapter 2, it is shown that for typical circumstances, the computational cost involved with the convolution is higher than utilizing the difference equation directly; thus, it seems senseless to even mention this approach. Readers are invited to refer to the next chapter, where the federal standard version of CELP is presented, with the advantages of the convolution method exposed.

* The Euclidean norm of $\mathbf{x} = [x_0 \quad x_1 \quad \cdots \quad x_{N-1}]^T$ is given by $\|\mathbf{x}\| = \left(\sum x_i^2\right)^{1/2}$; see Lancaster and Tismenetsky [1985].

11.5 POSTFILTER

As shown in Section 11.3, subjective quality of a CELP coder can be improved by incorporating a weighting filter into the error minimization procedure. This weighting filter reduces the noise components in spectral valleys but increases the noise components at spectral peaks. To further improve the subjective quality of the synthetic speech, a postfilter can be added so as to attenuate the components in the spectral valleys, thus enhancing the overall spectrum.

The frequency response of an ideal postfilter should follow the peaks and valleys of the spectral envelope of speech without giving an overall spectral tilt. In a predictive speech coder employing linear prediction, the synthesis filter has a frequency response that closely follows the spectral envelope of the input speech. Therefore, it is natural to derive the postfilter from the linear predictor. In this section, postfilter design is considered. It is shown that there are various forms of filtering topologies suitable for this purpose, with different levels of complexity; the choice of a particular filter is often decided by the desired level of quality and the amount of available resources.

Short-Term Postfilter

Consider a postfilter consisting of the bandwidth expanded LP synthesis filter

$$H_1(z) = \frac{1}{1 + \sum_{i=1}^{M} a_i \alpha^i z^{-i}} \tag{11.44}$$

with $0 < \alpha < 1$ a constant. Figure 11.15 shows the frequency response for this form of postfilter for various values of α (same LPC as in Example 11.1). Such a postfilter does reduce the perceived noise level. However, noise reduction is accompanied by muffled speech. This is due to the fact that the frequency response of this postfilter generally has a lowpass spectral tilt, as can be seen in Figure 11.15. To

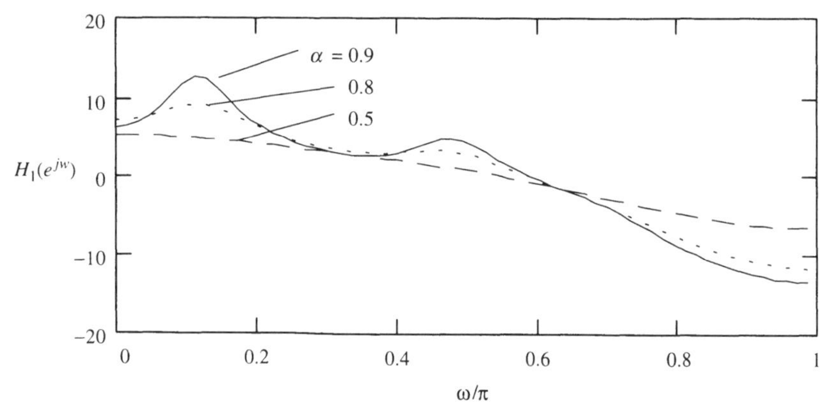

Figure 11.15 Examples of magnitude responses of an all-pole postfilter for different values of α.

reduce the spectral tilt of the all-pole postfilter described in (11.44), M zeros can be added so as to create the postfilter with system function

$$H_2(z) = \frac{1 + \sum_{i=1}^{M} a_i \beta^i z^{-i}}{1 + \sum_{i=1}^{M} a_i \alpha^i z^{-i}} \tag{11.45}$$

with $0 < \beta < \alpha < 1$. The magnitude of the frequency response in a logarithmic scale is given by

$$20 \log\left(\frac{1}{\left|1 + \sum_{i=1}^{M} a_i \alpha^i z^{-i}\right|}\right) - 20 \log\left(\frac{1}{\left|1 + \sum_{i=1}^{M} a_i \beta^i z^{-i}\right|}\right); \tag{11.46}$$

that is, it is the difference between the frequency responses of two bandwidth-expanded LP synthesis filters. The values of α and β are determined based on subjective listening test results. From Figure 11.15, we see that the response of (11.44) for $\alpha = 0.8$ has both spectral tilt and formant peaks (although greatly smoothed), while the response for $\alpha = 0.5$ has spectral tilt only. Hence, with $\alpha = 0.8$ and $\beta = 0.5$ in (11.46), we can remove the spectral tilt to a large extent by subtracting the response for $\alpha = 0.5$ from the response for $\alpha = 0.8$.

Even though the lowpass effect is reduced significantly with the addition of zeros (as in H_2), slight spectral tilt is still present. To further mitigate the lowpass effect, a first-order filter with a system function of $(1 - \mu z^{-1})$ can be added in cascade. A value of $\mu = 0.5$ has been found to provide good results. This filter is the same as the pre-emphasis filter (Chapter 4) and provides a highpass spectral tilt to compensate the lowpass effect. The overall system function of the postfilter is given by

$$H_3(z) = \left(1 - \mu z^{-1}\right) \frac{1 + \sum_{i=1}^{M} a_i \beta^i z^{-i}}{1 + \sum_{i=1}^{M} a_i \alpha^i z^{-i}}. \tag{11.47}$$

Figure 11.16 plots the magnitude responses of the three postfilters, where we can see that the spectral tilt of H_1 is reduced for H_2 and is further decreased for H_3, but

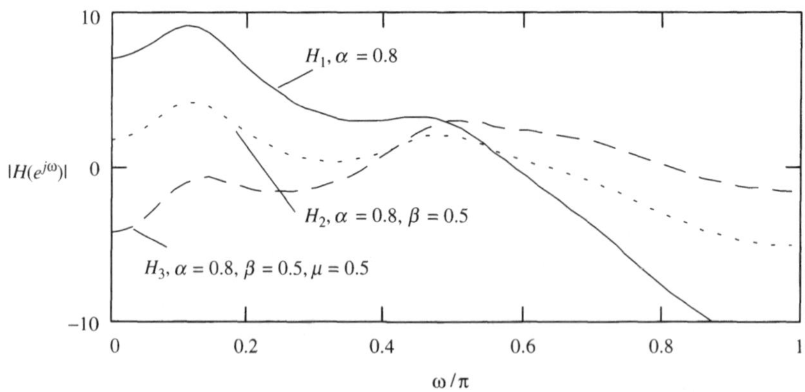

Figure 11.16 Some postfilter configurations and their magnitude responses.

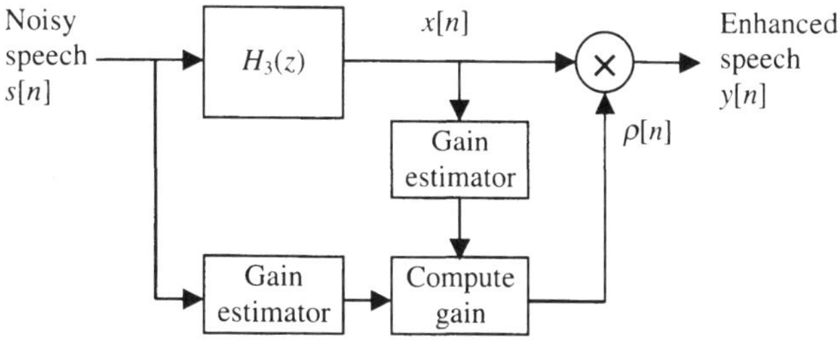

Figure 11.17 Block diagram of the postfilter.

the peaks and valleys of the original spectrum are largely preserved. The postfilter with system function given by (11.47) is included (with slight modification) by various standard CELP coders, such as the FS1016 (Chapter 12) and IS54 (Chapter 13). Since only the short-term LPCs are used, it is often referred to as the short-term postfilter.

Automatic Gain Control

To neutralize the power changes introduced by the postfilter, it is common to deploy a power normalization block within the architecture of the postfilter. Figure 11.17 shows the block diagram of the postfilter with automatic gain control. Power of the input noisy speech $s[n]$ and the filtered signal $x[n]$ are estimated with

$$\sigma_s^2[n] = \zeta\sigma_s^2[n-1] + (1-\zeta)s^2[n] \tag{11.48}$$

and

$$\sigma_x^2[n] = \zeta\sigma_x^2[n-1] + (1-\zeta)x^2[n], \tag{11.49}$$

respectively. Equations (11.48) and (11.49) act as first-order smoothers to the signals' power. A value of $\zeta = 0.99$ gives an adequate adaptation time for an 8-kHz sampling rate. For smaller values of ζ, faster adaptation is obtained; however, severe distortion can be introduced. The power estimates are updated on a sample-by-sample basis. For each sample, the gain

$$\rho[n] = \sqrt{\frac{\sigma_s^2[n]}{\sigma_x^2[n]}} \tag{11.50}$$

is calculated, and the enhanced speech is found with

$$y[n] = \rho[n]x[n]. \tag{11.51}$$

Thus, for each sample, a division and a square root operation are necessary to compute the gain. One simplification is given by

$$\sigma_s[n] = \zeta\sigma_s[n-1] + (1-\zeta)|s[n]|, \tag{11.52}$$

$$\sigma_x[n] = \zeta\sigma_x[n-1] + (1-\zeta)|x[n]|. \tag{11.53}$$

In this case, the magnitude is estimated instead of power. The scaling factor or gain can be calculated directly as the division

$$\rho[n] = \frac{\sigma_s^2[n]}{\sigma_x^2[n]}. \tag{11.54}$$

This scheme eliminates the need for the square root operation.

Adaptive Spectral Tilt Compensation

As stated earlier, the purpose of the $(1 - \mu z^{-1})$ term in (11.47) is to compensate the tilt of the transfer function so as to reduce the lowpass effect. Since the tilt is variable with the signal, it is possible to enhance the filter by making μ adaptive. In one proposition [Chen, 1995], μ is made proportional to the first reflection coefficient k_1:

$$\mu = 0.5\,k_1, \tag{11.55}$$

where k_1 can be derived from the LPCs a_i. For highly correlated voiced frames, $k_1 = R[1]/R[0]$ is close to one since $R[0] \approx R[1]$ (Chapter 4). Note that $R[0]$ and $R[1]$ are the autocorrelation values of the signal being analyzed. In this case, the resultant postfilter has essentially the same property as in (11.47). For unvoiced frames, however, the magnitude of k_1 tends to decrease, and k_1 might change from positive to negative; this is due to the fact that correlation among adjacent samples is weakened; also, the spectra of unvoiced frames tend to develop a high-pass spectral tilt. Therefore, it is better to either diminish the amount of compensation or even change to lowpass filtering in order to cancel the high-pass tilt. Figure 11.18 shows the magnitude responses of $H(z) = 1 - \mu z^{-1}$ for various values of μ. Subjectively, it is found that the proposed adaptation provides better enhancement to the quality of the processed speech when compared to a fixed coefficient.

Long-Term Postfilter

Detailed operation of the postfilter for the ITU-T G.728 LD-CELP coder (covered in Chapter 14) is described here. In this implementation, a long-term postfilter is incorporated to improve the subjective quality of the synthetic speech. Note that this coder utilizes a 20-sample frame, divided into four 5-sample subframes.

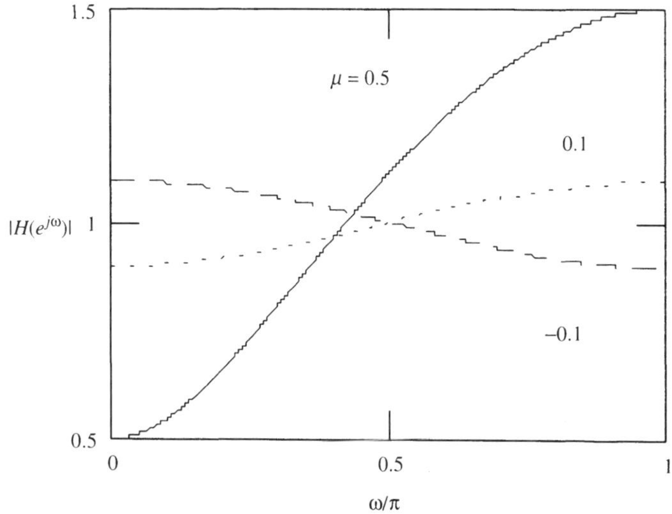

Figure 11.18 Plots of the magnitude of the system function $H(z) = (1 - \mu z^{-1})$ for three values of μ.

The postfilter has the same topology as in Figure 11.17, with the exception that $H_3(z)$ consists of the cascade connection of two filters: short-term filter $H_s(z)$ and long-term filter $H_l(z)$. The short-term filter has the same system function as (11.47), with prediction order $M = 10$, $\alpha = 0.75$, $\beta = 0.65$, and $\mu = 0.15\,k_1$. The coefficients a_i and k_1 are updated once per frame, with the update taking place in the first subframe.

The long-term postfilter is a comb filter with its spectral peaks located at multiples of the fundamental frequency of the speech to be processed. Its purpose is to attenuate frequency components between pitch harmonic peaks, enhancing the periodicity of the signal. System function of the filter is

$$H_l(z) = \frac{1}{1+b}\left(1 + bz^{-T}\right), \tag{11.56}$$

where the parameters b and T are updated once every four subframes, with the actual updates occurring at the third subframe.

Figure 11.19 shows the magnitude responses of (11.56) for several values of b and $T = 12$. Parameter b controls the amount of attenuation, and an adequate range is $b \in [0, 1]$. In the extreme case of $b = 0$, $H_l(z) = 1$ and the filter is allpass. As b increases, attenuation becomes higher and higher. The factor $1/(1 + b)$ normalizes the filter response in such a way that the peak magnitude is always equal to one.

Operations involved in the finding of the filter's parameters are summarized in Figure 11.20.

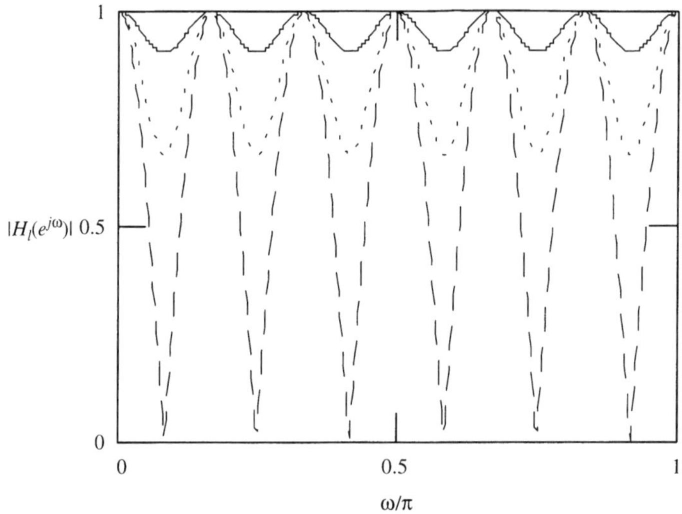

Figure 11.19 Magnitude responses of the long-term filter with $T = 12$ and $b = 0.05$ (*top*), 0.2 (*middle*), and 1 (*bottom*).

Pitch Period Estimation

The G.728 coder utilizes a highly efficient method for pitch period estimation. It first performs a coarse search on a reduced-resolution domain, followed by a search near the neighborhood around the coarse estimate. The final result is obtained by comparison with the past pitch period for multiplicity check (Chapter 2).

The prediction-error signal $e[n]$ is stored in the form shown in Figure 11.21; it is obtained by filtering the noisy speech $s[n]$ with the tenth-order prediction-error

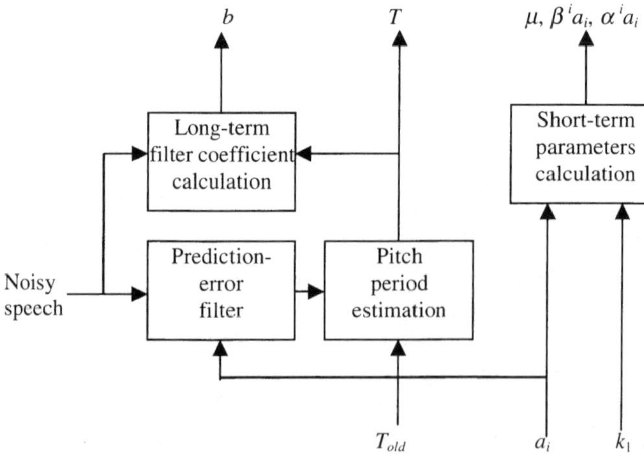

Figure 11.20 Operations involved in the computation of the postfilter's parameters.

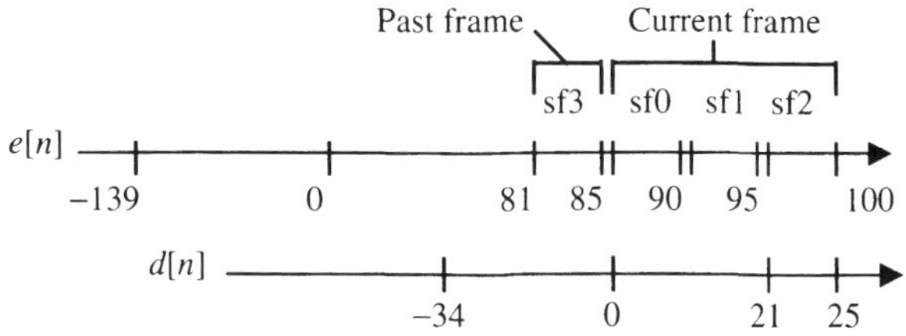

Figure 11.21 The prediction-error array (*top*) and its decimated version (*bottom*).

filter. Pitch period is found once per frame and is extracted at the third subframe (sf2). The pitch analysis window utilizes 100 samples and a range of pitch period from 20 to 140 samples. The overall procedure is illustrated in Figure 11.22.

The prediction error is decimated by a factor of four and is done with a third-order elliptic filter having a cutoff of 1 kHz, followed by down-sampling [Oppenheim and Schafer, 1989]. A first estimation is found in the decimated domain, where the autocorrelation

$$R_d[l] = \sum_{n=1}^{25} d[n]d[n-l]; \quad l = 5, 6, \ldots, 35, \tag{11.57}$$

is computed, with d denoting the decimated prediction-error signal. Note that the range of l corresponds to 20 to 140 samples in the original domain. The lag $l = T_d$ maximizes (11.57) and represents a coarse pitch period estimate.

The next step is to search around T_d in the original domain so as to yield a more accurate estimation. This is done by computing

$$R_e[l] = \sum_{n=1}^{100} e[n]e[n-l]; \quad l = 4T_d - 3 \text{ to } 4T_d + 3. \tag{11.58}$$

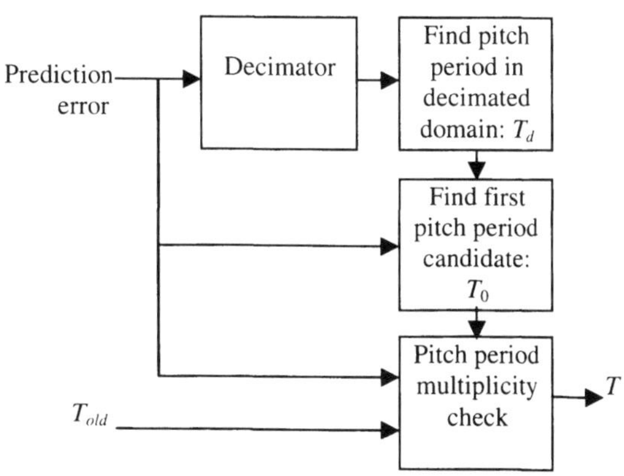

Figure 11.22 The pitch period estimation procedure.

The lag $l = T_0$ maximizes (11.58) and represents a first pitch period estimation with e denoting the prediction-error signal. This value may turn out to be a multiple of some fundamental pitch period. More processing is necessary to minimize this possibility.

First, T_0 is compared with T_{old} (the value from a past frame), if $T_0 \in [T_{old} - 6, T_{old} + 6]$; that is, the first estimate is in the neighborhood of T_{old}, then the desired result is settled with $T = T_0$. Otherwise, (11.58) is evaluated for $l = T_{old} - 6$ to $T_{old} + 6$, with $l = T_1$ the lag that maximizes the correlation within the interval. T_1 is a candidate for the final result. T_0 and T_1 are then compared using the normalized autocorrelation values:

$$R_i = \frac{\sum_{n=1}^{100} e[n]e[n - T_i]}{\sum_{n=1}^{100} e^2[n - T_i]}, \quad i = 0, 1. \tag{11.59}$$

The resultant autocorrelation values are clamped between 0 and 1; that is,

$$R_i \leftarrow \begin{cases} 0, & \text{if} \quad R_i \le 0 \\ 1, & \text{if} \quad R_i \ge 1, \quad i = 0, 1. \\ R_i, & \text{otherwise,} \end{cases} \tag{11.60}$$

Then the final pitch period T is given by

$$T = \begin{cases} T_0, & \text{if} \quad R_1 \le 0.4\, R_0, \\ T_1, & \text{otherwise.} \end{cases} \tag{11.61}$$

Long-Term Filter Coefficient Calculation

Assuming that the noisy speech is stored in the array $s[n]$, $n = -239$ to 5, with the current subframe residing in $n = 1$ to 5, then the normalized autocorrelation value

$$R_T = \frac{\sum_{n=-99}^{0} s[n]s[n - T]}{\sum_{n=-99}^{0} s^2[n - T]} \tag{11.62}$$

is calculated, which is a periodicity measure of the signal. The long-term filter coefficient is found with

$$b = \begin{cases} 0, & \text{if} \quad R_T < 0.6, \\ 0.15\, R_T, & \text{if} \quad 0.6 \le R_T \le 1, \\ 0.15, & \text{if} \quad R_T > 1. \end{cases} \tag{11.63}$$

In general, the closer R_T is to one, the more periodic is the speech waveform. For $R_T < 0.6$, which corresponds roughly to unvoiced frames, $b = 0$, implying that the long-term postfilter is turned off. For $0.6 \le R_T \le 1$, the postfilter is turned on, with the degree of comb filtering determined by R_T. For $R_T > 1$, b is limited to 0.15 so as to avoid excessive filtering.

11.6 SUMMARY AND REFERENCES

CELP is essentially an improved LPC coder, with the incorporation of a pitch synthesis filter to avoid the strict voiced/unvoiced classification, and an excitation codebook to preserve phase information. Problems observed in LPC coders, such as poor performance in the reproduction of background noise, are greatly rectified, leading to a far more robust and sturdy system. The superiority is a direct result of a more sophisticated speech production model, as well as the closed-loop analysis-by-synthesis approach contributing toward optimal adaptation to many different signal types.

At the algorithm level, the CELP encoder is an excellent illustration of how different strategies applied to the computation of various signals of the system have a great impact on computational complexity. Careful choice of methodologies and thorough understanding of their effects are indispensable for successful deployment in practice.

The core idea of CELP was first presented in Schroeder and Atal [1985]. Since the largest computational burden is the excitation codebook search process, many ideas have been proposed in the literature to tackle the problem; see Trancoso and Atal [1986] for some discussion on fast search methods. The search of the excitation codevector is in a sense a vector quantization process, where the index of the best codevector is found based on error criteria. Hence, it is possible to use the various complexity reduction techniques found in different suboptimal VQ schemes (Chapter 7) so as to speed up the search process; in Ahmed and Al-Suwaiyel [1993], several structures are proposed for the excitation codebook to achieve substantial computational reduction. It is possible to improve performance further by training the fixed excitation codebook using a representative set of speech signals. See Exercise 11.11 for a training algorithm based on a gradient-descent approach.

Unlike audio coding [Painter and Spanias, 2000], where structure of the coder is designed around the various well-known psychoacoustic phenomenon, such as signal masking and absolute threshold, speech coding seldom utilizes any perceptual modeling. This is partly because of the high computational demand involved with the approach. On the other hand, there is indication that perception of speech is very different from other nonspeech signals; in fact, there is evidence of a brain specialization where an isolated part of the brain is in charge of speech [Moore, 1997]. The CELP coder presented in this chapter takes advantage of simple psychoacoustic facts, which are found in the perceptual weighting filter as well as the postfilter, where the spectrum is shaped adaptively to improve the subjective quality. A more general form of perceptual weighting filter is used in some CELP coders, such as LD-CELP covered in Chapter 14. The idea of a postfilter is described in Chen and Gersho [1987], with a more complete work given in Chen and Gersho [1995], where additional techniques and experimental data are presented.

The CELP coder has been highly prosperous in practice and has spawned a series of standardized coders based on the same fundamental principles. Many of these coders are included in the next few chapters of this book. Despite its widespread acceptance, CELP is not free of limitations. One of the criticisms is that it does not behave well for audio signals, in general, but rather is highly specialized and adapted to speech signals. This leads to high efficiency in the representation of speech, but poor performance when presented with music signals having complex spectrum shapes. Designing an efficient coder that works well for both speech and general audio is indeed one of the goals that speech coding researchers crave to attain.

EXERCISES

11.1 Consider the analysis-by-synthesis loop described in Figures 11.5 and 11.8. We are given the parameters:

L: Size of excitation codebook,

N_1: Number of operations required by the pitch synthesis filter per excitation codevector,

N_2: Number of operations required by the (modified) formant synthesis filter per excitation codevector,

N_3: Number of operations required by the perceptual weighting filter per excitation codevector.

Further assume that the number of operations required by other components of the loop is negligible. Compare the total number of operations needed to perform one complete codebook search (i.e., L passes through the loop) for the case of Figure 11.5 and compare to that of Figure 11.8. Which scheme is more efficient?

11.2 By counting the number of additions and multiplications, show that for the state-save method, the number of sums and number of products are as shown in Table 11.2, with N being the length of the subframe and M the prediction order (short-term); while the computational cost for the zero-input zero-state method is as shown in Table 11.3. To count the number of multiplications, assume that $a_i r^i$, $i = 1$ to M are readily available, that is, no additional operations are involved with these modified coefficients.

11.3 An alternative zero-input zero-state method is shown in Figure 11.23. Write down the difference equations relating the signals. Using the same parameter values as in Table 11.1, calculate the numbers of sums and products. Is this approach more efficient? If so, explain.

TABLE 11.2 Computational Cost for the State-Save Method

Signal	#Sums (per Subframe)	#Products (per Subframe)
d	N	N
y	MN	MN
Total	$(M+1)N$	$(M+1)N$

TABLE 11.3 Computational Cost for the Zero-Input Zero-State Method

Signal	#Sums (per Subframe)	#Products (per Subframe)
$d1$	$N-T$	$N-T$
$d2$	0	N
d	N	0
$y1$	MN	MN
$y2$	MN	MN
$y3$	$(M-1)N$	MN
y	$2N$	0
Total	$3(M+1)N-T$	$(3M+2)N-T$

11.4 Plot the number of products per excitation codebook search for the state-save method and the zero-input zero-state method, with $M = 10$, $N = 60$, and $T = 50$, as a function of L, where L ranges from 8 to 1024 (3 to 10 bits codebook). For what range of L is the zero-input zero-state method more efficient? Repeat for $T = 80$.

11.5 From the signal flow graph of Figure 11.14, find out the number of sums (#sums), the number of products (#products), and the number of divisions (#divisions) required to perform one complete codebook search.

11.6 From the system function of the postfilter (11.47), find out the difference equation relating the filter's input/output. Using the difference equation, plot the first 30 samples of the impulse response of the filter when $\alpha = 0.8$, $\beta = 0.5$, $\mu = 0.5$, and the set of LPCs from Example 11.1.

11.7 Use some input speech signal to test the power estimation equation (11.48) for different values of ζ in the interval [0, 1]. What conclusion can be reached as far as the adaptation rate of the power estimator as a function of ζ? Repeat for the estimator specified by (11.52).

11.8 For the gain equations (11.50) and (11.54), a low signal level might cause numerical instability due to the division involved. Devise a mechanism to avoid this type of situation by detecting the signal level.

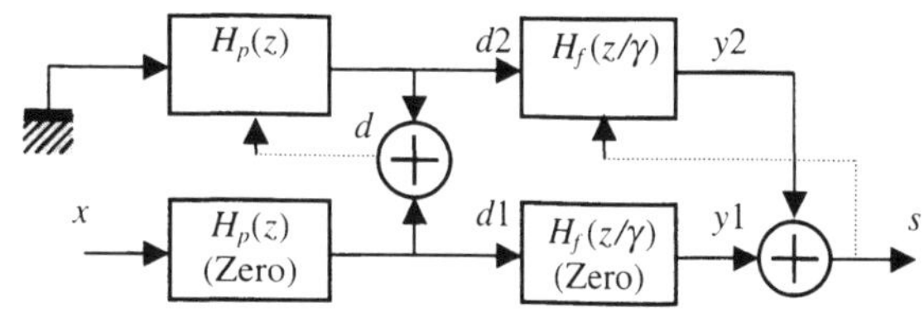

Figure 11.23 Block diagram showing an alternative zero-input zero-state method.

11.9 Assuming that the zero-state response is calculated through the convolution sum approach, modify the signal flow graph of Figure 11.14 so as to accommodate this alternative technique. Note that the impulse response of the filter cascade must be computed before performing any convolution operation.

11.10 For the postfilter of the G.728 coder:

(a) Plot the magnitude response of the short-term filter using the LPC of Example 11.1 and $k_1 = 0.8$.

(b) Plot the magnitude response of the long-term filter with $T = 30$ and $b = 0.15$.

(c) Plot the magnitude response of the cascade connection between the two filters.

11.11 Excitation codebook training based on *gradient-descent*. The purpose of this exercise is to develop a technique to train the excitation codebook so as to improve performance. Given the function f, the gradient is an N-dimensional vector given by

$$\nabla f = \left[\frac{\partial f}{\partial x_0} \quad \frac{\partial f}{\partial x_1} \quad \cdots \quad \frac{\partial f}{\partial x_{N-1}}\right]^T$$

with x_i being the variables affecting f. The gradient has a very important property: if we move along the gradient direction from any point in N-dimensional space, the function value increases at the fastest rate [Rao, 1996]; hence, the direction of the gradient is called the direction of *steepest ascent*. The negative of the gradient vector denotes the direction of *steepest descent*. Thus, any method that makes use of the gradient vector can be expected to give the minimum point faster than one that does not make use of the gradient vector.

Based on the principle, follow the steps below to design a training algorithm for CELP.

(a) In encoding, after the optimal codevector is located, show that the derivatives of the error energy $\varepsilon^{(l)}$—the function we seek to minimize—with respect to the elements of the excitation codevector $v^{(l)}[n]$ are

$$\frac{\partial \varepsilon^{(l)}}{\partial v^{(l)}[n]} = 2g^{(l)} \sum_{k=n}^{N-1} h[k-n]\left(y1^{(l)}[k] - u_o[k]\right) \tag{11.64}$$

for $n = 0$ to $N - 1$.

(b) Once the derivatives are known, we can travel in the direction of negative gradient with

$$v_{\text{new}}^{(l)}[n] = v^{(l)}[n] - \mu \frac{\partial \varepsilon^{(l)}}{\partial v^{(l)}[n]} \tag{11.65}$$

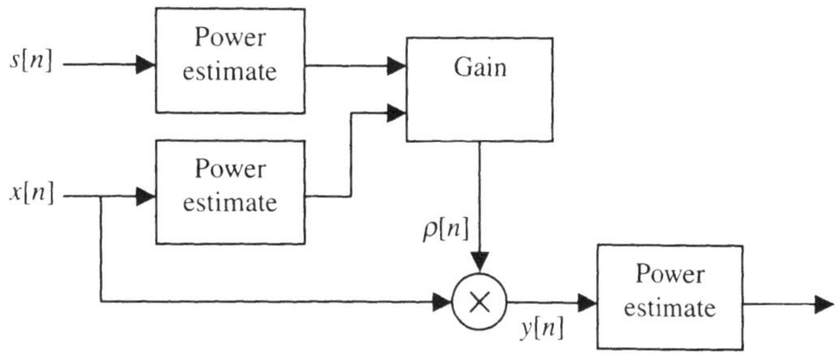

Figure 11.24 A system for automatic gain control tests.

where μ is a constant known as the step-size parameter. This constant determines how fast the training is going to be. A low μ implies slow training while high μ can lead to instability.

The training algorithm works as follows: for each input speech subframe from the training set, parameters of the CELP model are found; the gradient with respect to the chosen excitation codevector is calculated, which is then updated through gradient-descent. After presenting the training material for a long period of time, the excitation codebook is optimized, with an expected improvement in performance.

Prepare a detailed step-by-step procedure for excitation codebook training based on gradient-descent. Include all equations.

11.12 The automatic gain control mechanisms proposed for the postfilter can be tested by utilizing the system of Figure 11.24. Implement this system by following the equations presented in Section 11.5. If the proposition works, the power of $y[n]$ should be close to that of $s[n]$. For test purposes, steady sine waves can be used for both $s[n]$ and $x[n]$. Perform the test for the two described schemes of automatic gain control.

11.13 Assume that the pitch period T is uniformly distributed inside the interval [20, 147]. Calculate the average computational cost (per codebook search) with $M = 10$, $N = 60$, and $L = 512$ for the state-save method and the zero-input zero-state method.

CHAPTER 12

THE FEDERAL STANDARD VERSION OF CELP

In 1984, the U.S. Department of Defense (DoD) initiated a program to develop a new secure voice communication system to supplement the existing FS1015 LPC coder. Between 1988 and 1989, the 4.8-kbps CELP coder, jointly developed by the DoD and Bell Labs, was selected. Numerous tests confirmed the superiority of the coder, providing reasonable quality and robustness in acoustic noise, channel errors, and tandem coding conditions [Campbell et al., 1991]. The coder became an official federal standard in 1991 [National Communications System, 1992] and is known as FS1016.

Besides the basic principles of CELP coding, the FS1016 contains features and modifications to improve both the speech quality as well as the computational efficiency. Even though this coder has largely been replaced by the more recent MELP standard (Chapter 17), which produces almost the same quality at half the bit-rate, the clever design of the FS1016 set a milestone in speech coding development. Its core structure has exerted a strong influence on other CELP-based algorithms, as we will see in future chapters.

It is expected that readers have knowledge of the essential ideas of CELP coding, such as the material presented in Chapter 11. In this chapter, the attempt to improve the long-term predictor by utilizing a closed-loop analysis-by-synthesis approach is explained; this gave birth to the concept of the adaptive codebook. The incorporation of the adaptive codebook to the framework of CELP is shown; the structure of the stochastic codebook with fixed random samples is examined, followed by codebook search techniques. The chapter ends with an overview of the operations of the encoder and decoder.

12.1 IMPROVING THE LONG-TERM PREDICTOR

CELP algorithms rely on the long-term predictor or pitch synthesis filter to take advantage of the periodicity of the (voiced) speech signal so as to achieve efficient encoding. The parameters of the long-term predictor—pitch period (or lag) and long-term gain (or long-term LPC)—are necessary for the algorithm to operate, and they are estimated during the actual encoding process.

As indicated in Chapter 11, a long-term LP analysis procedure can be performed on the short-term prediction-error signal (Chapter 4). In this approach, the parameters are found to minimize the energy of the overall prediction error. The resultant parameters reflect the statistical property of the particular signal subframe; however, the goal of the procedure does not coincide exactly with CELP, since for this coder, the main purpose is the minimization of the weighted difference between the input speech and the synthetic speech. Hence, improvement is possible by using an estimation technique that directly targets the weighted difference.

Since the objective of the CELP encoder is to reduce the energy of the perceptually weighted difference between the input speech and the synthetic speech, in principle it is possible to augment the contribution of the long-term predictor by determining its parameters so as to minimize the mentioned weighted difference signal. Using the same notation as in Chapter 11, the energy of the weighted difference for the rth subframe is written as

$$\begin{aligned}\varepsilon_r^{(l)} &= \sum_{n=0}^{N-1}\left(u_r[n] - y_r^{(l)}[n]\right)^2 \\ &= \sum_{n=0}^{N-1}\left(u_r[n] - y1_r^{(l)}[n] - y2_r[n] - y3_r[n]\right)^2. \end{aligned} \tag{12.1}$$

To minimize the above quantity, we need to find the optimal set of parameters: excitation vector $v^{(l)}$, excitation gain $g^{(l)}$, pitch period T, and long-term gain b, with the assumption that the parameters of the formant synthesis filter are already known. Although these variables can be obtained by exhaustively searching for all gain, excitation, and the long-term predictor's parameters, the procedure becomes very computationally intensive, and suboptimal solutions are often used.

One way of reducing the search complexity is by obtaining the long-term predictor's parameters (pitch period and gain) and the excitation codevector (including gain) in two steps. First, we assume zero excitation gain and calculate the long-term predictor's parameters such that the error is minimized. Next, the long-term predictor is held constant and the optimal excitation plus gain is searched. The problem can be stated as follows: find b and T so that the sum of squared error

$$J_r = \sum_{n=0}^{N-1}\left(u_r[n] - y2_r[n] - y3_r[n]\right)^2 \tag{12.2}$$

is lowered as much as possible. Minimization can be conducted in the standard way of differentiation: find the derivative of J with respect to b and equate that to zero. Appendix E contains all the details of the procedures; only highlights of the results are provided here.

The derivative of J can be found from the derivative of $y2$, given by the convolution sum

$$\frac{\partial y2_r[n]}{\partial b} = \sum_{k=0}^{n} h[k] \frac{\partial}{\partial b} d2_r[n-k]; \quad 0 \le n \le N-1, \tag{12.3}$$

with $h[n]$ being the impulse response of the modified formant synthesis filter.

The trick is to note that in order to find the derivative of $d2$, it is necessary to express it as a function of the samples from the past—the $(r-1)$st subframe and further into the past, since $d2$ for the present subframe is not yet known. In fact, the rth subframe depends on b and T—the set of parameters for which we are searching.

From Chapter 11, we have

$$d2_r[n] = -b \cdot d2_r[n-T]; \quad 0 \le n \le N-1, \tag{12.4}$$

which is the zero-input response of the pitch synthesis filter. To express the current subframe of $d2$ in terms of the past samples, it is necessary to consider the range of T. For instance, if $T \ge N$, then no modification is needed for (12.4), since all $d2$ samples on the right-hand side of the equation have negative time index and hence belong to the past. This is not quite the case when $N/2 \le T < N$, where

$$d2_r[n] = \begin{cases} -bd2_r[n-T]; & 0 \le n \le T-1, \\ b^2 d2_r[n-2T]; & T \le n \le N-1; \end{cases} \tag{12.5}$$

and for $N/3 \le T < N/2$, we have

$$d2_r[n] = \begin{cases} -bd2_r[n-T]; & 0 \le n \le T-1, \\ b^2 d2_r[n-2T]; & T \le n \le 2T-1, \\ -b^3 d2_r[n-3T]; & 2T \le n \le N-1. \end{cases} \tag{12.6}$$

Note that (12.5) and (12.6) are derived from (12.4) with the simple purpose of setting the time index of the samples to negative values on the right-hand side of the equation.

From (12.4), (12.5), and (12.6), we see that for shorter and shorter values of the pitch period, the expression of $d2$ becomes more and more complex. Appendix E shows that a closed-form solution for b can be obtained for the case of $T \ge N$. For $N/2 \le T < N$, it becomes a cubic expression. And when $N/3 \le T < N/2$, it is so complex that it isn't even worth considering, since the computation load is too high. Hence, the proposed method for finding the long-term predictor's parameters is not very practical after all.

In this section, an attempt to improve the long-term predictor is described, but it ultimately leads to a dead end because of the computational burden. Even though the initiative is sound, and the performance should be much better than using the open-loop long-term LP analysis procedure, it is practically difficult, if not impossible, to implement. Readers might want to ask: What is the point of evaluating an inadequate approach? Well, throughout the history of research and development, a plethora of ideas and proposals have appeared; only a few of these propositions made it to mainstream engineering, with the rest left as a reminder of the various paths that researchers had examined. Evaluating some of the failed attempts helps us to comprehend why such a system is deployed in the way it is. As a matter of fact, ideas in this section generated the concept of the adaptive codebook: the main component that improved the quality of many practical CELP coders. The whole deal is like the old Chinese saying: "Failure is the mother of success."

12.2 THE CONCEPT OF THE ADAPTIVE CODEBOOK

In the last section, we saw that our attempt to find the long-term predictor's parameters so as to minimize the weighted difference can get rather complex, when the pitch period is less than the length of the subframe. The complexity is due to the operation of the pitch synthesis filter: the way the zero-input response $d2$ (12.4) is defined.

The concept of the adaptive codebook was developed as a modification to reduce complexity, but still utilize the same principle to minimize the weighted difference through a closed-loop analysis-by-synthesis approach. In other words, some "cheating" to overcome the computational barrier is employed. In this proposition, the operation of the pitch synthesis filter is redefined with

$$d2_r[n] = -b\begin{cases} d2_r[n-T]; & 0 \le n < T, \\ d2_r[n-2T]; & T \le n < 2T, \\ d2_r[n-3T]; & 2T \le n < 3T, \\ \vdots & \vdots \end{cases} \tag{12.7}$$

The effect is illustrated in Figure 12.1 for a typical case of $T < N$. Obviously, the adaptive codebook definition does not coincide with the original pitch synthesis

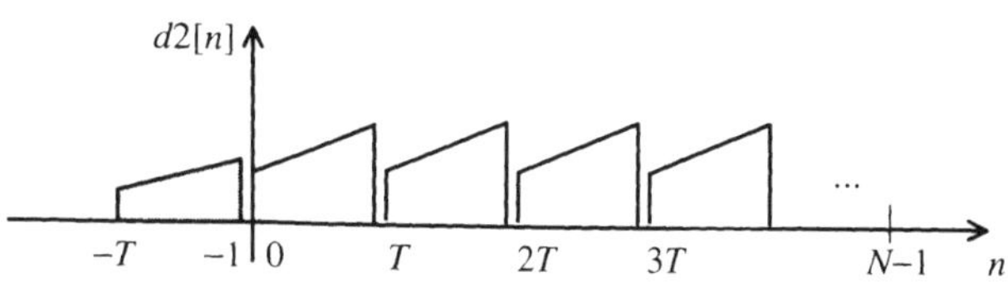

Figure 12.1 Extracting a codevector from the adaptive codebook when $T < N$: only the samples from $-T$ to -1 are used and are periodically replicated to form the codevector.

filter. For $T \geq N$, the two definitions match; for $T < N$, however, the response following the adaptive codebook definition is produced by a periodic extension of the past, multiplied by the long-term gain. By "cheating" in this manner, the zero-input response (with unit scaling) of the current subframe is completely determined from the history, thus eliminating the burden associated with the higher power of b found in the concept of the regular pitch synthesis filter ((12.5) and (12.6)). Of course, the two definitions still match when $b = 1$. According to this reformulation, b can be solved by searching through the values of T using simple procedures.

The limitation of the adaptive codebook approach is that the pitch pulses in a subframe maintain the same amplitude from one pulse to another when $T < N$. However, it has been observed that the procedure is subjectively very similar to the original formulation of the pitch synthesis filter. That is, in terms of subjective quality, little is lost or gained in using the modified method.

Operation of the Adaptive Codebook

The reason why the formulation is named adaptive codebook is due to the fact that the search procedure for the optimal pitch period can be considered as the evaluation of a codebook with overlapping codevectors. That is, adjacent codevectors have common samples, where each is indexed by the pitch period T. The codebook is adaptive because it is changed from subframe to subframe.

Consider the structure of the adaptive codebook shown in Figure 12.2. The codebook is represented as a 1-D array with T_{max} elements, with T_{max} being the maximum value of the pitch period. For convenience the elements of the codebook are denoted as $d[-T_{max}]$ to $d[-1]$, and the subframe index r is dropped by now.

Codevectors derived from this codebook are denoted by

$$\mathbf{d2}_o^{(t)} = [d2_o^{(t)}[0] \cdots d2_o^{(t)}[N-1]]^T; \quad t = T_{\min} \text{ to } T_{\max}.$$

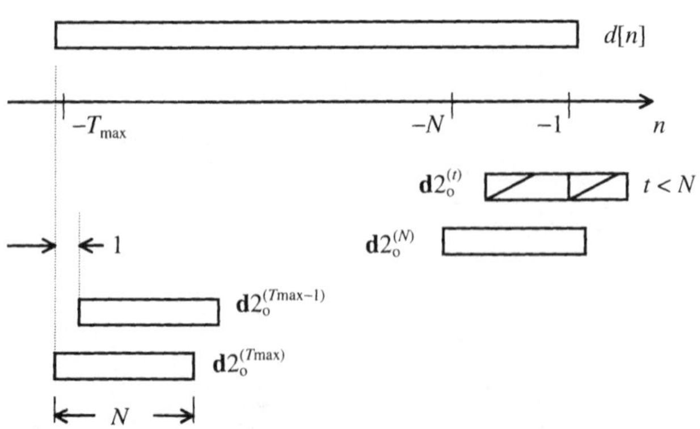

Figure 12.2 Illustration of the structure of an adaptive codebook.

Equation (12.7) can now be used to define the elements of the codevector as follows:

$$d2_o^{(t)}[n] = \begin{cases} d[n-t]; & 0 \le n < t, \\ d[n-2t]; & t \le n < 2t, \\ d[n-3t]; & 2t \le n < 3t, \\ \vdots & \vdots \end{cases} \tag{12.8}$$

if $t < N$, and

$$d2_o^{(t)}[n] = d[n-t] \tag{12.9}$$

if $t \ge N$, for $0 \le n \le N-1$. Thus, the codebook can be searched for all index t to yield the best codevector $\mathbf{d2}_o^{(t)}$ together with the optimal long-term gain b. Once these variables are known, the scaled codevector

$$\mathbf{d2}^{(t)} = b \cdot \mathbf{d2}_o^{(t)} \tag{12.10}$$

represents the optimal zero-input response of the modified pitch synthesis filter. The search procedure is similar to the one described in the last chapter for excitation codebook search and is outlined in Section 12.5.

Codebook Update

After $\mathbf{d2}^{(t)}$ is located, the total response of the pitch synthesis filter is

$$d[n] = d1^{(l)}[n] + d2^{(t)}[n]; \quad n = 0 \text{ to } N-1, \tag{12.11}$$

where $d1^{(l)}[n]$ is the zero-state response found by searching through the excitation codebook (or stochastic codebook, Section 12.4). The adaptive codebook is then updated with this signal according to

$$d[n] \leftarrow d[n+N]; \quad n = -T_{\max}, -T_{\max}+1, \ldots, -1. \tag{12.12}$$

From the point of view of an array, the operation is basically a left shift of N samples. The codebook is now ready to be searched in the next subframe.

A Summary

The adaptive codebook contains a history of the total response of the pitch synthesis filter; codevectors within the codebook are overlapped. The pitch period (t) indexes the codevector containing the best block of excitation from the past for use in

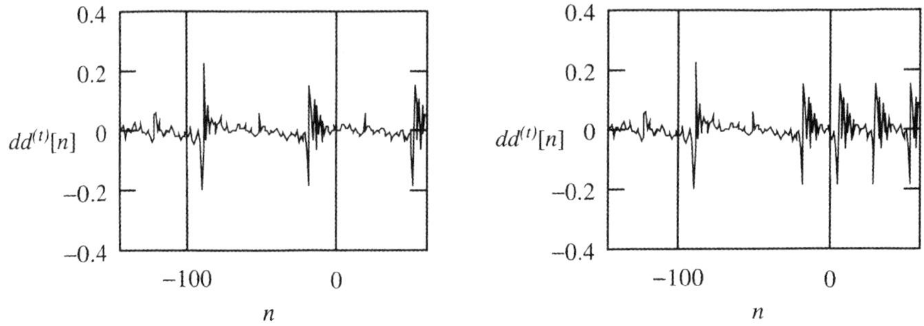

Figure 12.3 Examples of signals involved with the adaptive codebook. Samples from $n = -147$ to -1 form the codebook, with the actual codevector located from $n = 0$ to 59. *Left* : $t = 70$. *Right* : $t = 24$.

the present. For t less than the subframe length N, a full vector does not exist and the short vector is replicated (periodic extension) to the full vector length to form the codevector.

For the FS1016, $N = 60$, $T_{max} = 147$, with the index t ranging from 20 to 147, leading to a total of 128 values encodable using 7 bits. Figure 12.3 shows some examples of codevectors derived from the adaptive codebook, where we can clearly see the periodically extended sequence when $t < 60$.

12.3 INCORPORATION OF THE ADAPTIVE CODEBOOK TO THE CELP FRAMEWORK

In the last section, operation of the long-term predictor was modified to that of an adaptive codebook. Here, we consider the problem of assembling the adaptive codebook into the framework of a CELP encoder, already described in Chapter 11. An adaptive codebook can be incorporated directly into the block diagram of a standard CELP encoder, with the signal relations shown in Figure 12.4. The excitation codebook, as it is known in Chapter 11, is renamed the stochastic codebook; at this moment we assume that it contains samples with fixed values originating from a white noise source.

The fundamental principles of the encoder are the same as described in Chapter 11, except that the long-term predictor's parameters—pitch period and long-term gain—are determined in a closed-loop analysis-by-synthesis basis. That is, the pitch period t (or adaptive codebook index) and the long-term gain b (or adaptive codebook gain) are found to minimize

$$J^{(t)} = \sum_{n=0}^{N-1} \left(u[n] - y2^{(t)}[n] - y3[n]\right)^2. \tag{12.13}$$

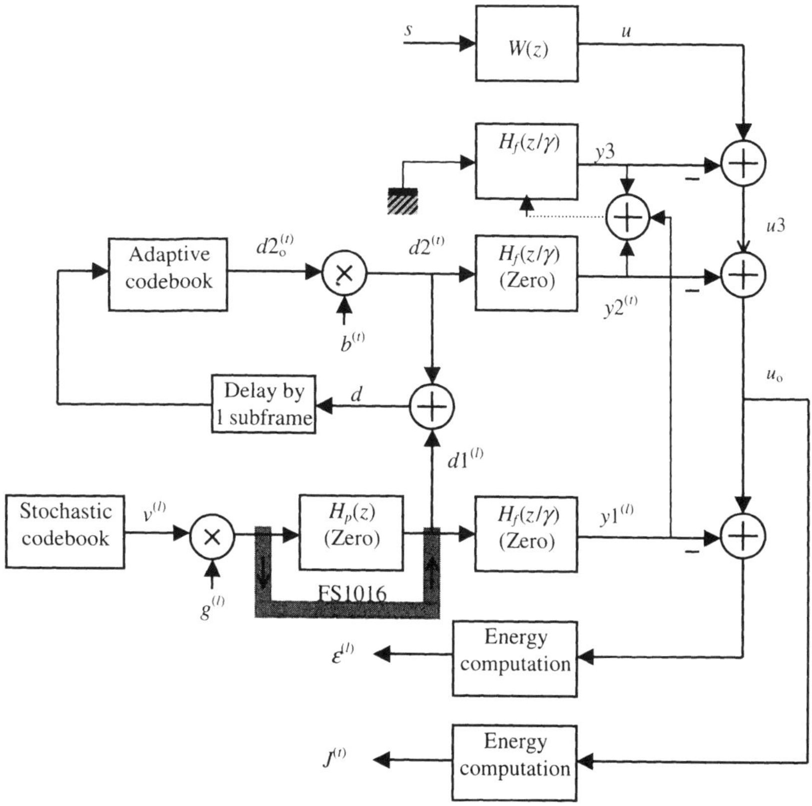

Figure 12.4 Signals involved in a CELP encoder with adaptive codebook. For the FS1016 coder, the pitch synthesis filter is bypassed.

Once the parameters of the adaptive codebook are known, one can search the stochastic codebook so that

$$\varepsilon^{(l)} = \sum_{n=0}^{N-1} \left(u[n] - y1^{(l)}[n] - y2^{(l)}[n] - y3[n] \right)^2 \tag{12.14}$$

is minimized. This is exactly the two-step suboptimal approach described in Section 12.1, where the adaptive codebook is searched first assuming the gain of the stochastic codebook to be zero; afterward, the stochastic codebook is examined to locate its parameters.

Further simplification to the encoder is possible by noting that the pitch synthesis filter ($H_p(z)$) with no initial condition has no effect when the pitch period is greater than the subframe length ($T > N$). Observe this fact from the difference equation describing its operation:

$$d1^{(l)}[n] = 0; \quad -T \leq n \leq -1, \tag{12.15}$$

$$d1^{(l)}[n] = g^{(l)}v^{(l)}[n] - b^{(l)}d1^{(l)}[n-T]; \quad 0 \leq n \leq N-1. \tag{12.16}$$

Thus, if $T > N$, $d1^{(l)}[n] = g^{(l)}v^{(l)}[n]$ for $0 \leq n < N$. That is, the pitch synthesis filter has no effect at all since the output is identical to the input. For $T \leq N$, $d1^{(l)}[n]$ contains sample values that are the result of addition between two random samples separated T apart, resulting in another random sequence with slightly different statistical property (Problem 12.1).

In practice, voiced excitation is largely due to the contribution of the adaptive codebook or, in the original formulation, the zero-input response of the pitch synthesis filter. On the other hand, the encoding process is an analysis-by-synthesis procedure: it is always possible to find an optimal excitation sequence that minimizes the sum of squared error. Thus, for simplicity, the pitch synthesis filter is removed from the stochastic codebook search loop, and the scaled adaptive codevector is added to the scaled stochastic codevector directly; the resultant sequence is used as excitation for the modified formant synthesis filter. This new proposition is already indicated in Figure 12.4, where the $H_p(z)$ block is bypassed. It has been observed that this method is subjectively very similar to the original formulation.

The equations governing these signals are already described in Chapter 11 and previous sections of this chapter.

12.4 STOCHASTIC CODEBOOK STRUCTURE

The standard excitation codebook for the CELP algorithm, as explained in Chapter 11, consists of L N-dimensional codevectors organized essentially as an $L \times N$ matrix. This type of structure is referred to as *nonoverlapping* and is illustrated in Figure 12.5. One problem with this codebook structure is the size of memory required to store the codevectors. The amount of memory locations needed to store the entire codebook is given by NL.

To overcome this problem, samples of the codebook are stored in a one-dimensional array, where most of the N samples of two consecutive vectors are common. The structure is known as *overlapping* codebook. An overlapping codebook with a shift value of S is shown in Figure 12.6. One can easily show that for an S-shift codebook, the amount of memory required is $S(L-1)+N$. Therefore, an overlapping codebook offers substantial reduction in memory requirements. Note that the standard codebook can be considered as an N-shift overlapping codebook, where $S = N$.

Denoting the entire codebook as a single array, we have

$$v[n]; \quad n = 0, \ldots, S(L-1)+N-1.$$

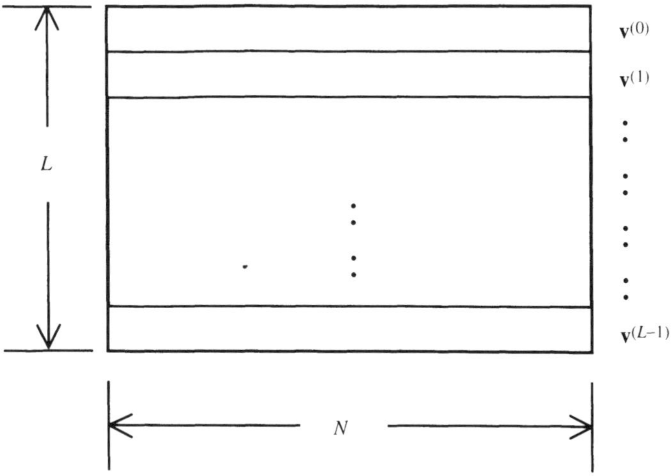

Figure 12.5 Structure of a nonoverlapping stochastic codebook.

Then each codevector can be identified with

$$
\begin{aligned}
v^{(0)}[n] &= v[n + (L-1)S],\\
v^{(1)}[n] &= v[n + (L-2)S],\\
&\vdots\\
v^{(L-1)}[n] &= v[n],
\end{aligned}
\tag{12.17}
$$

for $n = 0$ to $N-1$. It is straightforward to show that

$$v^{(l+1)}[n] = v^{(l)}[n - S]; \quad S \le n \le N - 1. \tag{12.18}$$

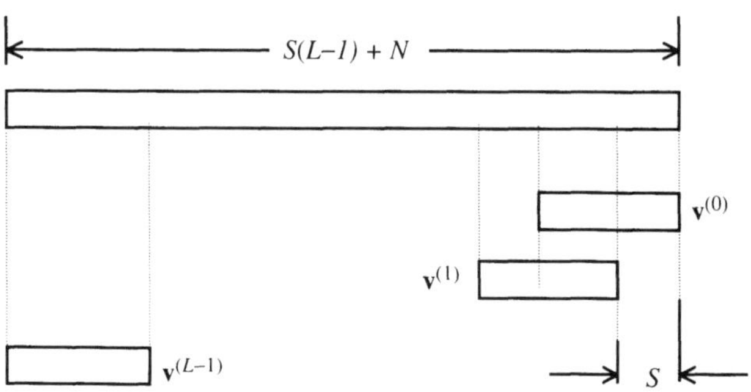

Figure 12.6 Structure of an overlapping stochastic codebook.

Subjectively, no significant difference is found between overlapping and nonoverlapping codebooks. This is expected since in principle the codebooks contain uncorrelated white noise samples, and shifted white noise sequences are also uncorrelated, satisfying the fundamental randomness assumption.

For the FS1016 coder, 9 bits is reserved to index the stochastic codebook, leading to $L = 2^9 = 512$ codevectors; the codebook is overlapped with a shift of $S = 2$, which sometimes is referred to as a "−2 shift" since the codevectors are ordered from right to left. With a codevector dimension of $N = 60$, the overlapping codebook requires a total of 1082 locations, an order of magnitude saving when compared with the nonoverlapping codebook ($NL = 30720$).

Another important advantage of overlapping codebook is the computational saving through the application of recursive convolution. As shown in Chapter 2, recursive convolution allows huge computational savings since the zero-state response of a given excitation codevector can be found from the response associated with the shifted version of the excitation. See Exercise 12.2 for numerical results on computational advantages.

So far, we haven't worried about the actual content of the codebook and have mentioned that the codebook samples are obtained from a white noise source. The FS1016 utilizes a special form of random samples, derived from a zero-mean unit-variance white Gaussian source, center clipped at 1.2 and ternary level quantized to $\{-1, 0, +1\}$. The National Communications System [1992] gives the actual 1082 codebook elements, which are plotted in Figure 12.7. Approximately 77% of the samples are zeros. This type of codebook is compact, causes little degradation in speech quality relative to other types of codebooks, and significantly reduces computational burden since products during convolution calculation are eliminated completely. (See Exercise 12.6 for an algorithm that takes advantage of these features.) In fact, experiments show that a codebook having a high percentage of zero samples performs almost the same as a codebook with less percentage of zeros [Davidson and Gersho, 1986].

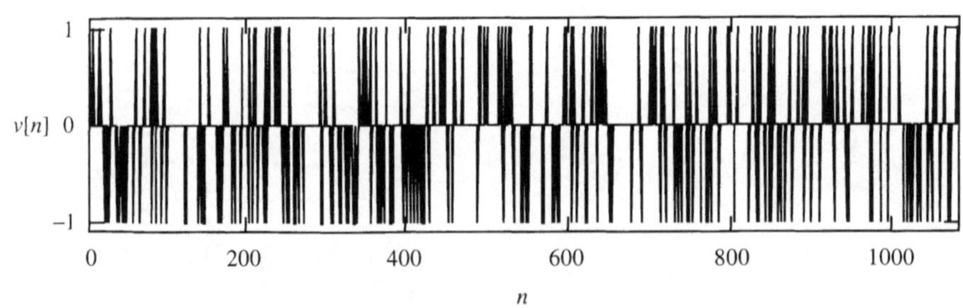

Figure 12.7 Plot of the elements of the FS1016 stochastic codebook. Data from [National Communications System, 1992], Table 11.

12.5 ADAPTIVE CODEBOOK SEARCH

The basic idea and structure of the adaptive codebook are discussed in previous sections. Here, we study the procedure to search for the optimal adaptive codevector, the use of fractional delay, and encoding of related parameters.

Integer Pitch Period

As explained in Section 12.3, the adaptive codebook is searched in a closed-loop analysis-by-synthesis manner, where the index of the best codevector, $t = T_{min}$, $T_{min} + 1, \ldots, T_{max}$, is located. Seven bits is allocated to this index, leading to $2^7 = 128$ values ranging from $T_{min} = 20$ to $T_{max} = 147$. The procedure to search the adaptive codebook is based on the same principles presented in Chapter 11 for the excitation codebook search. Only highlights are provided here; it is left as an exercise for readers to verify the validity of the equations.

The zero-state response of the modified formant synthesis filter is given by

$$\mathbf{y2}^{(t)} = \mathbf{H} \cdot \mathbf{d2}^{(t)} = b^{(t)}\mathbf{H} \cdot \mathbf{d2}_o^{(t)}, \tag{12.19}$$

with $\mathbf{H}$ being the impulse response matrix of the synthesis filter. Equation (12.13) is rewritten as

$$J^{(t)} = \|\mathbf{u} - \mathbf{y2}^{(t)} - \mathbf{y3}\|^2, \tag{12.20}$$

which is our target for minimization. Defining

$$\mathbf{u3} = \mathbf{u} - \mathbf{y3}, \tag{12.21}$$

we have

$$J^{(t)} = \|\mathbf{u3} - b^{(t)}\mathbf{H} \cdot \mathbf{d2}_o^{(t)}\|^2. \tag{12.22}$$

It can be shown that the optimal adaptive codebook gain that minimizes (12.22) is

$$b^{(t)} = \frac{\mathbf{u3}^T\mathbf{H} \cdot \mathbf{d2}_o^{(t)}}{\|\mathbf{H} \cdot \mathbf{d2}_o^{(t)}\|^2}, \tag{12.23}$$

and the optimal codebook index is the one that maximizes

$$Q^{(t)} = \frac{\left(\mathbf{u3}^T\mathbf{H} \cdot \mathbf{d2}_o^{(t)}\right)^2}{\|\mathbf{H} \cdot \mathbf{d2}_o^{(t)}\|^2} \tag{12.24}$$

for $t = T_{min}$ to T_{max}. Since the adaptive codebook is essentially an overlapping codebook with unit shift, recursive convolution can be incorporated to achieve efficient computation (Chapter 2).

Fractional Pitch Period

Fractional pitch period is already introduced in Chapters 2 and 4. Its introduction reduces both the reverberant distortion related to pitch multiplication, as well as the roughness of speakers with short pitch period. Noise perceived in synthetic speech is lowered since performance associated with long-term prediction is increased, which decreases the noisy stochastic excitation component.

To obtain the adaptive codevector with fractional delay, the FS1016 specifies an interpolation* technique based on a 40-point Hamming windowed sinc function†, defined as follows:

$$w_o(n) = 0.54 + 0.46\cos\left(\frac{2\pi n}{40}\right), \tag{12.25}$$

$$w(n,f) = w_o(n+f)\frac{\sin((n+f)\pi)}{(n+f)\pi}, \tag{12.26}$$

for $n = -20, -19, \ldots, 19$, with f being the fractional part of the pitch period. Only five fractional values are used: $f = \frac{1}{4}, \frac{1}{3}, \frac{1}{2}, \frac{2}{3}, \frac{3}{4}$. Figure 12.8 shows plots of the

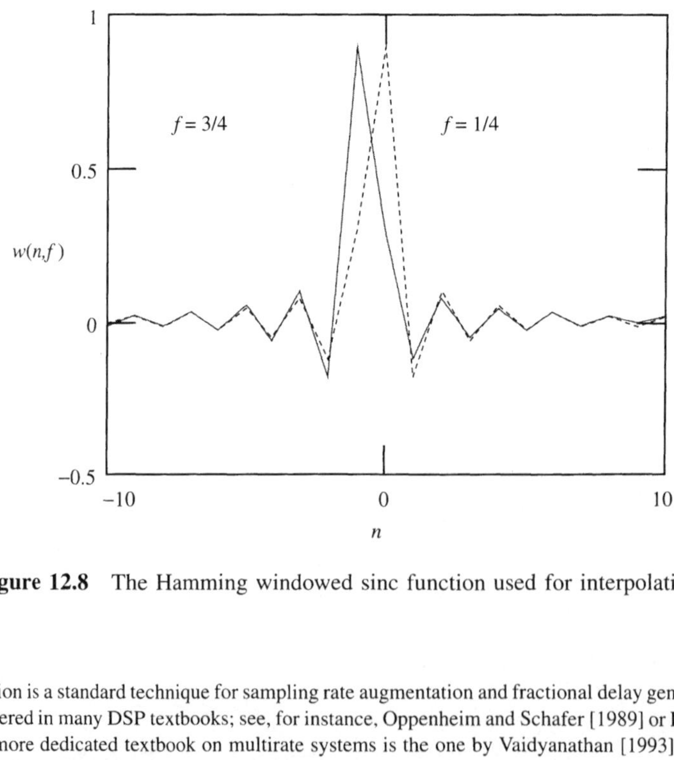

Figure 12.8 The Hamming windowed sinc function used for interpolation.

*Interpolation is a standard technique for sampling rate augmentation and fractional delay generation. The topic is covered in many DSP textbooks; see, for instance, Oppenheim and Schafer [1989] or Defatta et al. [1988]. A more dedicated textbook on multirate systems is the one by Vaidyanathan [1993].

† $\text{sinc}(x) \equiv \sin(\pi x)/(\pi x)$. See Stremler [1990].

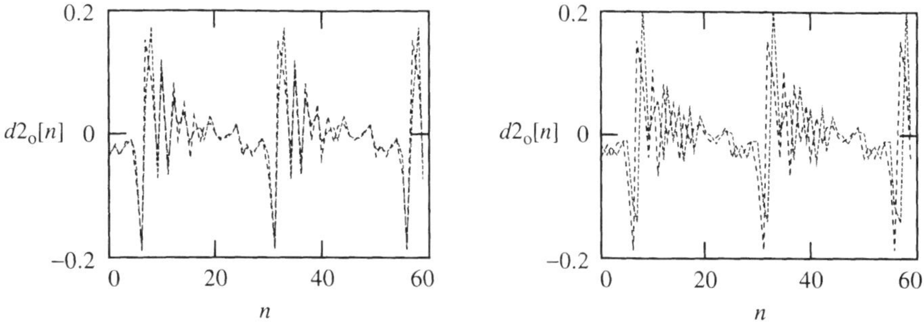

Figure 12.9 Adaptive codevector at $t = 25$. The signal is derived from the same codebook samples as in Figure 12.3. Two fractional delay values are considered: $\frac{1}{4}$ (*left*) and $\frac{3}{4}$ (*right*). The original codevector with no fractional delay is plotted for comparison (dotted line).

interpolation window. To obtain the codevector associated with the pitch period $t + f$, where t is the integer part while f is the fractional part, we first consider the signal

$$dd^{(t)}[n] = \begin{cases} d[n]; & -147 \leq n < 0, \\ d2_o^{(t)}[n]; & 0 \leq n < 60, \end{cases} \tag{12.27}$$

then

$$d2_o^{(t+f)}[n] = \sum_{k=-20}^{19} w(k,f)dd^{(t)}[n - t + k], \tag{12.28}$$

for $n = 0$ to 59. The above equation essentially creates fractionally delayed versions of the codevectors. Figure 12.9 shows some examples. After obtaining the codevectors with fractional delay, (12.24) can be used directly to choose the best codevector and (12.23) to find the gain.

Encoding of Pitch Period

FS1016 utilizes a total of 256 pitch period values in the interval [20, 147]. Within the interval, different resolutions are assigned in the following manner: $\frac{1}{3}$ for [20, $25\frac{2}{3}$] and [34, $79\frac{2}{3}$], $\frac{1}{4}$ for [26, $33\frac{3}{4}$], and 1 for [80, 147]. Note that the finest resolution is in the range of [26, $33\frac{3}{4}$], since for typical female speakers, it is more likely to fall inside this interval. The scheme is efficient in the sense that resources are allocated to where it is needed most.

After finding the integer part t and fractional part f of the pitch period, they are mapped to 8 bits through table lookup.

Subframe Encoding Strategy

FS1016 utilizes different numbers of bits to encode the pitch period, with the number depending on the position of the subframe. For the first and third subframes, 8 bits are allocated and the encoding scheme is as explained previously. For the second and fourth subframes, however, only 6 bits are employed. These 6 bits indicate a relative shift with respect to the pitch period of prior subframe, ranging from 31 codes lower to 32 codes higher, both integer and fractional values considered altogether.

The scheme provides higher encoding efficiency (smaller number of bits), as well as reduces computational load (shorter search range); the advantages come with negligible reduction in the quality of the synthetic speech since, in most instances, pitch periods of adjacent subframes do not differ significantly.

Encoding of Adaptive Codebook Gain

Gain of the adaptive codebook is encoded using 5 bits, with the quantized values ranging from −1 to 2.

12.6 STOCHASTIC CODEBOOK SEARCH

The stochastic codebook search shares identical techniques with the adaptive codebook. Equations involved in the process are summarized as follows.

Zero-state response of the modified formant synthesis filter is given by

$$\mathbf{y1}^{(l)} = g^{(l)}\mathbf{H} \cdot \mathbf{v}^{(l)}. \tag{12.29}$$

Equation (12.14) is rewritten as

$$\varepsilon^{(l)} = \left\|\mathbf{u} - \mathbf{y1}^{(l)} - \mathbf{y2} - \mathbf{y3}\right\|^2 = \left\|\mathbf{u}_o - g^{(l)}\mathbf{H}\mathbf{v}^{(l)}\right\|^2, \tag{12.30}$$

where

$$\mathbf{u}_o = \mathbf{u} - \mathbf{y2} - \mathbf{y3}. \tag{12.31}$$

The optimal stochastic codebook gain that minimizes (12.30) is

$$g^{(l)} = \frac{\mathbf{u}_o^T\mathbf{H} \cdot \mathbf{v}^{(l)}}{\|\mathbf{H} \cdot \mathbf{v}^{(l)}\|^2}, \tag{12.32}$$

and the optimal codebook index is the one that maximizes

$$P^{(l)} = \frac{\left(\mathbf{u}_o^T\mathbf{H} \cdot \mathbf{v}^{(l)}\right)^2}{\|\mathbf{H} \cdot \mathbf{v}^{(l)}\|^2} \tag{12.33}$$

for $l = 0$ to $L-1$, where $L = 512$ is the size of the stochastic codebook. As indicated in Section 12.4, the stochastic codebook is overlapping with a shift of 2; hence, recursive convolution can be used to achieve high efficiency. Also, the ternary nature of the codebook elements allows the convolution to be implemented without multiplication.

Modified Stochastic Codebook Gain

The FS1016 coder, while able to provide fairly good speech quality, has actually one major weakness: the stochastic codebook with its ternary quantized samples introduces noisy components to the synthetic speech, diminishing the achievable quality. These noisy components cannot be completely removed by the synthesis filter and postfilter. This limitation is indeed a target of improvement for other CELP-based coders, such as the IS54 covered in Chapter 13; as shown in that chapter, the IS54 coder utilizes a more sophisticated excitation codebook to achieve higher quality.

Within the context of FS1016, it is possible to improve the subjective quality by modifying the gain of the stochastic codebook. The technique consists of adaptively attenuating the stochastic codebook gain when the long-term predictor is efficient. This increases the relative adaptive codebook gain by attenuating the stochastic codebook gain; subjective quality is improved because this reduces roughness and quantization noise in sustained voiced segments. When long-term prediction is inefficient, the stochastic codebook gain is increased, thus providing a more subjectively pleasing match between the unvoiced speech segments of the input and synthetic speech. Efficiency of long-term prediction can be measured with the normalized cross-correlation

$$R = \frac{(\mathbf{u} - \mathbf{y3})^T (\mathbf{u} - \mathbf{y3} - \mathbf{y2}^{(t)})}{\|\mathbf{u} - \mathbf{y3}\|}. \tag{12.34}$$

The more efficient the long-term predictor, the shorter the norm of the vector $(\mathbf{u} - \mathbf{y3} - \mathbf{y2}^{(t)})$, leading to a smaller R. A scale factor is defined with

$$sf(R) = \begin{cases} 0.2; & |R| < 0.04, \\ 1.4\sqrt{|R|}; & |R| > 0.81, \\ \sqrt{|R|}; & \text{otherwise.} \end{cases} \tag{12.35}$$

Figure 12.10 shows the plot of the above function. The modified gain g' is given by

$$g' = sf(R) \cdot g^{(l)}. \tag{12.36}$$

Thus, for low R (long-term predictor is highly efficient), the original stochastic codebook gain is attenuated. On the other hand, when R is high, the original gain is amplified. Equation (12.35) is obtained empirically and confirmed to deliver higher

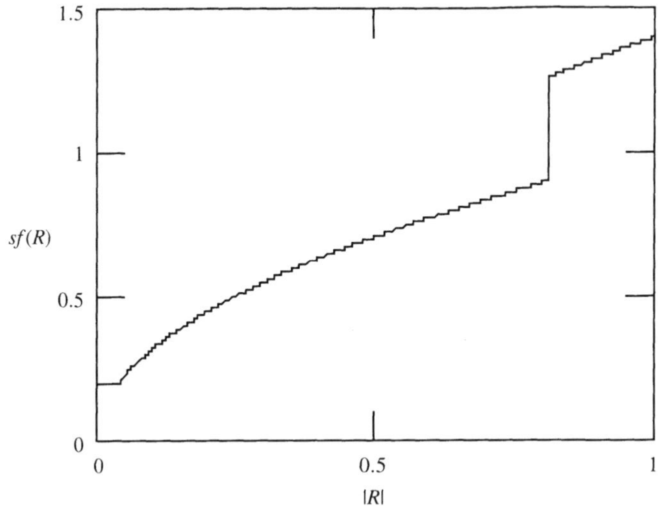

Figure 12.10 Scale factor as a function of the normalized cross-correlation used to modify the stochastic codebook gain.

subjective quality through listening tests. Since the modification is introduced after completing the stochastic codebook search loop, the increase in computation is insignificant.

Finally, gain of the stochastic codebook is quantized with 5 bits.

12.7 ENCODER AND DECODER

This section provides an overview of the operations of the encoder and decoder for the FS1016. Figure 12.11 shows the block diagram of the encoder. Several blocks are already discussed in previous sections and are not repeated here.

LP Analysis

LP analysis is performed once per frame; each frame consists of 240 samples having a duration of 30 ms. Eleven autocorrelation values are calculated from the frame using a Hamming window. The autocorrelation values are solved to obtain ten LPCs. The resultant coefficients are bandwidth expanded with a constant of $\gamma = 0.994$ (Chapter 4). The LPCs are quantized and interpolated for use by each 60-sample subframe (Chapter 8).

Perceptual Weighting Filter

A weighting factor of $\gamma = 0.8$, together with the quantized and interpolated LPCs are used for the perceptual weighting filter (Chapter 11). That is, its coefficients are updated every subframe.

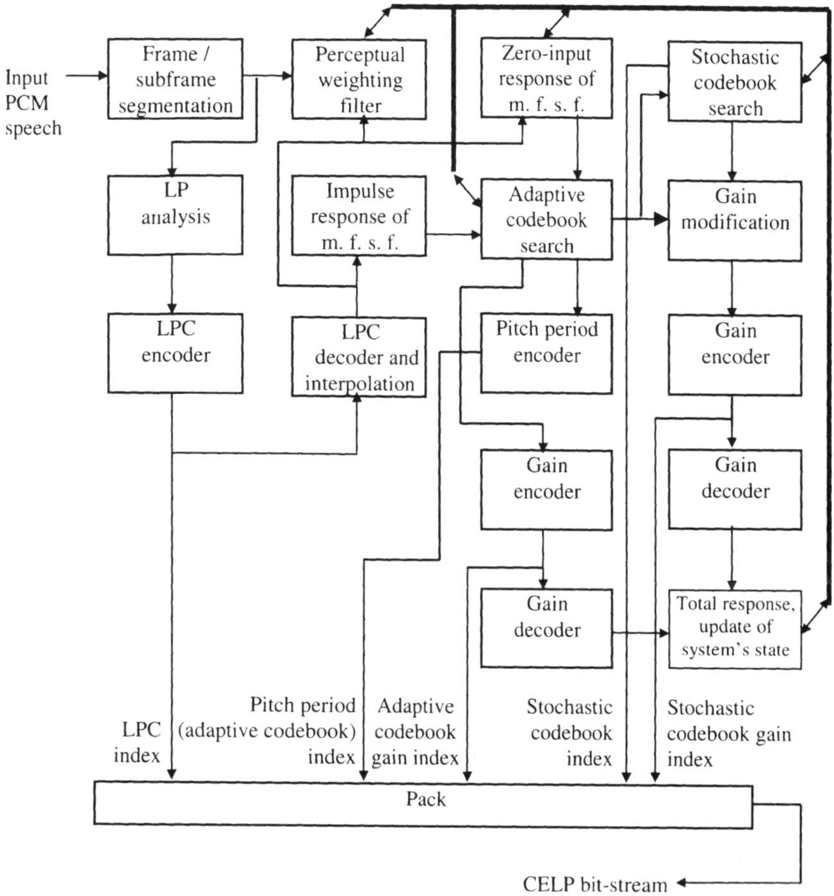

Figure 12.11 Block diagram of the FS1016 CELP encoder (m. f. s. f. means modified formant synthesis filter).

Total Response

This block computes the total excitation (*d*, Figure 12.4) and total response of the formant synthesis filter ($y1 + y2 + y3$), which are stored as history of the system to be used by the next subframe.

Bit Allocation

Table 12.1 summarizes the bit allocation scheme of the FS1016 coder. The synchronization bit alternates between zero and one from frame to frame, with the first transmitted frame starting at zero. Four bits per frame are assigned to error protection, with focus on the adaptive codebook index, since it plays the most important role in the quality of the synthetic speech. Finally, 1 bit is reserved for future

TABLE 12.1 Bit Allocation for the FS1016 CELP Coder[a]

Parameter	Number per Frame	Resolution	Total Bits per Frame
LPC	10	3, 4, 4, 4, 4, 3, 3, 3, 3, 3	34
Pitch period (adaptive codebook index)	4	8, 6, 8, 6	28
Adaptive codebook gain	4	5	20
Stochastic codebook index	4	9	36
Stochastic codebook gain	4	5	20
Synchronization	1	1	1
Error correction	4	1	4
Future expansion	1	1	1
Total			**144**

[a] Data from Campbell et al. [1991], Table 1.

upgrades of the coder. As we can see, a total of 144 bits are allocated per frame; with a frame length of 30 ms, the bit-rate is 4800 bps.

Decoder Operations

Figure 12.12 shows the block diagram of the decoder. The stochastic codevector addressed by the bit-stream index is scaled by the quantized gain value. The index

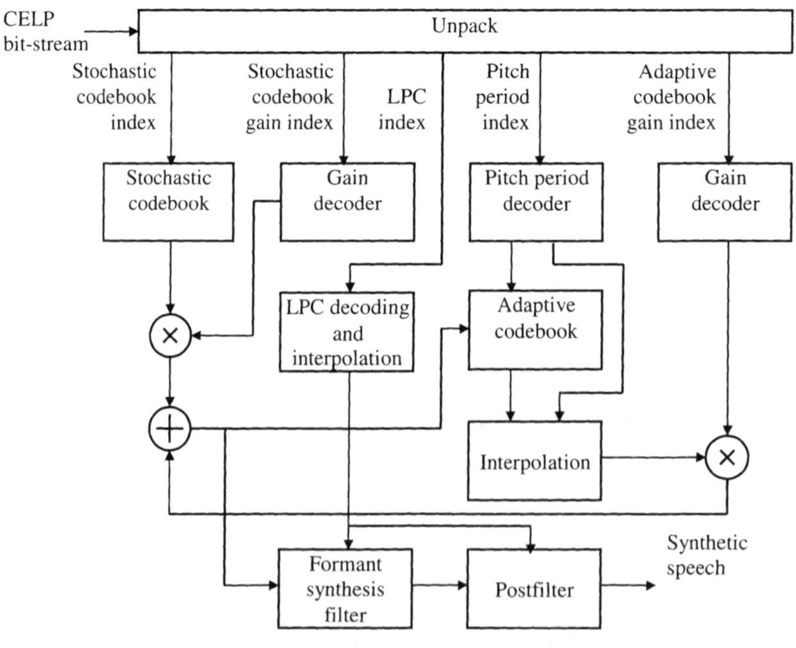

Figure 12.12 Block diagram of the FS1016 CELP decoder.

of the pitch period is used to find the integer part and the fractional part; if the fractional part is nonzero, interpolation is performed on the adaptive codevector. The recovered adaptive codevector is scaled by the quantized gain and added to the scaled stochastic codevector, serving as the update to the adaptive codebook as well as input excitation to the formant synthesis filter. See Chapter 11 for postfilter structure.

12.8 SUMMARY AND REFERENCES

Principles of the FS1016 CELP coder are described in this chapter. The idea of the adaptive codebook forms the core of many CELP-based standard coders since a closed-loop analysis-by-synthesis procedure produces far better quality than open-loop methods, such as long-term LP analysis. Due to the overlapping nature of the codebook, recursive convolution can be incorporated to yield high computational efficiency. The resultant quality can be further elevated with fractional delay, where the codevector is interpolated to increase time resolution.

The stochastic codebook is also overlapping in nature, allowing the use of recursive convolution. The ternary-valued samples provide adequate quality and permit further reduction in computation since multiplication is eliminated during the convolution sum. This is a perfect example of how imposing a certain structure on the codebook helps reduce complexity, with little loss of performance.

In essence, the adaptive codebook contains the predictable part of the excitation, that is, the component that can be obtained from the past; while the stochastic codebook contains the unpredictable component, or innovation. The scheme is highly efficient and well adapted to the nature of speech signals.

The FS1016 operating at 4.8 kbps definitely preserves far more naturalness of the original speech signal than the FS1015 LPC coder (Chapter 9) operating at 2.4 kbps. But the speech quality still contains many artifacts and in certain instances is considered inadequate to be used for general purpose telephone communications. Listening tests reveal that its quality is roughly equivalent to the MELP coder (Chapter 17), which operates at 2.4 kbps.

Major contributions of the FS1016 CELP coder can be summarized as follows:

- Demonstration of low bit-rate capability for the CELP algorithm.
- Introduction of fractional delay in the adaptive codebook search.
- Use of overlapping fixed codebook with ternary-valued samples to achieve high efficiency.

Technical details of the FS1016 are well documented [National Communications System, 1992]. The idea of modifying the stochastic codebook gain to increase subjective quality was proposed in Shoham [1991] with additional results. More theoretical results concerning CELP development is found in LeBlanc [1992]. See Chapter 16 for variations in adaptive codebook implementation.

EXERCISES

12.1 Given the WSS white noise process $x[n]$ with autocorrelation function (Chapter 3)

$$R_x[l] = \sigma_x^2 \delta[l],$$

find the autocorrelation function of the new process

$$y[n] = \alpha x[n] + \beta x[n - T],$$

where α and β are real constants and T is an integer. Prove that

$$R_y[l] = (\alpha^2 + \beta^2)R_x[l] + \alpha\beta R_x[l - T] + \alpha\beta R_x[l + T].$$

What can be said about the statistical property of $y[n]$?

12.2 This exercise considers the computational costs of three methods for the calculation of the L zero-state responses of the modified formant synthesis filter during CELP encoding. These methods are the difference equation, convolution, and recursive convolution. See Chapters 2 and 11 for background.

(a) For the difference equation method, show that

$$\#\text{sums} = \#\text{products} = L \cdot M \cdot N,$$

with M being the prediction order and N the subframe length.

(b) In the convolution sum, each response is found by performing an independent convolution operation with the input codevector. Show that

$$\#\text{sums} = L(N - 1)N/2 + (M - 2)(M - 1)/2 + (N - M)(N - 1),$$
$$\#\text{products} = LN(N + 1)/2 + (M - 2)(M - 1)/2 + (N - M)M.$$

In the above relations, the cost involved with finding the N-sample impulse response from the LPC is also accounted for.

(c) For recursive convolution, show that

$$\begin{aligned}\#\text{sums} &= (N - 1)N/2 + (M - 2)(M - 1)/2 + (N - M)(N - 1) \\ &\quad + S(L - 1)(2N - S - 1)/2, \\ \#\text{products} &= N(N + 1)/2 + (M - 2)(M - 1)/2 + (N - M)M \\ &\quad + S(L - 1)(2N - S + 1)/2.\end{aligned}$$

(d) Using parameters of the FS1016 coder, find the cost involved with the three methods. Which one is the best? For recursive convolution, consider the cases of $S = 1$ and $S = 2$.

12.3 The interpolation equation presented in Section 12.5 implies that the same procedure must be applied for every sample ($n = 0$ to 59). This is true only when $t > N$. For $t < N$, however, the codevector is obtained through periodic extension; thus, the interpolation equation need not be applied to all samples. Based on this observation, design a computationally efficient procedure to perform interpolation.

12.4 Elaborate a step-by-step procedure for the adaptive codebook search, including all equations and decision strategies. Take into account the integer and fractional parts of the pitch period, as well as the number of subframes. Consider the following cases of search strategies:

(a) Full search: Every possible codevector is computed with the best returned.

(b) Suboptimal search: The integer part of the pitch period is first located, followed by a fractional refinement near its neighbor.

Estimate the computational cost in each case.

12.5 Count all period values available for the FS1016 coder to prove that there are a total of 256 distinct values.

12.6 To compute the convolution sum

$$y[n] = h[n] * v[n] = \sum_{k=0}^{N-1} h[n-k]v[k],$$

with $h[n]$ the impulse response of a causal system ($h[n] = 0$ if $n < 0$), we can rely on the following procedure:

```
for n ← 0 to N − 1
    y[n] ← 0
    for k ← 0 to N − 1
        if n − k ≥ 0
            y[n] ← y[n] + h[n − k] v[k]
        else break
```

When $v[n]$ is ternary-valued, we can actually save some computation by using the procedure below:

```
for n ← 0 to N − 1
    y[n] ← 0
    for k ← 0 to N − 1
        if n − k ≥ 0
            if v[k] = 0 continue
            else
                if v[k] = 1
                    y[n] ← y[n] + h[n − k]
            else
                    y[n] ← y[n] − h[n − k]
```

Thus, no multiplication is required. However, note that more comparison is necessary. Traditionally, multiplication is much more expensive to deploy than comparison, and hence a codebook with ternary-valued samples is very attractive.

Extend these ideas to recursive convolution and elaborate the step-by-step procedure following a similar approach.

12.7 Consider a pitch period encoding scheme where the differences between the values of consecutive subframes are transmitted. Give some advantages and disadvantages of the proposition.

12.8 We learned that an overlapping codebook has the advantage of low memory cost. This cost can be further reduced by utilizing a circular overlapping structure. In this case a codevector is extracted with

$$v^{(l)}[n] = v[(n + (L - 1 - l)S) \bmod P]$$

for $n = 0$ to $N - 1, l = 0$ to $L - 1$, and P the number of elements in the v array. The inclusion of the modulo operator allows the elements of the codevector to be read from the beginning of the v array in case the boundary is passed. Show that $P = S \cdot L$. This is an interesting result since the number of elements in the v array is independent of the dimension of the codevector. What happens when $N < L$ and $N \geq L$? Compare the memory cost of a circular overlapping codebook with that of an ordinary overlapping codebook for the case of the FS1016. Is recursive convolution applicable in this case? This type of structure is adopted by the IS96 coder, covered in Chapter 18.

12.9 This exercise is concerned with storage of the stochastic codebook.

(a) It is said that samples of the stochastic codebook have three values, $\{-1, 0, 1\}$; hence, each sample can be represented with 2 bits since $2^2 = 4 > 3$. How many bits are required to store the whole codebook?

(b) Assume that two samples of the codebook are jointly encoded. In this case there are a total of $3^2 = 9$ possibilities: $\{-1, -1\}, \{-1, 0\}, \{-1, 1\}, \{0, -1\}, \{0, 0\}, \{0, 1\}, \{1, -1\}, \{1, 0\}$, and $\{1, 1\}$. How many bits are needed for each sample pair? How many bits for the whole codebook?

(c) Repeat (b) for the cases when three, four, five, and six samples are jointly encoded. Which scheme is most economical in storage?

CHAPTER 13

VECTOR SUM EXCITED LINEAR PREDICTION

The vector sum excited linear prediction (VSELP) coder is derived from the principles of a standard CELP coder (Chapter 11); indeed, it is a CELP coder with a particular codebook structure having reduced computational cost. There are at least three standard coders based on the same principles of VSELP, all originally developed by Motorola Inc. These standards are:

- TIA IS54. Standardized by the TIA in 1989 for time division multiple access (TDMA) digital cellular telephony in North America. It is a part of Interim Standard 54 (IS54) and operates at 7.95 kbps.
- ETSI GSM 6.20. This was created in order to double the capacity of the GSM cellular system. The coder works at 5.6 kbps.
- RCR STD−27B. This was standardized by the Research and Development Center for Radio Systems (RCR) for TDMA digital cellular telephone service in Japan.

In this chapter, the focus is placed on the study of the TIA IS54 coder. Like the FS1016 (Chapter 12), the IS54 utilizes the exact same principle of the adaptive codebook. Excitation vectors from the stochastic codebook, however, are obtained through linear combination of a number of fixed basis vectors—hence the name of vector sum excitation. The VSELP coder was designed to achieve the highest possible quality with reasonable computational complexity while providing robustness to channel errors, essential requirements for cellular telephony applications.

To understand the material in this chapter, readers need to have a general knowledge of the operation of CELP coding, such as the material in Chapter 11. Also, there are some common architectural features between the IS54 and FS1016; these were already presented in Chapter 12 and will not be repeated here. Concepts from linear algebra are used throughout the chapter and are covered in Appendix F. The chapter begins with the core encoding structure of the VSELP coder, followed by the structure of the stochastic codebook and search methodologies. Gain quantization is explained next, followed by an analysis of the structure of the encoder and decoder. Performance of the coder and comparison among the various standards are included in the summary section.

13.1 THE CORE ENCODING STRUCTURE

Structurally, the IS54 algorithm closely resembles the FS1016 coder. The difference lies in the form and structure of the stochastic codebook. In the FS1016 coder, an overlapping codebook with ternary-quantized random samples is used, whereas for the IS54 coder, two separate codebooks are used to generate the space of candidate vectors. Thus, the codebook search procedure of the FS1016 corresponds to two codebook searches in the IS54. Figure 13.1 shows the core encoding structure. For this coder, a frame consists of 160 samples, and a subframe contains 40 samples.

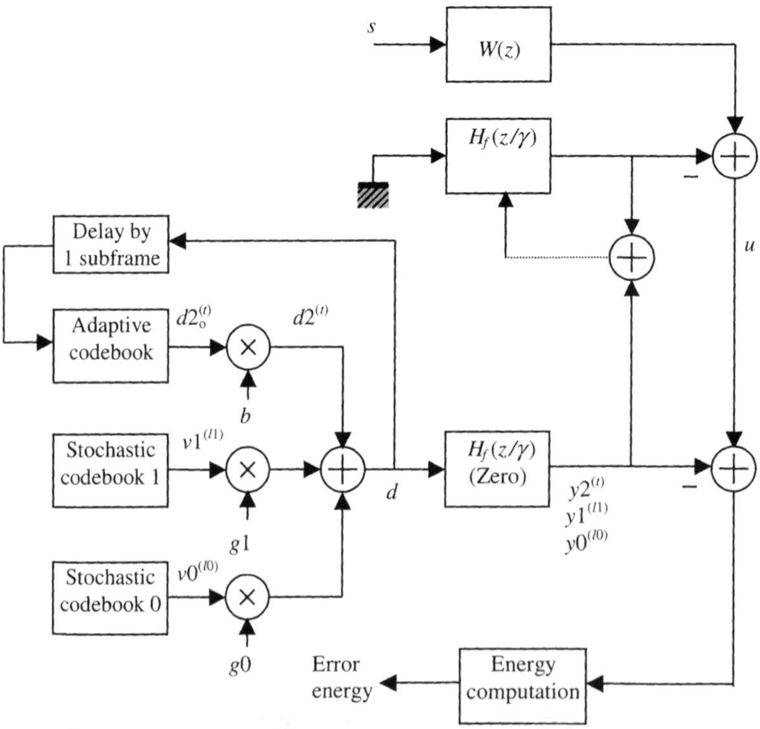

Figure 13.1 Signals involved in the IS54 encoder.

A total of $2^7 = 128$ codevectors are included in each stochastic codebook, with each codevector having 40 elements.

Basis Vectors

The 2^7 codevectors of each stochastic codebook are spanned from seven basis vectors. For codebook 1 we write

$$\mathbf{v1}^{(l1)} = \sum_{i=0}^{6} \theta_i^{(l1)} \mathbf{a1}_i \tag{13.1}$$

for $l1 = 0, \ldots, 2^7 - 1$; where

$$\theta_i^{(l1)} = \begin{cases} 1; & \text{if } i\text{th bit of } l1 = 1, \\ -1; & \text{if } i\text{th bit of } l1 = 0, \end{cases} \tag{13.2}$$

for $i = 0$ to 6. Hence, each codevector is constructed as a linear combination of the seven basis vectors: $\mathbf{a1}_0$ to $\mathbf{a1}_6$, with the weights of the linear combination defined by θ. In linear algebra terms, the space of the codevectors is spanned by seven basis vectors (Appendix F). At this point, we assume that the elements of the basis vectors are white noise samples.

Note that if the bits forming the index $l1$ are complemented (logically inverted, 1 to 0, and vice versa), the resultant codevector becomes the negative of the original. Therefore, for every codevector, its negative is also a codevector in the codebook. These pairs are called complementary codevectors and can be explored to improve computational efficiency. The chosen structure requires minimum storage (low memory cost) and enables fast search techniques.

Robustness against channel errors is also elevated when compared to the FS1016 coder. For instance, a single bit error in the codebook index changes the sign of only one of the basis vectors; the resulting codevector is still similar to the original.

Zero-State Response

During codebook search, the zero-state response of the synthesis filter for each of the codevectors must be found. In principle, 128 convolution operations must be done per codebook. However, due to the special structure imposed, a great deal of simplification is possible. From Figure 13.1,

$$\mathbf{y1}^{(l1)} = g1 \cdot \mathbf{H} \cdot \mathbf{v1}^{(l1)} \tag{13.3}$$

with $\mathbf{H}$ being the impulse response matrix. Substituting (13.1),

$$\mathbf{y1}^{(l1)} = g1 \sum_{i=0}^{6} \theta_i^{(l1)} \mathbf{H} \cdot \mathbf{a1}_i = g1 \sum_{i=0}^{6} \theta_i^{(l1)} \mathbf{f1}_i, \tag{13.4}$$

with

$$\mathbf{f1}_i = \mathbf{H} \cdot \mathbf{a1}_i; \quad i = 0, \ldots, 6 \tag{13.5}$$

being the filtered basis vectors of the codebook. Similarly, for codebook 0,

$$\mathbf{y0}^{(l0)} = g0 \sum_{i=0}^{6} \theta_i^{(l0)} \mathbf{f0}_i. \tag{13.6}$$

Therefore, for each stochastic codebook, only seven convolution operations are needed since the rest of the responses are spanned from the filtered basis vectors.

13.2 SEARCH STRATEGIES FOR EXCITATION CODEBOOKS

Different methods to search the three excitation codebooks—adaptive, stochastic 1, and stochastic 0—are described in this section.

Sequential Search: The FS1016 Style

Following a similar approach as for the FS1016, the three codebooks are searched sequentially in the following manner.

- *Adaptive Codebook.* This codebook is searched so that

$$J^{(t)} = \left\| \mathbf{u} - \mathbf{y2}^{(t)} \right\|^2 \tag{13.7}$$

  is minimized, where

$$\mathbf{y2}^{(t)} = \mathbf{H} \cdot d2^{(t)} = b \cdot \mathbf{H} \cdot \mathbf{d2}_o^{(t)}. \tag{13.8}$$

  For this coder, $20 \leq t \leq 147$ is encoded with 7 bits. Thus, the adaptive codebook is searched with the assumption that $g1$ and $g0$ are equal to zero.
- *Stochastic Codebook 1.* This codebook is searched so that

$$\varepsilon 1^{(l1)} = \left\| \mathbf{u} - \mathbf{y1}^{(l1)} - \mathbf{y2}^{(t)} \right\|^2 \tag{13.9}$$

  is minimized, with $\mathbf{y1}$ given by (13.3). That is, stochastic codebook 1 is searched after knowing the parameters of the adaptive codebook, with the assumption that $g0 = 0$.
- *Stochastic Codebook 0.* This codebook is searched so that

$$\varepsilon 0^{(l0)} = \left\| \mathbf{u} - \mathbf{y0}^{(l0)} - \mathbf{y1}^{(l1)} - \mathbf{y2}^{(t)} \right\|^2 \tag{13.10}$$

is minimized, with

$$\mathbf{y0}^{(l0)} = g0 \cdot \mathbf{H} \cdot \mathbf{v0}^{(l0)}. \tag{13.11}$$

Thus, stochastic codebook 0 is searched after knowing the parameters of the adaptive codebook as well as stochastic codebook 1.

Based on the description given before, standard codebook search procedures such as the one specified for the FS1016 can be applied with little modification. The outcomes, of course, are suboptimal in the sense that the resultant performance is far from the global optimum, obtainable only through a simultaneous search of the three codebook indices and the three codebook gains. Since the global optimum is expensive to reach, an alternative suboptimal approach is described next, which is better than the explained sequential search method.

An Improved Procedure: Joint Gain Optimization

It is possible to improve the performance of the coder by using this alternative procedure. First, the adaptive codebook is searched and the index t determined. Next, stochastic codebook 1 is searched to determine the index $l1$. Then stochastic codebook 0 is searched to arrive at a value for the index $l0$. Finally, the three gain values are found jointly to minimize the total error; hence, the gain values are left "floating" until all codebook indices are obtained.

How can this be done? From Chapter 12 we learned that in order to search an excitation codebook, its index completely fixes the gain, and vice versa; therefore, changing the gain value would likely deteriorate the quality. The approach in IS54 is to orthogonalize the vectors before examination, so that sequential search can be done independently from prior results; after the directions (codebook indices) of the vectors are known, the gains are found jointly from a VQ codebook. Details of the technique are described in the next two sections.

13.3 EXCITATION CODEBOOK SEARCHES

The improved search procedure, briefly explained in the last section, is fully described here. Concepts from linear algebra are necessary to understand some of the topics. Readers are invited to read Appendix F for a review.

Adaptive Codebook Search

The optimal index for the adaptive codebook is found by maximizing

$$P^{(t)} = \frac{\left(\mathbf{u}^T\mathbf{H} \cdot \mathbf{d2}_o^{(t)}\right)^2}{\left\|\mathbf{H} \cdot \mathbf{d2}_o^{(t)}\right\|^2}, \tag{13.12}$$

for $t = 20$ to 147. See Chapter 12 for the derivation of the above equation.

Some Linear Algebra Results

Before continuing with the rest of the excitation codebooks, two results from linear algebra are derived.

Lemma 13.1. Given the vectors $\mathbf{x}$, $\mathbf{y}$, and $\mathbf{z}$, we have

$$\|\mathbf{x}-\mathbf{y}-\mathbf{z}\|^2 = \|\mathbf{x}-\mathbf{y}\|^2 + \|\mathbf{x}-\mathbf{z}\|^2 - \|\mathbf{x}\|^2 \tag{13.13}$$

if $\mathbf{y}$ and $\mathbf{z}$ are orthogonal. See Exercise 13.1 for the proof of the above result.

Lemma 13.2. If $\mathbf{x}$ and $\mathbf{y}$ are orthogonal to $\mathbf{z}$,

$$\mathbf{x}^{\mathrm{T}}\mathbf{z} = \mathbf{y}^{\mathrm{T}}\mathbf{z} = 0,$$

then $\alpha\mathbf{x} + \beta\mathbf{y}$ is also orthogonal to $\mathbf{z}$,

$$(\alpha\mathbf{x} + \beta\mathbf{y})^T\mathbf{z} = \alpha\mathbf{x}^T\mathbf{z} + \beta\mathbf{y}^T\mathbf{z} = 0,$$

with α and β two scalars. This result can be extended in a straightforward manner to the sum of any number of vectors.

Search of Stochastic Codebook 1

Since the index of the adaptive codebook is known, we can search stochastic codebook 1 so as to minimize

$$\|\mathbf{u} - \mathbf{y}1^{(l1)} - \mathbf{y}2^{(t)}\|^2. \tag{13.14}$$

As explained in last section, the gain values are left "floating"; hence, error expression (13.14) cannot be evaluated, in general, unless there is a way to separate the error contribution between the adaptive codebook ($\mathbf{y}2$) and the stochastic codebook 1 ($\mathbf{y}1$). Consider what happens if $\mathbf{y}1^{(l1)}$ is orthogonal to $\mathbf{y}2^{(t)}$ for all values of $l1$. From Lemma 13.1,

$$\|\mathbf{u} - \mathbf{y}1^{(l1)} - \mathbf{y}2^{(t)}\|^2 = \|\mathbf{u} - \mathbf{y}1^{(l1)}\|^2 + \|\mathbf{u} - \mathbf{y}2^{(t)}\|^2 - \|\mathbf{u}\|^2. \tag{13.15}$$

The sum of squared error is minimized when the first two positive norm terms on the right-hand side are minimized, which are related to the adaptive codebook and the first stochastic codebook. Also note from (13.15) that minimization of a particular norm term can be done independently from the other; thus, error contributions from the two excitation sources are separated.

The next obvious question is how to obtain a $\mathbf{y}1$ vector that is orthogonal to $\mathbf{y}2$. This can be done by following the same principle of the Gram–Schmidt algorithm (Appendix F). Denoting $\mathbf{y}1'$ as the vector derived from $\mathbf{y}1$ that is orthogonal to $\mathbf{y}2$, we seek to minimize

$$\varepsilon 1'^{(l1)} = \|\mathbf{u} - \mathbf{y}1'^{(l1)}\|^2, \tag{13.16}$$

with

$$\mathbf{y1}'^{(l1)} = g1' \sum_{i=0}^{6} \theta_i^{(l1)} \mathbf{f1}_i' = g1' \mathbf{y1}_o'^{(l1)}. \tag{13.17}$$

To obtain the response $\mathbf{y1}'$, it is necessary to orthogonalize the $\mathbf{y1}$ vector with respect to $\mathbf{y2}$; since there are a total of 2^7 $\mathbf{y1}$ vectors, the proposition is rather expensive. Since the $\mathbf{y1}$ vectors are spanned by the seven filtered basis vectors $\mathbf{f1}$, it is only necessary to orthogonalize these seven vectors. From Lemma 13.2, the linear combination of these vectors will also be orthogonal to $\mathbf{y2}$. The orthogonalization is done by a procedure similar to the Gram–Schmidt algorithm:

$$\mathbf{f1}_i' = \mathbf{f1}_i - \frac{\mathbf{y2}^{(t)T}\mathbf{f1}_i}{\mathbf{y2}^{(t)T}\mathbf{y2}^{(t)}}\mathbf{y2}^{(t)} \tag{13.18}$$

for $i = 0$ to 6. The codebook is searched in such a way that

$$P1^{(l1)} = \frac{\left(\mathbf{u}^T\mathbf{y1}_o'^{(l1)}\right)^2}{\left\|\mathbf{y1}_o'^{(l1)}\right\|^2} \tag{13.19}$$

is maximized.

Search of Stochastic Codebook 0

In this case we seek to minimize

$$\left\|\mathbf{u} - \mathbf{y0}^{(l0)} - \mathbf{y1}^{(l1)} - \mathbf{y2}^{(t)}\right\|^2. \tag{13.20}$$

Following the same reasoning as before, we orthogonalize $\mathbf{y0}$ with respect to $\mathbf{y1}'$ and $\mathbf{y2}$, leading to the conclusion that the error expression we need to deal with is

$$\varepsilon0'^{(l0)} = \left\|\mathbf{u} - \mathbf{y0}'^{(l0)}\right\|^2, \tag{13.21}$$

where

$$\mathbf{y0}'^{(l0)} = g0' \sum_{i=0}^{6} \theta_i^{(l0)} \mathbf{f0}_i' = g0' \mathbf{y0}_o'^{(l0)} \tag{13.22}$$

and

$$\mathbf{f0}_i' = \mathbf{f0}_i - \frac{\mathbf{y2}^{(t)T}\mathbf{f0}_i}{\mathbf{y2}^{(t)T}\mathbf{y2}^{(t)}}\mathbf{y2}^{(t)} - \frac{\mathbf{y1}_o'^{(l1)T}\mathbf{f0}_i}{\mathbf{y1}_o'^{(l1)T}\mathbf{y1}_o'^{(l1)}}\mathbf{y1}_o'^{(l1)}, \tag{13.23}$$

which is the corresponding orthogonalization expression. The codebook is searched in such a way that

$$P0^{(l0)} = \frac{\left(\mathbf{u}^T \mathbf{y}0_o^{\prime(l0)}\right)^2}{\left\|\mathbf{y}0_o^{\prime(l0)}\right\|^2} \tag{13.24}$$

is maximized.

About Orthogonalization

From a geometrical point of view, the search procedure finds the angle or direction of the adaptive codevector that minimizes the error. Length of the codevector is left floating since the final excitation vector is formed by the addition of two more constituent vectors (from stochastic codebooks) that could possess components in the same direction as for the selected adaptive codevector.

To search stochastic codebook 1, we want to orthogonalize the codevectors with respect to the selected adaptive codevector; since the direction of the adaptive codevector is already known, further searching into the same direction is not likely to be productive. By orthogonalizing, we focus the search on all other directions except the one of the adaptive codevector.

For stochastic codebook 0, the principle is similar: look into those directions not specified by the adaptive codevector and the first stochastic codevector. In this manner, three directions (indices) are determined from the three excitation codebooks; lengths (gains) of these codevectors are found jointly so as to minimize the final error (Section 13.4).

Fast Search Procedure Based on Gray Code

Structures of the stochastic codebooks for the VSELP coders are designed to allow high efficiency. Here, the procedure for the stochastic codebook is described. As shown earlier, search procedures for the two stochastic codebooks are virtually identical after orthogonalization, both look into the maximization of the quantity

$$P^{(l)} = \frac{\left(C^{(l)}\right)^2}{G^{(l)}} = \frac{\left(\mathbf{u}^T \mathbf{y}^{(l)}\right)^2}{\left\|\mathbf{y}^{(l)}\right\|^2}. \tag{13.25}$$

For codebook 1, $l = l1$ and $\mathbf{y} = \mathbf{y}1_o^{\prime}$; while for codebook 0, $l = l0$ and $\mathbf{y} = \mathbf{y}0_o^{\prime}$. A full search requires evaluating all values of $P^{(l)}$ for all index values l. Due to the structure of the codebook, it is possible to search in a rather efficient manner. Note that

$$C^{(l)} = \mathbf{u}^T \mathbf{y}^{(l)} = \sum_{i=0}^{6} \theta_i^{(l)} \mathbf{u}^T \mathbf{f}_i^{\prime} \tag{13.26}$$

and

$$G^{(l)} = \|\mathbf{y}^{(l)}\|^2 = \sum_{i=0}^{6}\sum_{j=0}^{6}\theta_i^{(l)}\theta_j^{(l)}\mathbf{f}_i'^T\mathbf{f}_j' \tag{13.27}$$

with $\mathbf{f}' = \mathbf{f}0'$ or $\mathbf{f}1'$ being the orthogonalized filtered basis vector. Since

$$\theta_i^{(l)}\theta_j^{(l)} = 1 \tag{13.28}$$

for $i = j$, we have

$$G^{(l)} = \sum_{i=0}^{6}\mathbf{f}_i'^T\mathbf{f}_i' + 2\sum_{i=1}^{6}\sum_{j=0}^{i-1}\theta_i^{(l)}\theta_j^{(l)}\mathbf{f}_i'^T\mathbf{f}_j'. \tag{13.29}$$

Consider now two indices l_1 and l_2 that differ only in 1 bit at position k ($k = 0$ to 6). That is,

$$\theta_k^{(l_1)} = -\theta_k^{(l_2)} \tag{13.30}$$

and

$$\theta_i^{(l_1)} = \theta_i^{(l_2)}, \quad i \neq k. \tag{13.31}$$

From (13.26),

$$\begin{aligned} C^{(l_2)} &= \mathbf{u}^T\mathbf{y}^{(l_2)} \\ &= \sum_{i=0}^{6}\theta_i^{(l_2)}\mathbf{u}^T\mathbf{f}_i' \\ &= \sum_{i=0}^{6}\theta_i^{(l_1)}\mathbf{u}^T\mathbf{f}_i' + 2\theta_k^{(l_2)}\mathbf{u}^T\mathbf{f}_k'. \end{aligned}$$

Thus,

$$C^{(l_2)} = C^{(l_1)} + 2\theta_k^{(l_2)}\mathbf{u}^T\mathbf{f}_k'. \tag{13.32}$$

That is, the correlation C at index l_2 can be found from the corresponding value at index l_1, plus additional values that can be calculated with little effort. Similarly from (13.29),

$$\begin{aligned} G^{(l_2)} &= \sum_{i=0}^{6}\mathbf{f}_i'^T\mathbf{f}_i' + 2\sum_{i=1}^{6}\sum_{j=0}^{i-1}\theta_i^{(l_2)}\theta_j^{(l_2)}\mathbf{f}_i'^T\mathbf{f}_j' \\ &= \sum_{i=0}^{6}\mathbf{f}_i'^T\mathbf{f}_i' + 2\sum_{i=1}^{6}\sum_{j=0}^{i-1}\theta_i^{(l_1)}\theta_j^{(l_1)}\mathbf{f}_i'^T\mathbf{f}_j' + 4\sum_{i=0, i\neq k}^{6}\theta_i^{(l_2)}\theta_k^{(l_2)}\mathbf{f}_i'^T\mathbf{f}_k' \end{aligned}$$

or

$$G^{(l_2)} = G^{(l_1)} + 4\sum_{i=0, i\neq k}^{6}\theta_i^{(l_2)}\theta_k^{(l_2)}\mathbf{f}_i'^T\mathbf{f}_k'. \tag{13.33}$$

TABLE 13.1 Four-Bit Gray Code and Comparison with Binary Code

Gray	Binary
0000	0000
0001	0001
0011	0010
0010	0011
0110	0100
0111	0101
0101	0110
0100	0111
1100	1000
1101	1001
1111	1010
1110	1011
1010	1100
1011	1101
1001	1110
1000	1111

If codebook search is structured such that the index of a successive codeword differs from the previous one in only one bit position, (13.32) and (13.33) can be used to update $C^{(l)}$ and $G^{(l)}$ in a very efficient manner. Sequencing of the codewords in this way is accomplished using the Gray code. Table 13.1 shows the four-bit Gray code; note how the sequential code differs only by one bit position. See Exercises 13.2 and 13.3 for conversion methods between binary and Gray codes.

Note that complementary indices produce the same values for (13.25); hence, only half of the codevectors need to be evaluated ($2^6 = 64$). The sign of $C^{(l)}$ is checked to determine which of the complementary codevectors yields a positive gain:

$$g^{(l)} = \frac{C^{(l)}}{G^{(l)}}. \tag{13.34}$$

That is, if $C^{(l)}$ is positive then l is the desired codeword; otherwise the first complement (logical inverse) of l is selected as the codeword.

13.4 GAIN RELATED PROCEDURES

Gains of the excitation codebooks—adaptive, stochastic 1, and stochastic 0—are jointly optimized and vector quantized with 8 bits, which is performed once per subframe. The procedure to quantize the excitation gains and other related procedures are described here.

Energy of the Speech Frame

Energy of the speech frame is found and quantized once per frame using 5 bits at a step size of 2 dB. The quantized energy value is denoted as E_s. For each of the four subframes, energy of the speech frame is interpolated using the following rules:

- Subframe 0: Use energy of previous frame: $E_{s,\text{ previous}}$,
- Subframe 1: Use $\sqrt{E_{s,\text{ previous}} E_{s,\text{ current}}}$, the geometric mean between past and present.
- Subframe 2 and 3: Use $E_{s,\text{ current}}$.

By knowing the energy of the speech subframe, an estimate of the energy of prediction error (excitation) can be found with

$$E_e = E_s \prod_{i=1}^{10} \left(1 - k_i^2\right), \tag{13.35}$$

with k_i denoting the reflection coefficients of the current subframe. The above relation comes from the linear prediction framework (Chapter 4), where the prediction gain is given by

$$PG = E_s/E_e. \tag{13.36}$$

Excitation Gains: Selection of Parameters for Quantization

For excitation codebook gains, there are a total of three values to consider: b, $g1$, and $g0$. However, these gains are not quantized directly. The first parameter for quantization is the energy offset G, defined by

$$E_o = G \cdot E_e, \tag{13.37}$$

with E_o being the actual excitation energy (measured directly from the signal) and G representing an adjustment factor acting on the estimated value E_e.

Note that the complete excitation codevector is given by

$$\mathbf{d} = b \cdot \mathbf{d2}_o^{(t)} + g1 \cdot \mathbf{v1}^{(l1)} + g0 \cdot \mathbf{v0}^{(l0)}. \tag{13.38}$$

Let's define

$$R_0 = \mathbf{v0}^{(l0)T}\mathbf{v0}^{(l0)}, \tag{13.39}$$

$$R_1 = \mathbf{v1}^{(l1)T}\mathbf{v1}^{(l1)}, \tag{13.40}$$

$$R_2 = \mathbf{d2}_o^{(t)T}\mathbf{d2}_o^{(t)}, \tag{13.41}$$

representing the energy of each codevector. Then

$$P_2 = \frac{b^2 R_2}{E_o} \tag{13.42}$$

is the fractional energy contribution of the adaptive codevector; note that $0 \le P_2 \le 1$. The parameter P_2 is selected as the second parameter for quantization.

The third parameter for quantization is

$$P_1 = \frac{g1^2 R_1}{E_o}, \tag{13.43}$$

representing the fractional energy contribution of the codevector from the first stochastic codebook.

The three-dimensional vector $\{G, P_2, P_1\}$ is quantized with a codebook having 256 vectors (8 bits). In order to find $\{b, g1, g0\}$ from $\{G, P_2, P_1\}$, the following relations are derived:

$$b = \sqrt{\frac{G \cdot E_e \cdot P_2}{R_2}}, \tag{13.44}$$

obtained by substituting (13.37) in (13.42);

$$g1 = \sqrt{\frac{G \cdot E_e \cdot P_1}{R_1}}, \tag{13.45}$$

obtained by substituting (13.37) in (13.43); and

$$g0 = \sqrt{\frac{G \cdot E_e(1 - P_2 - P_1)}{R_0}} \tag{13.46}$$

since $1 - P_2 - P_1 > 0$ is the fractional energy contribution of stochastic codebook 0.

Gain Codebook Search

The codevector from the gain codebook that minimizes the total weighted error for the subframe is chosen. This error measure is

$$\varepsilon^{(l)} = \left\| \mathbf{u} - \mathbf{y2}^{(l)} - \mathbf{y1}^{(l)} - \mathbf{y0}^{(l)} \right\|^2 \tag{13.47}$$

with l denoting the index of the gain codebook. Thus, for $l = 0$ to 255, $\varepsilon^{(l)}$ is evaluated with the gain codevector $\{G^{(l)}, P_2(l), P_1(l)\}$, and the index producing the lowest error is selected. Expanding, we find

$$\begin{aligned}\varepsilon^{(l)} &= \mathbf{u}^T\mathbf{u} - 2\mathbf{y0}^{(l)T}\mathbf{u} - 2\mathbf{y1}^{(l)T}\mathbf{u} - 2\mathbf{y2}^{(l)T}\mathbf{u} \\ &\quad + 2\mathbf{y0}^{(l)T}\mathbf{y1}^{(l)} + 2\mathbf{y0}^{(l)T}\mathbf{y2}^{(l)} + 2\mathbf{y1}^{(l)T}\mathbf{y2}^{(l)} \\ &\quad + \mathbf{y0}^{(l)T}\mathbf{y0}^{(l)} + \mathbf{y1}^{(l)T}\mathbf{y1}^{(l)} + \mathbf{y2}^{(l)T}\mathbf{y2}^{(l)}.\end{aligned} \tag{13.48}$$

Since **u** is constant with respect to l, it can be eliminated with no effect on the final outcome. Redefine an error quantity with

$$\varepsilon'^{(l)} = \sum_{i=1}^{9} \gamma_i^{(l)}, \tag{13.49}$$

where

$$\gamma_1^{(l)} = -2\mathbf{y0}^{(l)T}\mathbf{u} = -2\sqrt{\frac{G^{(l)}\left(1 - P_2^{(l)} - P_1^{(l)}\right)E_e}{R_0}}\mathbf{y0}_o^T\mathbf{u} \tag{13.50}$$

and

$$\mathbf{y0}_o = \mathbf{y0}/g0, \tag{13.51}$$

$$\mathbf{y1}_o = \mathbf{y1}/g1, \tag{13.52}$$

$$\mathbf{y2}_o = \mathbf{y2}/b \tag{13.53}$$

are the filtered excitation vectors with unity gain. The remaining eight factors in (13.48) are

$$\gamma_2^{(l)} = -2\mathbf{y1}^{(l)T}\mathbf{u} = -2\sqrt{\frac{G^{(l)}P_1^{(l)}E_e}{R_1}}\mathbf{y1}_o^T\mathbf{u}, \tag{13.54}$$

$$\gamma_3^{(l)} = -2\mathbf{y2}^{(l)T}\mathbf{u} = -2\sqrt{\frac{G^{(l)}P_2^{(l)}E_e}{R_2}}\mathbf{y2}_o^T\mathbf{u}, \tag{13.55}$$

$$\gamma_4^{(l)} = 2\mathbf{y0}^{(l)T}\mathbf{y1}^{(l)} = 2E_eG^{(l)}\sqrt{\frac{P_1^{(l)}\left(1 - P_2^{(l)} - P_1^{(l)}\right)}{R_1R_0}}\mathbf{y0}_o^T\mathbf{y1}_o, \tag{13.56}$$

$$\gamma_5^{(l)} = 2\mathbf{y0}^{(l)T}\mathbf{y2}^{(l)} = 2E_eG^{(l)}\sqrt{\frac{P_2^{(l)}\left(1 - P_2^{(l)} - P_1^{(l)}\right)}{R_2R_0}}\mathbf{y0}_o^T\mathbf{y2}_o, \tag{13.57}$$

$$\gamma_6^{(l)} = 2\mathbf{y1}^{(l)T}\mathbf{y2}^{(l)} = 2E_eG^{(l)}\sqrt{\frac{P_2^{(l)}P_1^{(l)}}{R_2R_1}}\mathbf{y1}_o^T\mathbf{y2}_o, \tag{13.58}$$

$$\gamma_7^{(l)} = \mathbf{y0}^{(l)T}\mathbf{y0}^{(l)} = \frac{E_e}{R_0}G^{(l)}\left(1 - P_2^{(l)} - P_1^{(l)}\right)\mathbf{y0}_o^T\mathbf{y0}_o, \tag{13.59}$$

$$\gamma_8^{(l)} = \mathbf{y1}^{(l)T}\mathbf{y1}^{(l)} = \frac{E_e}{R_1} G^{(l)} P_1^{(l)} \mathbf{y1}_o^T \mathbf{y1}_o, \tag{13.60}$$

$$\gamma_9^{(l)} = \mathbf{y2}^{(l)T}\mathbf{y2}^{(l)} = \frac{E_e}{R_2} G^{(l)} P_2^{(l)} \mathbf{y2}_o^T \mathbf{y2}_o. \tag{13.61}$$

It is left as an exercise for readers to verify the accuracy of the above relations.

A Summary

Gain quantization as designed for the IS54 coder has the following advantages:

- The VQ scheme works well for all signal levels since the energy of the speech frame is quantized separately. With the average energy eliminated, the three gains can be quantized efficiently.
- The values of G, P_2, and P_1 are well behaved and highly suitable for quantization purposes. For instance, P_1 and P_2 have the limited range of [0, 1]. On the other hand, most values of G are less than or equal to one.
- As long as the energy of the speech frame is correct, the decoded signal energy will not be much greater than the intended level. This is due to the nature of the gain codebook, with most of the values of their elements (G, P_2, and P_1) less than one. Thus, transient instability will not occur while decoding the subframe. This is a desirable property since minor channel errors will not have a big impact on signal quality.

Using a set of speech data, the gain codebook for VQ can be designed using the GLA, explained in Chapter 7. See Exercise 13.4 for methods to speed up the codebook search.

13.5 ENCODER AND DECODER

Operations of the encoder and decoder for the IS54 VSELP standard are described in this section. Figure 13.2 shows the block diagram of the encoder. For LP analysis, ten LPCs are obtained at every frame interval of 20 ms (160 samples), which are quantized and interpolated to be used by each subframe (Chapter 8). A total of 38 bits are transmitted per frame for information regarding the LPC. Functions of the rest of the blocks are very much self-explanatory.

Table 13.2 summarizes the bit allocation scheme of the IS54 coder. A total of 159 bits are allocated per frame, leading to a bit-rate of 7950 bps. In the original standard, an additional 5050 bps are utilized for error protection and synchronization, bringing the total bit-rate to 13 kbps.

Figure 13.3 shows the block diagram of the decoder, which has the mission of reconstructing the excitation vector and passing it through the synthesis filter. A postfilter (see Chapter 11) is incorporated at the end to improve the subjective quality of the synthetic speech.

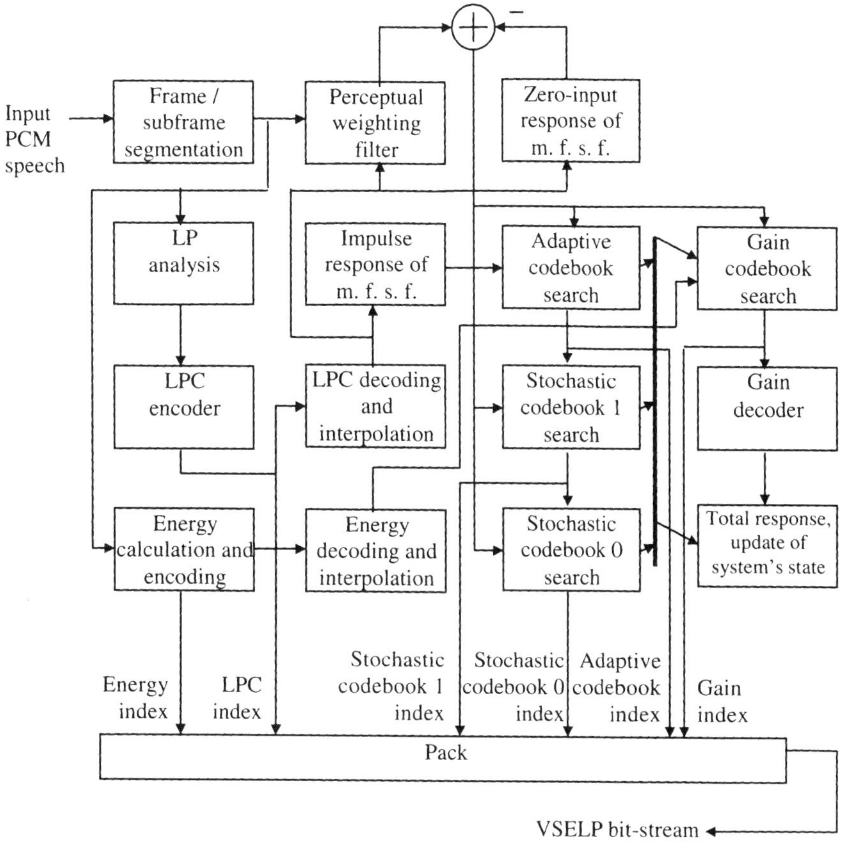

Figure 13.2 Block diagram of the IS54 VSELP encoder (m. f. s. f. means modified formant synthesis filter).

TABLE 13.2 Bit Allocation For the TIA IS54[a] VSELP Coder

Parameter	Number per Frame	Resolution	Total Bits per Frame
LPC	10	6,5,5,4,4,3,3,3,3,2	38
Adaptive codebook index	4	7	28
Stochastic codebook 1 index	4	7	28
Stochastic codebook 0 index	4	7	28
Frame energy	1	5	5
Gain index	4	8	32
Total			**159**

[a]Data From Macres [1994], Table 2.

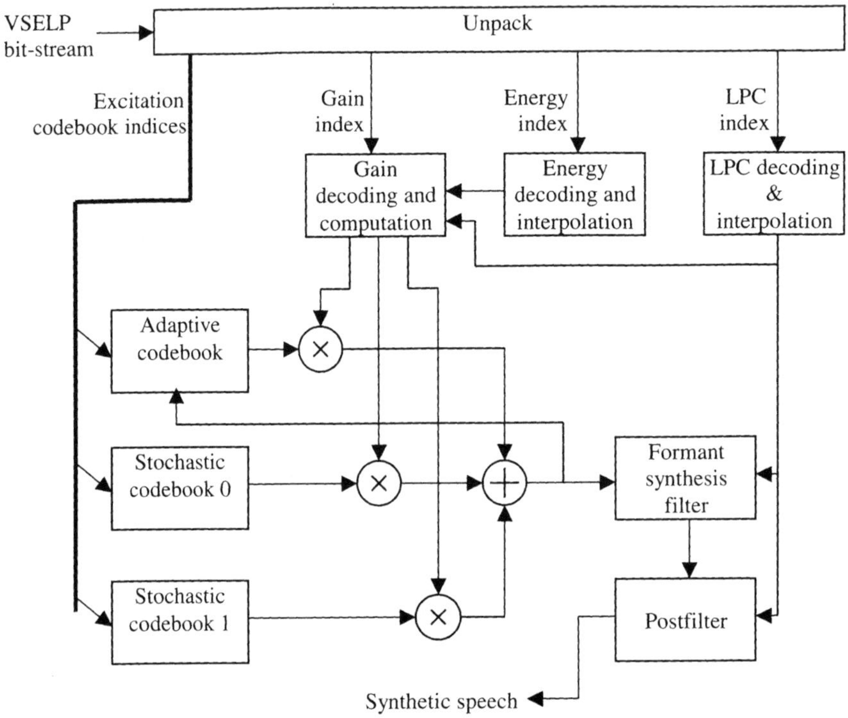

Figure 13.3 Block diagram of the IS54 VSELP decoder.

13.6 SUMMARY AND REFERENCES

Eminent features of VSELP are presented in this chapter. It is shown that the coder is designed to have reduced computational load—fast search is realizable with the excitation codebooks. Limited memory cost is achieved with a finite set of basis vectors, providing also high robustness against channel errors. The IS54 outperforms other standard CELP coders in quality, such as the FS1016 [Cox, 1995], although it operates at a higher bit-rate of 7.95 kbps. Thus, the IS54 is well suited for cellular telephony applications.

As indicated in Section 13.1, the basis vectors from which the stochastic codevectors are spanned contain white noise elements. It is possible to elevate the performance by further training the system using a large amount of speech data, with the objective of tuning the elements of the basis vectors in such a way that the total weighted error is minimized. For the case of the IS54, the optimal basis vectors are computed by solving the 14 basis vectors $\times$ 40 samples/basis vector $=$ 560 simultaneous equations, which are the results of taking the partial derivatives of the total weighted error with respect to each sample of each basis vector and setting them equal to zero. See Gerson and Jasiuk [1991] for additional information, where an increase of 13.41 to 14.05 dB in weighted segmental SNR after 16 iterations was reported.

Major contributions of the IS54 coder can be summarized as follows:

- First standardized medium bit-rate coder based on CELP.
- Introduction of the concept of adaptive codebook, enabling closed-loop optimization of long-term prediction parameters.
- Efficient implementation through the use of separate fixed codebooks with codevectors spanned by a small number of basis vectors.
- Joint quantization of the gains associated with the excitation via VQ.

The other two VSELP standards—GSM 6.20 and STD-27B—are based on similar architecture, but with a differing number of stochastic codebooks and/or basis vectors. The coder works essentially the same way if only one stochastic codebook is present, leading to lower implementational cost and bit-rate with sacrifice in quality. Main features of the VSELP algorithm are described in Gerson and Jasiuk [1991]. See Macres [1994] for the actual implementation on a DSP platform. In DeMartino [1993], a quality measurement report is presented where three coders—GSM 6.10, IS54, and STD-27B—are compared. The conclusion is that the IS54 is better than the other two coders, providing higher perceptual quality.

The IS54 coder was standardized by the TIA in approximately a year, beginning in early 1989. Due to insufficient time, the amount of testing was minimal and the coder suffers from a number of performance deficiencies that were never identified during testing, such as quality with background noise or music and tandem coding. The TIA attempted to repair the coder, but in 1994, discussion began about a total replacement [Cox, 1995]. The quest culminated in 1996 with the IS641 ACELP standard, operating at 7.4 kbps and described in Chapter 16.

EXERCISES

13.1 Prove Lemma 13.1 by expanding the norm term and using the fact that

$$\mathbf{y}^T\mathbf{z} = \mathbf{z}^T\mathbf{y} = 0$$

since $\mathbf{y}$ and $\mathbf{z}$ are orthogonal.

13.2 Let $g_{N-1}\cdots g_2g_1g_0$ denote a codeword in the N-bit Gray code, and let $b_{N-1}\cdots b_2b_1b_0$ designate the corresponding binary number, where the subscripts 0 and $N-1$ denote the least significant and most significant digits, respectively. Then the ith digit g_i can be obtained with

$$g_i = b_i \oplus b_{i+1}; \quad 0 \leq i \leq N-2,$$
$$g_{N-1} = b_{N-1},$$

with $\oplus$ denoting the exclusive or operation [Sandige, 1990]. Using the above relations, find the 3-bit Gray code from the corresponding binary numbers.

13.3 To convert from Gray code to binary, start with the most significant digit and proceed to the least significant digit. Then

$$\begin{aligned} b_i &= g_i, \quad \text{if number of 1's preceding } g_i \text{ is even,} \\ b_i &= g_i', \quad \text{otherwise,} \end{aligned}$$

with g' denoting the logically inverted version of g. Using the rule, convert the 3-bit Gray code to binary.

13.4 The excitation gains are vector quantized as a three-dimensional vector containing the elements G, P_2, and P_1. To find the best codevector from the codebook, an error expression containing nine sum terms is used. These sum terms are denoted as γ_1 to γ_9. Note from the expression for γ_1 that the quantity

$$-2\sqrt{G^{(l)}\left(1 - P_2^{(l)} - P_1^{(l)}\right)}$$

must be evaluated for all index values l. The expression can be precomputed in an off-line fashion and stored in a codebook, recovered by the same index l. This will save valuable processing time when the gain codebook is searched; the price to pay is additional storage space. Following the same principle, evaluate the expressions for γ_2 to γ_9 and find out the terms that can be computed off-line and stored in a codebook.

13.5 Once the optimal gain codevector is found, the elements $\{G, P_2, P_1\}$ must be converted to $\{b, g1, g0\}$. From the relations between these two sets of parameters, we can see that some quantities can be precomputed and stored in a codebook so as to save computational cost. For instance, in order to calculate b, we can precompute the quantity $(GP_2)^{1/2}$ and store them in a codebook. What can we do for $g1$ and $g0$?

13.6 Consider a VSELP coder without stochastic codebook 0; that is, the excitation is comprised of the adaptive codevector and one stochastic codevector. To quantize the excitation gains, a two-dimensional VQ is used, where the vector under consideration is $\{G, P_2\}$, with

$$b = \sqrt{\frac{G \cdot P_2 \cdot E_e}{R_2}}$$

and

$$g1 = \sqrt{\frac{G(1 - P_2)E_e}{R_1}}.$$

Find the error expression for this case that one can rely on to search the gain codebook. Express the answer as a function of $G^{(l)}$, $P_2^{(l)}$, E_e, R_1, R_2, $\mathbf{y1}_o$, $\mathbf{y2}_o$, and $\mathbf{u}$ (see Section 13.4).

13.7 Gerson and Jasiuk [1991] propose using the "pitch prefilter" with system function

$$H(z) = \frac{1}{1 - bz^{-T}}$$

during decoding. The excitation signal going into the formant synthesis filter is first processed by this filter. The purpose of the filter is, as its name implies, to enhance the pitch periodicity of the excitation signal. How would you choose the parameters b and T of this filter? Justify your answer.

13.8 Find out the computational costs associated with the stochastic codebook search with and without the Gray code ordering scheme. How much is gained by deploying the Gray code ordering?

CHAPTER 14

LOW-DELAY CELP

In the process of speech encoding and decoding, delay is inevitably introduced. Loosely defined, delay is the amount of time shift between the speech signal at the input of the encoder with respect to the synthetic speech at the output of the decoder, when the output of the encoder is directly connected to the input of the decoder. For schemes such as PCM and ADPCM (Chapter 6), the speech signal is encoded on a sample-by-sample basis: a few bits are found for each sample with the result transmitted immediately; the delay associated with these schemes is negligible. For many speech coders, such as CELP (Chapter 11), high compression ratio is achieved by processing the signal on a frame-by-frame basis, thus requiring a buffering procedure consuming typically 20 to 30 ms, depending on the length of the frame. It is this buffering process associated with most low bit-rate coders that augment the overall delay. See Chapter 1 for a precise definition of coding delay.

Delay is an important concern for real-time two-way conversations, and it basically can be thought of as the time the sound takes to travel from speaker to listener. For an excessively large delay, that is, above 150 ms, the ability to hold a conversation is impaired. The parties involved begin to interrupt or "talk over" each other because of the time it takes to realize the other party is speaking. When delay becomes high enough, conversations degrade to a half-duplex mode, taking place strictly in one direction at a time; hence, the lower the delay the better. Low delay is also highly desirable for typical telephone networks, since delay aggravates echo problems: the longer an echo is delayed, the more audible and annoying it is to the talker. Even though echo cancelers are normally incorporated, high delay makes the job of echo cancellation more difficult.

There is always a price to pay for a certain attribute, and for the case of low delay there is no exception. The low-delay constraint is in conflict with other desirable properties of a speech coder, such as low bit-rate, high quality, reduced computational cost, and robustness against channel errors. Therefore, delay reduction with minimum degradation of the good properties has been a great challenge to speech coding researchers.

This chapter is devoted to the ITU-T G.728 LD-CELP coder, standardized in 1992. At a bit-rate of 16 kbps, it is perhaps the most successful low-delay coder available. Core techniques of the coder were developed mainly by Chen while at AT&T Bell Labs [Chen 1990, 1991, 1995; Chen et al., 1990, 1991, 1992]. Even though the coder is based on the same principles as CELP, it utilizes many unconventional techniques to achieve low delay. In Section 14.1, strategies to achieve low delay are explained. Basic operational principles of the LD-CELP coder are described in Section 14.2, while issues related to LP analysis are given in Section 14.3. Excitation codebook structure and search procedures are covered in Section 14.4; the technique for backward gain adaptation is given in Section 14.5; operations of the encoder and decoder are covered in Section 14.6, followed by the algorithm for excitation codebook training in Section 14.7. A brief summary is given in the last section.

14.1 STRATEGIES TO ACHIEVE LOW DELAY

In this section, we analyze the most important strategies adopted by the G.728 LD-CELP coder to achieve low delay while maintaining low bit-rate.

Strategy 1. Reduce frame length to 20 samples. From Chapter 1 we know that the biggest component of coding delay is due to buffering at the encoder and is directly linked to the length of the frame selected for analysis. Therefore, an obvious solution is to reduce the length of the frame. Like conventional CELP, the LD-CELP coder partitions the input speech samples into frames that are further divided into subframes; each frame consists of 20 samples, containing four subframes of five samples each. As we will see later, encoding starts when five samples are buffered (one subframe); this leads to a buffering delay of 0.625 ms, producing a coding delay in the range of 1.25 to 1.875 ms. These value are much lower than conventional CELP, having a buffering delay of 20 to 30 ms.

Strategy 2. Recursive autocorrelation estimation. The first step for finding the LPCs is to calculate the autocorrelation values, which can be done using conventional techniques such as Hamming windowing (nonrecursive); due to the short frame length, the scheme gets highly computationally expensive and inefficient since the windows become overlapping for consecutive frames. To simplify, recursive methods can be employed. The LD-CELP coder utilizes the Chen windowing method (Chapter 3) to estimate the autocorrelation values, which in principle is a hybrid technique, combining both recursive and nonrecursive approaches.

Strategy 3. External prediction. Since the signal frame is relatively short, its statistical properties tend to be close to the near past or near future. It is therefore possible to estimate the LPCs entirely from the past and apply these coefficients to the current frame, which is the definition of external prediction (Chapter 4). By using external prediction, the encoder does not have to wait to buffer the whole frame (20 samples) before analysis; instead, the LPCs are available at the instant that the frame begins; encoding starts as soon as one subframe (five samples) is available. This is in high contrast to conventional CELP, where the LPCs are derived from a long frame, with the resultant coefficients used inside the frame (internal prediction).

Strategy 4. Backward adaptive linear prediction. Conventional speech coders use forward adaptation in linear prediction, where the LPCs are derived from the input speech samples; the coefficients are then quantized and transmitted as part of the bit-stream. The LD-CELP coder utilizes backward adaptation, with the LPCs obtained from the synthetic speech. By doing so, there is no need to quantize and transmit the LPCs since the synthetic speech is available at both the encoder and decoder side, thus saving a large number of bits for transmission. Note that this is a necessity to achieve low bit-rate, since otherwise the quantized LPCs must be transmitted at short frame intervals, leading to unacceptably high bit-rate. A disadvantage of the approach is its vulnerability toward channel errors: any error will propagate toward future frames while decoding.

Strategy 5. High prediction order. The LD-CELP coder utilizes a short-term synthesis filter with a prediction order equal to 50; no long-term predictor is employed. This design choice is due to the following reasons.

- In general, two options are available for capturing the periodicity of voiced frames: a long-term predictor combined with a short-term predictor (with a typical order of 10), or a short-term predictor with high prediction order, for instance, a value of 50.
- Long-term prediction with forward adaptation (i.e., parameters of the predictors are obtained from the input speech signal and are quantized and transmitted to the decoder) is not an option, since the number of bits required to carry information regarding the long-term predictor's parameters at short frame length would elevate the resultant bit-rate to prohibitive levels, ruining the low bit-rate goal.
- Long-term prediction with backward adaptation (i.e., parameters of the predictor are extracted from the synthetic speech; these parameters are not required to be transmitted since the decoder has access to the same synthetic speech signal) is possible, since no extra bit allocation is necessary. However,

it was found that backward block adaptation was extremely sensitive to channel errors and seemed to be inherently unstable [Chen, 1995]. Thus, the long-term predictor is abandoned.

- Pitch period of female speech is typically less than 50 samples. By using a short-term predictor with a prediction order of 50, it is possible to reproduce female speech with high quality. In addition, prediction gain practically saturates when the order is beyond 50; that is, further increasing the prediction order beyond 50 does not improve the quality much and only increases the complexity.
- High prediction order would create a burden for forward adaptation schemes since the LPCs must be quantized and transmitted; increasing the order implies an increasing number of bits for transmission. However, this is not an issue for LD-CELP, since backward adaptation is used with no need to encode the LPCs.
- Without the long-term predictor, the coder becomes less speech-specific, since no pitch quasiperiodicity is assumed. This feature improves the performance for nonspeech signals, like voice-band signaling tones found in most telecommunication systems and/or music.

Strategy 6. Backward excitation gain adaptation. Excitation gain is updated once every subframe (five samples) by using a tenth-order adaptive linear predictor in the logarithmic-gain domain. The coefficients of this log-gain predictor are updated once every four subframes by performing linear prediction analysis on previous logarithmic gain values.

By using a log-gain predictor of order ten, the predicted gain will be based on ten past gain values having a time span of $10 \cdot 5 = 50$ speech samples. This in turn allows the exploitation of pitch periodicity remaining in the excitation gain sequence for those female voices with a pitch period under 50 samples. Such a scheme is better at predicting the excitation gain for female voices. As a result, better coding efficiency can be achieved.

By making the excitation gain backward adaptive, the current gain value is derived from the information embedded on previously quantized excitation, and there is no need to send any bits to specify the excitation gain, since the decoder can derive the same gain in the same manner. Transmission of the highly redundant gain information is thus eliminated.

14.2 BASIC OPERATIONAL PRINCIPLES

After browsing the most distinguishing features of LD-CELP from the previous section, we are now ready for the basic operational principles. In LD-CELP, only five samples are needed to start the encoding process. On the other hand, only the excitation signal is transmitted: the predictor coefficients are updated by performing LP analysis on previously quantized speech. Thus, the

LD-CELP coder is basically a backward adaptive version of the conventional CELP coder. The essence of CELP, which is the analysis-by-synthesis codebook search, is retained. Figure 14.1 shows the general structures of the LD-CELP encoder and decoder.

The basic operational principle follows the conventional CELP algorithm, except that only the index to the excitation codebook is transmitted. The operation of the LD-CELP algorithm can be summarized as follows.

- Like conventional CELP, samples of speech are partitioned into frames and subdivided into subframes. In LD-CELP, each frame consists of 20 samples, which contains four subframes of five samples each. Since LP analysis is performed in a backward adaptive fashion, there is no need to buffer an entire frame before processing. Only one subframe (five samples) needs to be stored before the encoding process begins.

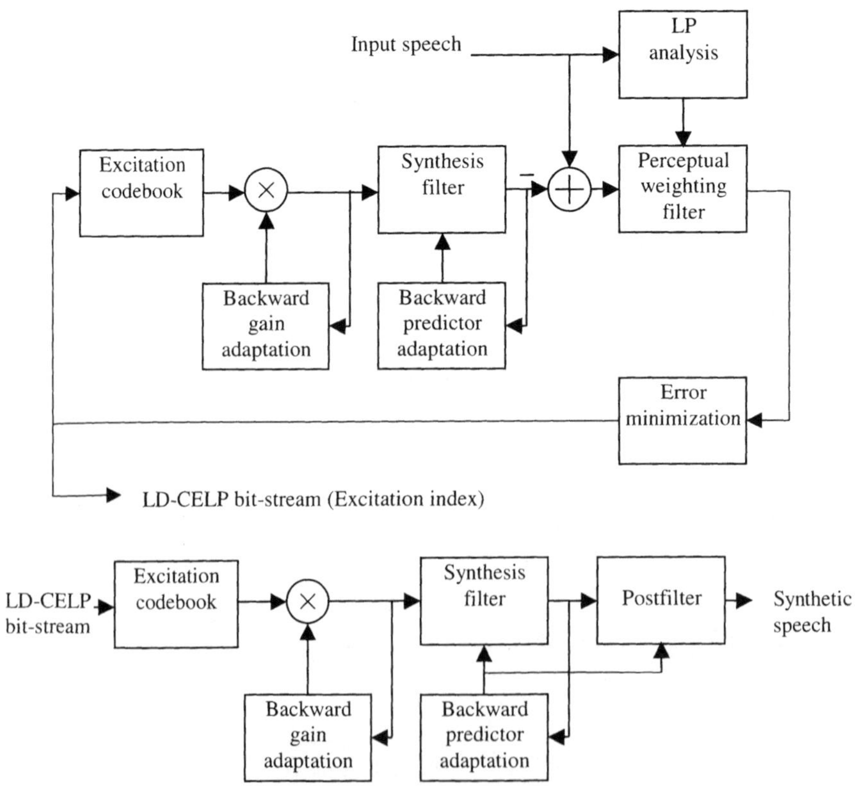

Figure 14.1 LD-CELP encoder (*top*) and decoder (*bottom*).

- The perceptual weighting filter has ten linear prediction coefficients derived from the original speech data. The filter is updated once per frame. The current frame's coefficients are obtained from previous frames' samples.
- The synthesis filter corresponds to that of a 50th-order AR process. Its coefficients are obtained from synthetic speech data of previous frames. The filter is updated once per frame. The zero-input response of the current frame can be obtained from the known initial conditions.
- Excitation gain is updated every subframe, with the updating process performed with a tenth-order adaptive linear predictor in the logarithmic-gain domain. The coefficients of this predictor are updated once per frame, with the LP analysis process done on gain values from previous subframes.
- The excitation sequence is searched once per subframe, where the search procedure involves the generation of an ensemble of filtered sequences: each excitation sequence is used as input to the formant synthesis filter to obtain an output sequence. The excitation sequence that minimizes the final error is selected.
- In the decoder, the initial conditions are restored. The synthetic speech is generated by filtering the indicated excitation sequence through the filter without any perceptual weighting. Since the excitation gain and the LPCs are backward adaptive, there is no need to transmit these parameters. A postfilter can be added to further enhance the output speech quality.

Functionality of each block is presented in the next sections.

14.3 LINEAR PREDICTION ANALYSIS

The LD-CELP algorithm requires multiple LP analysis procedures to be performed during operation since different sets of LPCs are used in various parts of the coder. These are:

- *Synthesis Filter.* This is a 50th-order filter; its coefficients are obtained by analyzing the synthetic speech in a backward adaptive fashion. The same procedure is performed in the encoder and decoder. In the decoder, the resultant coefficients are also used by the postfilter.
- *Perceptual Weighting Filter.* This is a tenth-order filter; its coefficients are obtained by analyzing the original input speech. This filter is available only in the encoder.
- *Backward Excitation Gain Adaptation.* The gain is obtained through backward adaptation, where a tenth-order predictor is applied in the logarithmic-gain domain. Coefficients of the predictor are obtained by analyzing the past gain terms. The operation is the same for the encoder and the decoder. Details are given in Section 14.5.

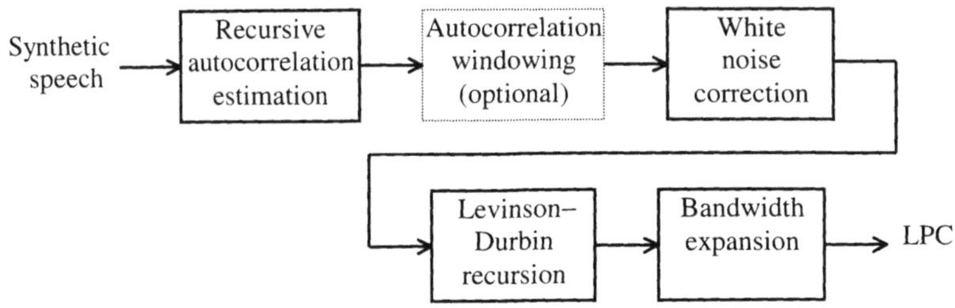

Figure 14.2 Procedure to obtain the LPCs for the synthesis filter.

Synthesis Filter

The system function of the filter is

$$H(z) = \frac{1}{1 + \sum_{i=1}^{50} a_i z^{-i}}. \tag{14.1}$$

The LPCs a_i, $i = 1, \ldots, 50$, are obtained by backward adaptation, where the input to LP analysis is the synthetic speech signal. The procedure of LP analysis is summarized in Figure 14.2. A set of autocorrelation coefficients is first estimated from the synthetic speech signal: $R[l]$, $l = 0, \ldots, 50$; the estimation is based on the Chen windowing procedure with parameters

$$\alpha = (3/4)^{1/40} = 0.992833749, \quad L = 35.$$

As explained in Chapter 3, the Chen window is hybrid in nature and consists of a recursive part and a nonrecursive part. It is highly efficient and provides good accuracy in practice.

Backward LP analysis is inherently more unstable than forward adaptation, and many of the techniques discussed in Chapter 4 for the alleviation of ill-conditioning are necessary to "tame" the system. Spectral smoothing by windowing the autocorrelation coefficients is optional; however, its application is highly recommended since system divergence (state of the decoder does not follow the encoder) had been observed for certain synthetic signals when this block is not present [Chen, 1995].

White noise correction is applied next with $\lambda = 257/256$. After using the Levison–Durbin recursion, the LPCs are bandwidth expanded with $\gamma = 253/256 \approx 0.9883$.

The LPCs are updated once every frame, or four subframes (20 samples). Within the frame–subframe structure, the update occurs at the third subframe. This scheme is illustrated in Figure 14.3. Note that the LPCs found from the samples before subframe 0 of the current frame are used by subframes 2 and 3 of the current frame,

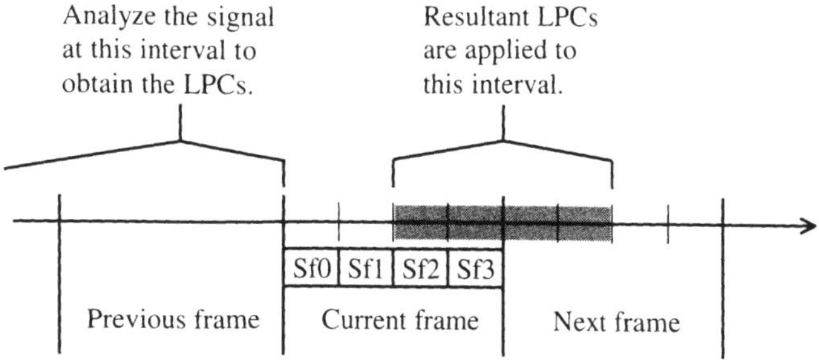

Figure 14.3 Illustration of LPC adaptation scheme for the synthesis filter.

and subframes 0 and 1 of the future frame. This scheme allows a more even distribution of computational activity during the various cycles of the encoding process and thus facilitates real-time implementation. This can be seen by observing that even though the autocorrelation values are available at the first subframe of each frame, computations may require more than one subframe worth of time. And since the Levinson–Durbin recursion is quite demanding, by updating at the third subframe, plenty of time is available to complete the task.

Perceptual Weighting Filter

The system function of the perceptual weighting filter is

$$W(z) = \frac{A(z/\gamma_1)}{A(z/\gamma_2)} = \frac{1 + \sum_{i=1}^{10} b_i \gamma_1^i z^{-i}}{1 + \sum_{i=1}^{10} b_i \gamma_2^i z^{-i}}. \tag{14.2}$$

The nominal values for (γ_1, γ_2) are (0.9, 0.6) and have been found to give good subjective quality. When compared with the form of weighting filter given in Chapter 11, (14.2) is more general, allowing more control over the spectral characteristics. Note that the prediction order is equal to 10, which is quite different from conventional CELP, where the order of the synthesis filter is equal to that of the weighting filter. Experimentally, a weighting filter with an order of 50 was found to produce occasional artifacts and was thus abandoned from consideration [Chen, 1995].

The LPCs b_i, $i = 1$ to 10, are derived from the original input speech; use of synthetic speech is avoided since it contains quantization errors. LP analysis follows a similar approach as for the synthesis filter, with

$$\alpha = (1/2)^{1/40} = 0.982820598, \quad L = 30$$

being the window parameters; white noise correction is the same as for the synthesis filter. The LPCs b_i in (14.2) are the output of the Levinson–Durbin module. Similar to the synthesis filter, the perceptual weighting filter is also updated once per frame, and the updates also occur at the third subframe.

14.4 EXCITATION CODEBOOK SEARCH

Like conventional CELP, the zero-input response is first subtracted from the input speech to obtain the target sequence. The target sequence is then used as reference during the excitation codebook search, where the codevector capable of generating the sequence as close as possible (in a sum of squared error sense) to the reference is selected. In this section, various techniques for the excitation codebook search are analyzed. Due to architectural differences with conventional CELP, LD-CELP requires different methodologies to improve efficiency.

The Analysis-by-Synthesis Loop

Figure 14.4 shows the encoding loop of the LD-CELP coder; unlike conventional CELP, the excitation gain is known during encoding of the current subframe, which is obtained through prediction from past values. Due to this fact, some computational saving is obtainable. Consider the alternative scheme shown in Figure 14.5. This new scheme has the excitation gain block moved out of the loop. After the zero-input response is subtracted from the original speech, the resultant sequence is divided by the excitation gain to generate the target sequence. In this way, there is no need to multiply all vectors of the excitation codebook by the gain. Using similar reasoning as for conventional CELP, it is possible to reposition the perceptual weighting filter as shown in Figure 14.6, leading to an additional cut in computational cost.

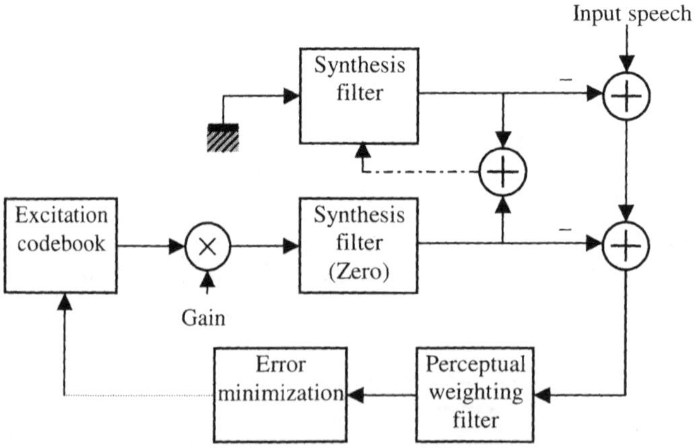

Figure 14.4 Analysis-by-synthesis loop of the LD-CELP encoder.

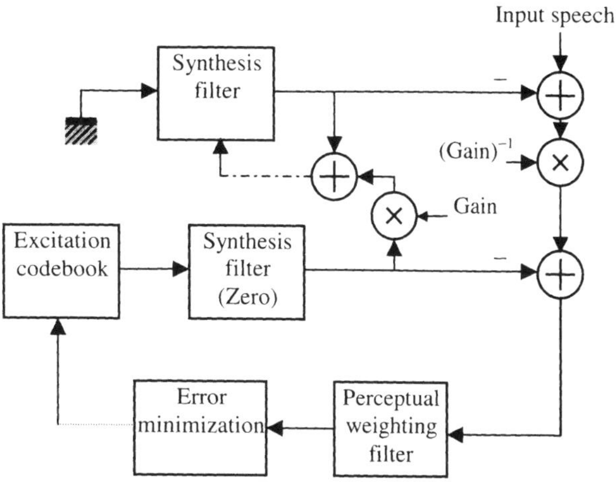

Figure 14.5 An equivalent analysis-by-synthesis loop.

Excitation Codebook Structure

The excitation codebook uses a shape-gain structure, where a 7-bit shape codebook forms from five-dimensional vectors together with a 3-bit gain codebook, constituting the 10-bit codebook. This configuration reduces computational complexity since filtering operations with the excitation sequences are done only 2^7 times.

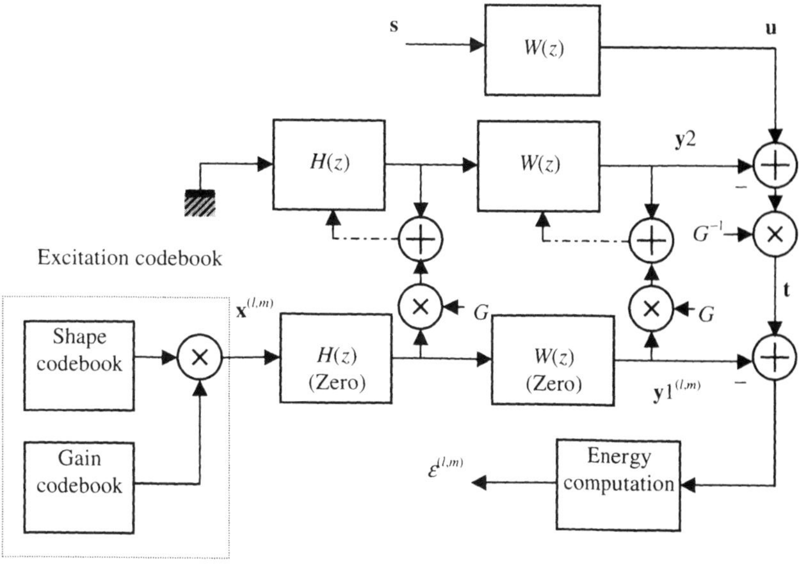

Figure 14.6 Another equivalent analysis-by-synthesis loop with reduced complexity.

The rest of the zero-state responses are found by multiplication with the gain values located in the gain codebook. Note also that the shape-gain approach reduces the storage requirement, since the gain codebook has relatively few components.

The exact values of the codebook elements are specified in [ITU, 1992]. For the gain codebook, 3 bits lead to a total of 8 different values. Two bits of the gain codebook address different magnitudes while the last bit contains the sign of the actual gain. Hence, there are four different magnitudes resulting in a total of eight gain values; half of them are positive and the other half negative. Denoting these values by $g^{(m)}$, $m = 0, \ldots, 7$, they are specified with

$$\begin{aligned} g^{(0)} &= 33/64; \\ g^{(m)} &= g^{(m-1)}7/4, \quad & m = 1, 2, 3; \\ g^{(m)} &= -g^{(m-4)}, & m = 4, 5, 6, 7. \end{aligned}$$

Denoting the five-element vectors of the shape codebook by $\mathbf{v}^{(l)}$, where $0 \le l \le 127$, there are a total of $2^{10} = 1024$ different five-element sequences from the excitation codebook, formed by the product between the shape codebook and gain codebook. Thus, each excitation sequence is denoted by $g^{(m)}\mathbf{v}^{(l)}$. These sequences are filtered by the cascade connection of formant synthesis filter and perceptual weighting filter to produce the zero-state response $\mathbf{y1}^{(l,m)}$ during encoding.

Objective of Excitation Codebook Search

For a given subframe, the zero-state response of the filters cascade is

$$\mathbf{y1}^{(l,m)} = g^{(m)}\mathbf{H}\mathbf{v}^{(l)} \tag{14.3}$$

with $l = 0, \ldots, 127$; $m = 0, \ldots, 7$; and $\mathbf{H}$ the impulse response matrix of the filters cascade. The objective is to find l and m so that

$$\varepsilon^{(l,m)} = \left\|\mathbf{t} - \mathbf{y1}^{(l,m)}\right\|^2 \tag{14.4}$$

is minimized, with

$$\mathbf{t} = \frac{1}{G}(\mathbf{u} - \mathbf{y2}) \tag{14.5}$$

being the target vector; $\mathbf{u}$ is the perceptually weighted input speech vector, and $\mathbf{y2}$ is the zero-input response vector of the filters cascade. Expanding (14.4), we find

$$\varepsilon^{(l,m)} = \mathbf{t}^T\mathbf{t} - 2\mathbf{t}^T\mathbf{y1}^{(l,m)} + \mathbf{y1}^{(l,m)T}\mathbf{y1}^{(l,m)}. \tag{14.6}$$

The first term, $\mathbf{t}^T\mathbf{t}$, is constant with respect to l and m. Hence, we need only consider the alternative error expression:

$$\begin{aligned} \varepsilon'^{(l,m)} &= -2\mathbf{t}^T\mathbf{y1}^{(l,m)} + \mathbf{y1}^{(l,m)T}\mathbf{y1}^{(l,m)} \\ &= -2g^{(m)}\mathbf{t}^T\mathbf{H}\mathbf{v}^{(l)} + g^{(m)2}\left\|\mathbf{H}\mathbf{v}^{(l)}\right\|^2. \end{aligned} \tag{14.7}$$

Propositions for Complexity Reduction

In actual encoding, (14.7) is evaluated for every value of l and m. Note that $\mathbf{H}$ is constant for four subframes; hence, the following proposals lead to computational savings:

- Compute $\mathbf{H}\mathbf{v}^{(l)}$, $l = 0, \ldots, 127$, and store the results for use by the four subframes. This requires $5 \cdot 128 = 640$ memory locations.
- Compute $\|\mathbf{H}\mathbf{v}^{(l)}\|^2$, $l = 0, \ldots, 127$, and store the results for use by the four subframes. This requires 128 memory locations.
- Compute $g^{(m)2}\|\mathbf{H}\mathbf{v}^{(l)}\|^2$, $m = 0, \ldots, 7$; $l = 0, \ldots, 127$, and store the results for use by the four subframes. This requires $8 \cdot 128 = 1024$ memory locations.
- Compute $g^{(m)}\mathbf{H}\mathbf{v}^{(l)}$, $m = 0, \ldots, 7$; $l = 0, \ldots, 127$, and store the results for use by the four subframes. This requires $5 \cdot 8 \cdot 128 = 5120$ memory locations.

The above proposals work by removing redundancies during the codebook search, since the same computation is repeated for four subframes. The third and fourth propositions have the drawback of requiring excessive storage, which might be prohibitive in practice due to high memory cost.

A Quantization Approach for Gain Search

For every index l, the obvious way to find the gain index m is by evaluating (14.7) for all m. The disadvantage of the approach is that (14.7) must be evaluated for every index m. To speed up the process, $gg^{(m)} = 2g^{(m)}$ and $g2^{(m)} = g^{(m)2}$ are often precalculated and stored in a codebook. The quantity $gc^{(m)}$ represents the midpoint level between adjacent gain levels, given by

$$gc^{(m)} = \tfrac{1}{2}\left(g^{(m)} + g^{(m+1)}\right); \quad m = 0,\ 1,\ 2,\ 4,\ 5,\ 6, \tag{14.8}$$

and is included to assist the search process.

Differentiating (14.7) with respect to the gain and equating to zero, the optimal gain value, denoted by g, is given by

$$g = \frac{\mathbf{t}^T\mathbf{H}\mathbf{v}^{(l)}}{\left\|\mathbf{H}\mathbf{v}^{(l)}\right\|^2} = \frac{C^{(l)}}{D^{(l)}}, \tag{14.9}$$

where

$$C^{(l)} = \mathbf{t}^T \mathbf{H} \mathbf{v}^{(l)} \tag{14.10}$$

and

$$D^{(l)} = \left\| \mathbf{H} \mathbf{v}^{(l)} \right\|^2. \tag{14.11}$$

Note that (14.7) can be expressed by

$$\varepsilon'^{(l,m)} = -gg^{(m)} C^{(l)} + g2^{(m)} D^{(l)}. \tag{14.12}$$

Therefore, in principle, we can quantize the optimal gain value (14.9) by selecting the codebook value $g^{(m)}$, $m = 0$ to 7, that is the closest. The procedure, however, requires one division operation that is not desirable for fixed-point implementation.

Quantization can basically be implemented with a series of comparisons with the quantizer cell boundaries (Chapter 5). For instance, to quantize a positive gain value g, one can rely on the following steps:

```
if g < gc^(0)
        m ← 0
else if g < gc^(1)
        m ← 1
else if g < gc^(2)
        m ← 2
else m ← 3
```

The same objective can be achieved without performing the division (14.9) to get g; instead we use the following code:

```
if C^(l) < gc^(0) D^(l)
        m ← 0
else if C^(l) < gc^(1) D^(l)
        m ← 1
else if C^(l) < gc^(2) D^(l)
        m ← 2
else m ← 3
```

In this latter code, division is avoided; for each m one multiplication is required. Multiplication is very often the preferred approach in a fixed-point environment since it can be implemented far more efficiently.

The Official Excitation Codebook Search Procedure of the ITU-T G.728 Coder

The code specified below is described in the original document of the ITU. It is assumed that $D^{(l)}$, $l = 0, \ldots, 127$, and that $\mathbf{t}^T\mathbf{H}$ are precomputed before entering. The outcomes are the two indexes, $l_{\min}$ and $m_{\min}$, pointing to the optimal excitation vector. This algorithm is very efficient and well suited for fixed-point implementation.

```
1.  min ← ∞
2.  for l ← 0 to 127
3.      C^(l) ← (t^T H) v^(l)
4.      if C^(l) < 0 goto 12
5.      if C^(l) < gc^(0) D^(l)
6.          m ← 0; goto 19
7.      if C^(l) < gc^(1) D^(l)
8.          m ← 1; goto 19
9.      if C^(l) < gc^(2) D^(l)
10.         m ← 2; goto 19
11.     else m ← 3; goto 19
12.     if C^(l) > gc^(4)
13.         m ← 4; goto 19
14.     if C^(l) > gc^(5)
15.         m ← 5; goto 19
16.     if C^(l) > gc^(6)
17.         m ← 6; goto 19
18.     else m ← 7
19.     epsilon ← -gg^(m) C^(l) + g2^(m) D^(l)
20.     if epsilon < min
21.         min ← epsilon
22.         l_min ← l; m_min ← m
23. return l_min, m_min
```

14.5 BACKWARD GAIN ADAPTATION

The excitation gain in the LD-CELP algorithm is backward adaptive and is updated once per subframe. The adaptation process is based on a linear predictor of order ten working in the logarithmic domain.

Principles of Operation

Given the excitation vector $\mathbf{x} = g \cdot \mathbf{v}$ and its scaled version

$$\mathbf{e}_r = G_r \cdot \mathbf{x}_r, \tag{14.13}$$

where r is the subframe index and G the excitation gain, then

$$\sigma_{\mathbf{e},r} = G_r \sigma_{\mathbf{x},r}, \tag{14.14}$$

where $\sigma_{\mathbf{x}}$ denotes the root mean square (rms) value of the vector $\mathbf{x}$ (standard deviation). Taking the logarithm of both sides of (14.14) gives

$$\log \sigma_{\mathbf{e},r} = \log G_r + \log \sigma_{\mathbf{x},r}. \tag{14.15}$$

Prediction

The variables in (14.15) can be associated with parameters in an AR model in the following manner:

- Log $\sigma_{\mathbf{e},r}$: an AR signal.
- Log G_r: the predicted AR signal.
- Log $\sigma_{\mathbf{x},r}$: prediction error.

Within the context of LP, prediction is found via a linear combination of past samples of the AR signal. That is,

$$\log G_r = -\sum_{i=1}^{10} c_i \log \sigma_{\mathbf{e},r-i}, \tag{14.16}$$

where c_i is the ith LPC for the current frame. These LPCs are changed only once per frame and are obtained by performing LP analysis on the past log-gain values $\sigma_{\mathbf{e},r}$, up to and including the past frame. Additional insight into the prediction procedure can be obtained by substituting (14.15) in (14.16):

$$\log G_r = \sum_{i=1}^{10} c_i \log G_{r-i} + \sum_{i=1}^{10} c_i \log \sigma_{\mathbf{x},r-i}. \tag{14.17}$$

The above equation describes a linear time-invariant (LTI) system whose input is $\sigma_{\mathbf{x},r}$ with output G_r. The system function is

$$H(z) = \frac{\sum_{i=1}^{10} c_i z^{-i}}{1 - \sum_{i=1}^{10} c_i z^{-i}}. \tag{14.18}$$

Assuming that the filter is stable, the associated impulse response will eventually decay to zero; this is highly desirable since, in the presence of channel error, the corresponding effect on the excitation gain will eventually decay to zero due to the decaying impulse response. Hence, this backward gain adaptation algorithm is robust to channel errors.

Example 14.1 From the discussion, it follows that the linear predictor is found to predict $\log \sigma_{\mathbf{e},r}$, with $\log G_r$ being the prediction for the rth frame. Therefore, the predictor is designed so that

$$\log G_r \approx \log \sigma_{\mathbf{e},r}, \tag{14.19}$$

and the prediction is based on the linear combination of the values $\sigma_{\mathbf{e},r-1}, \ldots, \sigma_{\mathbf{e},r-10}$. If the prediction is perfect, G_r is equal to $\sigma_{\mathbf{e},r}$; from (14.15) it follows that $\log \sigma_{\mathbf{x},r} = 0$, or $\sigma_{\mathbf{x},r} = 1$. In practice, however, prediction errors occur, and $\sigma_{\mathbf{x},r}$ will be different from 1. Figure 14.7 shows an example of a typical situation in gain adaptation, where the rms value of the original speech subframe is increased at $r = r_1$ and decreased at $r = r_2$.

Assuming stationary conditions where the rms value of the original speech subframe is constant, the prediction is perfect, or near perfect; hence, $\sigma_{\mathbf{x},r} \approx 1$ for $r < r_1$. At $r = r_1$, the rms value of the speech subframe increases to a new level; the predicted value $\log G_{r1}$ cannot respond instantaneously. The analysis-by-synthesis procedure during excitation codebook search finds the best excitation vector whose

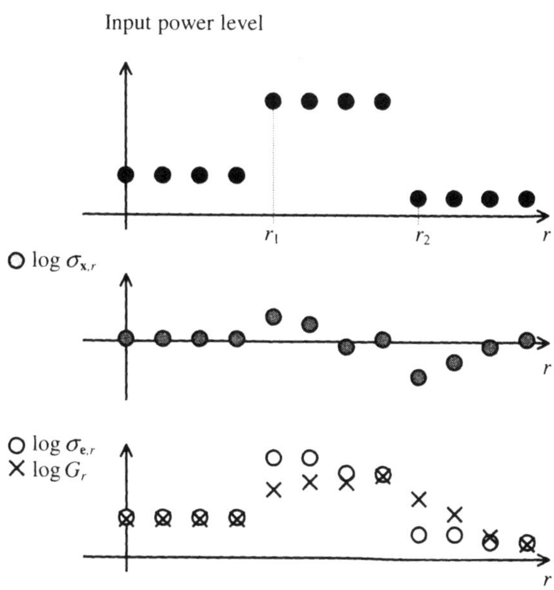

Figure 14.7 Example of gain adaptation.

rms value $\sigma_{\mathbf{x},r1}$ is greater than one so as to compensate the power increase. A similar situation happens at $r = r_2$ when the input power decreases.

Note from Figure 14.7 that without gain adaptation, the highly redundant gain sequence $\sigma_{\mathbf{e},r}$ must be sent. With gain adaptation, however, the sequence $\sigma_{\mathbf{x},r}$ is transmitted as part of the excitation vector. The dynamic range of $\sigma_{\mathbf{x},r}$ is far lower than $\sigma_{\mathbf{e},r}$. Coding efficiency is therefore increased by transmitting $\sigma_{\mathbf{x},r}$ instead of $\sigma_{\mathbf{e},r}$. We also see that in order for the scheme to work properly, the excitation vectors must have different rms values; this is effectively solved for the LD-CELP encoder, where the excitation codebook has a gain-shape structure.

Implementation

The actual backward gain adaptation mechanism of the G.728 coder is described here. Figure 14.8 shows the block diagram. The scaled excitation vector of the rth subframe is delayed by one subframe ($\mathbf{e}_{r-1}$), with its rms value calculated ($\sigma_{\mathbf{e},r-1}$), and the decibel value found ($20 \log \sigma_{\mathbf{e},r-1}$). A log-gain offset value of 32 dB is subtracted from the actual log-gain value so that

$$\delta_{r-1} = 20 \log \sigma_{\mathbf{e},r-1} - 32 \text{ dB}. \tag{14.20}$$

The offset value is meant roughly to equal the average excitation gain level for voiced frames. By removing the redundant value of the mean, the zero-mean sequence δ_r is created. Use of a zero-mean sequence has a numerical advantage in fixed-point processing, since the range of most variables involved in LP analysis is lowered, leading to a more efficient usage of the limited resolution.

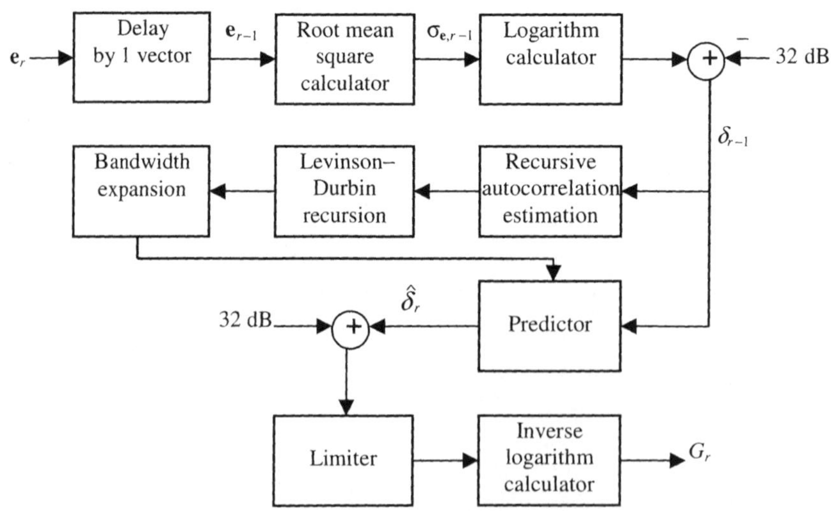

Figure 14.8 Backward excitation gain adaptation.

The same Chen windowing procedure, as for the synthesis filter and perceptual weighting filter, is applied here, with the input variables being δ_r. The window parameters are

$$\alpha = (3/4)^{1/8}, \quad L = 20.$$

After obtaining the LPCs through Levinson–Durbin recursion (prediction order of ten), the coefficients are bandwidth expanded with the constant $\gamma = 29/32$; such bandwidth expansion makes the gain adapter more robust to channel errors. Denoting the LPCs after bandwidth expansion as c_i, $i = 1$ to 10, we find

$$\hat{\delta}_r = -\sum_{i=1}^{10} c_i \delta_{r-i} \tag{14.21}$$

is the predicted value of δ_r. The offset value of 32 dB is added back to the predicted value, with the result clipped between 0 and 60 dB. This is performed by the limiter block, where gain values that are excessively large or small are set to the biggest and smallest limits, respectively. The gain limiter output is then fed to the inverse logarithm calculator, converting the gain back to the linear domain. Inclusion of the limiter ensures that the gain in the linear domain is between 1 and 1000.

The gain predictor is updated once every four subframes, with the updates taking place at the second subframe.

14.6 ENCODER AND DECODER

Operations of the encoder and decoder for the G.728 standard are described in this section.

Decoder

Block diagram of the decoder is essentially the one shown in Figure 14.1. The LD-CELP bit-stream consists of the excitation codebook index transmitted 10 bits every subframe (0.625 ms), leading to a bit-rate of 16 kbps. The recovered excitation codevector is scaled and passed through the synthesis filter and postfilter to yield the synthetic speech signal.

Mechanisms for gain and predictor adaptation are already explained in previous sections. Details of postfilter are covered in Chapter 11. Note that the postfilter utilizes a tenth-order short-term filter whose LPCs are obtained from the same LP analysis procedure as for the synthesis filter. All that is needed is to pause the 50th-order Levinson–Durbin recursion at order 10, get a copy of the coefficients, and resume the recursion from order 11 to 50.

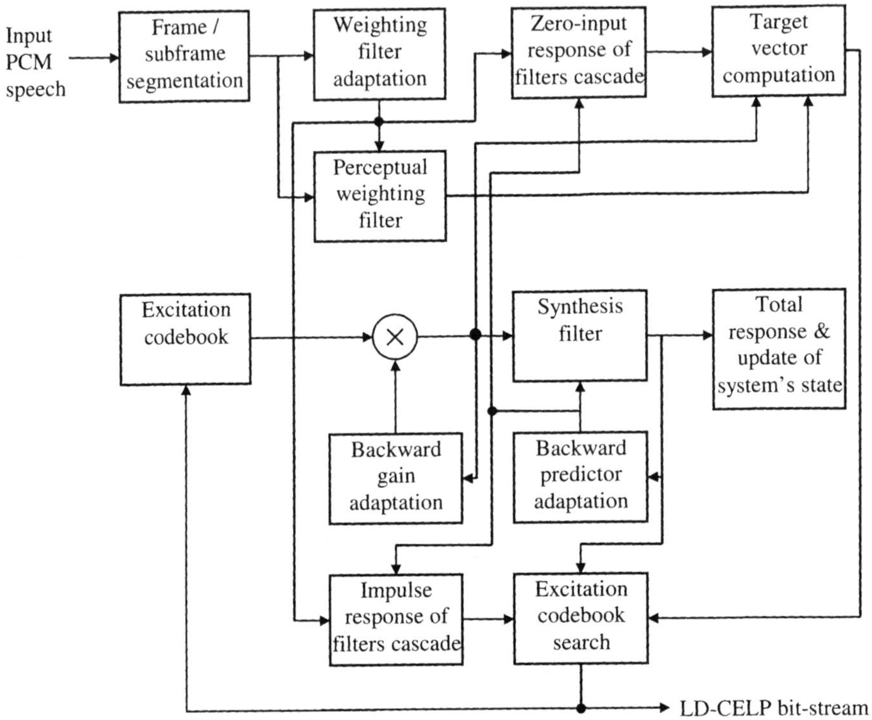

Figure 14.9 Block diagram of the G.728 encoder.

Encoder

Block diagram of the encoder appears in Figure 14.9. The input speech signal is passed through the perceptual weighting filter, with the output combined with the zero-input response and the gain to produce the target vector used during excitation codebook search. Prior to this step, the impulse response of the cascade filters is computed from the LPCs of these filters. The optimal codebook indices are transmitted as the bit-stream, which is the only information sent to the decoder.

Bit-Stream Synchronization

The LD-CELP bit-stream consists of 10 bits of the excitation index transmitted every 0.625 ms, leading to a bit-rate of 16 kbps. In practice, it is necessary for the decoder to know the boundaries of the received 10-bit codebook indices, so that the system states can be updated accordingly. Such synchronization information can be made available to the decoder by adding extra synchronization bits, leading to a bit-rate increase. Suppose that a rate increase is undesirable; it is still possible to send synchronization information without increasing the bit-rate.

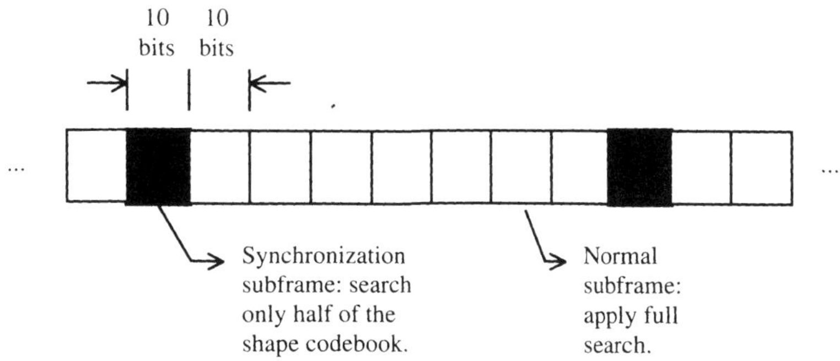

Figure 14.10 Illustration of a synchronization scheme where 1 bit is sent every 8 subframes.

Figure 14.10 illustrates one such scheme, where one synchronization bit is sent for a certain number of subframes. Since the coding algorithm has a basic adaptation cycle of four indices, it is reasonable to select a number of subframes in between synchronization that is a multiple of 4. The example in Figure 14.10 synchronizes once every eight subframes; in practice, synchronizing every 16 subframes is a good trade-off between robustness and quality.

Ten bits are sent every subframe, regardless of whether it is normal or not. If it is a normal subframe, the ten transmitted bits represent the index of the excitation codebook. Otherwise, one bit is allocated for synchronization (equal to 0 or 1), 6 bits for the shape codebook, and 3 bits for the gain codebook. Since the shape codebook is addressed by 7 bits, only half of the codevectors are searched during synchronization transmission. That is, the search range for the shape codebook is cut in half during synchronization. If this is done infrequently, such as once every 16 subframes, the speech quality degradation is essentially negligible. In this manner, synchronization is transmitted with no bit-rate modification.

14.7 CODEBOOK TRAINING

This section describes a training procedure to optimize the shape codebook and gain codebook of the LD-CELP coder originally outlined in Chen [1990]. Given the training data set

$$s_r[n]; \quad n = 0, \ldots, 4; \quad r = 0, \ldots, N_t - 1,$$

which represents an actual speech signal for training, with N_t being the size of the training data set (total number of five-sample subframes), the LD-CELP encoder can be used to encode the data set, resulting in the total error sum

$$D = \sum_{r=0}^{N_t - 1} \varepsilon_r^{(l_r, m_r)}, \tag{14.22}$$

which can be minimized by choosing the appropriate codebook contents. Note that the subscript r is introduced to index a subframe.

Optimal Shape Codevectors

Expanding (14.22) and using (14.3), we find

$$D = \sum_{r=0}^{N_t-1} \left\| \mathbf{t}_r - \mathbf{y}1^{(l_r, m_r)} \right\|^2 = \sum_{r=0}^{N_t-1} \left\| \mathbf{t}_r - \mathbf{H}_r \mathbf{v}^{(l_r)} g^{(m_r)} \right\|^2, \tag{14.23}$$

where $\mathbf{H}_r$ is the impulse response matrix for the cascade connection between the two filters. Differentiating the above equation with respect to $\mathbf{v}$, we find

$$\frac{\partial D}{\partial \mathbf{v}^{(l_r)}} = \sum_{r=0}^{N_t-1} -2\mathbf{t}_r \mathbf{H}_r g^{(m_r)} + 2\mathbf{H}_r \mathbf{H}_r^T \mathbf{v}^{(l_r)} \left(g^{(m_r)} \right)^2. \tag{14.24}$$

Equating to zero gives

$$\sum_{r=0}^{N_t-1} \mathbf{t}_r \mathbf{H}_r g^{(m_r)} = \sum_{r=0}^{N_t-1} \mathbf{H}_r \mathbf{H}_r^T \left(g^{(m_r)} \right)^2 \mathbf{v}^{(l_r)}. \tag{14.25}$$

The above equation indicates the requirements for the optimal shape codevectors $\mathbf{v}$.

Optimal Gain Codewords

Differentiating (14.23) with respect to g gives

$$\frac{\partial D}{\partial g^{(m_r)}} = \sum_{r=0}^{N_t-1} -2\mathbf{t}_r^T \mathbf{H}_r \mathbf{v}^{(l_r)} + 2g^{(m_r)} \left\| \mathbf{H}_r \mathbf{v}^{(l_r)} \right\|^2, \tag{14.26}$$

equating to zero results in the optimal gain codewords that minimize D:

$$g^{(m_r)} = \frac{\sum_{r=0}^{N_t-1} \mathbf{t}_r^T \mathbf{H}_r \mathbf{v}^{(l_r)}}{\sum_{r=0}^{N_t-1} \left\| \mathbf{H}_r \mathbf{v}^{(l_r)} \right\|^2}. \tag{14.27}$$

Optimization Procedure

The encoder is initialized appropriately and the signal subframes (from the training data set) are used as input to the encoder. As usual, all subframes are presented sequentially, the optimal indices are searched using the available excitation codebook, which can be initialized using random numbers.

Since there are a total of 128 possible values of l corresponding to 128 shape codevectors $\mathbf{v}^{(l)}$, and a total of 8 values of m or 8 different gain values, the relations for $\mathbf{v}$ (14.25) and g (14.27) must be accumulated separately and solved for the optimal codebook values at the end of the presentation of the whole training data set. This accumulation process is described by

$$\sum_{r:l_r=l} \mathbf{t}_r\mathbf{H}_r g^{(m_r)} = \sum_{r:l_r=l} \mathbf{H}_r\mathbf{H}_r^T \left(g^{(m_r)}\right)^2 \mathbf{v}^{(l_r)}; \quad l = 0, \ldots, 127, \tag{14.28}$$

$$g^{(m)} = \frac{\sum_{r:m_r=m} \mathbf{t}_r^T\mathbf{H}_r\mathbf{v}^{(l_r)}}{\sum_{r:m_r=m} \|\mathbf{H}_r\mathbf{v}^{(l_r)}\|^2}; \qquad m = 0, \ldots, 7. \tag{14.29}$$

The training procedure is summarized as follows. Each training subframe is presented sequentially to the encoder. Parameters of the encoder pertaining to the particular subframe are found. If $l_r = l$, the parameters of the rth subframe are used in the linear system described by (14.28) for the lth codevector. Similarly, if $m_r = m$, the parameters of the rth subframe are used in (14.29) for the mth codeword. After presenting all N_t subframes, a system of 128 equations results for the shape codebook; while for the gain codebook, a system of 8 equations is obtained. These equations are solved, resulting in a new excitation codebook. A new training epoch can start all over using the newly found codebooks. This process can be applied repeatedly up to the desired number of epochs. Using the described technique, Chen [1990] reported a SNR improvement on the order of 1 to 1.5 dB in actual LD-CELP coding using speech data outside the training set, with substantial gain in perceptual speech quality.

14.8 SUMMARY AND REFERENCES

In this chapter, the essence of the ITU-T G.728 LD-CELP coder is introduced. This coder achieves a one-way coding delay of less than 2 ms by using a short-length input buffer having only five samples, and by using backward adaptation for a 50th-order synthesis filter and the excitation gain. Speech quality is equivalent to or better than the 32-kbps G.726 ADPCM algorithm. Its unique design is also highly robust against channel errors without the help of any kind of forward error correction. Due to the lack of a pitch predictor, the G.728 can effectively model other signals besides speech with good accuracy; it has been shown that the coder is capable of passing all network information signals with little distortion.

Besides achieving the objective of low delay while maintaining toll quality speech at a relatively low bit-rate of 16 kbps, the parameters of the coder are selected carefully to facilitate fixed-point implementation; also, the different computational cycles are planned prudently to maintain a balanced load. These are important factors for the coder to be practicable. The level of complexity for the G.728 is higher than many low bit-rate coders: a trade-off is made to accomplish low delay. However, with the constant speed-up of microprocessors, the extra complexity will become less of an issue in the near future.

See Chen [1990] for inceptive ideas on LD-CELP, and a description of excitation codebook training. Fixed-point implementation issues are discussed in Chen et al. [1990] and Chen [1991]. A thorough description of the development and features of LD-CELP appears in Chen et al. [1991, 1992]. See Chen and Rauchwerk [1993] for the description of an 8-kbps LD-CELP coder. More updated surveys of low-delay coders and additional descriptions appear in Chen [1995], where the prime techniques are extended to the development of lower bit-rate coders, at 8, 6.5, and 4.67 kbps. Details of postfilter design are available in Chen and Gersho [1995]. The official standard is published in ITU [1992].

EXERCISES

14.1 Are the following statements true? Justify your answers.

- Minimum coding delay of a PCM system is equal to 0.125 ms.
- Minimum coding delay of the ETSI GSM 6.10 RPE-LTP coder (Chapter 10) is 40 ms.
- Minimum coding delay of the FS1016 CELP coder (Chapter 12) is 75 ms.
- Minimum coding delay of the TIA IS54 VSELP coder (Chapter 13) is 40 ms.

It is assumed that the encoder is transmitting in constant mode, with no processing delays.

Now, assume that the encoder is transmitting in *burst* mode; that is, all available bits of the compressed bit-stream for the frame are sent immediately to the decoder once encoding is completed. Recalculate the minimum coding delay for the above four cases.

14.2 Several multiplicative constants of the LD-CELP coder are specified as a fraction, like $\lambda = 257/256$ and $\gamma = 253/256$. What is the advantage of such numerical representation? *Hint:* The denominator is a power of 2.

14.3 During the excitation codebook search, the zero-state response $\mathbf{y1}^{(l,m)}$ must be evaluated for all possible combinations between l and m. For each value of l, $\mathbf{v}^{(l)}$ is filtered by the cascade filters; the resulting sequence is multiplied by one of the eight possible gain values to generate $\mathbf{y1}^{(l,0)}$ to $\mathbf{y1}^{(l,7)}$. This requires a total of 128 convolutions, plus $128 \cdot 8 = 1024$ products between a scalar and a vector to find all 1024 responses. It is possible to save computation by using a different configuration for the excitation codebook. Instead of separating into shape and gain, one of the gain values can actually be embedded into the shape codebook. For instance, the vectors in the shape codebook are now comprised by

$$g^{(0)}\mathbf{v}^{(l)};$$

then the gain codebook needs only seven values:

$$g^{(1)}/g^{(0)},\ g^{(2)}/g^{(0)},\ \ldots,\ g^{(7)}/g^{(0)}.$$

Study the proposition, count the total number of products, and compare to the original.

14.4 Design a bit-stream synchronization method for the G.728 coder based on the following features:

- One synchronization bit is transmitted every 16 subframes.
- Value of the synchronization bit is equal to one.

Take into account all possible real-world situations such as the occasional loss of synchronization.

14.5 For the codebook training technique described in Section 14.7, write down the step-by-step procedure with the assumption that the gain $g^{(m)}$ is known. That is, only the shape codevectors are modified through training.

14.6 Contrast the codebook training technique described in Section 14.7 with the regular GLA. Is it reasonable to assume that the total error sum is monotonically decreasing? Propose some stopping criterion to manage the training process.

CHAPTER 15

VECTOR QUANTIZATION OF LINEAR PREDICTION COEFFICIENT

Most speech coders standardized after 1994 utilize some sort of VQ for the LPC. This is because of the superior performance when compared with scalar quantization. Practical deployment of VQ is limited mainly by its inherent high complexity, and many structured schemes are often deployed. The methods described in this chapter deal exclusively with line spectral frequencies, which is the LPC representation of choice due to many favorable features as studied in Chapter 8.

The chapter begins with an investigation on interframe and intraframe correlation present among the LSF vectors; it is shown that the elements of the LSF vectors have a great deal of redundancy and can be explored for more efficient quantization. Three major techniques developed for LSF VQ, namely, split VQ, MSVQ, and PVQ, are described; very often in practice, these schemes are combined together to take advantage of the strength that each individual method provides. Detailed implementations of four LSF quantizers incorporated in various speech coding standards are given.

15.1 CORRELATION AMONG THE LSFs

The LSFs extracted from speech frames typically display a high level of redundancy, manifested as a dependency between elements of a frame (intraframe), as well as elements from different frames (interframe). Figure 15.1 shows the LSFs obtained from a female speech sequence, where tenth-order LP analysis is performed on 160-sample frames extracted with a Hamming window; a factor of 0.996 is applied for bandwidth expansion (Chapter 4). The data are gathered for a total of 174 frames. Note the dependency between LSF values of consecutive

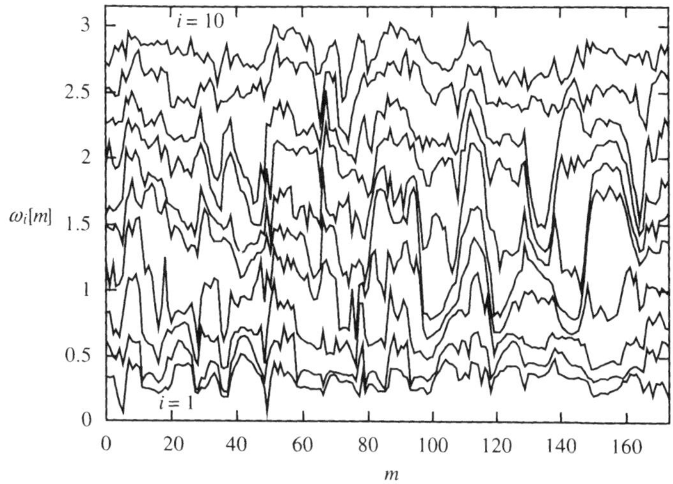

Figure 15.1 An example plot of LSFs obtained from consecutive frames of speech material.

frames, as well as LSF data within the same frame. Therefore, an encoding system that makes use of the correlation between elements will result in improved performance over those that do not consider these properties.

Since the means of the LSFs represent redundant information, they are often removed before using a prediction system. The mean in the present context refers to the time average of the LSFs values when they are concatenated frame-after-frame; its value can be estimated by analyzing a large amount of data, or by applying some time-adaptive scheme. Using the same data as in Figure 15.1, the means of the LSFs are estimated to be

$$m_1 = 0.291, \quad m_2 = 0.564, \quad m_3 = 0.851, \quad m_4 = 1.146, \quad m_5 = 1.444,$$
$$m_6 = 1.736, \quad m_7 = 2.033, \quad m_8 = 2.312, \quad m_9 = 2.595, \quad m_{10} = 2.882.$$

In some coders, the means are simply assumed to be

$$m_i = \pi.i/11, \quad i = 1 \text{ to } 10. \tag{15.1}$$

Outcomes of (15.1) are actually very close to the previous estimates obtained from real data and usually are satisfactory for practical applications.

To develop a correlation measure for the LSFs, consider the values

$$\omega_i[m]; \quad i = 1 \text{ to } 10; \quad m = 0, \ldots, N_t - 1$$

corresponding to a collection of LSF samples extracted from a speech source frame-after-frame; a total of N_t frames are considered. Then the mean-removed LSFs are

$$v_i[m] = \omega_i[m] - m_i. \tag{15.2}$$

The normalized correlation measure is defined by

$$R[i,j,k] = \frac{\sum_{m=0}^{N_t-k-1} v_i[m]v_j[m+k]}{\sqrt{\sum_{m=0}^{N_t-k-1}(v_i[m])^2 \sum_{m=0}^{N_t-k-1}(v_j[m+k])^2}}; \quad i,j = 1 \text{ to } 10; \quad k = 0,1,2,\ldots. \tag{15.3}$$

In (15.3), the correlation is measured between the ith LSF of a certain frame, with respect to the jth LSF k frames ahead. Thus, use $k = 0$ for intraframe measurement, and $k \neq 0$ for interframe calculation. Figure 15.2 shows some results on applying (15.3) to the data of Figure 15.1. For the case of $R[1,j,k]$, note how the correlation tends to be high for j near 1 and decreases as the index increases. Also, correlation is the highest for $k = 0$ and decreases as k increases. Similar observations are found for $R[6,j,k]$. These are expected results since the LSFs tend to have strong correlation with their neighbors that are close in time or frequency. Figure 15.3 shows additional results, where $R[1,j,k]$ is plotted as a function of k, with four values of j.

Researchers have used various schemes to explore the correlation among the LSFs so as to design more effective quantization systems. For instance, VQ can be used to effectively represent the LSF vector in a frame, hence taking advantage of the intraframe correlation. Moreover, predictive VQ can be deployed to explore the presence of interframe correlation. The former is often known as *memoryless*, since each LSF vector is quantized independently; while the latter is referred to as

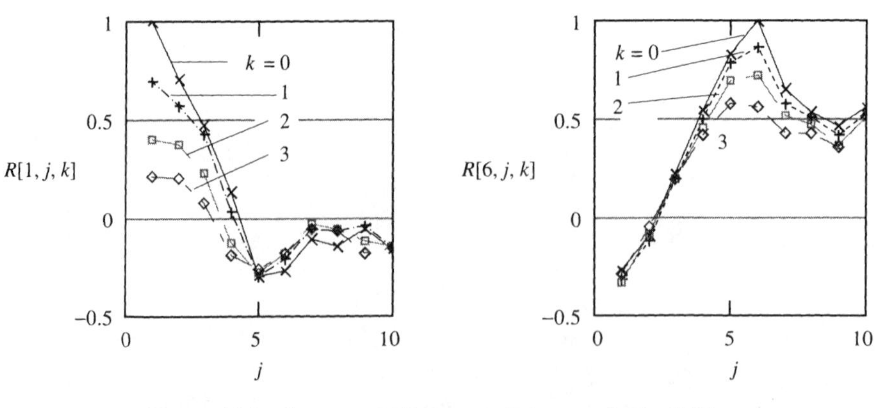

Figure 15.2 Some normalized correlation results for LSFs.

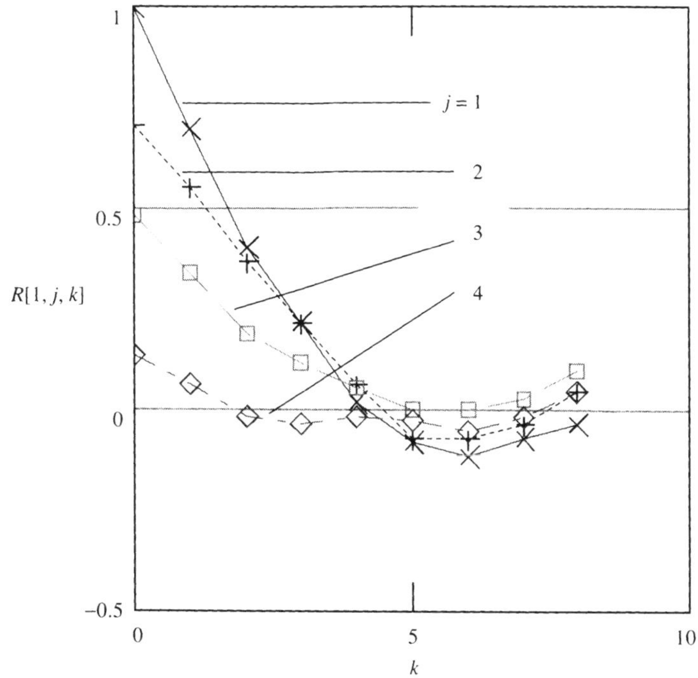

Figure 15.3 Additional normalized correlation results for LSFs.

memory-based: quantization of the current LSF vector depends on past outcomes. The strategies applied in practice are the topics for the rest of the chapter.

15.2 SPLIT VQ

Vector quantization (VQ) treats the whole set of LPCs as a vector, and distortion is minimized by finding the best vector, that is, the best combination of individual coefficients. Thus, the use of VQ is far more efficient than scalar quantization, especially for the case of LPCs, where transparent quantization is defined using spectral distortion—a measure dependent on the entire set of coefficients.

One of the earliest works in this area is by Buzo et al. [1980], where superiority of VQ over scalar quantization is demonstrated. Many variants have been proposed since then. Due to the fact that VQ is inherently more expensive to implement, most studies focused on cost reduction. Paliwal and Atal [1993] proposed the use of split VQ (Chapter 7) and showed that the resultant scheme outperformed the existent scalar quantization techniques, with moderate computational costs. Highlights of their work are presented in this section.

Distance Measure

Selection of a proper distance measure is the most important issue in the design and operation of a vector quantizer. Since transparent quantization of LPCs is defined using spectral distortion, the ideal distance measure in a VQ system is, obviously, the spectral distortion itself. However, due to the fact that SD is relatively complex to implement (Chapter 8), it is not suitable for direct deployment.

One of the simplest distance measures for the LSF vector $\mathbf{x}^T = [\omega_1, \omega_2, \ldots, \omega_M]$ and its quantized version is the squared Euclidean distance

$$d(\mathbf{x}, \hat{\mathbf{x}}) = (\mathbf{x} - \hat{\mathbf{x}})^T (\mathbf{x} - \hat{\mathbf{x}}) = \sum_{i=1}^{M} (\omega_i - \hat{\omega}_i)^2. \tag{15.4}$$

The squared Euclidean distance, however, is not optimum with respect to SD, in the sense that SD might not diminish even though the squared Euclidean distance is reduced. Of course, in the extreme case of zero Euclidean distance, SD is equal to zero as well. Due to the simplicity of (15.4) it is tempting to apply it in practice. The suboptimality of this distance measure is demonstrated in the following example.

Example 15.1 The relation between SD and squared Euclidean distance is shown here for a practical example. Using a fixed LSF vector as a reference, 1000 randomly generated LSF vectors are used to calculate two distances with respect to the reference vector. The first distance is the squared Euclidean and the second is

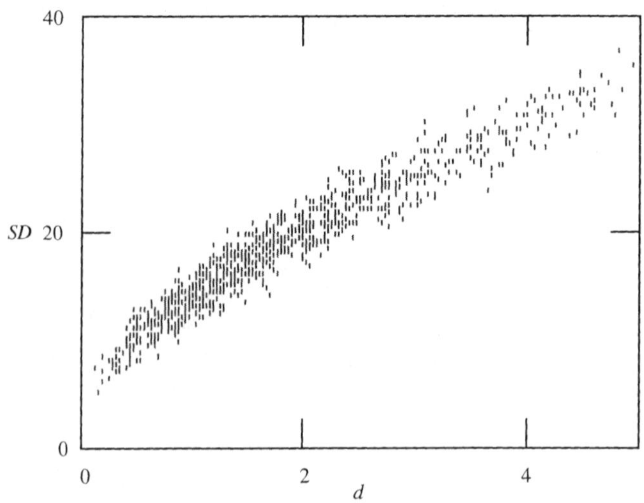

Figure 15.4 Spectral distortion (SD) versus squared Euclidean distance (d) for 1000 randomly generated LSF vectors.

the SD. In the computation of SD, the LSF vectors are transformed to the corresponding LPC vectors, and the spectral distortion is found using 200 sampling points. The results are plotted in Figure 15.4. In general, a small Euclidean distance corresponds to low SD. However, for a given Euclidean distance, a wide range of SD exists and vice versa. Thus, Euclidean distance is not optimum when minimization of SD is required, since its reduction does not necessarily lower the SD.

Squared Euclidean Distance Measure with Weighting

To improve the distance measure so as to capture more spectral information, a weighting scheme can be applied as follows:

$$\begin{aligned} d(\mathbf{x}, \hat{\mathbf{x}}) &= (\mathbf{x} - \hat{\mathbf{x}})^T \mathbf{W} (\mathbf{x} - \hat{\mathbf{x}}) \\ &= \sum_{i=1}^{M} w_i (\omega_i - \hat{\omega}_i)^2, \end{aligned} \tag{15.5}$$

where $\mathbf{W}$ is a diagonal matrix with the diagonal elements given by w_i; these elements might depend on $\mathbf{x}$. By selecting the weights w_i appropriately, it is possible to obtain a more optimized distance measure for the purpose of LSF quantization. In Paliwal and Atal [1993], the following weights are proposed:

$$w_i = \begin{cases} S(e^{j\omega_i})^{0.3}, & 1 \le i \le 8, \\ 0.64[S(e^{j\omega_i})]^{0.3}, & i = 9, \\ 0.16[S(e^{j\omega_i})]^{0.3}, & i = 10. \end{cases} \tag{15.6}$$

Note that $S(e^{j\omega})$ is the PSD defined by the set of LSFs $\{\omega_i\}$, and a system with a prediction order equal to ten is considered (Chapter 3). Hence, the weights $\{w_i\}$ vary with each LSF vector $\mathbf{x}$.

In this weighted Euclidean distance measure, the weight assigned to a given LSF is proportional to the PSD at a frequency value equal to that of the LSF. Thus, more weight is put on the peak of the PSD, which is important since the spectrum peaks play a higher role in auditory perception. Also note that for the last two weights ($i = 9$ and 10), fractional constants are applied so as to diminish their contributions toward the overall distance. Since the human ear cannot resolve differences at high frequencies as accurately as at low frequencies, more emphasis is put on the low- to mid-frequency regions. Therefore, the distance measure gives more weight to the LSFs corresponding to the high-amplitude formants than to those corresponding to the low-amplitude formants; the LSFs corresponding to the valleys in the PSD get the least weight. It is important to note that the weighting improves the perceptual performance; it does not necessarily reduce the spectral distortion.

Design of Split VQ

Many variables exist in a split VQ scheme. These variables are determined by answering some questions about what to expect from the system. Some of these can be answered by common sense, while others need experimentation. Questions and answers are recorded below.

- How many subvectors should be split?

  In split VQ, the input vector is divided into a number of subvectors, with each subvector quantized separately using regular VQ. The higher the number of subvectors, the lower the computational cost; the price to pay is a degradation in quality. A good trade-off for LSF quantization is to split into two subvectors.

- How many elements per subvector should there be?

  The input vector can be split into two parts in many ways: for instance, separately quantizing the first four elements and the last six elements, referred to as the (4, 6) scheme. Experimentally, it was found that the (4, 6) structure outperformed (3, 7), (5, 5), and (6, 4).

- How many bits per vector should there be?

  Experimentally, transparent quantization is achieved using 24 bits per vector together with the weighted Euclidean distance measure (15.5), where each subvector is encoded using 12 bits. The bit requirement is indeed a big leap forward when compared to traditional scalar quantizers (Chapter 8), which typically consume more than 30 bits/frame in order to achieve transparent quantization.

- How are the codebooks searched?

  The two codebooks in a (4, 6) scheme can be searched in many different ways. Full search, for instance, evaluates all possible combinations of subvectors from the two codebooks—a rather expensive proposition. By proceeding in a suboptimal sequential manner, the computational cost is greatly reduced. One approach is to search the first codebook in order to locate the first quantized subvector; this is done using the weighted Euclidean distance with the first four elements alone. Then the second codebook is searched so as to locate the second quantized subvector, this time using all 10 LSFs in the distance measure, with the first four elements fixed. To preserve the ordering of the LSFs, only those codevectors from the second codebook that satisfy $\omega_4 < \omega_5$ are considered. In this way, stability is guaranteed for the filter associated with the quantized LSF vector.

The split VQ scheme described here has influenced and inspired the design of LPC quantizers in generations to come. Several of these quantizers are described in the next sections.

15.3 MULTISTAGE VQ

MSVQ, as presented in Chapter 7, is a reduced-complexity suboptimal approach to VQ. Here, we study the work presented in LeBlanc et al. [1993], where MSVQ of the LSF is considered. The technique is included as part of the FS MELP standard (Chapter 17).

Codebook Design

Chapter 7 already discussed codebook design in MSVQ under squared Euclidean distance measure. Using the same notation, we seek to design the codebooks so as to minimize the distance sum

$$D = \sum_k (\mathbf{x}_k - \hat{\mathbf{x}}_k)^T \mathbf{W}_k (\mathbf{x}_k - \hat{\mathbf{x}}_k), \tag{15.7}$$

with $\mathbf{W}_k$ the diagonal weighting matrix associated with $\mathbf{x}_k$. Expanding (15.7), we find

$$\begin{aligned} D &= \sum_k (\mathbf{x}_k - \mathbf{B}_k\mathbf{y})^T \mathbf{W}_k (\mathbf{x}_k - \mathbf{B}_k\mathbf{y}) \\ &= \sum_k \mathbf{x}_k^T \mathbf{W}_k \mathbf{x}_k - 2\mathbf{y}^T \sum_k \mathbf{B}_k^T \mathbf{W}_k \mathbf{x}_k + \mathbf{y}^T \left(\sum_k \mathbf{B}_k^T \mathbf{W}_k \mathbf{B}_k \right) \mathbf{y}. \end{aligned} \tag{15.8}$$

Let's define

$$\mathbf{v} = \sum_k \mathbf{B}_k^T \mathbf{W}_k \mathbf{x}_k, \tag{15.9}$$

$$\mathbf{Q} = \sum_k \mathbf{B}_k^T \mathbf{W}_k \mathbf{B}_k, \tag{15.10}$$

$$D_o = \sum_k \mathbf{x}_k^T \mathbf{W}_k \mathbf{x}_k. \tag{15.11}$$

Substituting (15.9), (15.10), and (15.11) into (15.8) gives

$$D = D_o - 2\mathbf{y}^T\mathbf{v} + \mathbf{y}^T\mathbf{Q}\mathbf{y}. \tag{15.12}$$

Thus, the expression for the distortion takes the same form as in Chapter 7, and the design procedure can be applied with no modification.

One of the concerns in VQ of the LSF is the *outliers*—input vectors that are poorly represented in the training sequence and are quantized with a spectral distortion much larger than the average. The outlier performance can be improved significantly by appropriately weighting the distortion measure during training. The weighting matrix $\mathbf{W}_k$ in (15.7) is substituted by $\mathbf{W}'_k$, defined by

$$\mathbf{W}'_k = f(SD_k)\mathbf{W}_k, \tag{15.13}$$

where SD_k is the spectral distortion between $\mathbf{x}_k$ and $\hat{\mathbf{x}}_k$, and $f(\cdot)$ is a mapping function that can be selected according to a particular situation. One such function is

$$f(SD) = \begin{cases} 1, & SD \leq 1, \\ 1 + \lceil \eta(SD - 1) \rceil, & SD > 1, \end{cases} \tag{15.14}$$

with η a positive constant and $\lceil \cdot \rceil$ the ceiling function (returns the nearest integer greater than or equal to the enclosed variable). Therefore, the mapping function produces no changes to the weighting matrix for low *SD*. For $SD > 1$, however, the magnitudes of the original weights are augmented by a factor greater than one. Experimentally, it was found that the average *SD* was not disturbed significantly, but the percentage of outliers was decreased substantially [LeBlanc et al., 1993]. The typical range of η is between 2 and 100. For small η, reduction of the number of outliers is not very pronounced. For more effective control of the number of outliers, a higher value of η should be used. An excessively high value of η, however, can degrade the average performance of the final design.

Using MSVQ of the LSF parameters, with the codebooks designed with the explained procedures, it was found that the subjective performance, using 24 bits/frame, is equivalent to that of the LSF scalar quantizer for the FS1016 coder, which uses 34 bits/frame (Chapter 8). Thus, a much higher compression ratio is achieved. It is this superior performance that led to the inclusion of the MSVQ technique into the FS MELP standard. The reader must not forget that high performance is obtained with increasing implementational cost. However, cost associated with MSVQ is well within the reach of modern DSPs.

Other Implementational Issues

Let's ignore the case of MSVQ for a moment and consider an LSF-based LPC quantization system using optimal VQ. Figure 15.5 (*top*) shows the block diagram of such a system. The index output of the VQ encoder is transmitted to the VQ decoder, where a particular codevector, representing the quantized LSF vector is recovered. This quantized LSF vector is transformed back to LPC so as to yield the quantized LPC vector. Since all possible LSF codevectors are stored in the codebook of the VQ decoder, the LSF-to-LPC block can be eliminated by converting the LSF codevectors to LPC, and using the output index of the VQ encoder to directly recover a LPC vector, thus eliminating the computation associated with LSF-to-LPC conversion.

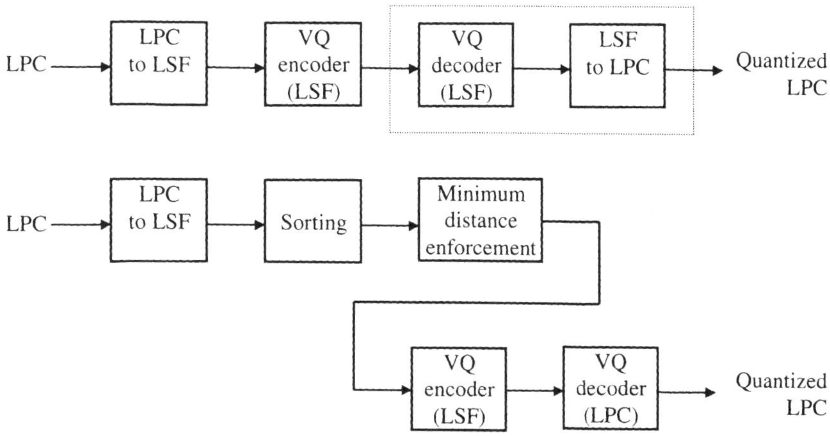

Figure 15.5 LSF-based LPC quantization using optimal VQ (*top*). Same system with increased robustness (*bottom*).

Assuming that the input LPC vectors are associated with stable synthesis filters, the corresponding LSF vectors have their elements sorted in ascending order (interlacing property, Chapter 8). This property is generally true if the numerical precision applied for the LPC-to-LSF conversion procedure is sufficiently high. Very often, numerical precision in practice is not what one desires due to resource limitation; for instance, one might not be able to use a floating-point processor due to the low price target of a product. Instead, all computation must be done using a cheaper fixed-point processor; in this case, accumulated errors during conversion may lead to LSF vectors with unsorted elements. Quantizing these LSF vectors using a VQ designed for sorted LSF vectors might lead to excessive quantization errors.

A low-complexity solution to the described problem is to deploy the system shown in Figure 15.5 (*bottom*). As we can see, a sorting block is incorporated to ensure that the LSF vectors are sorted in ascending order. These vectors are then passed to a minimum distance enforcement block, with the purpose of separating two elements that are too close to each other. That is, if the distance between any two elements is less than a certain threshold, they are modified in such a way that their distance is equal to the pre-established threshold. In this way, all input LSF vectors to the VQ encoder will satisfy the desired property.

The systems described in Figure 15.5 are based on an optimal VQ, where the codevectors are stored in one single codebook addressed by an index. This approach is computationally expensive and, in most instances, impractical. Let's go back into our MSVQ discussion. Figure 15.6 shows such a system: besides the preprocessing blocks prior to entering the encoder, the same blocks are deployed after the decoder. This is necessary since, for MSVQ, it is in general not possible to guarantee the ordering of the quantized vector; in fact, codevectors at the second stage and beyond are normally zero-mean vectors since they represent a population of error

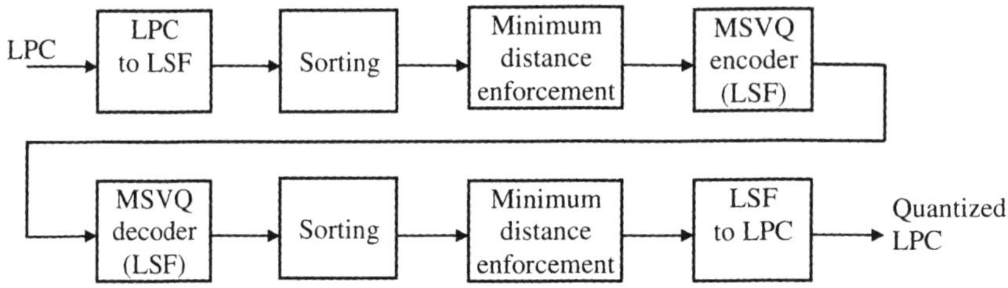

Figure 15.6 LSF-based LPC quantization using MSVQ.

vectors—the difference between input vectors and first-stage codevectors. On the other hand, codebook search in MSVQ is usually performed sequentially on a stage-by-stage basis. Based on the above factors, ordering and minimum distance between elements of the quantized vectors cannot be attained in general. Thus, the quantized LSF vectors at the output of the MSVQ decoder must be sorted and minimum distance maintained before converting back to LPC. In this way, the final LPC vectors are warranted to lead to stable synthesis filters.

LPC Quantization and Interpolation of the FS MELP Coder

This coder utilizes a frame length of 180 samples (22.5 ms); however, for LP analysis, a 200-sample (25 ms) Hamming window centered on the last sample in the current frame is used (Figure 15.7). A tenth-order LP analysis is performed. The LPCs are first transformed to LSFs, which are sorted (see Exercise 15.3 for ideas on sorting algorithms), and minimum distance between elements is enforced, where the minimum separation is equal to 50 Hz (see Exercise 15.4 for a suggested algorithm). The resulting vectors are quantized using MSVQ. The MSVQ codebooks consist of four stages of 128, 64, 64, and 64 codewords, respectively, resulting in a total of 25 bits/frame.

The MSVQ finds the codevectors that minimize the weighted squared Euclidean distance ((15.5) and (15.6)). Tree search is applied where the $M_a = 8$ best

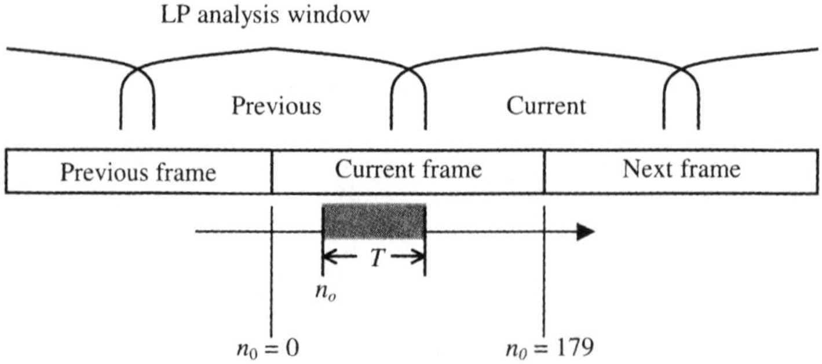

Figure 15.7 Positions of frame and LP analysis windows for the FS MELP coder.

codevectors from each stage are saved for use by the next stage. On the decoder side, the recovered LSF vectors are sorted and minimum distance is enforced; these vectors are then transformed to become the quantized LPC vectors.

During decoding for speech synthesis, the signal is generated on a pitch-period-by-pitch-period basis. The pitch period is determined from the input signal. For the pitch period that starts at time instant n_o, where $0 \leq n_o < 180$, the interpolation factor α is given by

$$\alpha = n_o/180, \tag{15.15}$$

and the LSF for the particular pitch period is obtained by linear interpolation with

$$\omega = (1 - \alpha)\omega_{\text{previous}} + \alpha\omega_{\text{current}}, \tag{15.16}$$

where ω_{previous} and ω_{current} represent the LSFs of the previous and current frames, respectively. Other parameters in the MELP model are interpolated in a similar manner. See Chapter 17 for more details.

15.4 PREDICTIVE VQ

As is evidenced from Section 15.1, LSF vectors from consecutive frames exhibit a high level of correlation, which can be explored efficiently using PVQ to improve the performance of LSF quantization. In this section, three PVQ-based schemes are described and are all part of standardized coders. The description is not limited to quantization, but also windowing and interpolation.

ITU-T G.723.1 MP-MLQ / ACELP

This coder utilizes a 30-ms (240-sample) frame divided into four 7.5-ms (60-sample) subframes. For each subframe, a Hamming window of length 180 centered on the subframe is applied; eleven autocorrelation coefficients are computed from the windowed data. A white noise correction factor of 1025/1024 is applied followed by spectral smoothing; LPCs are found by Levinson–Durbin conversion, which are bandwidth expanded. Thus, for every input frame, four sets of LPCs are computed, one for every subframe. The situation is illustrated in Figure 15.8. Among the four

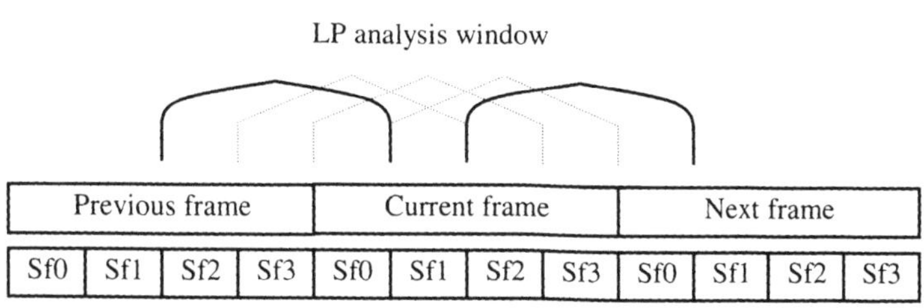

Figure 15.8 Positions of frame and LP analysis windows for the G.723.1 coder.

sets of LPCs that are obtained from a given frame, only the one corresponding to the last subframe is quantized and transmitted. The other three sets of LPCs associated with the first three subframes are only used for encoding operations, such as perceptual weighting.

LSF Quantization: PVQ with Split Structure

A block diagram of the LSF quantizer appears in Figure 15.9. The input LSF vector $\mathbf{x}[m]$ of the mth frame is first mean-removed:

$$\mathbf{v}[m] = \mathbf{x}[m] - \mathbf{m}, \tag{15.17}$$

leading to the zero-mean LSF vector $\mathbf{v}[m]$. Given the prediction $\mathbf{p}[m]$, the prediction-error vector is

$$\mathbf{e}[m] = \mathbf{v}[m] - \mathbf{p}[m], \tag{15.18}$$

with the prediction given by

$$\mathbf{p}[m] = \frac{12}{32}\hat{\mathbf{v}}[m-1]; \tag{15.19}$$

that is, it is a fixed first-order predictor that relies on the quantized mean-removed LSF vector to form the prediction. The prediction-error vector is quantized using split VQ, where the vector is split into three subvectors with dimension 3, 3, and 4.

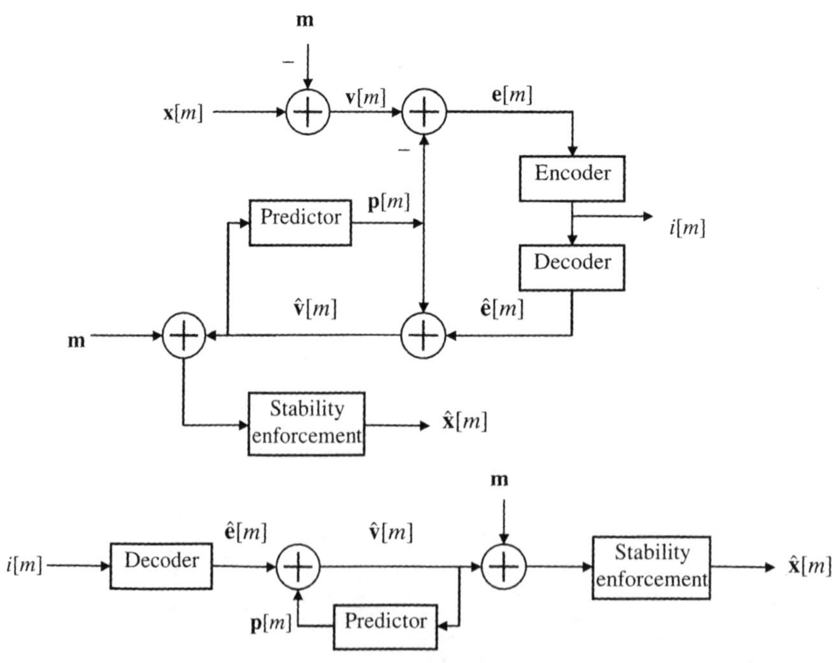

Figure 15.9 Encoder (*top*) and decoder (*bottom*) of the G.723.1 LSF quantizer.

The subvectors are quantized with three codebooks having the same size of 256, each addressed by 8 bits leading to a total of 24 bits/frame.

The objective of quantization is to locate the optimal codebook indices so that the weighted distance measure (15.5) is minimized. The weights are given by

$$w_i = \begin{cases} (\omega_2 - \omega_1)^{-1}, & i = 1, \\ (\min(\omega_i - \omega_{i-1}, \omega_{i+1} - \omega_i))^{-1}, & i = 2 \text{ to } 9 \\ (\omega_{10} - \omega_9)^{-1}, & i = 10. \end{cases} \tag{15.20}$$

Thus, more emphasis is placed on those LSFs that are situated close to each other. By utilizing these weights, sharp peaks in the spectrum can be represented more faithfully since they tend to have close-by LSFs. To enforce stability, the following procedure is applied:

```
1.  counter ← 0
2.  while counter < 10
3.      flag ← 0
4.      for i ← 1 to 9
5.          if ωi+1 − ωi < dmin
6.              avg ← (ωi + ωi+1)/2
7.              ωi ← avg − dmin/2
8.              ωi+1 ← avg + dmin/2
9.              flag ← 1
10.     counter++;
11.     if flag = 0 break
```

In the above pseudocode, a minimum distance of *dmin* is enforced between adjacent LSF values, with the official value being $dmin = 31.25$ Hz. The distance enforcement loop (Line 4 to Line 9) is entered a maximum of ten times. If the loop is passed with no modification to the LSF elements, the procedure is terminated (Line 11). If the LSF elements are still changed after the tenth exit to the loop, the LSF vector is considered invalid. The variable *flag* can be returned at the end of the procedure as a signal to the status of the operation.

During actual LSF decoding, if a decoded vector is invalid, it is discarded and replaced by the vector from the last frame.

LSF Interpolation

Linear interpolation is performed between the decoded LSF vector of the current frame ($\mathbf{x}$) and the past frame ($\mathbf{x}_{\text{past}}$) so as to obtain a different LSF vector for each subframe. This is done as follows:

$$\mathbf{x}_0 = 0.75\mathbf{x}_{\text{past}} + 0.25\mathbf{x}, \tag{15.21}$$

$$\mathbf{x}_1 = 0.5\mathbf{x}_{\text{past}} + 0.5\mathbf{x}, \tag{15.22}$$

$$\mathbf{x}_2 = 0.25\mathbf{x}_{\text{past}} + 0.75\mathbf{x}, \tag{15.23}$$

$$\mathbf{x}_3 = \mathbf{x}, \tag{15.24}$$

with $\mathbf{x}_0, \mathbf{x}_1, \mathbf{x}_2$, and $\mathbf{x}_3$ the LSF vectors of subframes 0, 1, 2, and 3, respectively.

ITU-T G.729 ACELP

This coder utilizes a 10-ms (80-sample) frame divided into two 5-ms (40-sample) subframes. A tenth-order LP analysis procedure is performed once per frame, with the autocorrelation values extracted using the asymmetric window:

$$w[n] = \begin{cases} 0.54 - 0.46\cos\left(\frac{2\pi n}{399}\right); & n = 0, \ldots, 199, \\ \cos\left(\frac{2\pi(n-200)}{159}\right); & n = 200, \ldots, 239. \end{cases} \tag{15.25}$$

There is a 5-ms look-ahead in LP analysis, meaning that 40 samples are needed from the future speech frame. Figure 15.10 illustrates the situation. The windowed samples are used to compute the autocorrelation values $R[l], l = 0, 1, \ldots, 10$. To avoid numerical problems for low-level input signals, the value of $R[0]$ has a low boundary of $R[0] = 1$.

Spectral smoothing is applied with the window

$$w_{ss}[l] = e^{-(2\pi 60 l/8000)^2/2}; \quad l = 1, \ldots, 10, \tag{15.26}$$

followed by white noise correction with a factor of 1.0001 (Chapter 4). The Levinson–Durbin algorithm is applied to solve for the LPCs.

Use of an asymmetric window coupled with a relatively short frame length of 10 ms is mainly motivated by coding delay issues. A shorter frame length is translated directly into lower delay. An asymmetric window with more weighting on the samples of the last subframe in the current frame implies less look-ahead during autocorrelation computation; that is, the need to buffer future samples is reduced, minimizing coding delay at the same time.

LSF Quantization: PVQ-MA with Switched Predictor and Multistage Split Structure

The G.729 coder utilizes 18 bits/frame to encode the LSF. The scheme is based on a combination of MSVQ, split VQ, and PVQ-MA with switched predictor. To quantize the prediction-error vectors within the framework of PVQ-MA, a two-stage

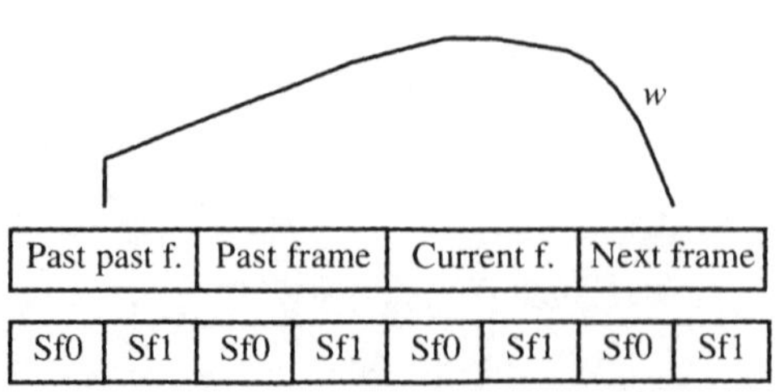

Figure 15.10 Positions of frame and LP analysis windows for the G.729 coder.

Table 15.1 Bit Allocation for LSF Quantization of the G.729 Coder[a]

Index	Function	Number of Bits
i_0	Predictor selection	1
i_1	First-stage codebook (10-D)	7
i_2	Second-stage codebook—low part (5-D)	5
i_3	Second-stage codebook—high part (5-D)	5

[a] Data from ITU [1996a], Table 8.

MSVQ is utilized; the first stage consists of a ten-dimensional quantizer, while the second stage is implemented as a split VQ with five-dimensional codebooks. Another important feature is the use of two different predictors, selected during encoding. Bit allocation of the quantizer is summarized in Table 15.1.

The switched prediction scheme allows the system to better adapt to different signal conditions, which is desirable since the LSFs display strong correlation during stationary segments, but mild correlation during transitions. To limit propagation of channel errors, the predictor is of the MA type. Prediction order was determined empirically and it was found that a fourth-order MA predictor forms a good compromise between performance and error control [Kataoka et al., 1994]. The use of a multistage and split VQ configuration effectively reduces implementational costs. The first stage is ten-dimensional, allowing exploitation of correlation among all vector components; at the second stage, these correlations are reduced, and the residue vector from the first stage is quantized efficiently using a split structure.

For comparison between residue vector from the first stage LSF vectors, the weighted Euclidean distance (15.5) is utilized, with

$$w_i = \begin{cases} \rho_i; & \text{if} \quad \omega_{i+1} - \omega_{i-1} - 1 > 0, \\ 10\rho_i(\omega_{i+1} - \omega_{i-1} - 1)^2 + 1; & \text{otherwise,} \end{cases} \tag{15.27}$$

for $i = 1$ to 10, and

$$\rho_i = \begin{cases} 1.2, & \text{if} \quad i = 5, 6, \\ 1.0, & \text{otherwise,} \end{cases} \tag{15.28}$$

with $\omega_0 = 0.04\pi$ and $\omega_{11} = 0.92\pi$.

During encoding and decoding, the following procedure is often invoked so as to maintain a minimum separation between the elements of the LSF vector. The minimum distance *dmin* is passed as a parameter to the procedure.

```
MIN_DISTANCE(ω1, ..., ω10, dmin)
for i ← 2 to 10
    if (ωi-1 > ωi - dmin)
        ωi-1 ← (ωi + ωi-1 - dmin)/2
        ωi ← (ωi + ωi-1 + dmin)/2
```

Decoding the LSF

Given the indices $i_0[m], i_1[m], i_2[m]$, and $i_3[m]$ of the mth frame, the vector

$$\hat{\mathbf{e}}[m] = \begin{bmatrix} \mathbf{y}1l_{i_1[m]} + \mathbf{y}2l_{i_2[m]} \\ \mathbf{y}1h_{i_1[m]} + \mathbf{y}2h_{i_3[m]} \end{bmatrix} \tag{15.29}$$

is constructed. It represents a scaled version of the quantized prediction-error vector. The vector $[\mathbf{y}1l^T|\mathbf{y}1h^T]^T$ is the 10-D codevector of the first-stage codebook, with $\mathbf{y}1l$ and $\mathbf{y}1h$ 5-D vectors. On the other hand, $\mathbf{y}2l$ and $\mathbf{y}2h$ denote the codevectors of the second-stage codebook, the low part and high part, respectively.

Minimum distance is enforced on the vector in (15.29) by calling MIN_DISTANCE twice; first with a value of $dmin = 0.0012$, then with a value of $dmin = 0.0006$. After this processing step, the quantized LSF vector is

$$\begin{aligned} \hat{\mathbf{x}}[m] &= \alpha^{(i_0[m])}\hat{\mathbf{e}}[m] + \mathbf{x}_p^{(i_0[m])}[m] \\ &= \alpha^{(i_0[m])}\hat{\mathbf{e}}[m] + \sum_{i=1}^{4} b_i^{(i_0[m])}\hat{\mathbf{e}}[m-i], \end{aligned} \tag{15.30}$$

where

$$\alpha^{(i_0)} = 1 - \sum_{i=1}^{4} b_i^{(i_0)} \tag{15.31}$$

and $b_i^{(i0)}$ are the MA coefficients of the i_0-th predictor. Initial values of the prediction-error vectors are given by

$$\hat{\mathbf{e}} = \begin{bmatrix} \frac{\pi}{11} & \frac{2\pi}{11} & \frac{3\pi}{11} & \cdots & \frac{10\pi}{11} \end{bmatrix}^T. \tag{15.32}$$

After computing (15.30), stability of the corresponding filter is ensured by performing the following operations:

1. Sort the elements of the LSF vector.
2. If $\hat{\omega}_1 < 0.005$ then $\hat{\omega}_1 \leftarrow 0.005$.
3. If $\hat{\omega}_{i+1} - \hat{\omega}_i < 0.0391$ then $\hat{\omega}_{i+1} \leftarrow \hat{\omega}_i + 0.0391$, for $i = 1$ to 9.
4. If $\hat{\omega}_{10} > 3.135$ then $\hat{\omega}_{10} \leftarrow 3.135$.

In the above procedure, $\hat{\omega}_i$ are the elements of the quantized LSF vector $\hat{\mathbf{x}}$. Figure 15.11 shows the block diagram of the decoder.

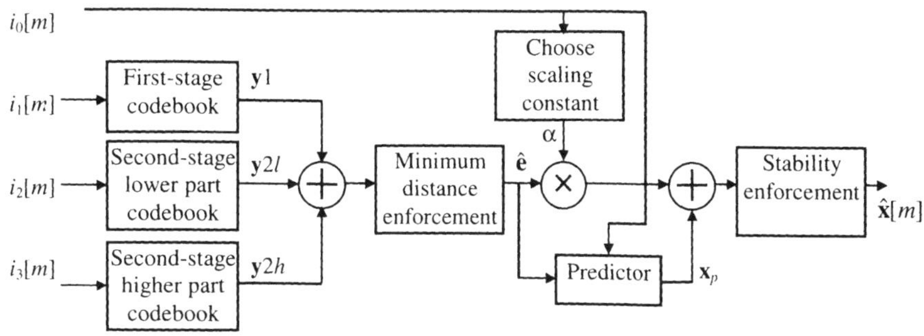

Figure 15.11 Block diagram of the LSF decoder for the G.729 coder.

Encoding the LSF

Given the input LSF vector of the mth frame,

$$\mathbf{x}[m] = [\omega_1[m] \quad \omega_2[m] \quad \cdots \quad \omega_{10}[m]]^T, \tag{15.33}$$

we have the prediction-error vector

$$\mathbf{r}^{(i_0)}[m] = \mathbf{x}[m] - \mathbf{x}_p^{(i_0)}[m]. \tag{15.34}$$

Note that the procedure is repeated for the two predictors: $i_0 = 0$ and $i_0 = 1$. The quantization target is the scaled prediction-error vector

$$\mathbf{e}^{(i_0)}[m] = \mathbf{r}^{(i_0)}[m]/\alpha^{(i_0)}. \tag{15.35}$$

The vector is quantized by searching the first-stage codebook to find the codevector that minimizes the Euclidean distance (no weighting); the resultant quantized vector is denoted by

$$\mathbf{e1}^{(i_0)}[m] = \begin{bmatrix} \mathbf{y1}l_{i_1}^{(i_0)} \\ \mathbf{y1}h_{i_1}^{(i_0)} \end{bmatrix}. \tag{15.36}$$

The next step is to search the second-stage low-part codebook. In this case, the vectors

$$\mathbf{e2}_{i_2}^{(i_0)}[m] = \begin{bmatrix} \mathbf{y1}l_{i_1}^{(i_0)} + \mathbf{y2}l_{i_2} \\ \mathbf{y1}h_{i_1}^{(i_0)} \end{bmatrix} \tag{15.37}$$

are constructed with $i_2 = 0, \ldots, 31$. Based on (15.37), the following quantized input vectors are found:

$$\hat{\mathbf{x}}\mathbf{2}_{i_2}^{(i_0)}[m] = \alpha^{(i_0)}\mathbf{e2}_{i_2}^{(i_0)}[m] + \mathbf{x}_p^{(i_0)}[m], \tag{15.38}$$

which are derived from (15.30). Minimum distance is enforced with $dmin = 0.0012$; the processed vectors are then applied in the weighted distance computation so as to compare with the input vector. The particular index i_2 that produces the smallest distance is selected. This concludes the search of the second-stage lower part codebook.

Using the selected first-stage codevector, and the second-stage lower part codevector, the high part of the second stage is searched. This is done by first constructing the vectors

$$\mathbf{e3}_{i_3}^{(i_0)}[m] = \begin{bmatrix} \mathbf{y1}l_{i_1}^{(i_0)} + \mathbf{y2}l_{i_2}^{(i_0)} \\ \mathbf{y1}h_{i_1}^{(i_0)} + \mathbf{y2}h_{i_3} \end{bmatrix} \tag{15.39}$$

for $i_3 = 0, \ldots, 31$. Both i_1 and i_2 are fixed from previous searches. Followed by calculation of the quantized input vectors similar to (15.38), the rest of the procedure is performed in an identical manner, with the ultimate goal of finding the optimal index i_3.

After the indices i_1, i_2, and i_3 are fixed, the same decoding procedure is applied to compute two quantized LSF vectors: $\hat{\mathbf{x}}^{(i_0)}[m], i_0 = 0$ and 1. These vectors are compared to the input vector with the one producing the lowest weighted distance selected. This final step determines the index i_0. Figure 15.12 shows the block diagram of the LSF encoder.

LSF Interpolation

Given the LSF vector $\mathbf{x}$, whose elements are in the angular frequency domain $\omega_i \in [0, \pi]$, it is first transformed by

$$\mathbf{u} = [\cos(\omega_1) \quad \cos(\omega_2) \quad \cdots \quad \cos(\omega_{10})]^T, \tag{15.40}$$

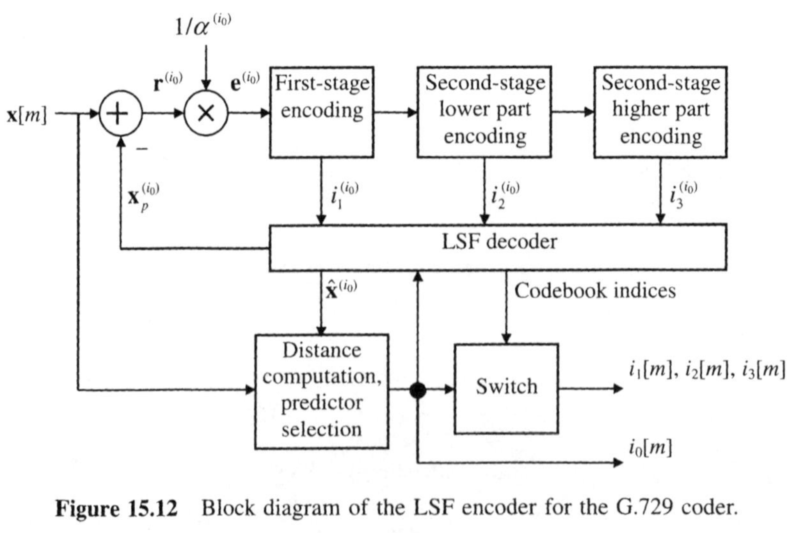

Figure 15.12 Block diagram of the LSF encoder for the G.729 coder.

where the ω_i are elements of $\mathbf{x}$. This vector is used directly by the second subframe. For the first subframe, we use

$$\mathbf{u}_0 = 0.5\,\mathbf{u}_{\text{past}} + 0.5\,\mathbf{u}. \tag{15.41}$$

The procedure is done for both the original LSF vector as well as the quantized version. Interpolation of either domain does not produce noticeable audible differences; the cosine domain, however, is sometimes convenient for practical implementation.

ETSI GSM EFR ACELP

This coder utilizes a 20-ms (160-sample) frame divided into four 5-ms (40-sample) subframes. A tenth-order LP analysis procedure is carried out twice for each 20-ms frame using two different asymmetric windows of length 30 ms (240 samples). Both procedures are performed for the same set of speech samples without using any samples from future frames. These two windows are defined by

$$w_1[n] = \begin{cases} 0.54 - 0.46\cos\left(\frac{\pi n}{159}\right); & n = 0, \ldots, 159, \\ 0.54 + 0.46\cos\left(\frac{\pi(n-160)}{79}\right); & n = 160, \ldots, 239, \end{cases} \tag{15.42}$$

and

$$w_2[n] = \begin{cases} 0.54 - 0.46\cos\left(\frac{2\pi n}{463}\right); & n = 0, \ldots, 231, \\ \cos\left(\frac{\pi(n-232)}{31}\right), & n = 232, \ldots, 239. \end{cases} \tag{15.43}$$

The windows are applied to 80 samples from a past frame in addition to the 160 samples of the current frame. Relative positions of the windows are illustrated in Figure 15.13. The windowed samples are used to compute the autocorrelation values $R[l], l = 0, 1, \ldots, 10$. The same processing steps as for the G.729 coder are used to obtain two sets of LPCs.

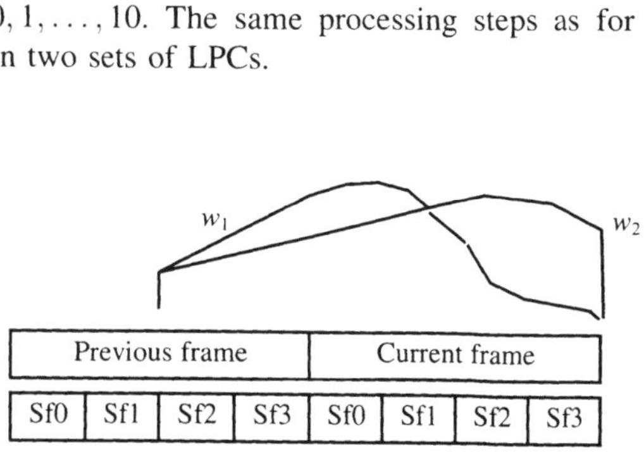

Figure 15.13 Positions of frame and LP analysis windows of the GSM EFR coder.

LSF Quantization: PVQ-MA with Split Matrix Quantization

Given the two sets of LPCs of the frame under consideration,

$$a_{1,i}; a_{2,i}; \quad i = 1, \ldots, 10,$$

where $a_{1,i}$ are obtained through w_1 while $a_{2,i}$ via w_2, they are first converted to two sets of LSFs:

$$\omega_{1,i}; \omega_{2,i}; \quad i = 1, \ldots, 10.$$

In vector notation,

$$\mathbf{x}_1^T = [\omega_{1,1}, \omega_{1,2}, \ldots, \omega_{1,10}], \qquad \mathbf{x}_2^T = [\omega_{2,1}, \omega_{2,2}, \ldots, \omega_{2,10}].$$

The mean of each individual LSF element is removed:

$$\mathbf{v}_1 = \mathbf{x}_1 - \mathbf{m}, \mathbf{v}_2 = \mathbf{x}_2 - \mathbf{m}, \tag{15.44}$$

with the prediction-error vectors

$$\mathbf{r}_1 = \mathbf{v}_1 - \mathbf{p}, \mathbf{r}_2 = \mathbf{v}_2 - \mathbf{p}, \tag{15.45}$$

where $\mathbf{p}$ is the predicted LSF vector for the current frame. The prediction is based on the prediction-error vector of the past frame, given by

$$\mathbf{p} = 0.65\hat{\mathbf{r}}_{2,\text{past}}, \tag{15.46}$$

where $\hat{\mathbf{r}}_{2,\text{past}}$ is the quantized second prediction-error vector of the prior frame. Thus, a first-order MA predictor is employed.

The two vectors $\mathbf{r}_1$ and $\mathbf{r}_2$ are joined together to form a matrix, which is divided into five submatrices of dimension 2×2: $\mathbf{R}_1$ to $\mathbf{R}_5$. This is illustrated as follows:

$$[\mathbf{r}_1|\mathbf{r}_2] = \begin{bmatrix} r_{1,1} & r_{2,1} \\ r_{1,2} & r_{2,2} \\ \vdots & \vdots \\ r_{1,10} & r_{2,10} \end{bmatrix} = \begin{bmatrix} \mathbf{R}_1 \\ \mathbf{R}_2 \\ \mathbf{R}_3 \\ \mathbf{R}_4 \\ \mathbf{R}_5 \end{bmatrix}. \tag{15.47}$$

Each submatrix consists of four elements and is quantized separately with 7, 8, $8 + 1$, 8, and 6 bits, respectively. The scheme is referred to as *split matrix* quantization, which in principle is a split VQ procedure. The third submatrix uses a 256-entry signed codebook; that is, 8 bits is allocated to the shape vectors and 1 bit is allocated to sign. In this way, more resources are allocated to the perceptually important frequency regions.

The objective of quantization is to find the vectors

$$\hat{\mathbf{x}}_1 = \hat{\mathbf{v}}_1 + \mathbf{m} = \hat{\mathbf{r}}_1 + \mathbf{p} + \mathbf{m}, \qquad \hat{\mathbf{x}}_2 = \hat{\mathbf{v}}_2 + \mathbf{m} = \hat{\mathbf{r}}_2 + \mathbf{p} + \mathbf{m} \tag{15.48}$$

and to minimize the distance

$$d(\mathbf{x}_1, \mathbf{x}_2, \hat{\mathbf{x}}_1, \hat{\mathbf{x}}_2) = \sum_{i=1}^{10} w_{1,i}(\omega_{1,i} - \hat{\omega}_{1,i})^2 + w_{2,i}(\omega_{2,i} - \hat{\omega}_{2,i})^2, \tag{15.49}$$

with $w_{1,i}$ and $w_{2,i}$ the weighting factors depending on the LSF values. Derivation of the weighting factors is identical for the two vectors, defined by

$$w_i = \begin{cases} 3.347 - 4.377\Delta\omega_i, & \Delta\omega_i < 0.353, \\ 2.143 - 0.970\Delta\omega_i, & \text{otherwise}, \end{cases} \tag{15.50}$$

where

$$\Delta\omega_i = \omega_{i+1} - \omega_{i-1} \tag{15.51}$$

with $\omega_0 = 0$ and $\omega_{11} = \pi$. This weighting scheme puts more emphasis on those LSFs that are surrounded closely by others, while LSFs that are far separated have proportionately less amount of weight. This is reasonable since for those LSFs that are close together, higher precision is needed to represent them as faithfully as possible; for those LSFs that are far apart, higher error can be tolerated. Figure 15.14 illustrates the quantization procedure.

LSF Interpolation

Given the quantized LSF vectors $\hat{\mathbf{u}}_1$ and $\hat{\mathbf{u}}_2$ whose elements are in the angular frequency domain, the second subframe utilizes $\hat{\mathbf{u}}_1$ directly while the fourth subframe

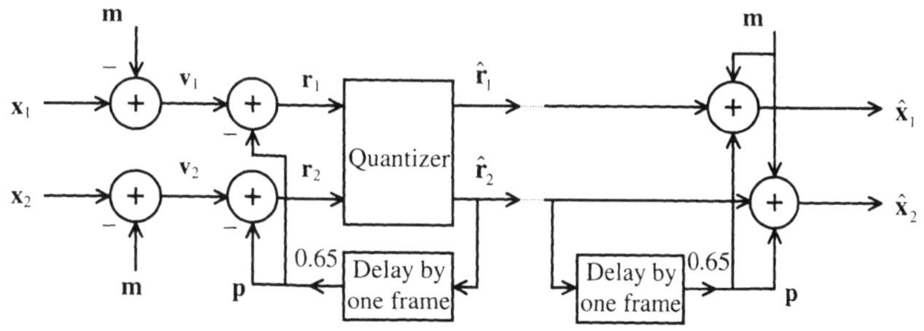

Figure 15.14 Illustration of LSF quantization for the GSM EFR coder.

utilizes $\hat{\mathbf{u}}_2$ directly. For the first and third subframes, linear interpolation is employed. First, the quantized LSF vectors are converted to

$$\mathbf{x}_1^T = [\cos(\hat{\omega}_{1,1}) \quad \cos(\hat{\omega}_{1,2}) \quad \cdots \quad \cos(\hat{\omega}_{1,10})],$$
$$\mathbf{x}_3^T = [\cos(\hat{\omega}_{2,1}) \quad \cos(\hat{\omega}_{2,2}) \quad \cdots \quad \cos(\hat{\omega}_{2,10})].$$

Then for the first and third subframes, we use

$$\mathbf{x}_0 = 0.5\mathbf{x}_{3,\text{past}} + 0.5\mathbf{x}_1 \tag{15.52}$$

and

$$\mathbf{x}_2 = 0.5\mathbf{x}_1 + 0.5\mathbf{x}_3, \tag{15.53}$$

respectively.

15.5 SUMMARY AND REFERENCES

This chapter captured some major trends in LPC quantization, which are all related to VQ of the LSF. These schemes have been widely deployed to many speech coding standards and are far superior to the prior generation of scalar quantizers. Success of these schemes is credited to the exploration of correlation among the LSFs, both interframe and intraframe.

One of the earliest work in VQ of LPCs is perhaps Buzo et al. [1980], with demonstration of the superiority of VQ over scalar quantization. Paliwal and Atal [1993] studied split VQ and concluded that LSF was better suited than LAR, and the resultant quantizer was as robust to channel errors as the scalar quantizers; transparent quantization was achieved with 24 bits/frame. LeBlanc et al. [1993] utilized MSVQ of the LSF and found that, at 24 bits/frame, performance is equivalent to the 34-bit/frame scalar quantizer of the FS1016. The same principle was later absorbed by the FS MELP standard, where a four-stage MSVQ at 25 bits/frame was utilized [McCree et al., 1997]. It is possible to prove that, at high resolution, the weighted Euclidean distance measure is optimum for LSF VQ, with a special form of weighting factors derived in Gardner and Rao [1995a, 1995b]. These principles are utilized in McCree and DeMartin [1997] to develop a switched PVQ scheme with multistage structure for LSF quantization at a rate of 20 bits/frame. The method outperformed the original quantizer of the FS MELP standard; together with other changes, the authors proposed a 1.6-kbps MELP coder.

Early works for PVQ in spectrum parameters quantization can be found in Shoham [1987] and Yong et al. [1988], where a predictor is selected according to certain criteria during encoding. Adaptive predictor and split VQ were proposed by Erzin and Cetin [1993]. The LSF quantizer of the G.729 was originally proposed in Kataoka et al. [1993, 1994] and later incorporated in the design of a CS-CELP coder [Kataoka et al., 1996]. A similar quantizer was investigated by Salami et al. [1994] and ultimately led to its inclusion as part of the G.729 standard [ITU, 1996a; Salami et al., 1998]. In Shoham [1999], a training algorithm is proposed to jointly optimize the predictors and residual codebook, leading to improvements with respect to the standardized quantizer. Use of an asymmetric window in LP analysis was investigated by Florencio [1993], which is very much favored by modern coders due to the cutback in algorithmic delay. Xydeas and Papanastasiou [1995] introduced the idea of split matrix quantization, where LSF vectors from consecutive frames are concatenated into a matrix to be quantized as a whole. Quantizer of the GSM EFR coder is described in Salami [1997a] and ETSI [1999], and that of the G.723.1 coder is found in ITU [1996b]. See Eriksson et al. [1999] for the "safety-net" PVQ design to improve robustness against channel errors as well as performance. Theoretical analysis of memoryless VQ and recursive coding of spectrum parameters can be found in Hedelin and Skoglund [2000] and Samuelsson and Hedelin [2001].

Quantization of LPCs is still an active research area where new algorithms with improved performance and less computational cost are proposed. Readers are referred to the current literature for the latest trend. Due to continuous technological advances, faster and more powerful processors are becoming accessible at low cost; some of the high-performance high-complexity algorithms that were prohibitive in the old days are becoming implementable. Therefore, in the near future, we will be able to witness the spawn of new generations of high-performance speech coders at very low bit-rate.

EXERCISES

15.1 Using some speech material, extract the LSF vectors based on 160-sample frames using a Hamming window. Calculate the normalized correlation $R[i, j, k]$ as defined in (15.3) for $i, j = 1$ to 10 and $k = 0$ to 3. Repeat the correlation calculation with the original LSF, that is, before the mean is removed. Compare the two sets of correlation values; explain the difference.

15.2 In split VQ of the LSF, and considering (3, 7), (5, 5), (4, 6), and (6, 4), using 24 bits and 12 bits per codebook, is there any difference in memory cost for these four schemes? What about computational cost using sequential search?

15.3 An elementary sorting method is known as bubble sort [Sedgewick, 1992]: keep passing through the array, exchanging adjacent elements, if necessary; when no exchanges are required on some pass, the array is sorted. The following pseudocode illustrates the idea:

```
BUBBLE_SORT(x[ ], N)
1. flag ← 1
```

```
2.  for i ← 0 to N − 2
3.          if x[i] > x[i + 1]
4.                  SWAP(x[i], x[i + 1])
5.                  flag ← 0
6.  if flag = 0 goto 1
```

The function SWAP at Line 4 exchanges two elements of the array. Verify that the algorithm does work. What is the best case and worst case computational cost associated with this method? See Cormen et al. [1990] for other sorting algorithms and performance analysis.

15.4 One suggested method for minimum distance enforcement for tenth-order LSF vector is the following:

```
MIN_DISTANCE(ω1, ... , ω10)
for i ← 1 to 9
        d ← ω_{i+1} − ω_i
        if d < dmin
                s1 ← s2 ← (dmin − d)/2
                if i = 1 and ω_i < dmin
                        s1 ← ω_i/2
                else if i > 1
                        temp ← ω_i − ω_{i−1}
                        if temp < dmin
                                s1 ← 0
                        else if temp < 2 * dmin
                                s1 ← (temp − dmin)/2
                if i = 9 and ω_{i+1} > π − dmin
                        s2 ← (π − ω_{i+1})/2
                else if i < 9
                        temp ← ω_{i+2} − ω_{i+1}
                        if temp < dmin
                                s2 ← 0
                        else if temp < 2 * dmin
                                s2 ← (temp − dmin)/2
                ω_i ← ω_i − s1
                ω_{i+1} ← ω_{i+1} + s2
```

Does the procedure deliver the expected results? What are the limitations, if any? The parameter *dmin* is the distance threshold between elements.

15.5 For the minimum distance enforcement routine of the G.723.1 coder, find out two situations with the input LSF values where the routine will return an invalid status: that is, the procedure finalizes and reports the condition where distance enforcement has not been completed.

15.6 Based on your knowledge of PVQ design (Chapter 7), summarize the procedure that you would apply to train the codebooks for the LSF quantizer of the G.729 coder, taking into account the particular structure with a switched predictor. Extend your method to the case where more than two predictors exist.

15.7 Plot the windows utilized by the G.729 and the GSM EFR coders for the autocorrelation calculation. Find the magnitude of the Fourier transform associated with the windows.

15.8 Using some speech material, extract the LSF vectors based on the rules outlined for the GSM EFR coder. With no quantization, apply the prediction scheme of the LSF quantizer to compute the prediction-error vectors. Calculate normalized correlation as defined in (15.3) for the LSF vectors as well as the prediction-error vectors. Is the prediction scheme effective in redundancy removal?

15.9 For the GSM EFR coder, plot the distance calculation weights as a function of the LSF difference (15.50). What conclusion is obtained?

15.10 One way to measure the effectiveness of the linear predictor is through the energy of the prediction error, and the associated prediction gain (Chapter 4). Using the windowing, LP analysis, and interpolation schemes of the G.729 coder, calculate the segmental prediction gain for some speech material. Repeat the measurement using a Hamming window and a rectangular window of the same length. Is the use of an asymmetric window effective in elevating the prediction gain?

15.11 Are the following statements true? Justify your answers.

(a) Minimum coding delay of the FS MELP coder is equal to 57.5 ms.

(b) Minimum coding delay of the G.723.1 coder is equal to 67.5 ms.

(c) Minimum coding delay of the G.729 coder is equal to 25 ms.

(d) Minimum coding delay of the GSM EFR coder is equal to 40 ms.

See Chapter 1 for definition of coding delay. It is assumed that the encoder is transmitting in constant mode, with no processing delays. Ignore the additional delay that the postfilter introduces for these coders during decoding.

Now, assuming that the encoder is transmitting in *burst* mode—that is, all available bits of the compressed bit-stream for the frame are sent immediately to the decoder once encoding is completed—recalculate the minimum coding delay for the above four cases.

15.12 Given the LSF vector

$$\mathbf{x} = [0.17, 0.33, 1.01, 1.14, 1.16, 1.57, 2.13, 2.54, 2.65, 2.95]^T,$$

show that the weight vectors to find the weighted Euclidean distance between $\mathbf{x}$ and its quantized version for various standardized coders are specified as follows:

(a) FS MELP:

$$\mathbf{w} = [1.741, 1.604, 1.101, 1.043, 1.035, 0.906, 0.805, 0.767, 0.486, 0.120]^T.$$

(b) G.723.1:

$$\mathbf{w} = [6.25, 6.25, 7.69, 50.0, 50.0, 2.44, 2.44, 9.09, 9.09, 3.33]^T.$$

(c) G.729:

$$\mathbf{w} = [7.33, 1.26, 1.36, 8.23, 5.10, 1.21, 1.01, 3.30, 4.48, 6.77]^T.$$

(d) GSM EFR:

$$\mathbf{w} = [1.90, 1.33, 1.36, 2.69, 1.73, 1.20, 1.20, 1.64, 1.75, 1.67]^T.$$

CHAPTER 16

ALGEBRAIC CELP

Algebraic CELP or ACELP was born as another attempt to reduce the computational cost of standard CELP coders. As explained in Chapter 11, excitation codebook search is the most intensive of all CELP operations. Throughout history, researchers have tried many techniques to bring the cost down to manageable levels; very often, the approach is to impose a certain structure for the fixed excitation codebook so as to enable fast search methods. For instance, the IS54 coder (Chapter 13) relies on a small number of basis vectors to generate the entire space of excitation codevectors; by doing so, efficient techniques can be applied to locate the optimal excitation sequence. The FS1016 coder (Chapter 12), on the other hand, utilizes an overlapping codebook with ternary-valued samples, allowing the usage of recursive convolution with no products.

Many ACELP-based standards follow the legacy of past CELP coders. The term "algebraic" essentially means the use of simple algebra or mathematical rules to create the excitation codevectors, with the rules being addition and shifting. Based on the approach, there is no need to physically store the entire codebook, resulting in significant memory saving. The original ideas were first introduced in Adoul and Lamblin [1987] and Adoul et al. [1987], two years after the landmark paper of Schroeder and Atal [1985]. It has been refined by many researchers, leading to at least five standardized coders:

- ITU-T G.723.1 Multipulse Maximum Likelihood Quantization (MP-MLQ)/ ACELP (1995).
- ITU-T G.729 Conjugate Structure (CS)–ACELP (1995).
- TIA IS641 ACELP (1996).

- ETSI GSM Enhanced Full Rate (EFR) ACELP (1996).
- ETSI Adaptive Multirate (AMR) ACELP (1999).

This chapter is devoted to the G.729 CS-ACELP coder. The most distinguishing features are use of the algebraic excitation codebook and conjugate VQ for the involved gains. The coder was developed originally for personal communication systems, digital satellite systems, and other applications such as packetized speech and circuit multiplexing equipment. The chapter starts with a revelation of the structure of the algebraic codebook, followed by the construction and search methodology of the adaptive codebook. Description of encoding and decoding operations, algebraic codebook search techniques, and gain quantization are given. The above-mentioned ACELP standards have been developed by the same group of researchers and therefore share many common attributes. The other four ACELP-based standards are studied in a separate section with the most salient features analyzed and compared.

The concept of ACELP has a great deal of influence on the direction of speech coding developments; study of its structure is necessary to grasp some of the puissant trends in this field.

16.1 ALGEBRAIC CODEBOOK STRUCTURE

As studied in Chapter 11, a CELP coder contains a fixed excitation codebook holding the codevectors that serve as input to the synthesis filter. The codebook is searched during encoding to locate the best codevector for a particular speech subframe. The G.729 coder has an algebraic structure for the fixed codebook (referred to from now on as the algebraic codebook) that is based on the *interleaved single-pulse permutation* design. In this scheme each codevector contains four nonzero pulses. Each pulse can have the amplitude of either 1 or -1 and can assume the positions given in Figure 16.1.

Each excitation codevector has 40 samples, which is the length of a subframe. The excitation codevector $v[n]$ is constructed by summing the four pulses according to

$$v[n] = \sum_{i=0}^{3} p_i[n] = \sum_{i=0}^{3} s_i \delta[n - m_i]; \quad n = 0, \ldots, 39, \tag{16.1}$$

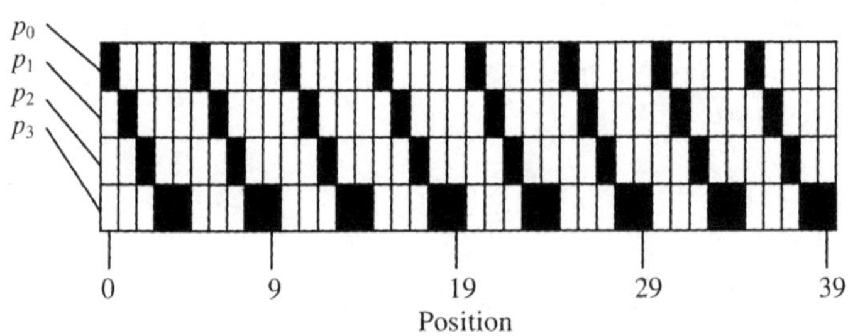

Figure 16.1 Positions of individual pulses in algebraic codebook for the G.729 coder, indicated by the black rectangles. Data from ITU [1996a], Table 7.

where $s_i = \pm 1$ is the sign of the pulse and m_i is the position. Thus, each pulse requires 1 bit per sign. For p_0, p_1, and p_2, 3 bits are needed for position; while for p_3, 4 bits are required. Hence, a total of 17 bits are needed to index the whole codebook.

The index to the codebook is therefore composed of two parts: sign (4 bits) and position (13 bits). Using a truncation function, $f(x) = x$ if $x > 0$ and $f(x) = 0$ otherwise. The sign index is given by

$$sindex = f(s_0) + 2f(s_1) + 4f(s_2) + 8f(s_3) \tag{16.2}$$

represented with 4 bits; while for position, we have

$$pindex = \frac{m_0}{5} + 8\left\lfloor\frac{m_1}{5}\right\rfloor + 64\left\lfloor\frac{m_2}{5}\right\rfloor + 512\left(2\left\lfloor\frac{m_3}{5}\right\rfloor + ms3\right) \tag{16.3}$$

represented with 13 bits, where

$$ms3 = \begin{cases} 0; & \text{if} \quad m_3 = 3, 8, \cdots, 38, \\ 1; & \text{if} \quad m_3 = 4, 9, \cdots, 39. \end{cases} \tag{16.4}$$

Note that (16.2) and (16.3) are the actual indices transmitted as part of the G.729 bit-stream. The advantage of this method is that no physical storage is required for the fixed codebook; however, the amount of bits allocated is relatively high when compared to other CELP standards, such as the FS1016.

16.2 ADAPTIVE CODEBOOK

The G.729 coder has some unique methodologies applied to the adaptive codebook, which are quite different from standards such as FS1016 or IS54. Figure 16.2 shows the signals involved in the encoder. When compared with the diagrams in Chapters 12 and 13, we see some major differences. In general, the G.729 is more complex with finer structure to boost the quality of synthetic speech. Due to advances in DSP technology, extra complexity can be tolerated. In fact, comparing the G.729 with previous generations of CELP coders, we can see the trend of increased complexity, affordable with the constant improvements in the hardware front. In this section we study the structure and search procedure of the adaptive codebook.

Perceptual Weighting Filter

G.729 utilizes a more sophisticated perceptual weighting filter than previous generations of CELP standards. The system function is

$$W(z) = \frac{A(z/\gamma_1)}{A(z/\gamma_2)} = \frac{1 + \sum_{i=1}^{10} a_i \gamma_1^i z^{-i}}{1 + \sum_{i=1}^{10} a_i \gamma_2^i z^{-i}}. \tag{16.5}$$

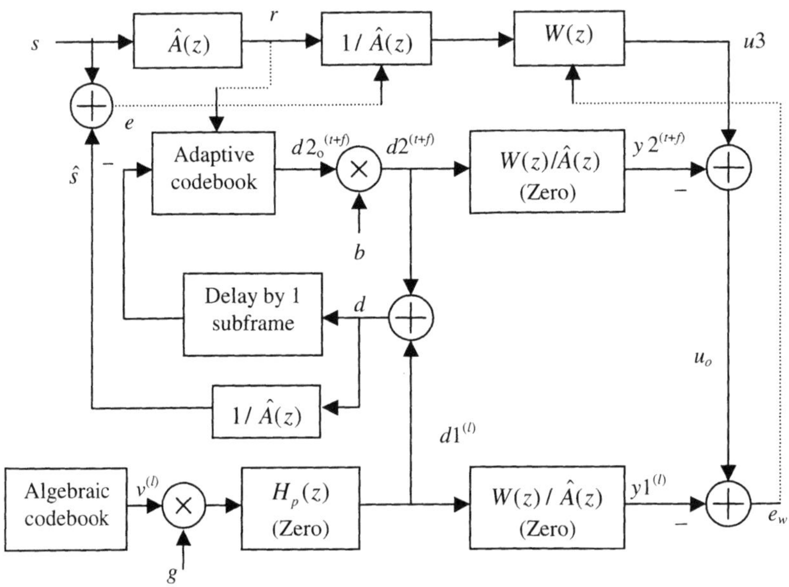

Figure 16.2 Signals involved in the G.729 encoder.

The LPC a_i are the unquantized coefficients and are updated once every subframe. For each subframe, the LPCs are obtained through interpolation in the LSF domain (Chapter 15).

The use of unquantized LPCs in the system function (16.5) allows a better capture of the original signal spectrum, leading to elevated subjective quality. This is in contrast to past generations of CELP coders, such as the FS1016, where quantized coefficients are used due to simplicity. As we have seen in Chapter 11, the perceptual weighting filter and the formant synthesis filter are merged together to form the modified formant synthesis filter. This technique does not apply to the G.729 coder, where modest reduction in computational cost is abandoned in favor of higher signal integrity.

The parameters γ_1 and γ_2 determine the frequency response of the filter and are made adaptive depending on the spectral characteristics. The adaptation is based on a spectrum flatness measure obtained through the first two RCs, as well as the closeness of the LSFs. There are two possible values for γ_1—0.94 or 0.98—and γ_2 pertains to the range [0.4, 0.7]. The rules for the determination of γ_1 and γ_2 are based on subjective tests and improvement in final speech quality has been reported [Salami et al., 1998].

Open-Loop Pitch Period Estimation

Open-loop pitch period estimation is the first step toward adaptive codebook search. The procedure is based on the weighted speech signal and is done once per frame,

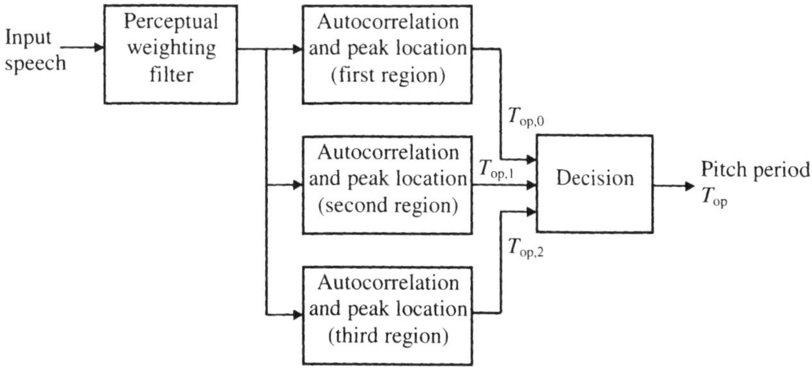

Figure 16.3 The open-loop pitch period estimation procedure for the G.729 encoder.

that is, every 10 ms. Figure 16.3 summarizes the major steps involved. The perceptually weighted speech, denoted by $u[n]$, is used in the following autocorrelation expression:

$$R[l] = \sum_{n=0}^{79} u[n]u[n-l]. \tag{16.6}$$

Note that 80 samples are involved corresponding to two subframes. Three peaks are separately found in three ranges of lag; these results are denoted by

$$R_0 = R[T_{op,0}] | R[l] \leq R[T_{op,0}], \quad 20 \leq l, T_{op,0} \leq 39, \tag{16.7}$$

$$R_1 = R[T_{op,1}] | R[l] \leq R[T_{op,1}], \quad 40 \leq l, T_{op,1} \leq 79, \tag{16.8}$$

$$R_2 = R[T_{op,2}] | R[l] \leq R[T_{op,2}], \quad 80 \leq l, T_{op,2} \leq 143. \tag{16.9}$$

The retained peak autocorrelation values are normalized according to

$$r_i = \frac{R_i}{\sqrt{\sum_n u^2[n - T_{op,i}]}}; \quad i = 0, 1, 2. \tag{16.10}$$

An overall winner is selected among these three outcomes by favoring the delays in the lowest range. The decision process is summarized as follows:

```
Top ← Top,0; r ← r0;
if r1 > 0.85r
        Top ← Top,1; r ← r1;
if r2 > 0.85r
        Top ← Top,2;
```

Thus, the procedure divides the lag range into three regions, with the final result (T_{op}) inclined toward the lower region, which is done to avoid choosing pitch multiples.

Search Range for Pitch Period

The pitch period or adaptive codebook index is encoded with 8 bits for the first subframe; for the second subframe, a relative value with respect to the first subframe is encoded with 5 bits. For the first subframe, a fractional pitch period is used with a resolution of $\frac{1}{3}$ in the range of $[19\,\frac{1}{3},\ 84\,\frac{2}{3}]$ and integers only in the range [85, 143]. For the second subframe, a delay with a resolution of $\frac{1}{3}$ is always used in the range

$$T_2 \in \left[round(T_1) - 5\frac{2}{3},\ round(T_1) + 4\frac{2}{3}\right],$$

with T_1 and T_2 the pitch period of the first and second subframes, respectively.

For each subframe, the pitch period is determined by searching through the adaptive codebook in such a way that the weighted mean-squared error is minimized. The search range is found from the open-loop estimate T_{op}: it is roughly $T_{op} \pm 3$ and $T_{op} \pm 5$ for the first and second subframes, respectively.

Signals Involved in Adaptive Codebook Search

After knowing the search range for adaptive codebook, the following signals are gathered before searching the codebook.

- Adaptive codebook array or excitation sequence.

  This is denoted as $d2[n]$. For $n < 0$, the array contains samples of past excitation (d in Figure 16.2). For $0 \le n < N$, where $N = 40$ is the subframe length, the array contains prediction error of the current subframe, obtained by filtering the input speech using the prediction-error filter. Coefficients of the filter are the quantized and interpolated LPCs, with the system function denoted by $\hat{A}(z)$. With this arrangement, the preliminary adaptive codevector for integer pitch period t is given by

$$d2'^{(t)}[n] = d2[n - t]; \quad 0 \le n < N. \tag{16.11}$$

  Note that this is different from the traditional adaptive codebook formulation, where the codevector repeats the t samples of $d2[n]$ in the range $n \in [-t, -1]$ when $t < N$ (Chapter 12). A drawback of repeating the t samples is that recursive convolution does not apply when $t < N$. To improve computational efficiency, $d2[n]$, $0 \le n < N$ is filled with prediction error so that recursive convolution can be utilized for all t. Also note that the codevectors (16.11) are preliminary, meaning that they are not the final version utilized for speech synthesis. Obviously, this approach is suboptimal since the codevectors during the search stage are not the same as the final versions used in speech synthesis—more on this issue later.

- Target sequence for adaptive codebook search.

  The target sequence $u3[n]$ is found using the perceptually weighted speech and the zero-input response of the cascade connection between perceptual weighting filter and formant synthesis filter. The system function of the formant synthesis filter is denoted by $1/\hat{A}(z)$, where the coefficients are the quantized and interpolated LPCs.

- Impulse response of the cascade connection between perceptual weighting filter and formant synthesis filter, denoted by $h[n]$.

Finding Integer Lag

Given the preliminary codevector $d2'^{(t)}[n]$, it is processed by the filter cascade to generate the output sequence $y2'^{(t)}[n]$; that is,

$$y2'^{(t)}[n] = d2'^{(t)}[n] * h[n] = \sum_{k=0}^{n} h[k]d2'^{(t)}[n-k] \tag{16.12}$$

for $t = T_{\min} - 4$ to $T_{\max} + 4$ and $n = 0$ to $N - 1$, with $T_{\min}$ and $T_{\max}$ found in the previous step. Note that the range of t has extended; this is done to enable interpolation. Normalized correlation is calculated using

$$R[t] = \frac{\sum_{n=0}^{39} u3[n]y2'^{(t)}[n]}{\sqrt{\sum_{n=0}^{39} \left(y2'^{(t)}[n]\right)^2}}. \tag{16.13}$$

The value of t that maximizes $R[t]$ within the interval $[T_{\min}, T_{\max}]$ is denoted as T, which is the integer lag sought.

Finding Fractional Lag

If it is the first subframe and the integer lag satisfies $T > 84$, the fractional part of the pitch period is set to zero. Otherwise, the fractions around T are evaluated by interpolating the correlation values (16.13) in the following way:

$$R_o(t,f) = \sum_{i=0}^{3} R[t-i]w13[3f+3i] + \sum_{i=0}^{3} R[t+i+1]w13[3-3f+3i] \tag{16.14}$$

for $f = 0$, $\frac{1}{3}$, and $\frac{2}{3}$. In actual implementation, five fractional values are considered: $f = -\frac{2}{3}, -\frac{1}{3}, 0, \frac{1}{3}$, and $\frac{2}{3}$. Since (16.14) only applies for $f \geq 0$, t must be decremented by one in order to evaluate negative f. For transmission, only three fractional values are used: $-\frac{1}{3}$, 0, and $\frac{1}{3}$. If $f = -\frac{2}{3}$: $T \leftarrow T-1, f \leftarrow \frac{1}{3}$. If $f = \frac{2}{3}$: $T \leftarrow T+1, f \leftarrow \frac{1}{3}$.

The interpolation weights $w13$ are obtained through a Hamming truncated sinc function at ± 11 and padded with zero at ± 12. See Exercise 16.5 for details.

Generation of Adaptive Codevector

Given t, f, and $d2$, the actual adaptive codevector used for speech synthesis is calculated with the following procedure:

```
Adaptive_cv(d2, t, frac)
n0 ← −t
frac ← −frac
if frac < 0
        frac ← frac + 3
        n0 ← n0 − 1
for n ← 0 to 39
        d2[n] ← 0
        for i ← 0 to 9
                d2[n] ← d2[n] + d2[n0 − i + n] · w31[frac + 3 · i]
                        + d2 [n0 + 1 + i + n] · w31 [3 − frac + 3 · i]
```

The parameter *frac* is equal to -1, 0, or 1 in the above code, corresponding to the fractional values of $-\frac{1}{3}$, 0, and $\frac{1}{3}$, respectively. The codevectors are found at $d2[n]$, $n = 0$ to 39. The interpolation weights $w31[n]$ are obtained from a Hamming truncated sinc function at ± 29 and padded with zeros at ± 30 (Exercise 16.5).

Note from the previous pseudocode that the initial values in the $d2$ array for $n \in [0, 39]$ are completely replaced. Recall that, before searching the adaptive codebook, these elements of the array are filled with prediction-error samples. Interpolation is done for all fractional values, including $f = 0$. Thus, the actual adaptive codevector—henceforth denoted as $d2_o^{(t+f)}[n] = d2[n], n = 0$ to 39, with $d2[n]$ the result of applying the interpolation routine Adaptive_cv—is in general not the same as the preliminary codevectors, defined in (16.11). In this manner, the encoder and the decoder are able to generate the exact same codevectors from t and f, since samples of the codevector are formed solely from past excitation.

Therefore, we have a search procedure where the integer and fractional lags are based on a set of preliminary codevectors. Once the lags are determined the actual codevector is found, which is different from the preliminary version. How accurate could this be? Well, as shown in the next example, the accuracy is quite good when $t \geq 40$. In this case, interpolating using (16.14) and identifying the correlation peak produces a similar outcome as first calculating the actual adaptive codevectors and later finding the correlation using an equation like (16.13). This is because the interpolation weights $w13$ and $w31$ are selected to achieve a close match. For $t < 40$, however, the process is inaccurate, since the preliminary codevectors obtained from the prediction-error samples are not the same as those found through interpolation of past samples. This limits the potential of the coder to represent speech with short pitch period, such as females or children. Nevertheless, the technique performs reasonably well, and when combined with the rest of the coder's components, the synthetic speech is of good quality in most instances.

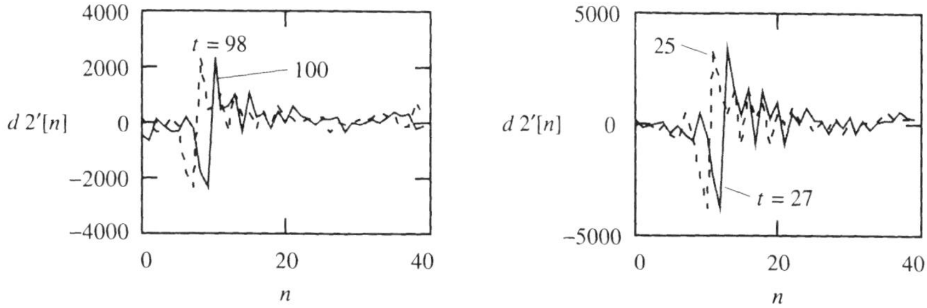

Figure 16.4 Examples of preliminary adaptive codevectors.

Example 16.1 Samples of the adaptive codebook at a certain encoding instance are recorded, denoted as $d2[n]$, with $n < 0$ the past excitation samples, while $n \in [0, 39]$ is the prediction error of the current subframe. Figure 16.4 shows some preliminary adaptive codevectors, where we can see that, for nearby pitch periods, the codevectors are shifted versions of each other.

By knowing the LPC of the formant synthesis filter, the zero-state responses are found for all preliminary codevectors from $t = 16$ to 150, which are then used to calculate the normalized correlation as given in (16.13), where a target sequence $u3[n]$ is available. Using these correlation values (at integer lags), the correlation at fractional lags are found using (16.14). Figure 16.5 shows the results. Note how the interpolated values $R_o(t, 0)$ are roughly equal to $R[t]$, with $R_o\left(t, \frac{1}{3}\right)$ and $R_o\left(t, \frac{2}{3}\right)$ positioned in between $R[t]$ and $R[t + 1]$. Under a full search criterion, the goal would be the evaluation of the $R_o(t,f)$ for all t and f so as to locate the one combination that provides the highest correlation.

Next, all codevectors for $t \in [20, 147]$ and $f = \{-\frac{1}{3}, 0, \frac{1}{3}\}$ using the procedure described in the pseudocode are found. Figure 16.6 shows some example waveforms, where we can see that they are related by a fractional time shift. Additional

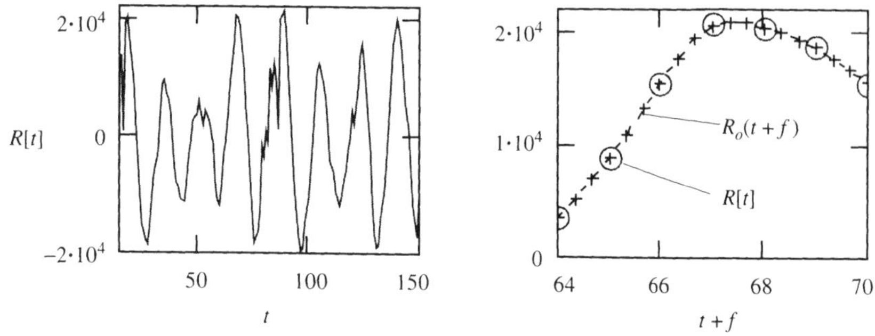

Figure 16.5 Examples of correlation values during adaptive codebook search. *Left*: For integer lags. *Right*: Plot of correlation at integer lag (o) and fractional lag (+).

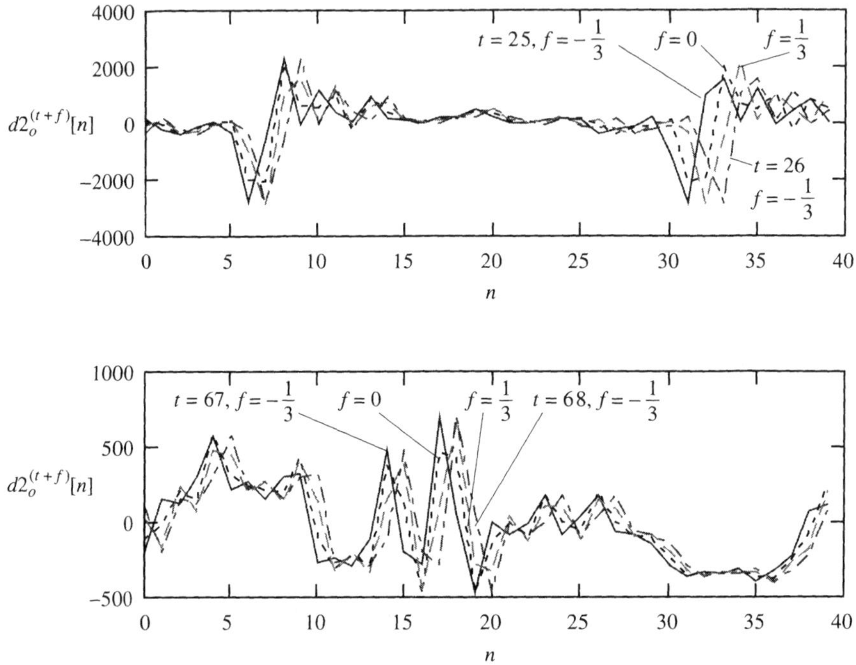

Figure 16.6 Examples of adaptive codevectors.

plots appear in Figure 16.7, where the actual and preliminary codevectors are compared (zero fractional lag). Note that the two versions are different, which is more pronounced when $t < 40$.

The correlation values are recalculated using the actual codevectors; this is done by modifying (16.13) slightly. A comparison appears in Figure 16.8, where we can see that for $t + f \geq 40$, the two sets of values match each other closely; for $t + f < 40$, however, the difference becomes obvious.

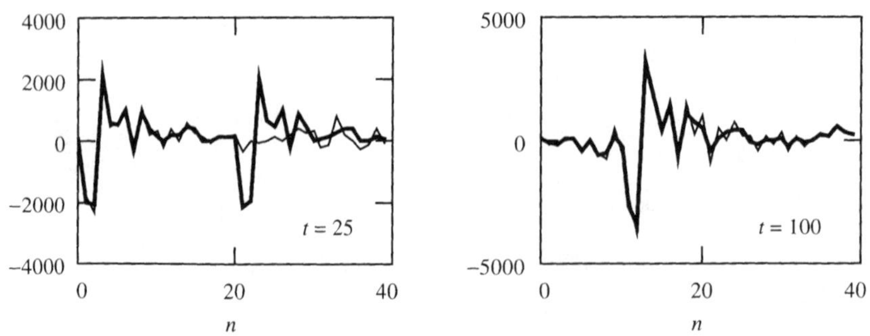

Figure 16.7 Some comparisons between actual codevectors (bold) and preliminary codevectors (fine).

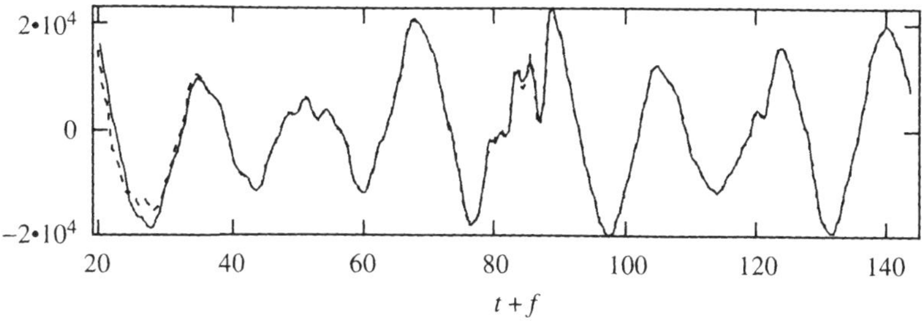

Figure 16.8 Comparison between the correlation values used for codebook search (R_o, solid) and the correlation values found with the actual codevectors (R'_o, dot).

A Summary

The G.729 adopts a strategy for the adaptive codebook search that is intended to be computationally efficient at some sacrifice on overall accuracy, especially for short pitch periods ($t + f < 40$). We can say that the approach is suboptimal in the sense that the best possible solution is not found, at the benefit of reduced computational burden. In practice, however, the method in combination with other components of the coder produces synthetic speech of high quality.

16.3 ENCODING AND DECODING

The G.729 follows the basic schemes for encoding and decoding as for a generic CELP coder (Chapter 11). Differences exist mainly because of the extra complexity introduced to achieve higher performance. These differences are explained in this section.

Signals Involved in Encoding and Filter-State Updates

For the G.729 coder, the perceptual weighting filter utilizes the original (unquantized) LPCs, while the formant synthesis filter employs the quantized LPCs. Hence, cascade connection of these filters does not produce the modified formant synthesis filter, as for the case of the generic CELP coder. The advantage of this approach is that, in theory, perceptual weighting is more accurate, since no distortion due to quantization is introduced. The price to pay is extra complexity. Due to the need to have the prediction-error signal (r) in the adaptive codebook search (Section 16.2), the configuration in Figure 16.9 is deployed. As explained later, this scheme provides some advantage in fixed-point implementation.

It is possible to further simplify the system by combining the top two layers of filters. Figure 16.10 summarizes the idea. Two filters—marked as #1 and #2—require special attention. For filter #1, its state is given by the synthesis error $e = s - \hat{s}$. Thus, output of the filter in response to r (prediction error) with initial

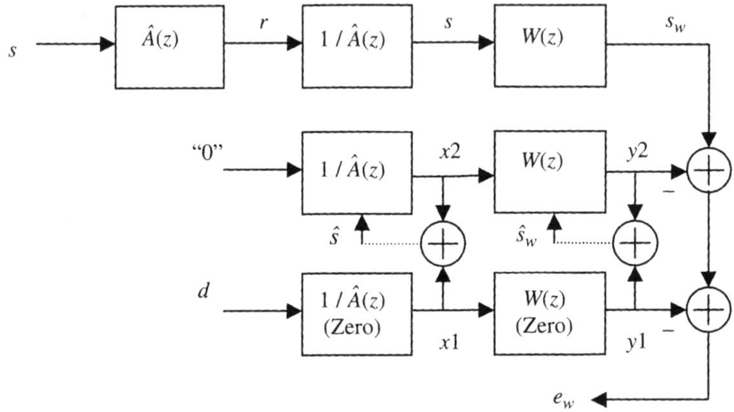

Figure 16.9 Signals involved in G.729 encoding: equivalent realization as in Figure 16.1.

state given by s (input speech) is equal to s; while output in response to zero input with initial state given by $\hat{s}$ (synthetic speech) is equal to $x2$ (Figure 16.9). Therefore, the overall response is given by $s - x2$. For filter #2, its state is given by the weighted synthesis error $e_w = s_w - \hat{s}_w$. Output of the filter in response to s with initial state given by s_w (weighted input speech) is equal to s_w; while output in response to $-x2$ with initial state $-\hat{s}_w$ is equal to $-y2$ (Figure 16.9). The total response is hence $s_w - y2$. We conclude that the two systems (Figures 16.9 and 16.10) are equivalent. These arguments are valid due to linearity of the underlying filters. The diagram in Figure 16.2 is essentially the same as that in Figure 16.10 with all components made explicit.

The advantage of this configuration is its higher performance under a fixed-point environment. In the generic approach and in Figure 16.9, zero-input response of the filter ($x2$) is calculated separately and is subtracted later from the perceptually weighted speech. The problem is that samples of the zero-input response decay rapidly toward zero, due to the fact that the filter is stable. Under fixed-point,

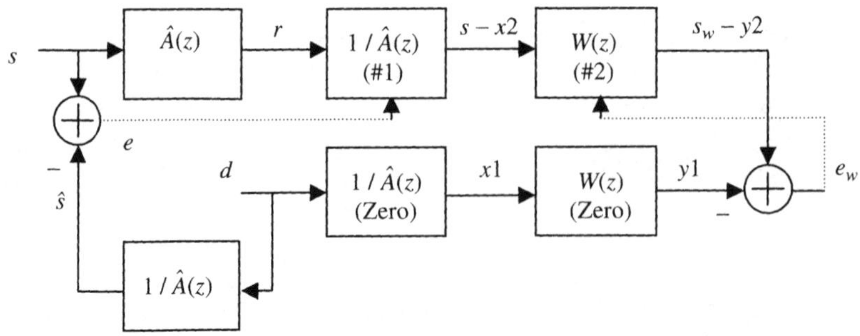

Figure 16.10 Signals involved in G.729 encoding: another equivalent realization.

low-magnitude samples cannot be represented faithfully, leading to relatively high quantization errors. The system in Figure 16.10 avoids computing the zero-input response separately. As we can see, the filters process at all time the relatively high-amplitude input speech. By initializing the filters appropriately, equivalent functionality is achieved. Therefore, no weak signals are computed in isolation, leading to higher precision for the final outcomes.

Overview of Encoding Operations

Figure 16.11 shows the block diagram of the G.729 encoder. The input speech signal is highpass filtered and segmented into 80-sample frames (10 ms), subdivided into two 40-sample subframes (see Exercise 16.6).

LP analysis is performed once per frame (Chapter 15). Two sets of LPCs are obtained at the end: original and quantized. Both sets are interpolated so as to apply to each individual subframe. The original LPCs (after interpolation) are used by the

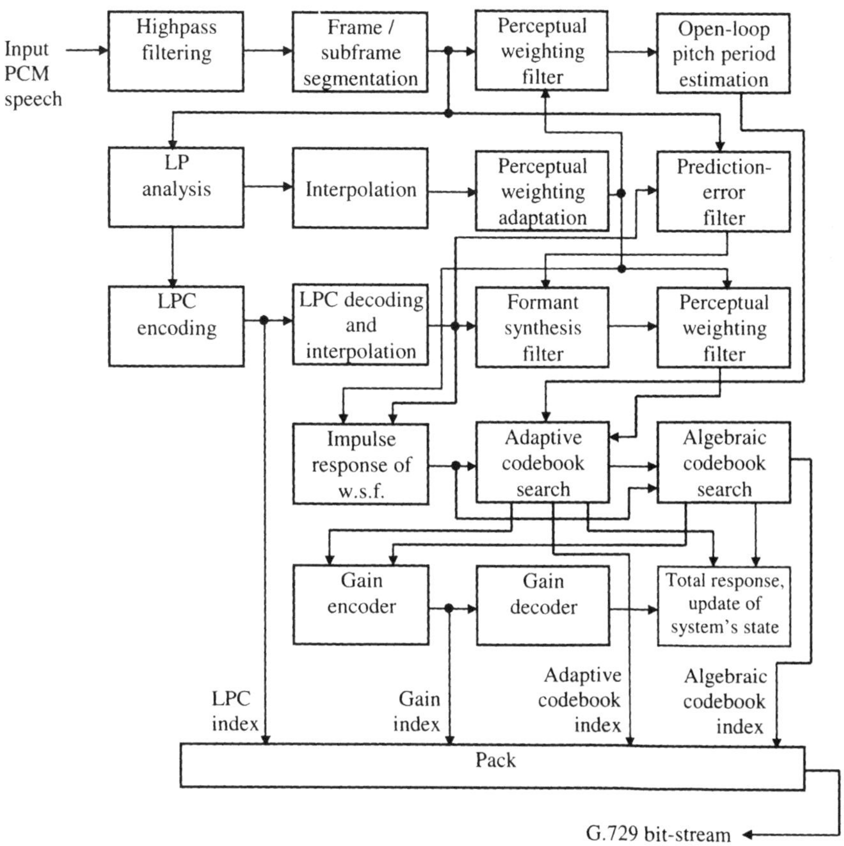

Figure 16.11 Block diagram of the G.729 encoder (w.s.f. means weighted synthesis filter).

TABLE 16.1 Bit Allocation for the G.729 Coder[a]

Parameter	Number per Frame	Resolution	Total Bits per Frame
LPC index	1	18	18
Pitch period (adaptive codebook index)	2	8, 5	13
Parity bit for pitch period	1	1	1
Algebraic codebook index	2	17	34
Gain index	2	7	14
Total			**80**

[a] Data from ITU [1996a], Table 1.

perceptual weighting filter. The input speech subframe is then filtered by it, with the output used in the open-loop pitch period estimation. The procedure for perceptual weighting adaptation and pitch period estimation is explained in Section 16.2.

The impulse response of the filter cascade between the perceptual weighting filter and the formant synthesis filter is calculated. The resultant filter with system function $W(z)/\hat{A}(z)$ is referred to as the weighted synthesis filter. Note that the formant synthesis filter utilizes the quantized LPCs $\hat{a}_i$, with system function

$$\frac{1}{\hat{A}(z)} = \frac{1}{1 + \sum_{i=1}^{10} \hat{a}_i z^{-i}}. \tag{16.15}$$

Target signal $u3$ for the adaptive codebook search is found using the scheme in Figure 16.2, with proper initialization for the filters. Computation of the adaptive codebook gain is identical to that of the FS1016, with the resultant gain bounded between 0 and 1.2. See Section 16.4 for the algebraic codebook search, and Section 16.5 for the gain quantization.

Table 16.1 summarizes the bit allocation scheme of the G.729 coder. A total of 80 bits are allocated per frame, leading to a bit-rate of 8000 bps. One parity bit is transmitted per frame for error detection.

Summary of Decoding Operations

Figure 16.12 shows the decoder. The algebraic codevector is found from the received index, specifying positions and signs of the four pulses, which is filtered by the pitch synthesis filter in zero state. Parameters of this filter are specified in Section 16.4. The adaptive codevector is recovered from the integer part of the pitch period and interpolated according to the fractional part of the pitch period. Algebraic and adaptive codevectors are scaled by the decoded gains and added to form the excitation for the formant synthesis filter. A postfilter is incorporated to improve subjective quality.

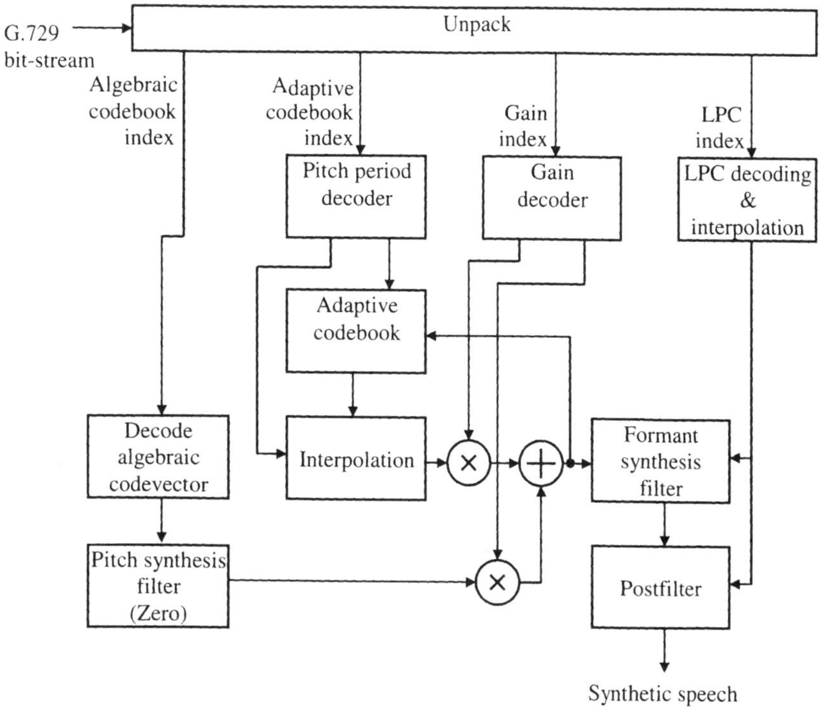

Figure 16.12 Block diagram of the G.729 decoder.

16.4 ALGEBRAIC CODEBOOK SEARCH

The different techniques for the algebraic codebook search are explained in this section.

Exhaustive Search

The 17-bit codebook is searched in such a way that (Chapter 11)

$$P^{(l)} = \frac{\left(\mathbf{u}_o^T \mathbf{H} \mathbf{v}^{(l)}\right)^2}{\mathbf{v}^{(l)T} \mathbf{H}^T \mathbf{H} \mathbf{v}^{(l)}} \tag{16.16}$$

is maximized, with $l = 0, \ldots, 2^{17} - 1$ being the index to the codebook; $\mathbf{u}_o = \mathbf{u}3 - \mathbf{y}2$ is the target vector; $\mathbf{H}$ is the impulse response matrix of the cascade connection between $H_p(z)$ and $W(z)/\hat{A}(z)$; and $\mathbf{v}^{(l)}$ is the excitation codevector corresponding to the index l. During the actual search, the vector

$$\mathbf{u}_h = \mathbf{H}^T \mathbf{u}_o = \begin{bmatrix} h[0] & h[1] & \cdots & h[39] \\ 0 & h[0] & \cdots & h[38] \\ \vdots & \vdots & \ddots & \vdots \\ 0 & 0 & \cdots & h[0] \end{bmatrix} \begin{bmatrix} u_o[0] \\ u_o[1] \\ \vdots \\ u_o[39] \end{bmatrix} \tag{16.17}$$

is precomputed and stored. Elements of the vector are

$$u_h[n] = \sum_{i=n}^{39} u_o[i]h[i-n]; \quad n = 0, \ldots, 39, \tag{16.18}$$

representing the correlation between the target vector and the impulse response. On the other hand,

$$\Phi = \mathbf{H}^T\mathbf{H} \tag{16.19}$$

is the autocorrelation matrix of the impulse response; the matrix is symmetric with elements

$$\phi[i,j] = \sum_{n=j}^{39} h[n-i]h[n-j]; \quad i = 0, \ldots, 39, j = i, \ldots, 39. \tag{16.20}$$

Based on (16.17) and (16.19), (16.16) can be rewritten as

$$P^{(l)} = \frac{\left(\mathbf{u}_h^T\mathbf{v}^{(l)}\right)^2}{\mathbf{v}^{(l)T}\Phi\mathbf{v}^{(l)}} = \frac{\left(C^{(l)}\right)^2}{G^{(l)}}. \tag{16.21}$$

In order to search the codebook and locate the optimal codevector, $P^{(l)}$ must be computed 2^{17} times to find the peak. Due to the structure of the codebook, computation of $P^{(l)}$ is not as complex as it appears from (16.21); after all, each codevector has only four nonzero samples. It is simple to verify that

$$C^{(l)} = \sum_{i=0}^{3} s_i^{(l)} u_h\left[m_i^{(l)}\right], \tag{16.22}$$

where s_i and m_i are the signs and positions of the nonzero pulses in the codevector. Similarly, the denominator of (16.21) is given by

$$\begin{aligned} G^{(l)} &= \sum_{j=0}^{3}\sum_{i=0}^{3} s_j^{(l)} s_i^{(l)} \phi\left[m_i^{(l)}, m_j^{(l)}\right] \\ &= \sum_{i=0}^{3} \phi\left[m_i^{(l)}, m_i^{(l)}\right] + \sum_{j=0}^{3}\sum_{i=0, i\neq j}^{3} s_j^{(l)} s_i^{(l)} \phi\left[m_i^{(l)}, m_j^{(l)}\right]. \end{aligned} \tag{16.23}$$

Taking into account that Φ is symmetric—$\phi[i,j] = \phi[j,i]$—we have

$$G^{(l)} = \sum_{i=0}^{3} \phi\left[m_i^{(l)}, m_i^{(l)}\right] + 2\sum_{j=0}^{2}\sum_{i=j+1}^{3} s_j^{(l)} s_i^{(l)} \phi\left[m_i^{(l)}, m_j^{(l)}\right]. \tag{16.24}$$

Thus, (16.22) and (16.24) can be utilized to compute $P^{(l)}$ for all l so as to locate the optimal codevector, and it must be done 2^{17} times since that is the size of the codebook; this number represents the amount of work required for exhaustive search.

Suboptimal Search

From (16.22) we see that $C^{(l)}$ will have a high probability of being maximized when the condition

$$s_i^{(l)} = \operatorname{sgn}\left(u_h\left[m_i^{(l)}\right]\right) \tag{16.25}$$

is satisfied for all combinations of i and l, where $\operatorname{sgn}(x) = \pm 1$ depending on the sign of x. In other words,

$$\operatorname{sgn}\left(v^{(l)}[n]\right) = \operatorname{sgn}(u_h[n]) \tag{16.26}$$

for all l and $n = 0, \ldots, 39$. That is, signs of the excitation codevector samples are set to be equal to that of the correlation sequence $u_h[n]$. By fixing the signs of each codevector sample according to $u_h[n], P^{(l)}$ needs to be evaluated only 2^{13} times, instead of 2^{17} times, since the 4 bits associated with the signs are effectively removed from the search loop. Under (16.25) or (16.26), (16.22) becomes

$$C^{(l)} = \sum_{i=0}^{3} \left| u_h\left[m_i^{(l)}\right] \right|. \tag{16.27}$$

Note that (16.20) can be modified to include the values of the sign as follows:

$$\phi'[i,j] = \begin{cases} \phi[i,j], & \text{if} \quad i = j, \\ \frac{1}{2}\operatorname{sgn}(u_h[i]) \operatorname{sgn}(u_h[j]) \phi[i,j], & \text{otherwise}. \end{cases} \tag{16.28}$$

Then (16.24) becomes

$$G^{(l)} = \sum_{i=0}^{3} \phi'\left[m_i^{(l)}, m_i^{(l)}\right] + \sum_{j=0}^{2} \sum_{i=j+1}^{3} \phi'\left[m_i^{(l)}, m_j^{(l)}\right]. \tag{16.29}$$

Therefore, (16.27) and (16.29) can be utilized to compute $P^{(l)}$, with the range of l a subset of the widest possible range $[0, 2^{17} - 1]$. Note that there is no guarantee that the codevector found in this way is the absolute best; however, in practice, the suboptimal search approach provides performance very close to exhaustive search.

The search loop is described by the following pseudocode. It is assumed that $u_h[n]$ and $\phi'[i,j]$ for all i, j, and n are known before entering the loop.

```
1.  peak ← 0
2.  for i0 ← 0 to 7
3.      for i1 ← 0 to 7
4.          for i2 ← 0 to 7
5.              for i3 ← 0 to 15
6.                  m0 ← 5i0; m1 ← 5i1 + 1; m2 ← 5i2 + 2;
7.                  if i3 > 7
8.                      m3 ← 5(i3 − 8) + 4;
9.                  else
10.                     m3 ← 5i3 + 3;
11.                 P ← C²[m0, m1, m2, m3]/G[m0, m1, m2, m3];
12.                 if P > peak
13.                     peak ← P
14.                     m'0 ← m0; m'1 ← m1; m'2 ← m2; m'3 ← m3
```

Note that the search time associated with the fourth pulse (Line 5) is the longest, since it has a total of 16 possible positions. In Line 11, (16.27) and (16.29) are utilized to find C and G for the four position values. Upon completion, the final positions m_0', m_1', m_2', and m_3' are returned as the results.

Suboptimal Search with Reduced Complexity

One of the biggest computational burdens for the suboptimal search method described before is the loop involved with the fourth pulse, which has 16 different positions. This loop is entered $8^3 = 512$ times. By limiting the number of times that this last loop is entered, a great deal of computation can be eliminated.

By keeping track of the relative contribution of the first three pulses toward the objective of maximizing (16.21), it is possible to make the (heuristic) decision not to enter the fourth search loop. This is done by setting a reference threshold to which that the contribution of the first three pulses is compared. This threshold is given by

$$C_{\text{th}} = C_{\text{avg}} + \left(C_{\max} - C_{\text{avg}}\right)\alpha, \tag{16.30}$$

where C_{avg} is the average correlation due to the first three pulses,

$$C_{\text{avg}} = \frac{1}{8}\left(\sum_{n=0}^{7} |u_h[5n]| + \sum_{n=0}^{7} |u_h[5n+1]| + \sum_{n=0}^{7} |u_h[5n+2]|\right); \tag{16.31}$$

$C_{\max}$ is the maximum correlation due to the first three pulses,

$$C_{\max} = \max_{n=0,\ldots,7} |u_h[5n]| + \max_{n=0,\ldots,7} |u_h[5n+1]| + \max_{n=0,\ldots,7} |u_h[5n+2]|; \tag{16.32}$$

and the parameter α controls the number of times the last loop is entered. Experimentally, it was found that $\alpha = 0.4$ provides performance very close to exhaustive search [Salami et al., 1998].

Note that for the previous suboptimal search technique, the last loop is entered $N = 2^{3+3+3} = 512$ times. For the present method and on an experimental basis, with $\alpha = 0.4$, the average value of N is 60 and only 5% of the time exceeds 90. To limit the worst case complexity, the value of N in the two subframes is limited to a maximum value of $N_{\max} = 180$. The first subframe is allowed a maximum of $N_1 = 105$, and the second subframe is left with $N_2 = N_{\max} - N_1$. Thus, during codebook search for a particular subframe, if the last loop is entered more than the maximum allowable times, the whole search process is stopped with the best results found so far returned. In practice only 1% of the subframes have to be stopped due to the limits being exceeded. Pseudocode for this improved search method is given below, where it is assumed that C_{th} is known before initiating the search process.

```
1.  peak ← 0; count ← 0;
2.  for i0 ← 0 to 7
3.          for i1 ← 0 to 7
4.                  for i2 ← 0 to 7
5.                          m0 ← 5i0; m1 ← 5i1+1; m2 ← 5i2 + 2;
6.                          C ← |uh[m0]| + |uh[m1]| + |uh[m2]|
7.                          if C < Cth
8.                                continue;
9.                          count++;
10.                         if count > MAX_N
11.                               goto end
12.                         for i3 ← 0 to 15
13.                               if i3 > 7
14.                                      m3 ← 5(i3 - 8) + 4
15.                               else
16.                                      m3 ← 5i3 + 4
17.                               C1 ← C + |uh[m3]|
18.                               P ← C1^2/G[m0, m1, m2, m3]
19.                               if P > peak
20.                                      peak ← P
21.                                      m0' ← m0; m1' ← m1; m2' ← m2; m3' ← m3;
22. end:
```

As we can see, a variable *count* is incorporated to track the number of times that the innermost loop is entered. If *count* is greater than the pre-established limit (*MAX_N*), the whole process is terminated immediately (Line 11). The contribution of the first three pulses is calculated (Line 6) and compared to the threshold (Line 7); if it is below threshold, the last loop will not be entered (Line 8); otherwise, the loop is entered if *count* is below the maximum allowable number.

Pitch Synthesis Filter

Many CELP-based speech coding standards, such as the FS1016 and the IS54, have eliminated the pitch synthesis filter in the filtering path during the excitation codebook search. This is done for simplicity reasons since at a pitch period greater than the subframe length, the effect of the filter is void (Chapter 12). In many instances, long-term periodicity can effectively be generated by the adaptive codebook alone, which is the motive for its exclusion in the filtering path.

Voices with low pitch period will have more than one pitch pulse in a subframe. In that situation, it is beneficial to introduce periodicity to the excitation codevector from the algebraic codebook so as to enhance the overall periodicity of the signal. This can effectively be generated through the pitch synthesis filter, with system function

$$H_p(z) = \frac{1}{1 + bz^{-T}}. \tag{16.33}$$

In using (16.33), one of the obvious questions is the determination of the parameters b and T. One particular scheme that works well within the context of G.729 is to use a rounded value of the pitch period (or adaptive codebook index) from the current subframe and the quantized adaptive codebook gain of the past subframe. This is due to the fact that, for the current subframe, the quantized adaptive codebook gain is not yet known. Thus,

$$b = -\hat{b}_{\text{past}} \tag{16.34}$$

with $\hat{b}_{\text{past}}$ being the quantized adaptive codebook gain of the past subframe, bounded by [0.2, 0.8].

For $T < 40$, the introduction of the pitch synthesis filter requires computing the signal

$$d1[n] = \begin{cases} v[n], & 0 \leq n \leq T - 1, \\ v[n] - bd1[n - T], & T \leq n \leq 39, \end{cases} \tag{16.35}$$

where $d1[n]$ is the output of the pitch synthesis filter. This signal is needed for the adaptive codebook update and is calculated only after the optimal excitation codevector $v[n]$ is located.

Addition of the pitch synthesis filter does not modify the general procedure for the excitation codebook search. If the impulse response of the weighted synthesis filter is denoted by $h_w[n]$, then the total impulse response is

$$h[n] = \begin{cases} h_w[n], & 0 \leq n \leq T - 1, \\ h_w[n] - bh_w[n - T], & T \leq n \leq 39, \end{cases} \tag{16.36}$$

since the pitch synthesis filter is in cascade with the weighted synthesis filter. Equation (16.36) represents the impulse response of $H_p(z)W(z)/\hat{A}(z)$.

16.5 GAIN QUANTIZATION USING CONJUGATE VQ

After locating the adaptive and algebraic codevectors, the associated gains are vector quantized; this is done based on a conjugate structure VQ system. The objective of quantization is the minimization of

$$J = \| \mathbf{u3} - b \cdot \mathbf{y2}_o - g \cdot \mathbf{y1}_o \|^2, \tag{16.37}$$

where:

- $\mathbf{u3}$ = Target vector for adaptive codebook search.
- $\mathbf{y2}_o$: Zero-state response of weighted synthesis filter to adaptive codevector (no scaling).
- $\mathbf{y1}_o$: Zero-state response of the cascade connection between pitch synthesis filter and weighted synthesis filter to algebraic codevector (no scaling).
- b: Adaptive codebook gain.
- g: Algebraic codebook gain.

The objective of quantization is to choose quantized values of b and g so as to minimize (16.37).

Quantization of Algebraic Codebook Gain

Given the codevector $\mathbf{d1}_r$, which is the algebraic codevector filtered by the pitch synthesis filter with no scaling, its power is calculated as the mean energy:

$$Ed_r = 10\log\left(\frac{1}{40}\sum_{n=0}^{39} d1_r^2[n]\right), \tag{16.38}$$

with r being the subframe index. Note that $\mathbf{d1}_r$ has only a few nonzero elements, and when the pitch period of the past subframe is greater than 40, $\mathbf{d1}_r$ is the same as the algebraic codevector with only four nonzero pulses. After scaling $\mathbf{d1}_r$ with the gain g_r, the power of the resultant vector becomes $Ed_r + 20\log g_r$. Let Eo_r be the mean–removed power of the scaled algebraic codebook contribution:

$$Eo_r = 20\log g_r + Ed_r - 30, \tag{16.39}$$

with 30 dB being the experimentally found mean. The gain g_r can then be expressed by

$$g_r = 10^{(Eo_r - Ed_r + 30)/20}. \tag{16.40}$$

This gain is written as

$$g_r = \gamma_r g'_r, \tag{16.41}$$

with γ a correction factor, and g' the predicted gain of the rth subframe. In actual implementation, it is the factor γ that is being quantized. For gain prediction, we consider first the power prediction:

$$Eo'_r = \sum_{i=1}^{4} b_i \hat{U}_{r-i}, \tag{16.42}$$

with $\{b_1, b_2, b_3, b_4\} = \{0.68, 0.58, 0.34, 0.19\}$ being the MA coefficients for the predictor, and $\hat{U}$ the quantized version of prediction error. That is,

$$U_r = Eo_r - Eo'_r. \tag{16.43}$$

The predicted gain is found by replacing Eo_r by its predicted value in (16.40):

$$g'_r = 10^{(Eo'_r - Ed_r + 30)/20}. \tag{16.44}$$

Substituting (16.39) and (16.44) in (16.43), one can show that

$$U_r = 20 \log \gamma_r. \tag{16.45}$$

The encoder of the gain quantizer is shown in Figure 16.13, where we can see that the quantized gain is given by

$$\hat{g}_r = \hat{\gamma}_r g'_r. \tag{16.46}$$

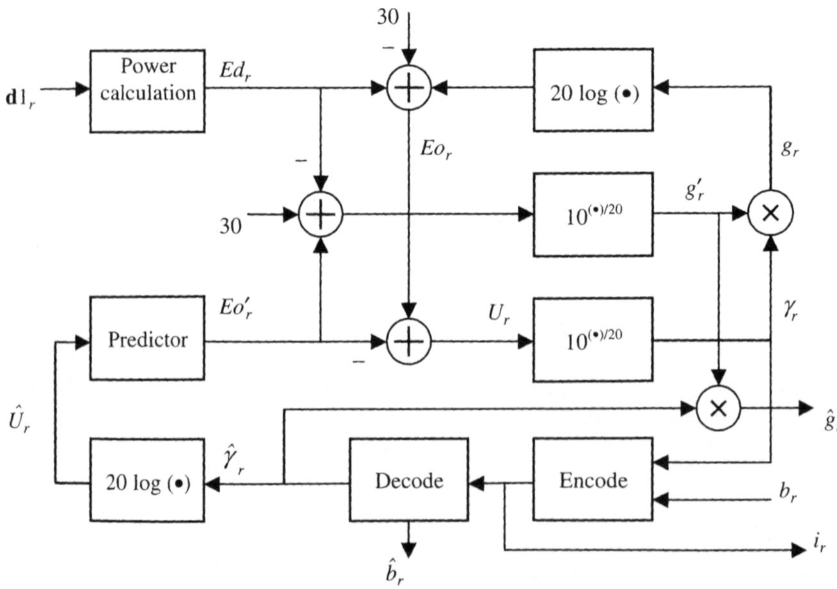

Figure 16.13 Quantization of algebraic codebook gain (G.729 encoder).

The quantization process is done by searching the available candidates in the codebook and constructing the quantized gain values according to (16.46). The particular value that minimizes (16.37) is selected as the final result. See Exercise 16.11 for the corresponding gain decoder.

VQ Structure and Search Technique

The adaptive codebook gain b and the correction factor γ are vector quantized using a two-stage conjugate-structured codebook. The first stage consists of a 3-bit 2-D codebook, and the second stage a 4-bit 2-D codebook. Denote the first stage codewords as

$$\mathbf{y}_{i_1}^{(1)} = \begin{bmatrix} y_{i_1,1}^{(1)} & y_{i_1,2}^{(1)} \end{bmatrix}^T; \quad i_1 = 1, \ldots, 8,$$

and the second stage codewords as

$$\mathbf{y}_{i_2}^{(2)} = \begin{bmatrix} y_{i_2,1}^{(2)} & y_{i_2,2}^{(2)} \end{bmatrix}^T; \quad i_2 = 1, \ldots, 16.$$

The first element of each codevector contains a fraction of the quantized adaptive codebook gain, while the second element represents a fraction of the quantized algebraic codebook gain correction factor. Given the indices i_1 and i_2, the quantized adaptive codebook gain is given by

$$\hat{b} = y_{i_1,1}^{(1)} + y_{i_2,1}^{(2)}, \tag{16.47}$$

and the quantized fixed codebook gain is given by

$$\hat{g} = g'\hat{\gamma} = g'\left(y_{i_1,2}^{(1)} + y_{i_2,2}^{(2)}\right). \tag{16.48}$$

A full search procedure can be applied where all candidates are plugged into (16.37) to evaluate the error, and afterwards the codewords with the lowest error are selected; this requires the evaluation of $8 \cdot 16 = 128$ times the error expression (16.37).

A suboptimal search technique that requires only $4 \cdot 8 = 32$ error computations using (16.37) is described. In this method, the optimal gains are derived from (16.37). Note that (16.37) can be written in the expanded form:

$$J = \mathbf{u3}^T\mathbf{u3} + b^2\mathbf{y2}_o^T\mathbf{y2}_o + g^2\mathbf{y1}_o^T\mathbf{y1}_o - 2b\mathbf{u3}^T\mathbf{y2}_o - 2g\mathbf{u3}^T\mathbf{y1}_o + 2bg\mathbf{y1}_o^T\mathbf{y2}_o. \tag{16.49}$$

Differentiating with respect to b and g and equating the results to zero leads to the following system of equations:

$$\begin{bmatrix} \mathbf{y2}_o^T\mathbf{y2}_o & \mathbf{y1}_o^T\mathbf{y2}_o \\ \mathbf{y1}_o^T\mathbf{y2}_o & \mathbf{y1}_o^T\mathbf{y1}_o \end{bmatrix} \begin{bmatrix} b \\ g \end{bmatrix} = \begin{bmatrix} \mathbf{u3}^T\mathbf{y2}_o \\ \mathbf{u3}^T\mathbf{y1}_o \end{bmatrix}, \tag{16.50}$$

which can easily be solved to obtain the optimal gain values.

The design of the conjugate VQ codebooks is such that the second elements (γ) of the first-stage codevectors are in general larger than the first elements (b). For the second codebook, the situation is reversed. This design allows the usage of a pre-selection technique to reduce computation.

Given the optimal g and therefore γ, the four first-stage codevectors with second element closest to the optimal value are selected. Using the optimal b, the eight second-stage codevectors with first element closest to the optimal value are selected. Hence, for each codebook, the best 50% candidate vectors are selected. This is followed by a search over the $4 \cdot 8 = 32$ possibilities, such that the combination of the two indices minimizes (16.37) or (16.49).

The suboptimal search method reduces the complexity without noticeable loss of quality compared to the full search. The codebooks can be designed using an iteration process in which one of the codebooks is optimized while the other is kept fixed.

16.6 OTHER ACELP STANDARDS

Salient features of other ACELP-based standards are described in this section; since all these were developed by the same group of researchers, they share many common features. Interested readers are referred directly to the indicated references for full details.

ITU-T G.723.1 MP-MLQ/ACELP

The G.723.1 coder was initiated for very-low-bit-rate videophone applications. It started with the screening of ten candidate coders in 1993. In 1994 two coders met the proposed requirements, and subsequently the two were merged to form a dual rate coder, with the draft of the standard finalized in 1995 [Cox, 1997].

Since low-bit-rate videophones only send a few image frames per second, the involved delay is relatively high. The researchers at the time felt that a 30-ms frame length would be acceptable. A longer frame is normally associated with lower bit-rate, since the speech samples can be analyzed and represented more efficiently. When compared with G.729, the bit-rate of 5.3 kbps (low rate) and 6.3 kbps (high rate) is lower, but the minimum coding delay of 67.5 ms is higher (8 kbps and 25 ms for the G.729; see Chapter 15—Exercise section).

The G.723.1 encoder utilizes the same analysis-by-synthesis principle of most CELP coders. LP analysis, LPC quantization, and LPC interpolation are covered in Chapter 15. For the two even subframes, pitch period is encoded with 7 bits; while for the two odd subframes, pitch period is encoded differentially from a previous value using 2 bits. The pitch synthesis filter has a five-tap structure with the coefficients vector quantized.

ACELP is the adopted method for low bit-rate in the G.723.1 standard. The algebraic codebook is addressed by 17 bits, with each codevector containing at

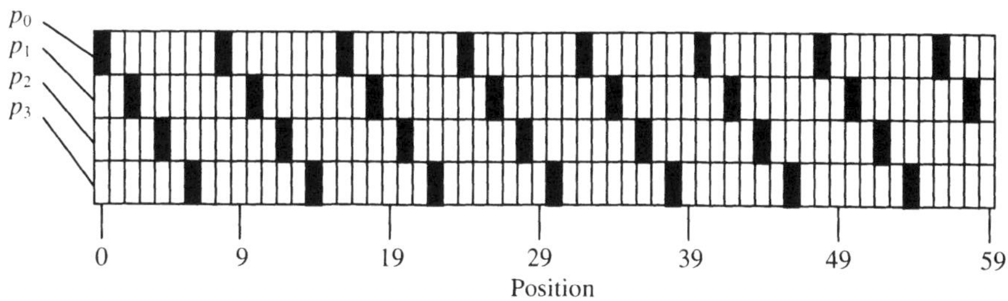

Figure 16.14 Positions of individual pulses in the algebraic codebook for the G.723.1 coder, indicated by the black rectangles. Data from ITU [1996b], Table 1.

most four nonzero pulses. Each pulse can have the amplitudes either of 1 or −1 and can assume the positions given in Figure 16.14. Each excitation codevector has 60 samples, which is the length of a subframe. The excitation codevector $v[n]$ is constructed by summing the four pulses according to

$$v[n] = \sum_{i=0}^{3} p_i[n] = \sum_{i=0}^{3} s_i \delta[n - m_i + grid], \tag{16.51}$$

where $s_i = \pm 1$ is the sign of the pulse, m_i are the positions according to Figure 16.14, and $grid = 0$ when even positions are used (just as in Figure 16.14), and $grid = 1$ when odd positions are utilized. Each pulse position is encoded with 3 bits and each pulse sign is encoded with 1 bit. This gives a total of 16 bits for the four pulses, plus one grid bit leading to a total of 17 bits.

High bit-rate excitation utilizes a similar principle but with more pulses. The scheme is referred to as *multipulse maximum likelihood quantization* (MP-MLQ). For even subframes, 6 pulses are used; for odd subframes, 5 pulses are used. Restriction on the pulse positions is that they can either be all odd or all even, selected with a grid bit. Positions, signs, and grid are determined using analysis-by-synthesis. Excitation gain is scalar quantized. Tables 16.2 and 16.3 show the bit-allocation scheme for the G.723.1 coder, resulting in the bit-rates of 6300 and 5266.67 bps.

TIA IS641 ACELP

Due to advances in speech coding technology, TIA launched a standardization process around 1995 to search for a substitute for the existent IS54 standard of the North American TDMA system. The process culminated in 1996 when the coder proposal submitted jointly by Nokia and the University of Sherbrooke was approved as the IS641 standard. The coder was found to be superior to IS54 not only under error-free conditions, but also in the presence of channel errors and background noise [Salami et al., 1997c].

The coder is based on the ACELP algorithm, with a source bit-rate of 7.4 kbps and a channel bit-rate of 5.6 kbps. It operates on speech frames of 20 ms, divided

TABLE 16.2 Bit Allocation for the G.723.1 Coder at High Bit-Rate[a]

Parameter	Number per Frame	Resolution	Total Bits per Frame
LPC index	1	24	24
Pitch period	4	7, 2, 7, 2	18
All the gains combined	4	12	48
Pulse positions	4	20, 18, 20, 18	73[b]
Pulse signs	4	6, 5, 6, 5	22
Grid bit	4	1	4
Total			**189**

[a] Data from ITU [1996b], Table 2.
[b] A special packaging technique is used to achieve this number.

into four 5-ms subframes. The extracted parameters are typical of CELP coders, with the bit-allocation scheme shown in Table 16.4. General operation of the coder is very similar to G.729. In fact, it uses the same algebraic codebook.

ETSI GSM EFR ACELP

In order to improve on the existent GSM full-rate coder (6.10 RPE-LTP) and half-rate coder (6.20 VSELP), ETSI started a prestudy phase in 1994, where a set of essential requirements was drafted and technical feasibility was assessed. The goals include high quality at error-free condition (with ITU-T G.728 LD-CELP as a reference) and to achieve significant improvement with respect to the existing GSM full-rate coder under severe error conditions, providing graceful degradation without annoying effects. The target coder was designated as *enhanced full rate* (EFR).

In addition, essential requirements were set for bit-rate, complexity, and delay. The same channel bit-rate of 22.8 kbps as for the existing full rate coder is used. The complexity was not to exceed that of the GSM half-rate coder, and coding delay was not to surpass that of the full-rate coder.

TABLE 16.3 Bit Allocation for the G.723.1 Coder at Low Bit-Rate[a]

Parameter	Number per Frame	Resolution	Total Bits per Frame
LPC index	1	24	24
Pitch period	4	7, 2, 7, 2	18
All the gains combined	4	12	48
Pulse positions	4	12	48
Pulse signs	4	4	16
Grid bit	4	1	4
Total			**158**

[a] Data from ITU [1996b], Table 3.

TABLE 16.4 Bit Allocation for the IS641 Coder[a]

Parameter	Number per Frame	Resolution	Total Bits per Frame
LPC index	1	26	26
Pitch period	4	8, 5, 8, 5	26
Gain index	4	7	28
Algebraic codebook index	4	17	68
Total			**148**

[a] Data from Salami et al. [1997c], Table 1.

Six candidate coders were submitted to the competitive EFR coder selection process launched by ETSI in 1995. After the first phase of testing, the ACELP algorithm jointly developed by Nokia and the University of Sherbrooke was selected in October. During 1996, verification tests for the EFR coder were completed, finalizing the new standard [Salami et al., 1997a].

The GSM EFR coder is based on ACELP and shares many common features, like its other ACELP cousins. It operates on 20-ms speech frames subdivided into four 5-ms subframes. In the encoder, the speech signal is analyzed with the parameters of the CELP model extracted. LP analysis, LPC quantization, and LPC interpolation are described in Chapter 15.

The fixed excitation codebook has an algebraic structure indexed by 35 bits. Each excitation codevector has 40 samples, which is the length of a subframe. The codevector is formed by a summation of 10 pulses. These pulses are basically shifted impulses with known position; they have unit magnitude and associated sign.

The index to the algebraic codebook contains information about the position and sign of each of the ten pulses. Three bits are allocated to the positions of each pulse, and 1 bit is used to encode the sign of two pulses. This leads to $10 \cdot 3 + 5 \cdot 1 =$ 35 bits for the codebook. Available positions of the pulses are summarized in Figure 16.15, where we can see that the ten pulses are separated into groups of

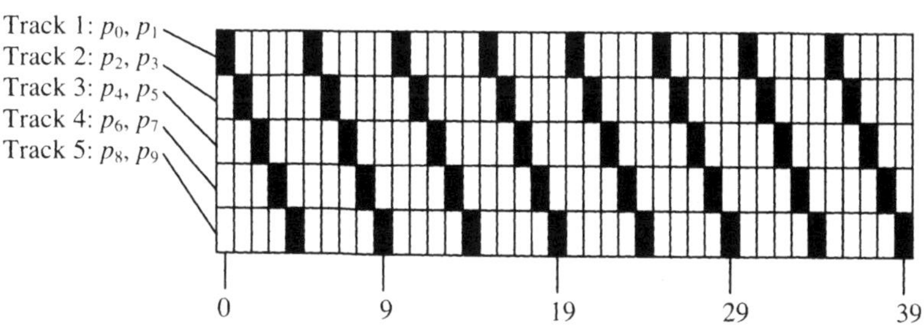

Figure 16.15 Positions of individual pulses in the algebraic codebook for the GSM EFR coder, indicated by the black rectangles. Data from Salami et al. [1997a], Table II.

TABLE 16.5 Examples of Pulses and Resultant Sequences from the First Track for the GSM EFR Coder

Position of p_0	Position of p_1	Sign	Resultant Sequence
0	5	+	
0	5	−	
5	0	+	
5	0	−	
5	5	+	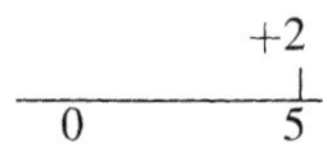
5	5	−	

two, called a track. Positions of the two pulses inside each track are encoded with 6 bits, or 8 positions per pulse, and one sign bit is assigned to each track. The sign bit indicates the sign of the first pulse, while the sign of the second pulse is determined by its relative position with respect to the first pulse. If the position of the second pulse is smaller, then it has opposite sign; otherwise it has the same sign as that of the first pulse. Table 16.5 uses a single track to show the various forms of signals encodable by the specified scheme. Note that two pulses in each track may overlap, resulting in a single pulse with amplitude +2 or −2.

The bit-allocation scheme is shown in Table 16.6, resulting in a source coding bit-rate of 12.2 kbps. For channel coding, a bit-rate of 10.6 kbps is used, resulting in a 22.8-kbps channel bit-rate.

TABLE 16.6 Bit Allocation for the GSM EFR Coder[a]

Parameter	Number per Frame	Resolution	Total Bits per Frame
LPC index	1	38	38
Pitch period	4	9, 6, 9, 6	30
Adaptive codebook gain	4	4	16
Algebraic codebook index	4	35	140
Algebraic codebook gain	4	5	20
Total			**244**

[a] Data from Salami et al. [1997a], Table 1.

ETSI AMR ACELP

The adaptive multirate (AMR) coder standardized by ETSI in 1999 pertains to the family of coders called *network-controlled multimode*. Precise definition for multimode coding is given in Chapter 1. Network control means that the coder responds to an external control signal to switch the data rate to one of a predetermined set of rates. The control signal is assumed to be remotely generated, typically in response to traffic levels in the network or in response to requests for signaling information.

The AMR coder consists of a family of fixed-rate coders, each having a different rate. There are a total of eight coders and all are based on the ACELP algorithm. The available bit-rates are 12.2, 10.2, 7.95, 7.40, 6.70, 5.90, 5.15, and 4.75 kbps. At the highest bit-rate of 12.2 kbps, the algorithm is identical to GSM EFR. At lower bit-rates, the available bits are diminished, and different schemes are used to allocate the bits to various parameters of the ACELP model [ETSI, 1999].

16.7 SUMMARY AND REFERENCES

Principles of ACELP are covered in this chapter, with detailed description of the G.729 coder and abbreviated recitals of four other well-known ACELP standards. It is shown that ACELP follows the fundamental structure of CELP, with a fixed-codebook design that allows fast search with no storage requirement. Particularly for the G.729, various innovations in the form of refined structure with added complexity are incorporated. The extra complexities essentially reflect the progress in speech coding development, where more sophistication is added to the coder framework for general improvement.

See Adoul and Lamblin [1987] and Adoul et al. [1987] for initial ideas on ACELP. Details of the G.729 coder are given in ITU [1996a] and Salami et al. [1998]; quality of the coder under various test conditions is summarized in Perkins et al. [1997]. A faster codebook search technique is described in Salami et al. [1997d, 1997e]; the resultant coder—together with other changes—is published separately by the ITU as G.729 Annex A, or G.729a, which is bit-stream compatible with the G.729. There are variations of the G.729 standard that operate at bit-rates other than 8 kbps. For instance, Annex D of the G.729 specifies a 6.4-kbps coder utilizing the same CS-ACELP principle [ITU, 1998b]. Details of the G.723.1 coder are given in ITU [1996b]; In Cox [1997], comparison is made between G.729, G.723.1, and G.729a. See Salami et al. [1997c] and Salami et al. [1997a] for descriptions of the IS641 and GSM EFR, respectively. Issues regarding implementation of the GSM EFR coder on a DSP platform are described in Du et al. [2000]. Details of the AMR coder are given in ETSI [1999]. Note that documentations and reference codes of most ETSI standards are available free of charge at their official website: www.etsi.org.

EXERCISES

16.1 For the G.729 coder, given the pulse positions $\{m_0, m_1, m_2, m_3\} = \{20, 31, 12, 18\}$, find *pindex*, the associated position index. Repeat for $\{0, 16, 33, 29\}$.

16.2 Given *pindex* (16.3) in the G.729 coder, specify the procedure to recover the four pulse positions m_0, m_1, m_2, and m_3.

16.3 For the G.729 coder, count the number of pitch period values available for the first subframe and the second subframe, and confirm the fact that they can be encoded using 8 and 5 bits, respectively.

16.4 For the G.729 coder, if the open-loop pitch period estimate is $T_{op} = 22$, what is the search range for the first subframe? If $T_1 = 23\ \frac{1}{3}$, what is the search range for the second subframe?

16.5 For the G.729 coder, the interpolation weights $w13[n]$, $n = 0$ to 12 and $w31[n]$, $n = 0$ to 30 are given by the following:

(a) $w13$: {0.901, 0.760, 0.424, 0.0841, −0.106, −0.121, −0.0476, 0.0163, 0.0312, 0.0157, 0, −0.00593, 0},

(b) $w31$: {0.899, 0.769, 0.449, 0.0959, −0.134, −0.179, −0.0849, 0.0370, 0.0955, 0.0689, 0, −0.0504, −0.0508, −0.0142, 0.0231, 0.0335, 0.0168, −0.00747, −0.0193, −0.0138, 0, 0.00940, 0.00903, 0.00238, −0.00366, −0.00503, −0.00241, 0.00105, 0.00278, 0.00215, 0}.

Gather some prediction error data from a real speech signal and perform an experiment similar to Example 16.1 so as to confirm some of the claims made in Section 16.2. The above numbers are rounded results; see ITU [1996a] for full precision specifications.

16.6 The G.729 coder specifies the following filter to process the input speech:

$$H(z) = \frac{0.46364 - 0.92725z^{-1} + 0.46364z^{-2}}{1 - 1.9059z^{-1} + 0.91140z^{-2}}.$$

Plot the magnitude response of the filter and describe its behavior. The above filter's coefficients are rounded results; see ITU [1996a] for full precision specifications.

16.7 Section 16.4 presented a suboptimal algebraic codebook search procedure for the G.729 coder. Extend the same idea to the G.723.1 coder, where the positions of the pulses for the algebraic codebook are given in Figure 16.14. Write down the procedure in the form of a pseudocode. Repeat for the GSM EFR coder, with the positions given in Figure 16.15.

16.8 Based on your knowledge of the G.729 standard, assess the validity of the statements below. Justify your answers.

(a) Algebraic codebook gain is always greater than or equal to zero.

(b) The exhaustive search procedure for the algebraic codebook, as explained in Section 16.4, cannot be implemented in practice, since the synthetic speech will suffer severe distortion.

If your assessment to the latter statement is true, redesign an exhaustive search procedure that will work well under the G.729 framework. Write down the pseudocode to perform the search.

16.9 Bozo proposed reducing the bit-rate of the G.729 coder by simply lengthening the frame length, say, for instance, from 10 ms to 12 ms, with all codebooks' parameters intact. Is Bozo's idea implementable? What would be the main problem?

16.10 For the ETSI GSM EFR ACELP coder, the positions (m) and signs (s) are given by

$$\{m_0 = 15, m_1 = 35, s_1 = 1\},$$
$$\{m_2 = 21, m_3 = 1, s_2 = -1\},$$
$$\{m_4 = 17, m_5 = 17, s_3 = 1\},$$
$$\{m_6 = 33, m_7 = 18, s_4 = 1\},$$
$$\{m_8 = 9, m_9 = 34, s_5 = -1\}.$$

Find out the pulses of the associated excitation sequence, specifying the positions and amplitudes.

16.11 Based on the block diagram of the gain encoder for the G.729 coder (Figure 16.13), draw the block diagram of the corresponding gain decoder, with inputs being the codevector **d**1 and index i, producing the quantized gain $\hat{g}$ at its output.

CHAPTER 17

MIXED EXCITATION LINEAR PREDICTION

The LPC coder, as studied in Chapter 9, uses a fully parametric model to efficiently encode the important information of the speech signal. The coder produces intelligible speech at bit-rates in the neighborhood of 2400 bps. However, the highly simplistic model in the core of the LPC coder generates annoying artifacts such as buzzes, thumps, and tonal noises; which is due to the many limitations of the model, thoroughly analyzed in Chapter 9.

The mixed excitation linear prediction (MELP) coder is designed to overcome some of the limitations of LPC. It utilizes a more sophisticated speech production model, with additional parameters to capture the underlying signal dynamics with improved accuracy. The essential idea is the generation of a mixed excitation signal as input to the synthesis filter, where the "mixing" refers to the combination of a filtered periodic pulse sequence with a filtered noise sequence. The benefits require an extra computational cost, practicable only with the powerful digital signal processors (DSPs) introduced in the mid-1990s.

Originally developed by McCree (see McCree and Barnwell [1995]) as a Ph.D. thesis, the MELP coder incorporated many research advances at the time, including vector quantization, speech synthesis, and enhancement to the basic LPC model. It was then refined and submitted as a candidate for the new U.S. federal standard at 2.4 kbps in 1996 [McCree et al., 1996] and officially became a federal standard in 1997 [McCree et al., 1997].

In this chapter, the speech production model that the MELP coder relies on is described and compared with LPC. Several fundamental processing techniques are analyzed next, followed by a detailed discussion on the encoder and decoder operations. The discussion is based on the core structure of the FS MELP coder, as described in the open literature. Readers must be aware that in order to implement a bit-stream-compatible version of the coder, consultation with official documentation is mandatory.

17.1 THE MELP SPEECH PRODUCTION MODEL

A block diagram of the MELP model of speech production is shown in Figure 17.1, which is an attempt to improve upon the existent LPC model. Comparing with the block diagram in Chapter 9, we can see that the MELP model is more complex. However, the two models do share some fundamental similarities; such as the fact that both rely on a synthesis filter to process an excitation signal so as to generate the synthetic speech. As explained later in the chapter, the MELP decoder utilizes a sophisticated interpolation technique to smooth out interframe transitions.

The main improvements of the MELP model with respect to the LPC model are highlighted below.

- A randomly generated period jitter is used to perturb the value of the pitch period so as to generate an aperiodic impulse train.

As mentioned in Chapter 9, one of the fundamental limitations in LPC is the strict classification of a speech frame into two classes: unvoiced and voiced. The MELP coder extends the number of classes into three: unvoiced, voiced, and *jittery voiced*. The latter state corresponds to the case when the excitation is aperiodic but not completely random, which is often encountered in voicing transitions. This jittery voiced state is controlled in the MELP model by the pitch jitter parameter and is essentially a random number. Experimentally, it was found that a period jitter uniformly distributed up to ±25% of the pitch period produced good results. The short isolated tones, often encountered in LPC coded speech due to misclassification of voicing state, are reduced to a minimum.

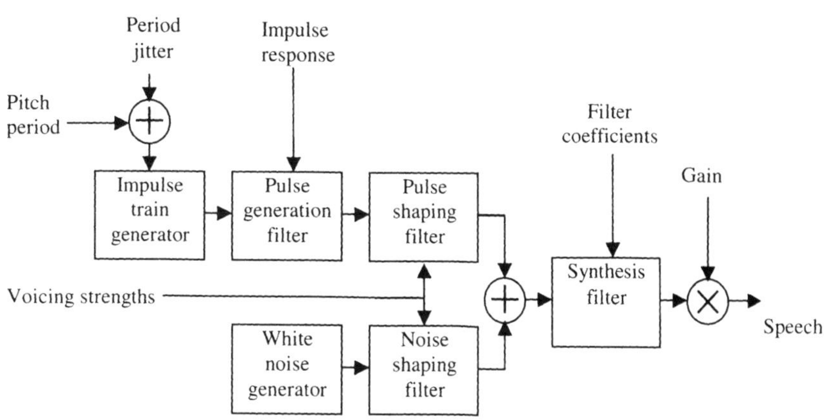

Figure 17.1 The MELP model of speech production.

Methods to determine the voicing states are explained in the next sections.

- Shape of the excitation pulse for periodic excitation is extracted from the input speech signal and transmitted as information on the frame.

In the simplest form of LPC coding, voiced excitation consists of a train of impulses; which differs a great deal from real world cases, where each excitation pulse possesses a certain shape, different from an ideal impulse. The shape of the pulse contains important information and is captured by the MELP coder through Fourier magnitudes (next section) of the prediction error. These quantities are used to generate the impulse response of the pulse generation filter (Figure 17.1), responsible for the synthesis of periodic excitation.

- Periodic excitation and noise excitation are first filtered using the pulse shaping filter and noise shaping filter, respectively; with the filters' outputs added together to form the total excitation, known as the *mixed* excitation, since portions of the noise and pulse train are mixed together.

This indeed is the core idea of MELP and is based on practical observations where the prediction-error sequence is a combination of a pulse train with noise. Thus, the MELP model is much more realistic than the LPC model, where the excitation is either impulse train or noise.

In Figure 17.1, the frequency responses of the shaping filters are controlled by a set of parameters called voicing strengths, which measure the amount of "voicedness." The responses of these filters are variable with time, with their parameters estimated from the input speech signal, and transmitted as information on the frame; procedural details are given in the next sections.

17.2 FOURIER MAGNITUDES

The MELP coder depends on the computation of Fourier magnitudes from the prediction-error signal to capture the shape of the excitation pulse, which basically are the magnitudes of the Fourier transform. These quantities are quantized and transmitted as information on the frames. The objective is to create, on the decoder side, a periodic sequence as close as possible to the original excitation signal. In this section, the procedure for Fourier magnitude computations is described; examples using real speech signals are included to illustrate the technique. Note that Fourier magnitudes are calculated only when the frame is voiced or jittery voiced.

Many properties of the discrete Fourier transform (DFT) and discrete-time Fourier transform (DTFT) are used to explain the topics in this section. Readers are referred to Oppenheim and Schafer [1989] for full coverage of the subjects. Efficient algorithms for the computation of the DFT, known as the fast Fourier transform (FFT), are also contained in the same reference.

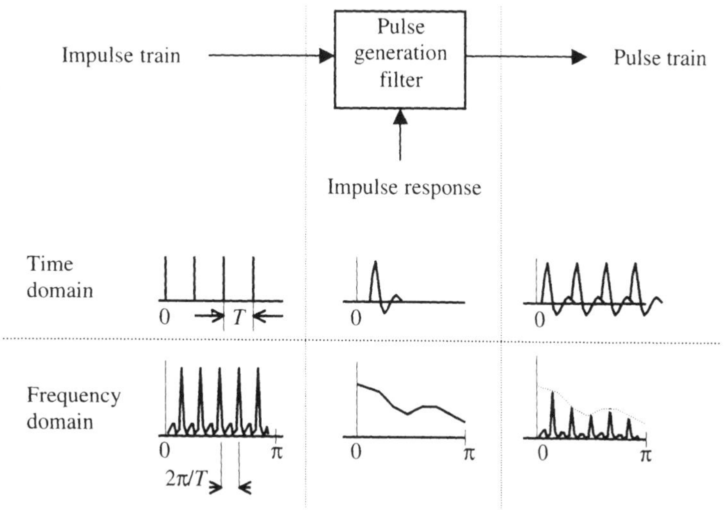

Figure 17.2 Illustration of signals associated with the pulse generation filter.

Pulse Generation Filter

The MELP model relies on the pulse generation filter (Figure 17.1) to produce periodic excitation. The idea is to find the impulse response of the filter during encoding and transmit the response to the decoder so as to generate the pulse train, used as the periodic excitation to the synthesis filter. Figure 17.2 illustrates the pulse generation process, where an impulse train of period T excites the filter to obtain a pulse train at its output. Looking at the magnitude of the Fourier transform, the spectrum of the pulse train is given by the product between the spectrum of the impulse train and the magnitude response of the filter. Therefore, by measuring the height of the peaks of the pulse train's magnitude spectrum, it is possible to get acquainted with the magnitude response of the pulse generation filter. From Figure 17.2, measurements of heights are done at frequency values of $\omega = 2\pi i/T$, $i = 1, 2, \ldots$; once the magnitude response is found, the impulse response of the pulse generation filter, and hence the shape of the pulse, is known.

During encoding, the magnitude of the DFT for the prediction error is found; peaks of the magnitude spectrum corresponding to the harmonics associated with the pitch frequency are measured. The peak values are the Fourier magnitudes and are transmitted to the decoder to construct the excitation pulse, or impulse response, of the pulse generation filter.

Fourier Magnitudes: Calculation and Quantization

The procedure used by the FS MELP coder for Fourier magnitude calculations and quantizations is presented next (Figure 17.3). Inputs to the procedure are the speech data (200 samples), the LPC, and the pitch period. It is assumed that the LPC and

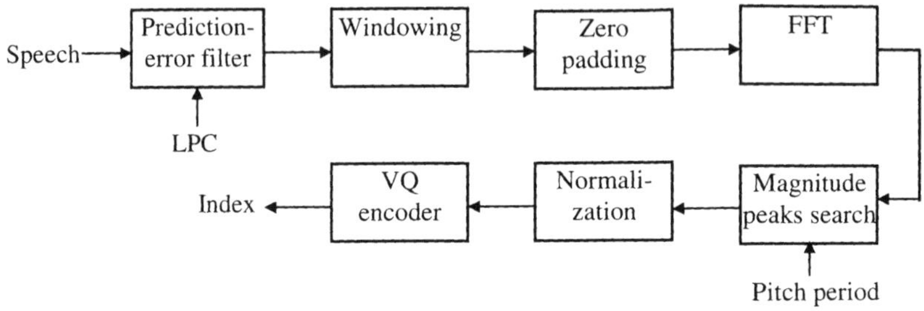

Figure 17.3 Block diagram of the Fourier magnitude calculation and quantization procedure.

pitch period are already extracted from the signal. Prediction error is first computed by passing the speech data through the prediction-error filter. A Hamming window multiplies the resultant 200-sample prediction-error sequence. The windowing step is standard in FFT deployment; its incorporation maintains spectral leakage to a minimum. The resultant 200-sample sequence is zero padded to a 512 sample, with the FFT calculated afterward, resulting in 512 complex-valued samples. Recall that the FFT is an efficient way to compute the DFT, with the DFT defined by

$$X[k] = \sum_{n=0}^{N_o-1} x[n]e^{-j(2\pi kn/N_o)}; \quad k = 0, \ldots, N_o - 1 \tag{17.1}$$

being the analysis equation of the transform, and

$$x[n] = \frac{1}{N_o}\sum_{k=0}^{N_o-1} X[k]e^{j(2\pi kn/N_o)}; \quad n = 0, \ldots, N_o - 1 \tag{17.2}$$

being the synthesis equation. We say that $x[n]$ and $X[k]$ form a DFT transform pair, and $N_o = 512$ is the length of the sequences.

Why is there a need to zero pad? And why is the length equal to 512? To answer these questions, it is necessary to realize that, under very general conditions, the sequence $X[k]$ in (17.1) is obtained by sampling the DTFT at regular interval, defined by

$$X(e^{j\omega}) = \sum_{n=-\infty}^{\infty} x[n]e^{-j\omega n}, \tag{17.3}$$

which is the spectrum of the signal corresponding to continuous frequency values. Thus, each value of k in (17.1) is associated with one particular frequency, given by

$$\omega_k = \frac{2\pi k}{N_o}; \quad k = 0, \ldots, N_o - 1. \tag{17.4}$$

Therefore, increasing the length of our original sequence from 200 to 512 samples with zero padding effectively increases the frequency resolution, resulting in higher accuracy; the process is equivalent to interpolation in the frequency domain. The number of 512 is selected as a trade-off between frequency resolution and computational cost and is a power of 2 because most FFT algorithms work under that constraint. In the present case, since the prediction-error signal is real-valued, the resultant DFT sequence (17.1) is symmetric, and only half of the samples need to be computed; that is, only 256 samples are passed to the next block. Operation of the magnitude peaks search block (Figure 17.3) is summarized as follows, where it is assumed that the pitch period T is known. The approach used for pitch period estimation is given in Section 17.4.

```
MAG_PEAKS_SEARCH(X[0 ... 255], T, Fmag[1 ... 10])
 1.  for i ← 1 to 10
 2.       freq ← round(512·i/T)
 3.       if freq > 255
 4.            Fmag[i] ← SMALLEST_MAG
 5.       else
 6.            peak ← 0
 7.            for j ← freq − 5 to freq + 5
 8.                 if j > 255 break
 9.                 if |X[j]| > peak
10.                      peak ← |X[j]|
11.            Fmag[i] ← peak
12.  return Fmag[1 ... 10]
```

The purpose of the above code is to search in the neighborhoods surrounding the frequency values $512i/T$, $i = 1, \ldots, 10$. These values correspond to the ten first harmonics associated with the pitch frequency. If the frequency value is above 255, that is, the harmonic is out of the evaluation range, the corresponding peak value, or Fourier magnitude, is set at SMALLEST_MAG (Line 4), which is a constant. In practice, SMALLEST_MAG is set to a small positive value. This latter case occurs when the pitch frequency is too high. If the frequency values are less than or equal to 255, the search for peaks is performed in the range round($512i/T$) $\pm$ 5; this range can be adjusted accordingly depending on the desired accuracy. At the end of the routine, ten peak magnitudes are obtained: $Fmag[i]$, $i = 1$ to 10. These ten values are the sought-after Fourier magnitudes. Thus, the Fourier magnitudes are the peak magnitude values of the harmonics related to the pitch frequency. Ten of these quantities are found to be sufficient for the present coding application.

The Fourier magnitude sequence $Fmag[i]$ is normalized according to

$$Fmag'[i] = \alpha \cdot Fmag[i], \tag{17.5}$$

with

$$\alpha = \left(\frac{1}{10}\sum_{i}(Fmag[i])^2\right)^{-1/2}. \tag{17.6}$$

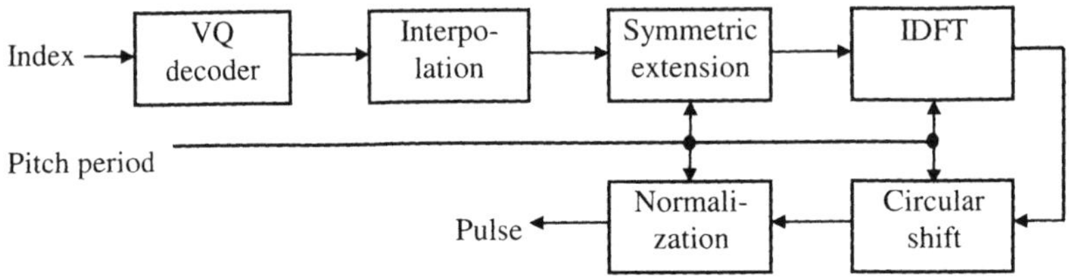

Figure 17.4 Block diagram of the excitation pulse generation procedure during decoding.

The resultant sequence $Fmag'[i]$ has an rms value of one and is vector quantized with an 8-bit VQ. (Procedures for VQ design are presented in Chapter 7.) The codebook is searched using a perceptually weighted Euclidean distance, with fixed weights that emphasize the low-frequency regions, since it is subjectively more important.

Pulse Excitation Generation

During decoding, the excitation signal is synthesized on a pitch-period-by-pitch-period basis; that is, if T is the pitch period under consideration, then a T-sample pulse is generated. A block diagram of the pulse generation procedure is shown in Figure 17.4. Given the decoded Fourier magnitudes, which are interpolated during MELP decoding (discussed later), the resultant sequence is denoted by $Fmag[i]$, $i = 1, \ldots, 10$; the following pseudocode summarizes the procedure.

```
PULSE_GENERATION(Fmag[1...10], T, y[0...T−1])
 1. Y[0] ← 0                    // DC component set to zero
 2. for k ← 11 to T−11          // Default components set to one
 3.        Y[k] ← 1
 4. for k ← 1 to 10             // Symmetric extension
 5.        Y[k] ← Fmag[k]
 6.        Y[T−k] ← Fmag[k]
 7. IDFT(Y[0...T−1], y[0...T−1], T)
 8. CIRCULAR_SHIFT(y[0...T−1], T, 10)
 9. NORMALIZE(y[0...T−1], T)
10. return y[0...T−1]
```

The routine takes the ten Fourier magnitudes and the pitch period as inputs to generate the T-sample array Y by setting $Y[0]$ to 0 and the rest of the elements to 1. The Y-array is used for the inverse DFT (IDFT), thus representing frequency-domain information. By setting the first element to zero, the DC component of the sequence obtained after IDFT is going to be zero. Next, the ten Fourier magnitudes are placed at $Y[1 \ldots 10]$ and $Y[T-10 \ldots T-1]$, resulting in a symmetric sequence. Symmetric extension is necessary for the generation of a real-valued sequence after IDFT. In Line 7, the IDFT is calculated with

$$y[n] = \frac{1}{T}\sum_{k=0}^{T-1} Y[k]e^{j2\pi nk/T}; \quad n = 0, \ldots, T-1. \tag{17.7}$$

In Line 8, the resultant sequence $y[n]$ is circularly shifted by ten samples and is normalized (Exercise 17.1) in Line 9 so that the y-array has unit rms value. The circular shift operation is specified by the following:

```
CIRCULAR_SHIFT(y, T, n)
for i ← 0 to T − 1
        x[i] ← y[MOD(i − n, T)]
for i ← 0 to T − 1
        y[i] ← x[i]
return y[0 ... T − 1]
```

The modulo operation MOD(n, N) is defined* by

```
MOD(n, N)
1. if n < 0
2.         n ← n + N
3.         if n < 0 goto 2
4. else
5.         n ← n − N
6.         if n ≥ N goto 5
7. return n
```

The purpose of circular shifting is to prevent abrupt changes at the beginning of the period.

Example 17.1 The procedures explained before are illustrated using a real speech signal. Figure 17.5 shows the prediction-error sequence used, with 200 samples and a pitch period roughly equal to 49. The magnitude plot of the 512-point DFT, corresponding to the windowed and zero-padded prediction-error signal, is also shown

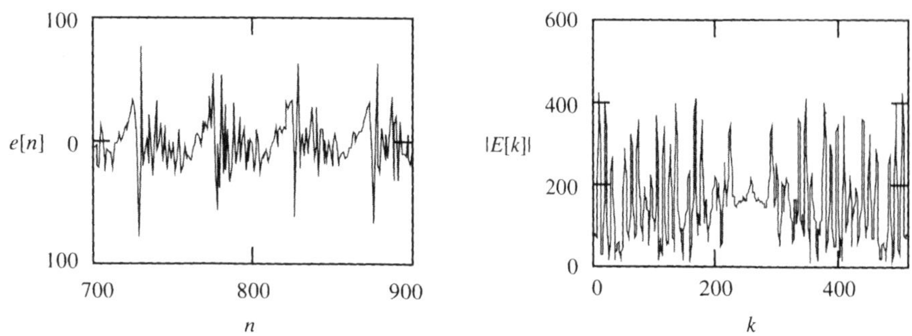

Figure 17.5 *Left*: A 200-sample prediction-error sequence. *Right*: Magnitude plot of the resultant 512-point DFT.

*When the input variable is negative, this definition of modulo operation might differ from other commonly used definitions found in calculators or math software. The present definition is necessary to implement the circular shift operation in a meaningful way.

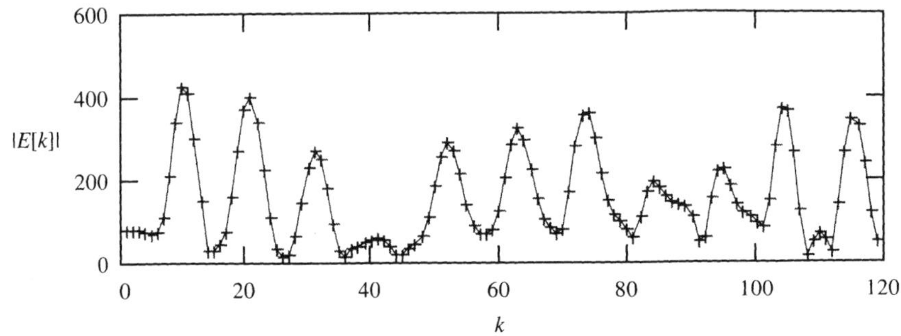

Figure 17.6 Expanded view of the magnitude plot of the DFT in Figure 17.5.

in the same figure. Since the pitch period is equal to 49, the magnitude peaks search occurs at $512i/49 \approx 10i$, for $i = 1$ to 10. Figure 17.6 shows an expanded view of the magnitude of the DFT, where we can see that the peaks are roughly located around $10i$. The peaks search operation leads to the following Fourier magnitude sequence:

$$\begin{aligned} Fmag[1] &= 422.5, & Fmag[2] &= 398.3, & Fmag[3] &= 269.4, \\ Fmag[4] &= 59.4, & Fmag[5] &= 287.8, & Fmag[6] &= 322.9, \\ Fmag[7] &= 358.0, & Fmag[8] &= 193.9, & Fmag[9] &= 226.7, \\ Fmag[10] &= 370.4. \end{aligned}$$

The above sequence is normalized and used to generate the excitation pulse. Quantization is not considered in this example.

To generate the excitation pulse, the normalized Fourier magnitude sequence is first symmetrically extended to 49 samples (Figure 17.7); then the inverse DFT is

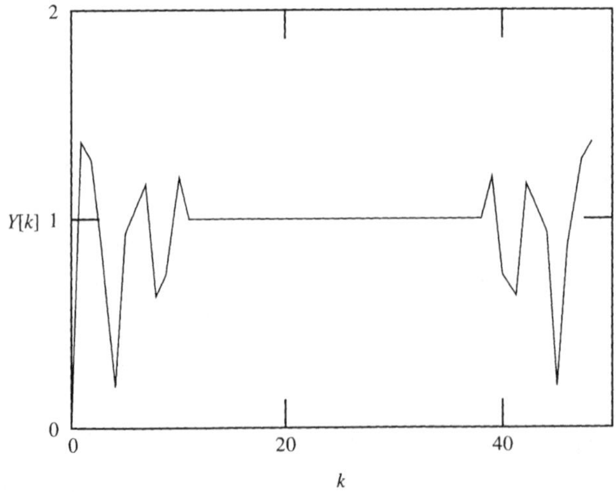

Figure 17.7 Symmetrically extended Fourier magnitude sequence.

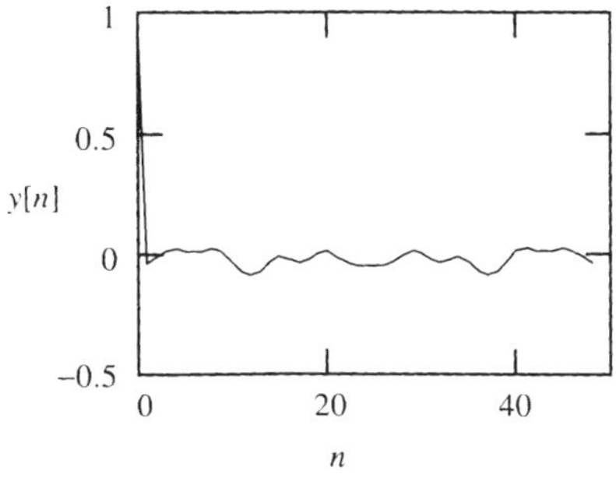

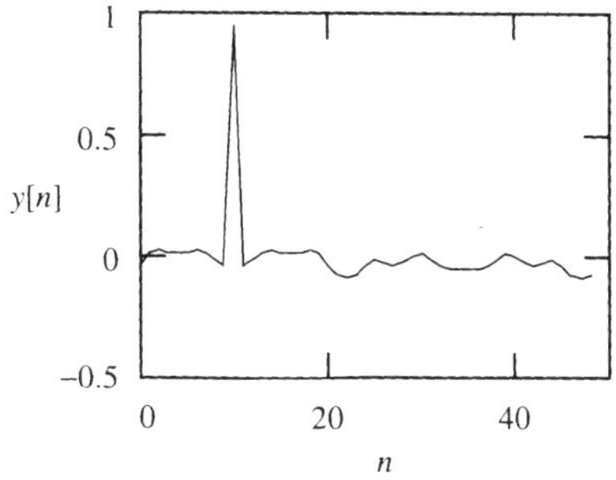

Figure 17.8 *Left*: Excitation pulse after IDFT. *Right*: The same pulse after circular shift.

calculated, leading to the sequence shown in Figure 17.8. The final excitation pulse is obtained after circular shifting. Figure 17.9 contrasts the original magnitude spectrum (prediction error), the magnitude spectrum of the pulse train generated by the Fourier magnitude, and the magnitude spectrum of an impulse train. These spectrums are obtained following the procedure of windowing using a 200-sample window, zero padding to 512 samples, and calculating the DFT. As we can see, frequency distribution of the pulse train is much closer to the original than the impulse train. In fact, the impulse train exhibits a spectrum having equal-height peaks; while for the pulse train, peaks of the first ten harmonics follow the shape of the original spectrum. Hence, incorporation of Fourier magnitudes adds naturalness and improves the overall quality.

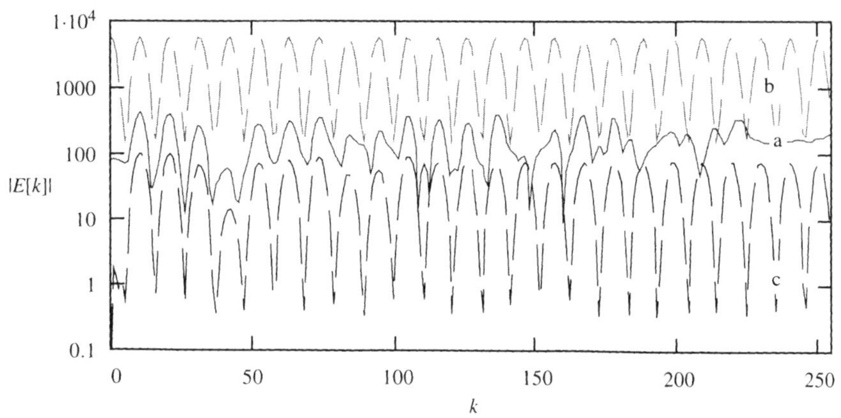

Figure 17.9 Comparison between the magnitude spectrum of (a) original prediction error, (b) impulse train, and (c) pulse train generated using Fourier magnitude.

17.3 SHAPING FILTERS

The MELP speech production model makes use of two shaping filters (Figure 17.1) to combine pulse excitation with noise excitation so as to form the mixed excitation signal. Responses of these filters are controlled by a set of parameters called voicing strengths; these parameters are estimated from the input signal. By varying the voicing strengths with time, a pair of time-varying filters results. These filters decide the amount of pulse and the amount of noise in the excitation, at various frequency bands.

In FS MELP, each shaping filter is composed of five filters, called the synthesis filters, since they are used to synthesize the mixed excitation signal during decoding. Each synthesis filter controls one particular frequency band, with passbands defined by 0–500, 500–1000, 1000–2000, 2000–3000, and 3000–4000 Hz. The synthesis filters connected in parallel define the frequency responses of the shaping filters. Figure 17.10 shows the block diagram of the pulse shaping filter, exhibiting the mechanism by which the frequency response is controlled. Denoting the impulse responses of the synthesis filters by $h_i[n]$, $i = 1$ to 5, the total response of the pulse shaping filter is

$$h_p[n] = \sum_{i=1}^{5} vs_i h_i[n], \tag{17.8}$$

with $0 \leq vs_i \leq 1$ being the voicing strengths. Equation (17.8) results from the fact that the synthesis filters are connected in parallel. The noise shaping filter, on the other hand, has the response

$$h_n[n] = \sum_{i=1}^{5} (1 - vs_i) h_i[n]. \tag{17.9}$$

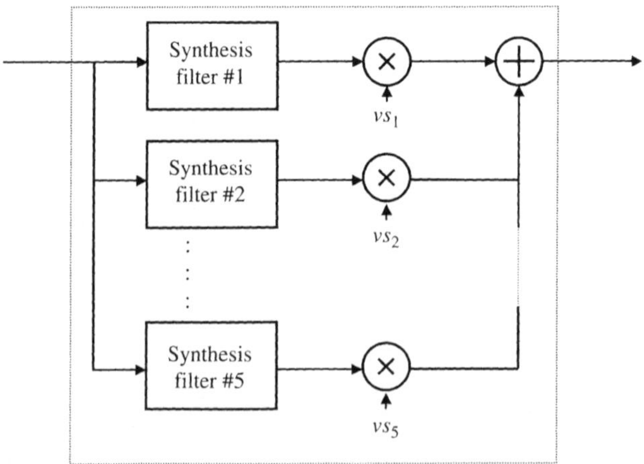

Figure 17.10 Block diagram of the pulse shaping filter.

Thus, the two filters complement each other in the sense that if the gain of one filter is high, then the gain of the other is proportionately lower, with the total gain of the two filters remaining constant at all times. As we will see later in the chapter, during MELP decoding, both pulse excitation and noise excitation are generated to have the same power level.

The synthesis filters are implemented as FIR with 31 taps. FIR filters are utilized due to the following reasons:

- *Linear phase.* A linear phase system simply means that the group delay is constant with frequency, which is a feature of FIR systems. A constant group delay for all frequency values will not distort the shape of a pulse passing through the system, since all components are delayed by the same amount, which is important for the processing of a pulse excitation sequence.
- *Frequency response can be changed with relative ease.* As we have already seen, the total response of the filter can be obtained by some scaling and sum operations, which can be done in practice at relatively low cost. In fact, by combining the synthesis impulse responses together, only one convolution is needed to compute the output, instead of doing it five times.
- *Interpolation can be done with relative ease.* To guarantee smoothness in transition between frames, the impulse responses of the shaping filters are interpolated during MELP decoding. The FIR nature of the filter allows linear interpolation of the impulse responses, or filter coefficients, without instability concerns, since stability is guaranteed for FIR systems. Figure 17.11 shows the impulse responses ($h_i[n]$, $i = 1$ to 5, $n = 0$ to 30) of the synthesis filters with the magnitude responses shown in Figure 17.12. See Exercise 17.2 for instructions on the design of these filters.

Figure 17.13 shows some example of the frequency responses of the shaping filters. As expected, when the gain of one band is maximized in one filter, the gain of the same band is minimized in the other filter. Gain of the parallel combination of the two filters is always constant.

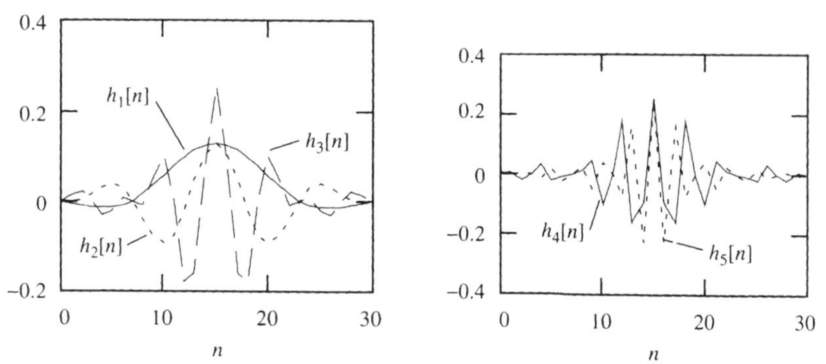

Figure 17.11 Impulse responses of the FS MELP synthesis filters.

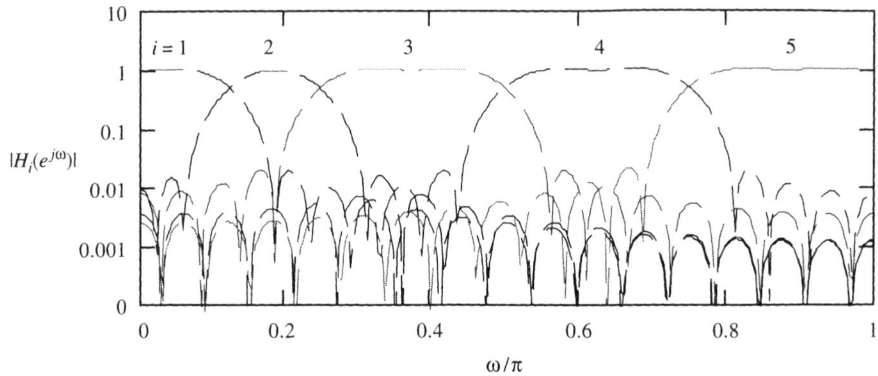

Figure 17.12 Magnitude responses of the FS MELP synthesis filters.

The function of the shaping filters become obvious: they determine how pulse-like or noise-like the mixed excitation is in each of the five frequency bands. Within the context of MELP coding, the two shaping filters are used only during decoding, where mixed excitation is generated.

17.4 PITCH PERIOD AND VOICING STRENGTH ESTIMATION

The FS MELP coder employs a sophisticated procedure to accurately estimate the pitch period and voicing strengths, since these parameters play a weighty role in the quality of the synthetic speech. In fact, fractional refinement is used throughout the encoding process. An analysis filter bank is utilized to separate the input signal into five bands, with the voicing strength found in each band. The operations described in this section are performed only during encoding, where the speech signal is analyzed with the required parameters extracted.

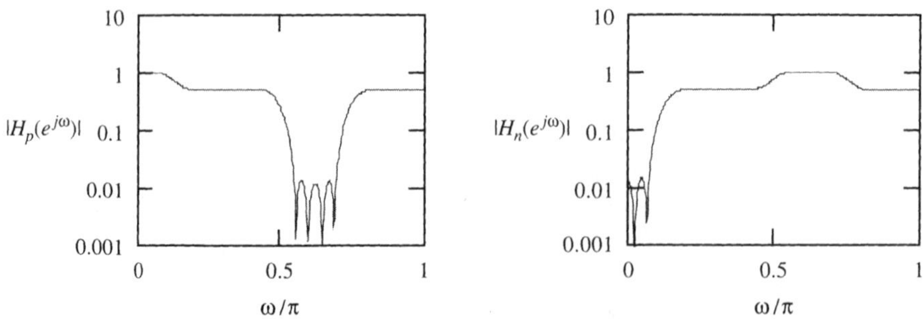

Figure 17.13 Magnitude response plots of the pulse shaping filter (*left*) and noise shaping filter (*right*) when $vs_1 = 1$, $vs_2 = vs_3 = 0.5$, $vs_4 = 0$, and $vs_5 = 0.5$.

Analysis Filters

The five analysis filters possess the same bandwidth settings as for the synthesis filters. Unlike the synthesis filters discussed in Section 17.3, these filters are implemented as sixth-order Butterworth [Oppenheim and Schafer, 1989]; that is, they are IIR. This design decision is based mainly on the relatively low computational cost associated with IIR configurations. For instance, sixth-order Butterworth requires 12 multiplications per output sample, whereas for a 31-tap FIR filter, 31 products are needed; and both configurations meet or exceed the required passband and stopband characteristics. Nonlinearity in phase response—typical of IIR systems—is of minor importance here, since the filters' outputs are used for the correlation computation. See Exercise 17.3 for the design of analysis filters.

Positions of Windows

Before continuing with pitch period estimation, the positions of some signal processing windows with respect to the signal frames are first described. Figure 17.14 summarizes the major windows utilized by the FS MELP coder. Like the FS1015 LPC coder, each frame is comprised of 180 samples. The positions of these analysis windows are selected to facilitate interpolation, since parameters of a given frame are interpolated between two different sets, calculated from the analysis windows

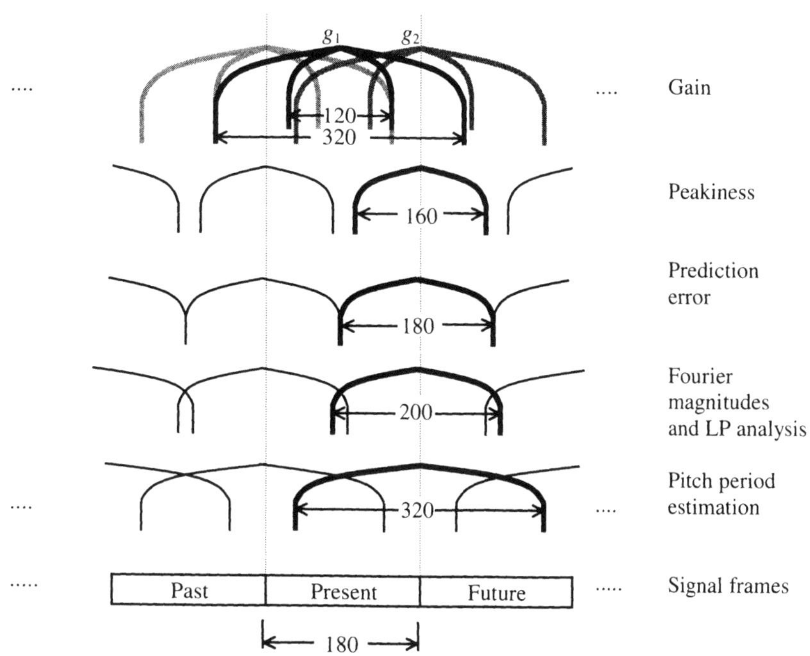

Figure 17.14 Positions of different windows with respect to the signal frames. Windows associated with the present frame are marked using bold lines.

centered at the beginning sample and end sample of the frame (gain parameter is an exception to this rule).

As we can see, some windows are overlapping and others are not. For most parameters (except gain), the analysis window corresponding to the present frame is centered on the last sample of the present frame. For instance, Fourier magnitudes utilize a 200-sample window (Section 17.2); this window consists of the last 100 samples of the present frame, and the first 100 samples of the future frame.

First-Stage Pitch Period Estimation

The input speech signal is filtered by the first analysis filter, with a passband of 0 to 500 Hz. For each 180-sample frame, the normalized autocorrelation

$$r[l] = \frac{c[0, l, l]}{\sqrt{c[0, 0, l]c[l, l, l]}}, \tag{17.10}$$

where

$$c[l, m, k] = \sum_{n=-\lfloor k/2 \rfloor - 80}^{-\lfloor k/2 \rfloor + 79} s[n + l]s[n + m], \tag{17.11}$$

is calculated for $l = 40, 41, \ldots, 160$. The resultant pitch period is the corresponding value of l for which $r[l]$ is maximized. Normalization in (17.10) is used to compensate for variations in signal energy; its introduction also simplifies voicing classification since the same threshold can be applied to all frames (low-energy frames excluded).

Autocorrelation calculation according to (17.11) utilizes 160 products. Taking into account the range of l, it is straightforward to verify that for each pitch period estimation operation, 320 consecutive samples of the signal are involved. The 320 consecutive samples constitute the pitch period estimation window. This window is based on 160 samples of the current frame, plus 160 samples of the next frame. The sample $s[0]$ in (17.11) corresponds to the first sample of the next frame. Denoting the integer pitch period from the above procedure as T, the following steps are used:

- The autocorrelation values $c[0, T+1, T]$ and $c[0, T-1, T]$ are calculated. If $c[0, T-1, T] > c[0, T+1, T]$, then $T \leftarrow T-1$; otherwise proceed to the next step. The purpose of this step is to verify the likelihood of the maximum autocorrelation point being located in $[T, T+1]$ or $[T-1, T]$; this is done by comparing the autocorrelation values at lags of $T-1$ and $T+1$.
- Compute the fractional pitch period:

$$\eta = \frac{c[0, T+1, T]c[T, T, T] - c[0, T, T]c[T, T+1, T]}{c[0, T+1, T](c[T, T, T] - c[T, T+1, T]) + c[0, T, T](c[T+1, T+1, T] - c[T, T+1, T])}. \tag{17.12}$$

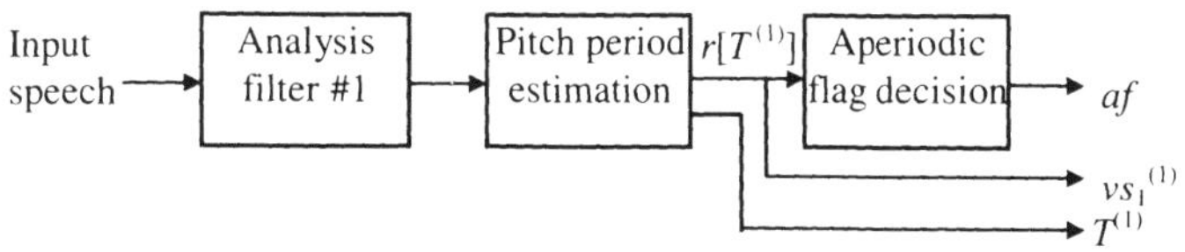

Figure 17.15 Illustration of first-stage pitch period estimation.

- Compute the normalized autocorrelation value:

$$r[T+\eta] = \frac{(1-\eta)c[0,T,T] + \eta c[0,T+1,T]}{\sqrt{c[0,0,T]((1-\eta)^2 c[T,T,T] + 2\eta(1-\eta)c[T,T+1,T] + \eta^2 c[T+1,T+1,T])}}. \tag{17.13}$$

The resultant real pitch period is denoted by $T^{(1)} = T + \eta$, referred to as the first-stage pitch period (Figure 17.15). The origin of the above formulas is from the Medan–Yair–Chazan method discussed in Chapter 2.

Low-Band Voicing Strength

Voicing strength of the low band (0–500 Hz, vs_1) is equal to the normalized autocorrelation value associated with $T^{(1)}$, given by (17.13). That is,

$$vs_1 = r\left[T^{(1)}\right]. \tag{17.14}$$

This value of voicing strength is subjected to modification later according to other properties of the signal.

Aperiodic Flag

The aperiodic flag depends on the low-band voicing strength:

$$af = \begin{cases} 1, & \text{if } vs_1 < 0.5, \\ 0, & \text{otherwise.} \end{cases} \tag{17.15}$$

For high voicing strength (above 0.5), the frame is strongly periodic and the flag is set to zero. For weak periodicity, the flag is set to one, signaling the MELP decoder to generate aperiodic pulses as voiced excitation (jittery voiced). The flag is transmitted as part of the MELP bit-stream using one bit.

Voicing Strength Determination for the Four High-Frequency Bands

Figure 17.16 shows the system used to estimate the voicing strengths of the four upper-frequency bands. Details are described below.

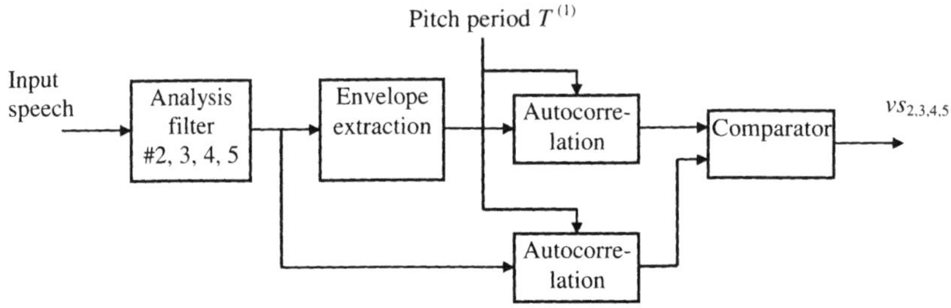

Figure 17.16 Illustration of bandpass voicing strengths estimation.

- Calculate $r_1 = r[T^{(1)}]$, where $r[\cdot]$ is the normalized autocorrelation given by (17.13), and $T^{(1)}$ is the first-stage real pitch period. The signal used to calculate the autocorrelation is the output from the corresponding bandpass filter.
- Calculate $r_2 = r[T^{(1)}]$. This time, however, the signal used is the envelope of the bandpass signal, where the envelope is obtained by full-wave rectification (absolute value of the samples), followed by lowpass filtering. The lowpass filter possesses a zero at $\omega = 0$ so as to cancel the DC component. Thus, the full-wave rectified signal is smoothed. In many instances, the envelope of the bandpass signal reflects better the underlying periodicity due to the fundamental pitch frequency; in fact, these envelopes tend to rise and fall with each pitch pulse. Autocorrelation analysis of these bandpass signal envelopes yields an estimate of the amount of pitch periodicity in the corresponding band. Figure 17.17 shows some example waveforms, where we can see that the envelope roughly displays the same periodicity due to the fundamental pitch period. The autocorrelation value is decremented by 0.1 to compensate for an experimentally observed bias.
- The voicing strength of the band is given by

$$vs = \max(r_1, r_2). \tag{17.16}$$

Repeating the above procedures for the four remaining filters leads to the voicing strengths vs_2, vs_3, vs_4, and vs_5. Thus, the voicing strength is determined by comparing the autocorrelation values obtained directly from the bandpass signal and the one obtained from the envelope of the signal itself, whichever is higher.

LP Analysis and Prediction Error

A tenth-order LP analysis is performed on the input speech signal using a 200-sample (25-ms) Hamming window centered on the last sample in the current frame (Figure 17.14). The autocorrelation method is utilized together with the Levinson–Durbin algorithm. The resultant coefficients are bandwidth-expanded with a constant of 0.994 (Chapter 4). The coefficients are quantized (Chapter 15) and used to calculate the prediction-error signal.

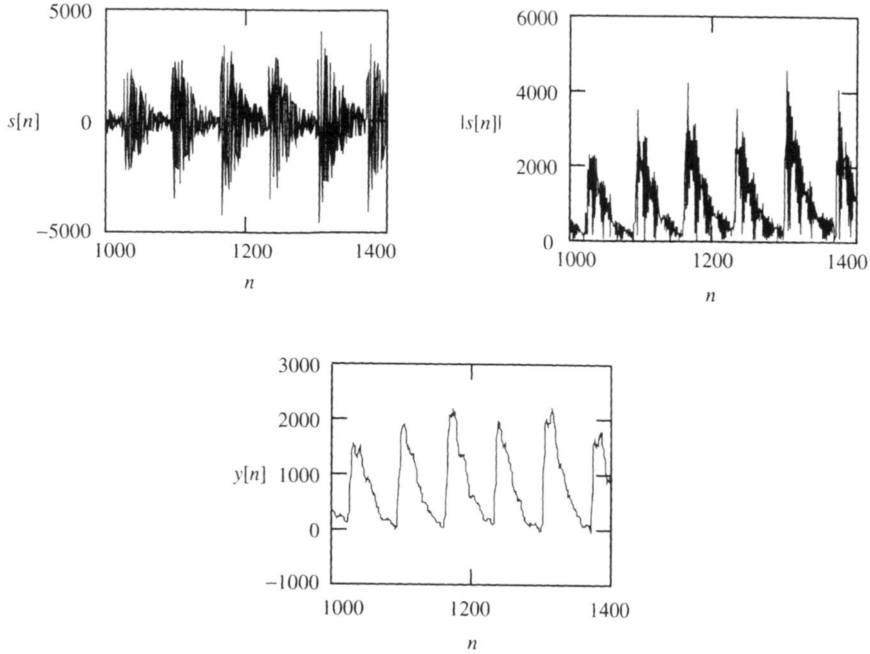

Figure 17.17 Bandpass output from the third analysis filter (1–2 kHz), obtained using a voiced portion of a speech waveform (*top left*). The full-rectified bandpass output (*top right*). The smoothed bandpass envelope (*bottom*).

Peakiness

Peakiness of the prediction-error signal is calculated over a 160-sample window centered on the last sample in the current frame (Figure 17.14). The peakiness value is defined by

$$p = \frac{\sqrt{\frac{1}{160}\sum_{n=-80}^{79} e^2[n]}}{\frac{1}{160}\sum_{n=-80}^{79} |e[n]|} \tag{17.17}$$

and is a measure of the "peaky" level of the signal. Generally, the "peaky" level refers to the presence of samples having relatively high magnitudes ("peaks") with respect to the average magnitude of a group of samples, which is effectively captured by (17.17). For high peakiness, the signal is expected to have outstanding "peaks." In linear algebra terms, peakiness is the ratio of the L2 norm to the L1 norm of the signal.

Example 17.2 Figure 17.18 shows some commonly encountered signals and their peakiness measures. As we can see, a single impulse has the highest peakiness. A train of scattered impulses also has a relatively high peakiness. The sinewave and

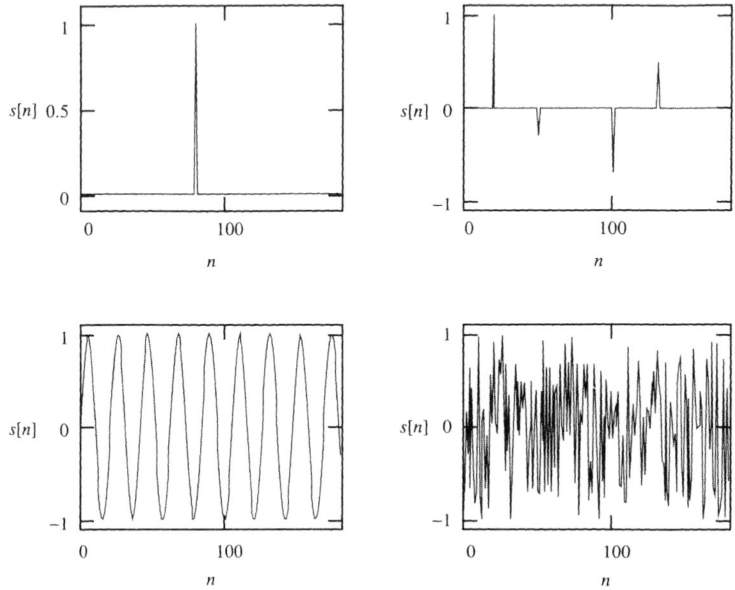

Figure 17.18 Some signals and their peakiness values. *Top left*: Impulse, $p = 13.4$. *Top right*: Some irregular impulses, $p = 7.3$. *Bottom left*: Sinewave, $p = 1.1$. *Bottom right*: White noise, $p = 1.2$.

white noise sequences have low peakiness due to the fact that no outstanding peaks are present in these signals.

Figure 17.19 plots the peakiness value of a periodic impulse train having a certain period. As we can see, peakiness grows monotonically with the period. When

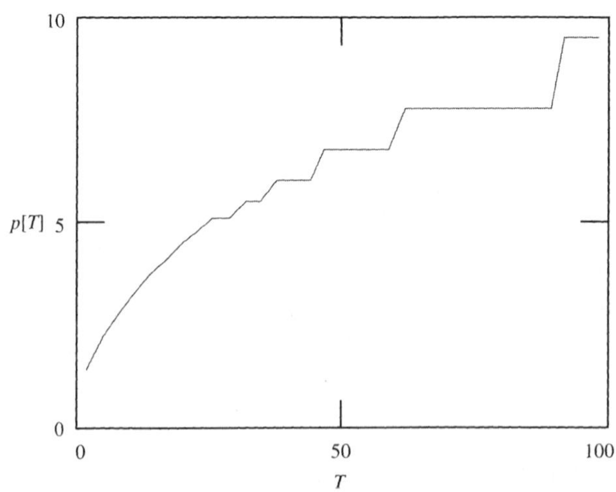

Figure 17.19 Peakiness of a uniformly spaced impulse train as a function of the period.

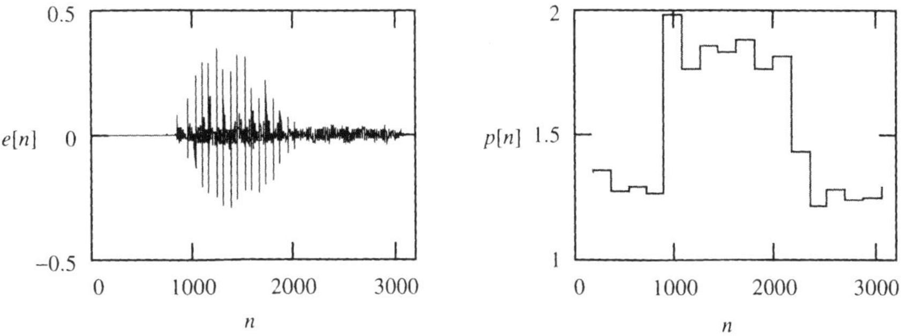

Figure 17.20 Prediction error obtained from a speech waveform (*left*) and peakiness measure applied to the prediction error (*right*).

peakiness of prediction error is calculated, it is expected that the measure is higher for voiced frames due to the quasiperiodic impulse train structure. In particular, its value can be maximized when the number of pulses in the frame becomes sparse, such as the case of transition frames. Therefore, peakiness measure is useful for voiced/unvoiced decision, as well as in the detection of transition.

Figure 17.20 shows an example of peakiness measure on a practical prediction-error signal. Note that peakiness is much higher between $n = 1000$ and 2000, where the signal is voiced. Peakiness is especially high for the frame containing $n = 1000$ due mainly to the transition nature, where a few excitation pulses are irregularly distributed within the frame.

Peakiness and Voicing Strengths

According to the value of peakiness, voicing strengths of the lowest three bands are modified according to the following:

- If $p > 1.34$ then $vs_1 \leftarrow 1$.
- If $p > 1.60$ then $vs_2 \leftarrow 1$ and $vs_3 \leftarrow 1$.

Thus, when peakiness is high, it overwrites some of the voicing strengths by setting them directly to a high value. As we will see later, each voicing strength is quantized to 0 or 1, transmitted using 1 bit.

A combination of peakiness and autocorrelation measures is highly effective in voicing state classification. Typical unvoiced frames have low peakiness and autocorrelation, leading to weak voicing strengths. For transition frames, however, peakiness is high with medium autocorrelation. For these frames, the relatively low autocorrelation sets the aperiodic flag to 1 (17.15), signaling the decoder to generate random periods. On the other hand, high peakiness sets the voicing strengths to their maximum values; the resultant conditions indicate the jittery voiced state. For voiced frames, peakiness is medium with high autocorrelation; this would set the aperiodic flag to zero, with high voicing strengths. The MELP coder relies on the aperiodic flag combined with voicing strengths to manage voicing state information.

Final Pitch Period Estimation

The prediction-error signal is filtered using a lowpass filter with 1-kHz cutoff, with the output used for pitch period estimation. The result is obtained by searching in a range surrounding the first-stage pitch period $T^{(1)}$. Reestimating the pitch period using prediction error yields a more accurate result, since the formant structure of the original speech signal is removed.

17.5 ENCODER OPERATIONS

After studying the core techniques in previous sections, we are ready to put pieces together so as to form the MELP encoder; a block diagram is shown in Figure 17.21. Functionality of some blocks is already explained and will not be repeated here.

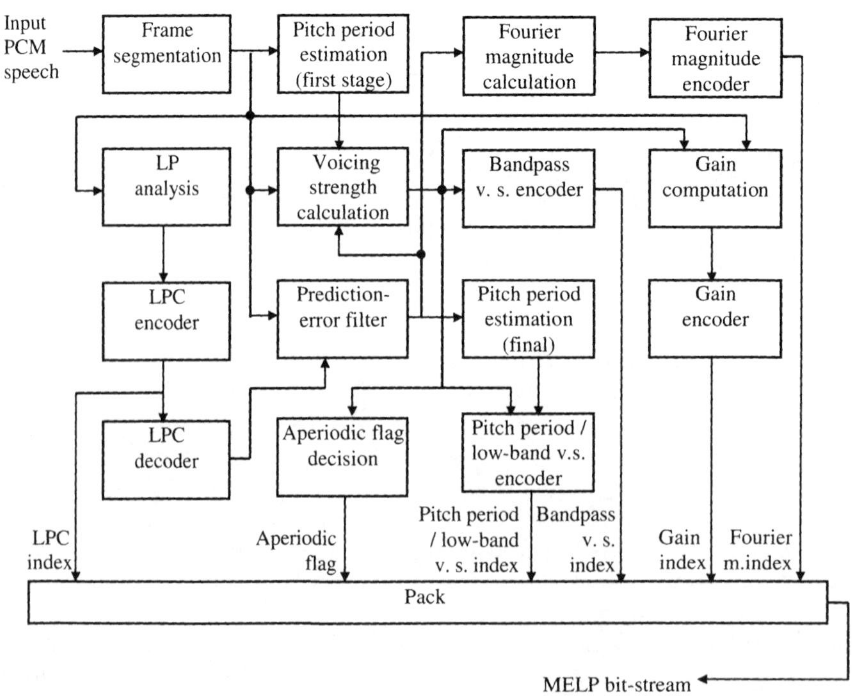

Figure 17.21 Block diagram of the MELP encoder.

Bandpass Voicing Strength Encoder

Voicing strengths of the four high-frequency bands are quantized according to the following pseudocode:

```
QUANTIZE_VS(vs1, ..., vs5)
 1. if vs1 ≤ 0.6                    // Unvoiced
 2.       for i ← 2 to 5
 3.             qvsi ← 0
 4. else                            // Voiced
 5.       for i ← 2 to 5
 6.             if vsi > 0.6
 7.                   qvsi ← 1
 8.             else
 9.                   qvsi ← 0
10.       if qvs2 = 0 and qvs3 = 0 and qvs4 = 0
11.             qvs5 ← 0
12. return qvs2, ..., qvs5
```

The procedure takes the voicing strength of the five bands as input. For the unvoiced case, which is determined by the magnitude of vs_1, the quantized voicing strengths (qvs_i) of the four high-frequency bands are set to zero (Lines 2 and 3). Otherwise, they are quantized to zero or one according to their magnitudes. The case of $(qvs_2, qvs_3, qvs_4, qvs_5) = (0, 0, 0, 1)$ is explicitly avoided in Lines 10 and 11. Finally, the four quantized voicing strengths are returned in Line 12.

Quantization of Pitch Period and Low-Band Voicing Strength

The final pitch period T and the low-band voicing strength vs_1 are quantized jointly using 7 bits. If $vs_1 \leq 0.6$, the frame is unvoiced and the all-zero code is sent. Otherwise, $\log T$ is quantized with a 99-level uniform quantizer ranging from $\log 20$ to $\log 160$. The resulting index is mapped to the appropriate code based on a table. The quantized vs_1 denoted as qvs_1 is equal to 0 for the unvoiced state and 1 for the voiced state. No separate transmission is necessary since the information can be recovered from the same table.

Gain Computation

The input speech signal gain is measured twice per frame using a pitch-adaptive window length. This length is identical for both gain measurements and is determined as follows:

- If $vs_1 > 0.6$, the window length is the shortest multiple of $T^{(1)}$ (first-stage, pitch period), which is longer than 120 samples. If this length exceeds 320 samples, it is divided by 2. This case corresponds to voiced frames, where pitch synchronization is sought during gain computation. By utilizing an integer multiple of pitch period, variation in the value of gain with respect to the position of the window is minimized.

- If $vs_1 \leq 0.6$, the window length is 120 samples. Thus, for unvoiced or jittery voiced frames, a default window length is used.

Gain calculation for the first window produces g_1 and is centered 90 samples before the last sample in the current frame (Figure 17.14). The calculation for the second window produces g_2 and is centered on the last sample in the current frame. The equation for gain calculation is

$$g = 10 \log_{10}\left(0.01 + \frac{1}{N}\sum_n s^2[n]\right), \tag{17.18}$$

where N is the window length and $s[n]$ the input speech signal. The range of n depends on the window's length as well as the particular gain (g_1 or g_2). The 0.01 factor prevents the argument from going too close to zero. If a gain measurement is less than 0, it is clamped to 0. The gain measurement assumes that the input signal range is $[-32768, 32767]$ (16 bits per sample).

Gain Encoder

Gain g_2 is quantized with a 5-bit uniform quantizer ranging from 10 to 77 dB. Gain g_1 is quantized with 3 bits using the pseudocode described below:

```
ENCODE_g1(g1, g2, g2,past)
 1. if |g2 - g2,past| < 5 and |g1 - (g2 + g2,past)/2| < 3
 2.        index ← 0
 3. else
 4.        gmax ← MAX(g2,past, g2) + 6
 5.        gmin ← MIN(g2,past, g2) - 6
 6.        if gmin < 10
 7.               gmin ← 10
 8.        if gmax > 77
 9.               gmax ← 77
10.        index ← UNIFORM(g1, gmin, gmax, 7)
11. return index
```

The procedure takes the gains g_1 and g_2 of the current frame and $g_{2,\text{past}}$ of the past frame. In Line 1 some conditions are verified to determine whether the frame is steady-state (slight change in energy). If the condition is met, zero is returned as the encoding result. Otherwise the frame is transitory and a seven-level uniform quantizer is used. The limits of the quantizer (*gmin*, *gmax*) are calculated in Lines 4 to 9; encoding through the uniform quantizer is performed in Line 10, resulting in the index set $\{1, 2, \ldots, 7\}$. Thus, a total of eight levels exist, deployable with 3 bits. The function UNIFORM in Line 10 takes g_1 as input and encodes it

TABLE 17.1 Bit Allocation for the FS MELP Coder[a]

Parameter	Resolution: Voiced	Resolution: Unvoiced
LPC	25	25
Pitch period/low-band voicing strength	7	7
Bandpass voicing strength	4	—
First gain	3	3
Second gain	5	5
Aperiodic flag	1	—
Fourier magnitudes	8	—
Synchronization	1	1
Error protection	—	13
Total	**54**	**54**

[a] Data from McCree et al. [1997], Table 1.

using a uniform quantizer (Chapter 5) having the input range [*gmin*, *gmax*] and seven codewords. The method is essentially an adaptive quantization scheme, where parameters of the past and future are utilized to improve efficiency.

Bit Allocation

Table 17.1 summarizes the bit allocation scheme of the FS MELP coder. As described in Chapter 15, the LPCs are quantized as LSFs using MSVQ. Synchronization is an alternating one/zero pattern. Error protection is provided for unvoiced frames only, using 13 bits. A total of 54 bits are transmitted per frame, at a frame length of 22.5 ms. A bit-rate of 2400 bps results.

17.6 DECODER OPERATIONS

Figure 17.22 shows the block diagram of the MELP decoder, where the bit-stream is unpacked with the indices directed to the corresponding decoder. Comparing with Figure 17.1, we can see that the speech production model is embedded within the structure of the decoder. Two additional filters are added along the processing path: the spectral enhancement filter taking the mixed excitation as input, and the pulse dispersion filter at the end of the processing chain. These filters enhance the perceptual quality of the synthetic speech.

Parameter Decoding and Interpolation

In MELP decoding, parameters from the bit-stream are unpacked and decoded according to the appropriate scheme. These parameters are LPC (LSF), pitch period/

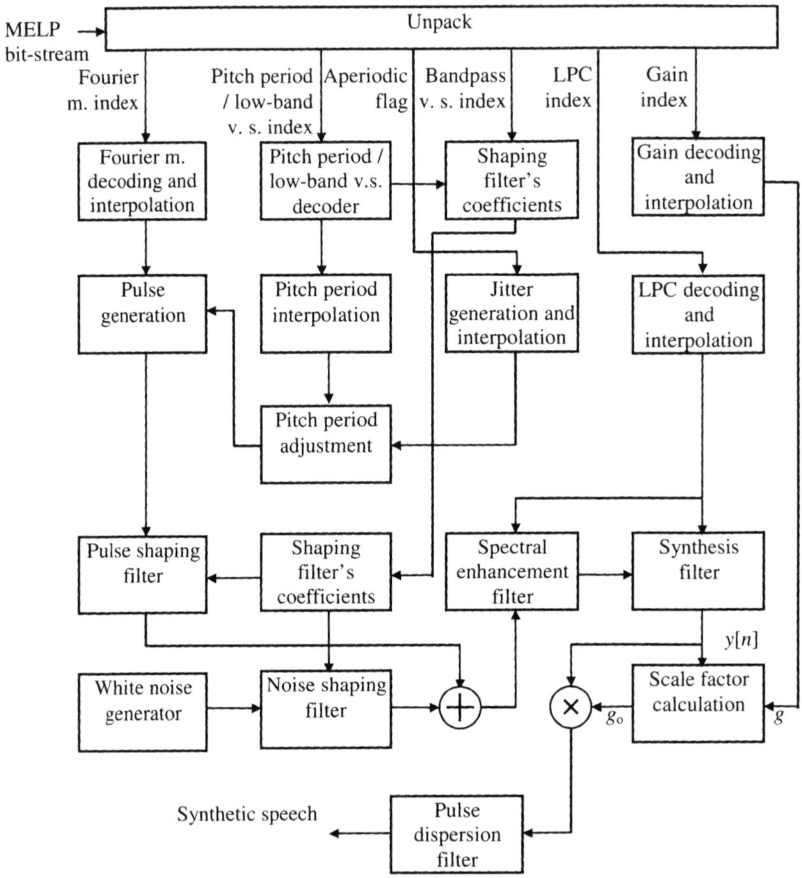

Figure 17.22 Block diagram of the MELP decoder.

low-band voicing strength, bandpass voicing strengths, gains (g_1 and g_2), aperiodic flag, and Fourier magnitudes. The mentioned parameters represent information on the frame. Most of them are interpolated linearly during speech synthesis.

For unvoiced frames (detectable from the pitch period/low-band voicing strength code), default values for some parameters are used. These include 50 for the pitch period, 0.25 for the jitter, all 1's for Fourier magnitudes, and all 0's for voicing strengths. Default values are necessary for unvoiced frames since linear interpolation is performed on a pitch-period-by-pitch-period basis during speech synthesis. Note that a new parameter—jitter—is introduced. Jitter is used only in the decoder to control the amount of randomness during aperiodic voiced excitation generation.

For voiced frames, the value of jitter is assigned according to: jitter $\leftarrow$ 0.25 if aperiodic flag is equal to one; otherwise, jitter $\leftarrow$ 0. In this case, pitch period is decoded from the bit-stream. After interpolation, the actual pitch period to use is given by

$$T = T_o(1 + \text{jitter} \cdot x), \tag{17.19}$$

with T_o being the decoded and interpolated pitch period, and x a uniformly distributed random number in the interval $[-1, 1]$. In this way, erratic periods are generated, simulating the conditions encountered in transition frames. The quantity $(T_o \cdot \text{jitter} \cdot x)$ is the period jitter described in Figure 17.1. Also note that the maximum perturbation to the pitch period is equal to $\pm 25\%$, a limit found to be adequate for most practical situations.

During speech synthesis, the signal is generated on a pitch-period-by-pitch-period basis. That is, at a given instant of time n_o, where $n_o \in [0, 179]$ pertains to the current frame, the set of parameters required for synthesis is determined from linear interpolation of data from the past frame and present frame. The interpolation factor α is given by

$$\alpha = n_o/180. \tag{17.20}$$

This interpolation factor is applied in a similar manner for the parameters LPC (LSF), pitch period, jitter, Fourier magnitudes, and shaping filters' coefficients. Even though voicing strengths can be interpolated, interpolating the shaping filters' coefficients (Section 17.3) results generally in lower computational cost (Exercise 17.6). As an example, the pitch period value to be used for synthesis is given by

$$T = (1 - \alpha)T_{\text{past}} + \alpha T_{\text{present}}, \tag{17.21}$$

with the result rounded to the nearest integer. The gain to use is given by the interpolation formulas:

$$g = \begin{cases} (1 - \alpha)g_{2,\text{past}} + \alpha g_{1,\text{present}}, & n_o < 90, \\ (1 - \alpha)g_{1,\text{present}} + \alpha g_{2,\text{present}}, & 90 \le n_o < 180. \end{cases} \tag{17.22}$$

See Exercise 17.7 for gain decoding.

For the case that $n_o + T > 180$, that is, the period under consideration crosses the frame boundary, the parameters are still interpolated using the same rule. For the next period, n_o is adjusted by subtracting 180 to reflect the coordinate of the new frame.

Mixed Excitation Generation

The T-sample pulse sequence generated from the Fourier magnitudes has unit rms value (Section 17.2). This sequence is filtered by the pulse shaping filter and added with the filtered noise sequence to form the mixed excitation. Noise is generated by a zero-mean uniform random number having unit rms value. Coefficients of the filters are interpolated pitch synchronously.

Spectral Enhancement Filter

This filter has the system function

$$H(z) = (1 - \mu z^{-1}) \frac{1 + \sum_{i=1}^{10} a_i \beta^i z^{-i}}{1 + \sum_{i=1}^{10} a_i \alpha^i z^{-i}}, \tag{17.23}$$

where the a_i are the linear prediction coefficients. The parameters μ, α, and β are made adaptive depending on the signal conditions. This filter is identical to the widely used postfilter in CELP coders—see Chapter 11 for details. It is stated without further comment that the sole purpose of the spectral enhancement filter, as its name implies, is to enhance the perceptual quality of the synthetic speech by accentuating the original spectral characteristics.

Synthesis Filter

This is a formant synthesis filter in direct form, with the coefficients corresponding to the interpolated LSFs.

Scale Factor Calculation

Power of the synthesis filter's output must equal the interpolated gain g (17.22) of the current period. Since the excitation is generated at an arbitrary level, a scaling factor g_o is calculated so as to scale the synthesis filter's output ($y[n]$, Figure 17.22) to produce the appropriate level. This is given by

$$g_o = \frac{10^{g/20}}{\sqrt{\frac{1}{T} \sum_n y^2[n]}}. \tag{17.24}$$

By multiplying g_o by $y[n]$, the resultant T-sample sequence will have a power of $(10^{g/20})^2$, or g dB.

Pulse Dispersion Filter

This is the last block in the decoding chain. The filter is a 65-tap FIR filter derived from a spectrally flattened triangle pulse (see Exercise 17.8). Figure 17.23 shows the impulse response and magnitude response of this filter. As we can see, it is almost an allpass filter, where changes in magnitude response are relatively small.

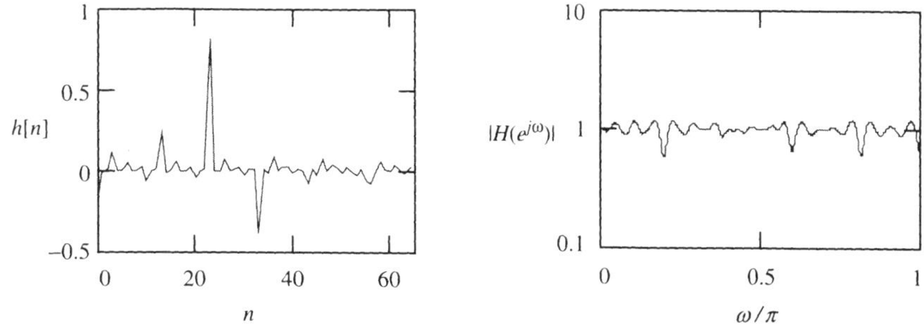

Figure 17.23 Impulse response (*left*) and magnitude response (*right*) of the pulse dispersion filter.

What is the purpose of this ineffectual looking filter? The explanation from McCree and Barnwell [1995] is that it improves the match of bandpass-filtered synthetic and natural speech waveforms in regions not containing a formant resonance, where the bandpass-filtered natural speech has a smaller peak-to-valley ratio than the synthetic speech. We can think of the function of the filter as to "stir" slightly the spectrum of the synthesis filter's output so as to improve naturalness; this is beneficial since mixed excitation is formed by combining the noise and pulse through a filter bank with fixed bandwidth.

17.7 SUMMARY AND REFERENCES

The theory and practice of the MELP coder are discussed in this chapter. It is shown that the speech production model of MELP is basically an improvement with respect to LPC, where additional parameters are introduced to add naturalness, smoothness, and adaptability to more diverse signal conditions. The MELP coder has achieved all these advantages without elevating the overall bit-rate, which is mainly due to the inclusion of vector quantization techniques. Much of the buzzy quality usually associated with LPC is eliminated by introducing frequency-dependent voicing strength, jittery voiced state with aperiodic excitation pulses, and enhancement filters to better match the synthetic waveform with the original.

MELP pertains to the family of parametric coders. When comparing with hybrid coders such as CELP, the bit-rate requirement is much lower since detailed description for the excitation is avoided. Instead, a coarse set of parameters is extracted to represent the excitation. Similar to LPC, it can also be classified as a source-controlled multimode coder, since different encoding schemes are applied depending on the properties of the signal.

Many details of the operations of the encoder and decoder are simplified to facilitate digestion of the chapter's material. Readers are referred to official documentation for a complete description, where additional techniques are included to improve robustness and accuracy under various practical situations. Like any coder, MELP is not perfect and many studies have been published with the aim of enhancement. See Unno et al. [1999] for improvement ideas, where transition noise, plosive reproduction, and low-pitch male speakers are addressed. In McCree and DeMartin [1998], the bit-rate is reduced to 1.7 kbps with alternative quantization techniques. In Stachurski et al. [1999], attempts were made to achieve near toll quality at 4.0 kbps.

The ideas of mixed excitation, where a noise component and a periodic component are combined together, have been investigated by many researchers in the past. The multiband excitation coder, for instance, replaces the single voiced/unvoiced classification of the LPC coder with a set of such decisions over harmonic intervals in the frequency domain [Griffin and Lim, 1988]. The idea was subsequently standardized by the International Maritime Satellite Corporation (Inmarsat) in 1990 at a bit-rate of 4.15 kbps. In Kondoz [1994], additional proposals are given to reduce the bit-rate to 2.4 kbps and below. However, when the model on which multiband excitation is based no longer fits the input signal, the resultant reproduction quality is greatly reduced. This is particularly true when there is music or noise mixed with the speech signal [Cox, 1995]. The MELP coder, on the other hand, behaves quite well even for certain classes of nonspeech signals. Its quality is mainly due to the robustness and flexibility of the underlying model, allowing good adaptation to a more general class of audio signals.

EXERCISES

17.1 Prove that the normalization step in the PULSE_GENERATION(...) routine is accomplished by multiplying each sample of the pulse sequence ($y[n]$) by the square root of the pitch period T. That is, $\sqrt{T} \cdot y[n]$, $n = 0, \ldots, T-1$ has an rms value of one. *Hint*: The decoded Fourier magnitude sequence has an rms value of 1.

17.2 This exercise concerns MELP synthesis filters design. These are 31-tap FIR filters, designed by windowing the impulse response of an ideal filter. Follow the procedures below to find the impulse responses of these filters.

(a) Synthesis filter #1: This is essentially a lowpass filter. For an ideal lowpass filter with a cutoff frequency of ω_p and unit passband gain, the impulse response is [Oppenheim and Schafer, 1989]

$$h_d[0] = \frac{\omega_p}{\pi},$$

$$h_d[n] = \frac{\sin(\omega_p n)}{\pi n}, \quad \text{if } n \neq 0.$$

Note that the impulse response is infinite in length. For finite length, the response is truncated using a window sequence $w[n]$ with $N = 31$

samples, centered around the origin:

$$h[n] = h_d\left[n - \left(\frac{N-1}{2}\right)\right] w\left[n - \left(\frac{N-1}{2}\right)\right]; \quad n = 0, 1, \ldots, N-1.$$

Some popular window sequences are the Hamming, Hanning, and Kaiser.

(b) Synthesis filters #2, 3, and 4: The same procedure can be applied. However, the impulse response of an ideal bandpass filter is utilized, which is given by

$$h_d[0] = \frac{\omega_{p2} - \omega_{p1}}{\pi},$$
$$h_d[n] = -\frac{\sin(\omega_{p1} n)}{\pi n} + \frac{\sin(\omega_{p2} n)}{\pi n}, \quad \text{if } n \neq 0.$$

(c) Synthesis filter #5: This filter is basically a highpass filter; hence, the impulse response of an ideal highpass filter is used, given by

$$h_d[0] = 1 - \frac{\omega_p}{\pi},$$
$$h_d[n] = -\frac{\sin(\omega_p n)}{\pi n}, \quad \text{if } n \neq 0.$$

After finding the five impulse responses $h_i[n]$, $i = 1$ to 5, add them up to confirm the fact that the result is an impulse. Thus, the five synthesis filters connected in parallel produce an allpass filter. Finally, plot the frequency responses for the five filters.

17.3 This exercise concerns MELP analysis filters design. These are sixth-order Butterworth filters. Design procedures are well documented and can be found in Oppenheim and Schafer [1989] or Bose [1993]. The system function is given by

$$H(z) = \frac{H_o\left(1 + \sum_{i=1}^{6} a_i z^{-i}\right)}{1 + \sum_{i=1}^{6} b_i z^{-i}},$$

with H_o a scaling constant; a_i and b_i are the filter's coefficients. Parameters of the filters, designed with an attenuation level of -3 dB at the cutoff frequency values, are given in Table 17.2. Plot the frequency responses of these filters and compare with the synthesis filters.

17.4 Explain the differences between the Medan–Yair–Chazan method of pitch period estimation as explained in Chapter 2 and that of Section 17.4.

TABLE 17.2 Parameters of the Analysis Filters

Parameter	Analysis Filter #1	Analysis Filter #2	Analysis Filter #3	Analysis Filter #4	Analysis Filter #5
H_o	$2.883 \cdot 10^{-5}$	$5.300 \cdot 10^{-3}$	$3.169 \cdot 10^{-2}$	$3.169 \cdot 10^{-2}$	$1.052 \cdot 10^{-3}$
a_1	6	0	0	0	−6
a_2	15	−3	−3	−3	15
a_3	20	0	0	0	−20
a_4	15	3	3	3	15
a_5	6	0	0	0	−6
a_6	1	−1	−1	−1	1
b_1	−4.485	−4.425	−1.847	1.847	2.979
b_2	8.529	8.798	2.631	2.631	4.136
b_3	−8.779	−9.953	−2.216	2.216	3.260
b_4	5.148	6.753	1.575	1.575	1.517
b_5	−1.628	−2.608	−0.6229	0.6229	0.3911
b_6	0.2166	0.4535	0.1978	0.1978	0.04336

17.5 Calculate and plot the peakiness measure of the pulse

$$h[n,T] = \begin{cases} 1, & 0 \le n < T, \\ 0, & \text{otherwise,} \end{cases}$$

as a function of T, where $T \in [2, 100]$. Peakiness is computed with 180 samples. What conclusion can be drawn?

17.6 It is mentioned that during MELP decoding, the coefficients of the shaping filters are interpolated. Confirm the fact that interpolating the voicing strengths produces the same results. Find out the computational cost involved in each case. Which approach offers the lowest cost?

17.7 This exercise concerns gains decoding.

(a) How is g_2 decoded given its index ig_2 from the bit-stream?

(b) Confirm the fact that g_1 can be decoded using the following pseudocode:

```
DECODE_g1(ig1, ĝ2, ĝ2,past)
if ig1 = 0
        ĝ1 ← (ĝ2 + ĝ2,past)/2
else
        gmax ← MAX(ĝ2,past, ĝ2) + 6
        gmin ← MIN(ĝ2,past, ĝ2) − 6
        if gmin < 10
              gmin ← 10
        if gmax > 77
```

```
            gmax ← 77
        ĝ1 ← UNIFORM_DECODE(ig1,gmin,gmax,7)
return ĝ1
```

17.8 This exercise concerns the design of a pulse dispersion filter. The procedures specified here are described in McCree and Barnwell [1995].

(a) Consider the triangle pulse described by

$$x[n] = \begin{cases} n/n_1, & \text{if} \quad 0 \le n < n_1, \\ (n - n_1)/(n_1 - n_2) + 1, & \text{if} \quad n_1 < n \le n_2, \\ 0, & \text{otherwise}, \end{cases}$$

where $0 < n_1 < n_2$ are integers.

(b) Find the DFT of the triangle pulse sequence, with $n_1 = 23$, and $n_2 = 33$, using 65 samples ($n = 0$ to 64).

(c) Denoting the resultant DFT sequence as $X[k]$, $k = 0$ to 64, set all samples to unit magnitude with

$$H[k] = X[k]/|X[k]|.$$

(d) Calculate the IDFT of $H[k]$, resulting in the desired impulse response sequence $h[n]$, $n = 0$ to 64. The impulse response represents the coefficients of the pulse dispersion filter. Plot the magnitude spectrum and compare with Figure 17.23.

17.9 In the FS1015 LPC coder, the LPC quantizer achieves good efficiency by transmitting a different number of linear prediction coefficients, depending on whether the frame is voiced or unvoiced. Why is this seemingly excellent approach not used in the FS MELP coder?

17.10 Applying the interpolation rule specified for decoding, show that when $T_{past} = 50$, $T_{present} = 60$, and $n_o = 0$ the sequence of pitch periods {50, 53, 56, 59} will be generated. Find the sequence of pitch periods when $T_{past} = 60$, $T_{present} = 70$, and $n_o = 38$.

17.11 Given the voicing strength values $\{vs_1, \ldots, vs_5\} = \{0.8, 0.6, 0.5, 0.9, 0.1\}$, find the quantized values qvs_2 to qvs_5. Repeat for {0.5, 0.4, 0.6, 0.9, 0.7} and {0.8, 0.2, 0.1, 0.3, 0.7}.

17.12 Write down the pseudocode capable of producing the length of the window used for gain computation. Inputs to this procedure are the low-band voicing strength and first-stage pitch period.

CHAPTER 18

SOURCE-CONTROLLED VARIABLE BIT-RATE CELP

In a conventional speech coder, the same number of bits is allocated to all speech frames regardless of the fact that a speaker is silent roughly 63% of the time in a two-way conversation. Furthermore, due to the dynamic nature of the speech signal, the number of bits required to represent the frames faithfully varies with time. By changing the bit-rate as a function of the signal properties, it is possible to yield an average bit-rate that is substantially less than the fixed bit-rate of a conventional coder. This is the principle of the *source-controlled variable rate*, where the coding algorithm responds to the time-varying local character of the speech signal to determine the data rate.

The principles of source-controlled variable bit-rate algorithm have been around since the early stage of speech coding development. The FS1015 LPC coder (Chapter 9) and the FS MELP coder (Chapter 17), for instance, pertain to this family. For these coders, parameters of the speech production model are encoded using different numbers of bits, depending on whether the frame is voiced or unvoiced.

This chapter is dedicated to the TIA IS96 variable bit-rate (VBR) CELP standard, in which the control mechanism is based on a background noise estimate and the energy of the signal. The coder was created to increase the capacity of a *code division multiple access* (CDMA)-based digital cellular system by decreasing the mutual interference among users. The IS96 algorithm is also known as QCELP, since it was initially designed by Qualcomm and later standardized by the TIA in 1993 [Gardner et al., 1993]. The chapter begins with a thorough description of the principles of adaptive rate decision, followed by LP analysis, LSF quantization, and operations of decoding and encoding. Since many details of CELP operations are covered extensively in previous chapters (Chapters 11, 12, 13, 14, and 16), topics related to decoding, encoding, and quantization are described only briefly. Readers are referred to TIA [1998] for a complete description of the IS96 standard.

18.1 ADAPTIVE RATE DECISION

The most critical component of a source-controlled variable rate coder is the mechanism by which the bit-rate is selected, since the efficiency and quality of the resultant algorithm depend heavily on the selected technique. This section describes the method adopted by the IS96 standard.

Available Bit-Rates

The IS96 coder dynamically selects one of four data rates every 20 ms (160-sample frames), depending on speech activity. The four bit-rates are:

- Full rate: 8.55 kbps.
- Half rate: 4 kbps.
- Quarter rate: 2 kbps.
- Eighth rate: 0.8 kbps.

Typically, active speech is coded at full rate, while silence and background noise are coded at lower rates.

Frame/Subframe Structure

IS96 follows the conventional CELP structure (Chapter 11), with the parameters of the speech production model extracted at regular time intervals. Depending on the bit-rate, a 20-ms frame is divided into subframes of different lengths. Two types of subframes are considered:

- *Pitch Subframe.* This is the interval within which the parameters of the pitch synthesis filter are extracted.
- *Codebook Subframe.* This is the interval where the fixed excitation codebook is searched to find the best codevector.

Lengths of the subframes are summarized in Table 18.1.

Bit-Rate Transition Rules

An adaptive algorithm is used to determine the bit-rate for each frame, which keeps a running estimate of the background noise energy and selects the bit-rate based on the difference between the background noise energy estimate and the current frame's energy. Figure 18.1 summarizes the major rules for transition between

TABLE 18.1 Subframe Lengths in Number of Samples for Different Bit-Rates[a]

Bit-Rate	Pitch Subframe	Codebook Subframe
Full	40	20
Half	80	40
Quarter	160	80
Eighth	—	160

[a] Data from TIA [1998], Table 2.4.1-1.

bit-rates, where the algorithm stays at a certain bit-rate at a given frame and switches to other values or remains the same for the next frame. It is only permitted to decrease by one rate per frame, that is, full → quarter is prohibited; if a request for full → quarter appears, it is converted to full → half. However, there is no restriction on bit-rate increases; thus, eighth → quarter, eighth → half, and eighth → full are all valid. These rules are designed to smooth the transitions so as to minimize the amount of perceivable artifacts, as well as to represent transients (sudden increases in energy) in the most faithful manner, since they play an important role in subjective quality.

Background Noise Estimate and Bit-Rate Decision

The first variable that the algorithm relies on for bit-rate decision is the autocorrelation at zero lag, or energy of the windowed speech signal, written as

$$Es = \sum_{n=0}^{159} (w[n]s[n+60])^2, \tag{18.1}$$

where $s[n]$ denotes the input speech signal, with the shift of 60 reflecting the position of the LP analysis window (Section 18.2); $w[n]$ is the Hamming window

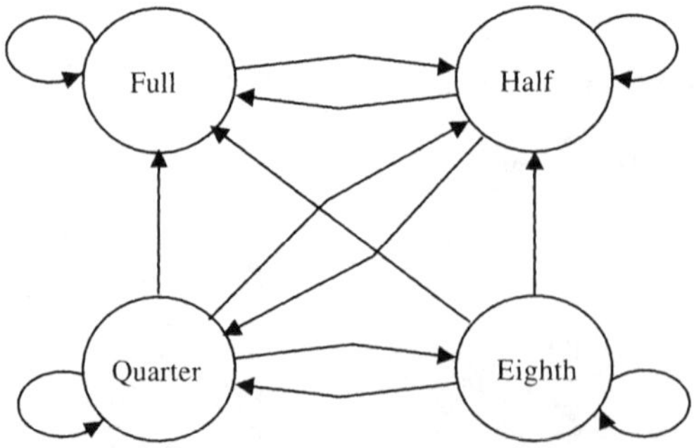

Figure 18.1 State diagram for bit-rate transitions.

sequence. It is important to note that the equations to follow assume 14-bit uniform PCM for the input speech, with a maximum range of ± 8031. The window, on the other hand, is assumed to have a peak magnitude of 1.

The background noise estimate of current frame is found with

$$BN = \min(Es_{\text{past}},\ BN_{\max},\ \max(\alpha \cdot BN_{\text{past}},\ BN_{\text{past}} + 1)) \tag{18.2}$$

where $\min(x, y, z)$ returns the minimum of x, y, and z; and $\max(a, b)$ returns the maximum of a and b. Es_{past} and BN_{past} correspond to past frame quantities. $BN_{\max}$ represents the upper limit of the background noise estimate and is a parameter of the algorithm; its value is set at $\approx 5.06 \cdot 10^6$. For initialization, BN is equal to $BN_{\max}$. Both terms inside the max( ) operator ($\alpha \cdot BN_{\text{past}}$ and $BN_{\text{past}} + 1$) increment the candidate for the noise estimate by a small amount for the next frame, with $\alpha = 1.00547$. The expression is based on the assumption that the noise of the next frame is almost the same as for the current frame, with a slight increase. In fact, the adaptation speed of the noise estimate is controlled by the constant α of the first term, as well as the addition of 1 in the second term. Changing these two constants modifies the adaptation speed. Why are the two terms inside the max( ) operator? This is because the incremental effect of $1.00547BN_{\text{past}}$ might be void for a low value of the noise estimate under a fixed-point environment (rounding), and the use of $BN_{\text{past}} + 1$ ensures that a higher number is available for the next round, since 1 is the smallest possible unit.

Example 18.1 Figure 18.2 shows some plots of the background noise estimate, where a typical speech waveform is utilized. The energy of the signal is calculated as in (18.1) with the noise estimate computed by (18.2). From the plots we see the following: BN is an estimate of the lowest energy level of the signal. The core assumption is that the background noise is equal to the lowest input energy at all times. Thus, BN continuously tracks the minimum in the energy level of the input signal. If the energy of the input signal is above BN, BN is increased slightly with time, with the adaptation speed dictated by the constants inside the max( ) operator in (18.2): α and 1. These two constants can be modified if a different adaptation speed is desired. Unless there is a new low in the input energy level so that $BN[m] = Es[m-1]$, the inclusion of the max( ) operator ensures that the value of BN is always on the upside.

Decision Thresholds for Bit-Rate Decision

Based on the noise estimate, a threshold equation is defined with

$$Th_i = \begin{cases} \alpha_{i,1}BN^2 + \alpha_{i,2}BN + \alpha_{i,3}; & BN \le 1.6 \cdot 10^5, \\ \alpha_{i,4}BN^2 + \alpha_{i,5}BN + \alpha_{i,6}; & \text{otherwise}, \end{cases} \tag{18.3}$$

for $i = 1$, 2, and 3. The threshold functions are designed in such a way that $Th_3 > Th_2 > Th_1$ for $BN \in [0, BN_{\max}]$, which is satisfied except at the very end of

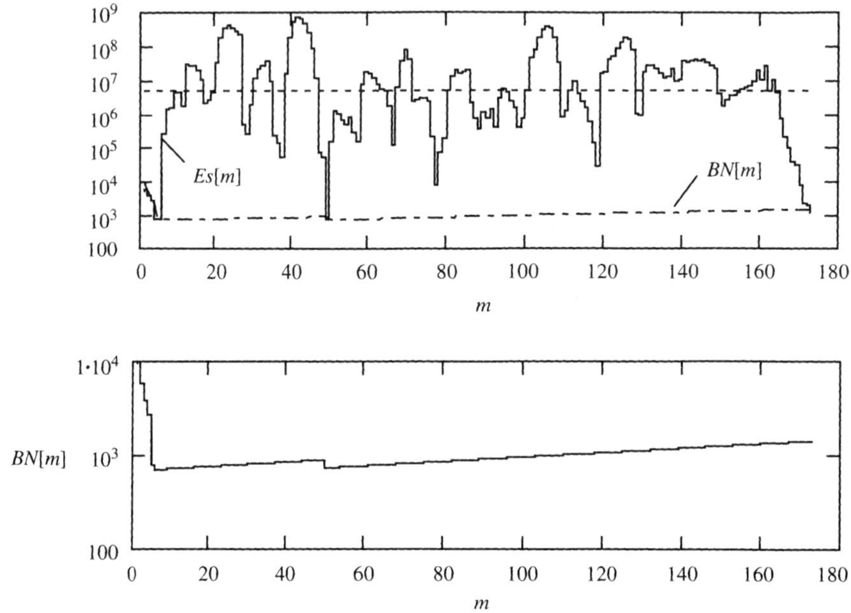

Figure 18.2 Example of signal energy *Es* and background noise estimate *BN*.

the range: $BN \approx BN_{\max}$. The α constants have the values $\alpha_{1,1} \approx -5.54 \cdot 10^{-6}$, $\alpha_{1,2} \approx 4.05$, $\alpha_{1,3} \approx 362$, $\alpha_{1,4} \approx 9.04 \cdot 10^{-8}$, $\alpha_{1,5} \approx 3.54$, $\alpha_{1,6} \approx -6.21 \cdot 10^{4}$, $\alpha_{2,1} \approx -1.53 \cdot 10^{-5}$, $\alpha_{2,2} \approx 8.75$, $\alpha_{2,3} \approx 1.14 \cdot 10^{3}$, $\alpha_{2,4} \approx -1.99 \cdot 10^{-7}$, $\alpha_{2,5} \approx 4.94$, $\alpha_{2,6} \approx 2.24 \cdot 10^{5}$, $\alpha_{3,1} \approx -3.96 \cdot 10^{-5}$, $\alpha_{3,2} \approx 18.9$, $\alpha_{3,3} \approx 3.35 \cdot 10^{3}$, $\alpha_{3,4} \approx -4.84 \cdot 10^{-7}$, $\alpha_{3,5} \approx 8.63$, and $\alpha_{3,6} \approx 6.46 \cdot 10^{5}$. Figure 18.3 plots the three thresholds as a function of the background noise estimate. To find the desired bit-rate for the current frame, the energy *Es* is compared to the thresholds. If *Es* is greater than the three thresholds, the full rate is selected; if *Es* is greater than only two thresholds, the half rate is selected; if *Es* is greater than only one threshold, the quarter rate is selected; otherwise the eighth rate is selected. It is necessary to note that the actual bit-rate for the frame depends on the past history, which is governed by the rules of Figure 18.1. Thus, the actual bit-rate is determined from the desired bit-rate in conjunction with the past bit-rate. This multithreshold scheme improves robustness and stability under noisy conditions.

Example 18.2 Figure 18.4 plots the energy *Es* of the 160-sample frames, together with the noise estimate *BN* and the thresholds Th_1, Th_2, Th_3 for the same signal as in Example 18.1. It shows that the three thresholds are proportional to the noise estimate, which tracks the bottom of the signal energy. Note also that $Th_3 > Th_2 > Th_1$. At the beginning, the signal energy is low and decreasing (first six frames), and *BN* is equal to *Es*. As the speech energy grows, *BN* increases slowly, until a low-energy interval corresponding to a short silent period appears near the 50th frame; at that

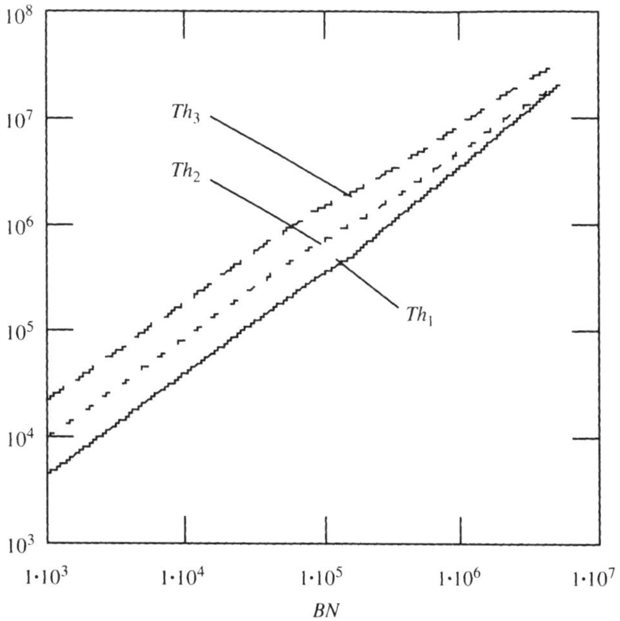

Figure 18.3 Plot of decision thresholds as a function of the background noise estimate.

point *BN* is lowered and increases slowly afterward. Figure 18.5 shows the desired bit-rate, *bit-rate'* [*m*], obtained when the energy is compared with the three thresholds. The coding scheme is: bit-rate = 0 for full, 1 for half, 2 for quarter, and 3 for eighth. Initially (first six frames), the eighth rate is the choice since the signal energy is lower than the three thresholds. This is again the case during the short

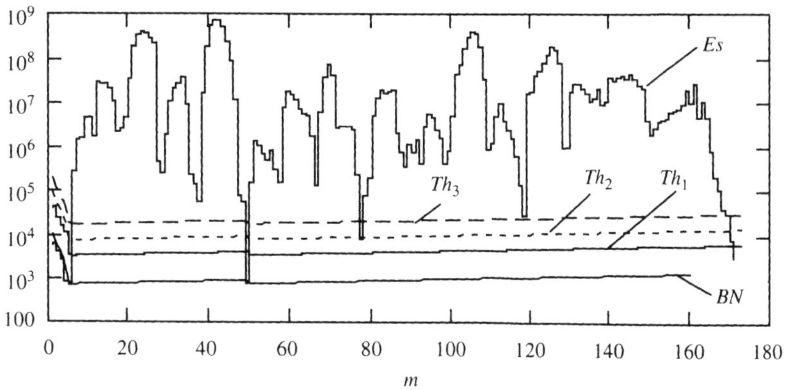

Figure 18.4 Example of signal energy, background noise estimate, and the three decision thresholds.

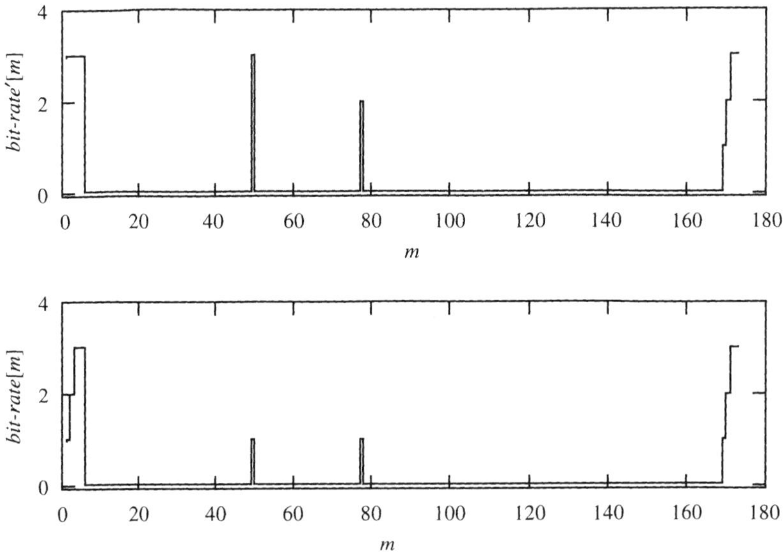

Figure 18.5 Desired bit-rate (*top*) and actual bit-rate (*bottom*) for the signal in Figure 18.4.

silence period near the 50th frame. Most of the frames, however, have a desired bit-rate of full. The incorporation of three thresholds effectively improves the robustness of the decision mechanism against noisy conditions, leading to a stable bit-rate decision outcome. The desired bit-rate is processed so as to produce the actual bit-rate, *bit-rate*[*m*], following the restriction rules for bit-rate changes. Note how the sudden alteration in bit-rate, such as full $\rightarrow$ eighth, is replaced by full $\rightarrow$ half.

Example 18.3 White noise is used to illustrate the behavior of the bit-rate selection algorithm. Figure 18.6 shows the situation, where the first few frames are low-energy noise followed by high-energy noise. Except at the transition, the noise is stationary. We can see how *BN* initially tracks the bottom of the signal energy and increases slowly after the transition. The three thresholds display similar behaviors. As time progresses, Th_2 and Th_1 surpass the signal energy. Figure 18.7 shows the actual bit-rate. Note that right after the transition, the half rate is selected, followed by an oscillatory zone where the rate changes between half and quarter; and as Th_1 moves higher, the bit-rate eventually switches to eighth.

Bit Allocation

At a bit-rate of full, half, and quarter, five sets of parameters are transmitted: LPC, pitch period, pitch gain, codebook index, and codebook gain. These parameters are typical of a CELP algorithm. Depending on the bit-rate, a different number of bits are used for encoding. At the full rate, additional bits are sent for error detection. At

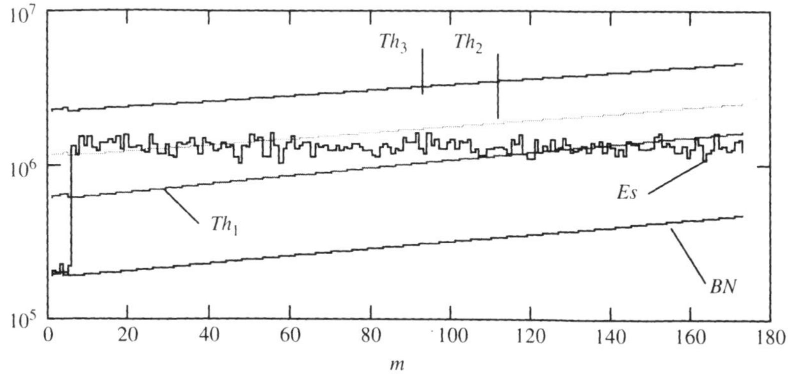

Figure 18.6 Example of signal energy, background noise estimate, and the three decision thresholds. The signal in this case consists of low-energy white noise switched to high-energy white noise.

a bit-rate of eighth, no parameters from the pitch synthesis filter are transmitted. The bit allocation scheme is summarized in Table 18.2.

A Summary

The technique adopted by the IS96 standard for bit-rate selection is based on the energy of the frames. In this method, a background noise estimate is found based on the lowest energy of the signal frame observed so far. This noise estimate is allowed to increase slowly with time if the signal energy remains above it. Three thresholds are defined on top of the noise estimate. The energy of the frame is compared to these thresholds so as to select one out of four different bit-rates.

The technique is an *open-loop* approach, since bit-rate selection is made directly from observations of parameters extracted from the input speech without assessing how the specific coding mode will perform for the particular frame. The advantage of using this method is its simplicity, when compared to a *closed-loop* method, where each coding mode must be evaluated with the best selected.

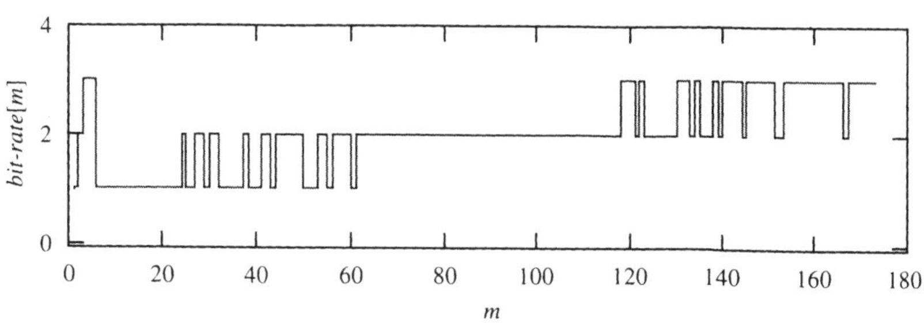

Figure 18.7 Actual bit-rate for the signal in Figure 18.6.

TABLE 18.2 Bit Allocation for the IS96 Coder[a]

Parameter	Number per Frame	Resolution	Total Bits per Frame
	Full Rate		
LPC	10	4	40
Pitch period	4	7	28
Pitch gain	4	3	12
Codebook index	8	7	56
Codebook gain	8	3	24
Parity check	11	1	11
Total			**171**
	Half Rate		
LPC	10	2	20
Pitch period	2	7	14
Pitch gain	2	3	6
Codebook index	4	7	28
Codebook gain	4	3	12
Total			**80**
	Quarter Rate		
LPC	10	1	10
Pitch period	1	7	7
Pitch gain	1	3	3
Codebook index	2	7	14
Codebook gain	2	3	6
Total			**40**
	Eighth Rate		
LPC	10	1	10
Codebook seed	1	4	4
Codebook gain	1	2	2
Total			**16**

[a] Data from TIA [1998], Table 2.4.1-2.

18.2 LP ANALYSIS AND LSF-RELATED OPERATIONS

This section describes the operations related to LP analysis, LSF quantization, and LSF interpolation.

LP Analysis

This coder utilizes a 20-ms (160-sample) frame, and a Hamming window of length 160 is applied to the signal for autocorrelation computation. The window is centered on the last full-rate pitch subframe (Figure 18.8). The windowed data

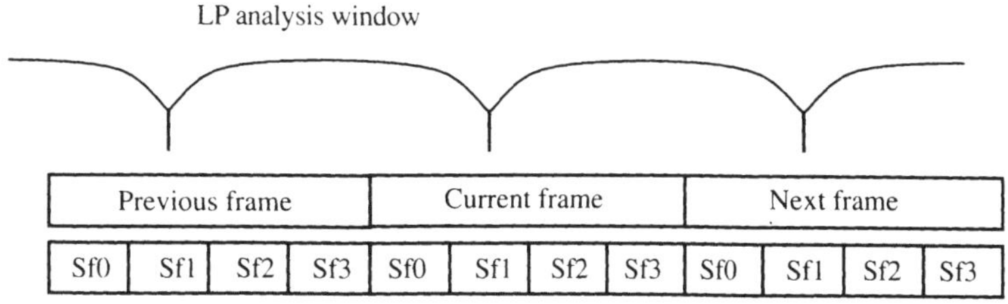

Figure 18.8 Positions of frame and LP analysis windows.

are used in the autocorrelation computation, where 11 values are obtained. The Levinson–Durbin algorithm is then applied, resulting in ten LPCs, which are bandwidth-expanded with a factor of 0.9883, and later converted to LSFs. The LSFs are mean-removed by subtracting the constant bias

$$\mu_k = k\pi/11; \quad k = 1, \ldots, 10, \tag{18.4}$$

representing the mean values of the LSFs when they are concatenated frame after frame.

LSF Quantization

At full rate, the mean-removed LSFs are quantized directly using ten different uniform scalar quantizers, with the same resolution of 4 bits. A predictive scalar quantizer or DPCM (Chapter 6) is deployed for quantization for all other rates. The scheme allows exploitation of redundancy present among the LSFs at consecutive frames (Chapter 15). Prediction is based on a scaled version of the LSF from the past frame, with the scale factor or prediction coefficient being 0.90625. The prediction error in the DPCM scheme is uniformly quantized at a resolution of 2 bits for half rate and 1 bit for quarter and eighth rates.

LSF Decoding and Interpolation

After decoding the LSFs using the appropriate scheme, stability is enforced with the following procedure; for simplicity, the input set of LSFs is denoted by ω_i, $i = 1$ to 10.

```
ω0 ← 0; i ← 0;
while i < 10
        if ω(i+1) − ω(i) < dmin
                ω(i+1) ← ω(i) + dmin
                i++
ω11 ← π
while i > 0
        if ω(i+1) − ω(i) < dmin
```

```
9.          ωi ← ωi+1 − dmin
10.         i--
```

The procedure first proceeds from low to high order. If the ordering or minimum distance requirement is not satisfied, the high-order LSF is replaced by the low-order LSF plus *dmin*, where $dmin = 0.02\pi$ corresponds to 80-Hz minimum separation (Line 4). By enforcing minimum distance between the LSFs, large peaks in the transfer function of the associated synthesis filter are eliminated. The procedure then proceeds from high to low order, achieving a similar goal but in reverse order. Note that two additional LSFs are created: $\omega_0 = 0$ and $\omega_{11} = \pi$.

After stability enforcement, the quantized LSFs of the current frame $\hat{\omega}_k$ are smoothed according to

$$\hat{\omega}_k \leftarrow \beta\hat{\omega}_{k,\text{past}} + (1 - \beta)\hat{\omega}_k. \tag{18.5}$$

At full rate, no smoothing is performed where $\beta = 0$. The level of smoothing is higher for the lower rates of quarter and eighth since quantization noise is higher for these cases and more smoothing is necessary to ensure good quality.

Finally, the LSFs are linearly interpolated to be used by the different subframes.

18.3 DECODING AND ENCODING

Decoding and encoding operations of the IS96 are described in this section.

Decoder Structure

The IS96 is based on the CELP algorithm (Chapter 11), with the overall speech synthesis or decoder model shown in Figure 18.9. First, a vector is taken from one of two sources depending on the bit-rate. For the eighth rate a pseudorandom vector is generated. For all other rates, a vector specified by an index is taken

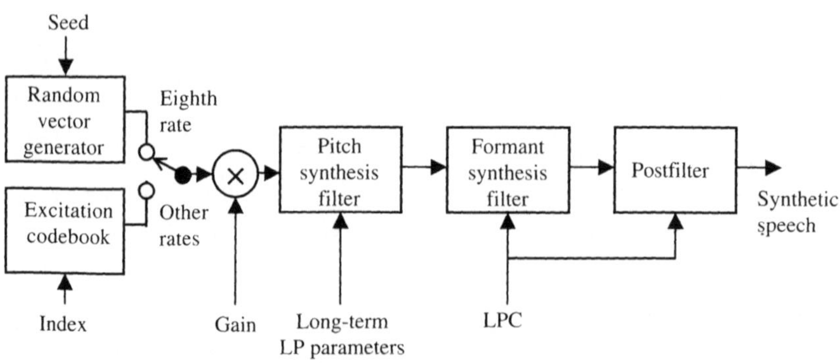

Figure 18.9 Block diagram of the IS96 decoder.

from the excitation codebook. The vector is multiplied by a gain term and passed through the cascade connection of the pitch synthesis filter, formant synthesis filter, and postfilter so as to generate the synthetic speech at its output.

Excitation Codebook Structure

The excitation codebook consists of $2^7 = 128$ codevectors and has an overlapping structure, where two consecutive codevectors are shifted by one sample (Chapter 12). To further save memory cost, the codebook is stored as a 128-element array, with the elements of the codevector obtained through circular shifting. That is, if the codebook index combined with the codevector length surpasses the boundary of the array, the last elements of the codevector are read from the beginning of the array. This design is highly compact and allows a fast codebook search through recursive convolution. The structure is referred to as a *circular overlapping* codebook. Denoting the entire codebook as a single array,

$$v[n]; \quad n = 0, \ldots, 127,$$

then each codevector is identified with

$$v^{(l)}[n] = v[(n + l)\text{mod}128]; \quad l = 0, \ldots, 127; \quad n = 0, \ldots, N_c - 1, \qquad (18.6)$$

where N_c is the length of the codevector, which depends on the bit-rate. Samples of the codebook take on discrete values in the set $\{-2.5, -2, -1.5, -1, 0, 1, 1.5, 2, 2.5, 3\}$, where 79% are zeros. A plot of all the samples of the excitation codebook array appears in Figure 18.10.

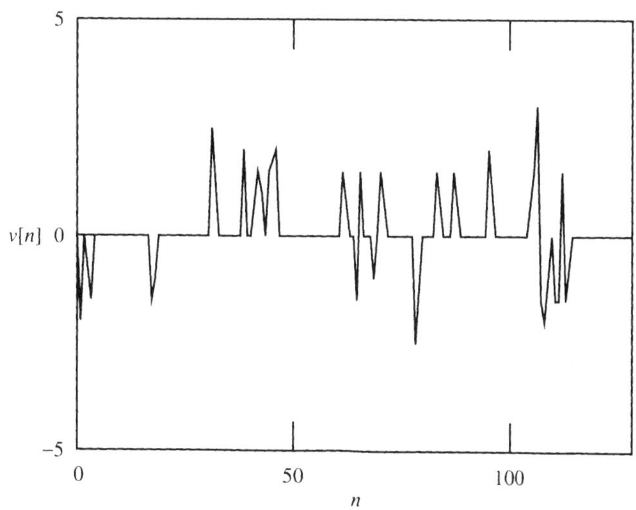

Figure 18.10 Plot of the excitation codebook array for the IS96 coder. Data from TIA [1998], Table 2.4.6.1-1.

Random Seed Generation

For frames encoded with the eighth bit-rate, a pseudorandom number generator is used to produce the excitation sequence. The sequence is generated using a 4-bit seed value, which is transmitted as information on the packet. Both the decoder and encoder use the same seed to generate the exact same excitation sequence.

Finding the Parameters of the Pitch Synthesis Filter

The pitch synthesis filter has a system function

$$H_p(z) = \frac{1}{1 + bz^{-T}}, \tag{18.7}$$

with b the pitch gain (or long-term gain) and T the pitch period. Both parameters must be determined for the pitch subframe under consideration. Note that these parameters are not found for the case of the eighth rate, where a default value of $b = 0$ is applied.

Like other CELP coders (i.e., IS54, Chapter 13), the IS96 utilizes a suboptimal two-step sequential procedure to locate the best parameters during encoding. In the first step, a long-term gain and pitch period are found, assuming zero excitation gain; and in the second step, an index to the excitation codebook with the associated gain is determined, with the long-term parameters fixed. An analysis-by-synthesis loop is applied to find the long-term parameters; this is done by searching through all possible combinations of pitch period and quantized gain values.

The pitch lag T is encoded with 7 bits and ranges from 17 to 143. The pitch gain b is represented by 3 bits and ranges from 0 to -2 in steps of -0.25 (uniform quantization). For each pitch subframe, the chosen parameters b and T are encoded using 3 and 7 bits, respectively.

Excitation Codebook Search

After the parameters of the pitch synthesis filter are found, the next step is to search the excitation codebook so as to get the optimal excitation codevector and the associated gain. The codebook search procedure is performed for every codebook subframe. Given the excitation codebook gain g and index l, they are encoded using 3 and 7 bits, respectively.

Gain magnitude is quantized using a DPCM technique, which operates on a codebook subframe basis regardless of the bit-rate. That is, it operates eight times during a full rate frame, four times during a half rate frame, two times during a quarter rate frame, and only once for an eighth rate frame.

18.4 SUMMARY AND REFERENCES

This chapter briefly covers the TIA IS96 VBR-CELP standard, with a description of its most salient features. It is shown that the coder follows the principles of an open-loop source-controlled variable bit-rate algorithm, with the bit-rate selected based on the energy of the signal frame. In particular, the energy of the background noise is monitored and used to adapt a set of three time-varying thresholds that remain above the noise level. Comparison between the energy of the frame and the thresholds determines the desired bit-rate. Other implementational details of IS96 are quite similar to its CELP cousins and hence are skipped for brevity.

There are many other proposed multimodal coders; see Das et al. [1995] for an overview. Discussion of issues related to variable bit-rate speech coding appears in Gersho and Paksoy [1993]; development of the QCELP algorithm is given in Gardner et al. [1993]. In Wang [1999], a higher bit-rate QCELP algorithm is described with improved performance. See TIA [1998] for technical descriptions of the IS96 standard.

EXERCISES

18.1 Given the desired bit-rate sequence $bit\text{-}rate'[m]$, write the pseudocode to convert it to the actual bit-rate sequence $bit\text{-}rate[m]$. All transition rules among bit-rates must be observed. *Hint*: Inputs to the procedure are $bit\text{-}rate'[m]$ and $bit\text{-}rate[m-1]$.

18.2 Using some speech waveforms, compute the energy per frame and the background noise estimate and plot out the resultant sequences. Modify the constants 1.00547 and 1 inside the max( ) operator in the expression for background noise estimate and replot the background noise estimate. You can either increase those constants or decrease them. What is the conclusion?

18.3 Implement the threshold equations for bit-rate decision and observe their behaviors using different types of waveforms. Based on a comparison between energy and threshold, find out the actual bit-rate assigned to the frames.

18.4 IS96 relies on the pseudorandom number generator described by

$$SD[n] = (521 \cdot SD[n-1] + 259) \bmod 2^{16}.$$

Based on the equation, generate several sequences (160 samples) using various initial conditions. Are the resultant sequences "white" enough? Justify your answer. Is it possible for this pseudorandom number generator to degenerate [Banks and Carson, 1984]? That is, under certain conditions, subsequent outcomes become the same number (constant output).

18.5 For the bit-rate transition rules, assume all aspects remain the same with the exception that full → quarter is permitted. Draw the state diagram for bit-rate transitions and write the pseudocode to convert the desired bit-rate sequence to the actual bit-rate sequence.

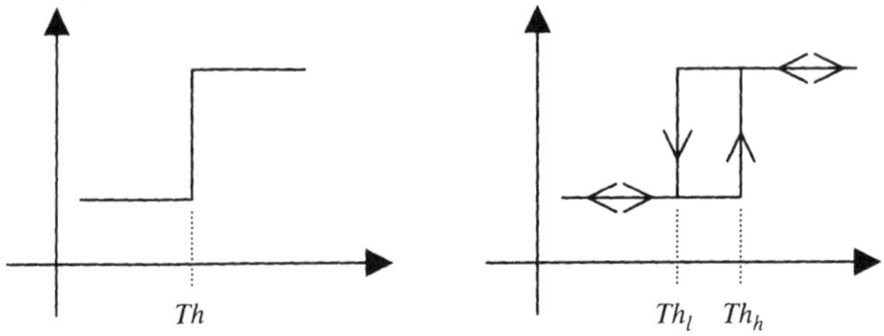

Figure 18.11 Transfer characteristics of a simple comparator (*left*) and a comparator with hysteresis (*right*).

18.6 Example 18.3 shows a situation where rapid oscillation is developed in the bit-rate pattern (Figure 18.7). This happens when the value of the threshold is near the background noise estimate, since a simple pass–fail comparison rule is deployed. The situation is undesirable since frequent switching between bit-rates can generate artifacts. One solution to the problem is to introduce hysteresis to the comparison rule, with the comparison based on two thresholds as illustrated in Figure 18.11. By properly adjusting the distance between the two thresholds in a comparator with hysteresis, it is possible to eliminate or reduce the oscillations at the output, leading to improved stability. Modify the IS96 scheme to incorporate hysteresis in the decision process. Generate some signals in a similar way as in Example 18.3 to test and contrast the cases with and without hysteresis.

CHAPTER 19

SPEECH QUALITY ASSESSMENT

Quality of synthetic speech is one of the most important attributes of a speech coder. At the very early development stage of digital communication, signal-to-noise ratio (SNR) is often the method of choice to evaluate the quality of waveform coding schemes such as PCM and ADPCM. When the FS1015 LPC coder was developed around 1984, it was observed that the synthetic speech was quite intelligible; however, it has a poor SNR value. The conclusion was that some new measure must be developed in order to handle these types of parametric coders. A series of subjective measures were invented to meet this need, where a selected group of human subjects was asked to listen to some speech material and evaluate them. An average score, representing the subjective quality, is obtained at the end of the section.

This chapter aims to present a survey of current speech coding evaluation practice. It begins with an overview of speech quality, its scope and measuring conditions. Early objective measures are described next, followed by some standard subjective test procedures. Recent developments in objective measures are discussed in the last section, with some description on standardization efforts. The material in this chapter is intended for general information only; no detail algorithmic descriptions are provided. Interested readers are referred to the references for additional information.

19.1 THE SCOPE OF QUALITY AND MEASURING CONDITIONS

There are many dimensions in quality perception, some of them are described as follows.

- *Intelligibility.* Whether the underlying message of the speech signal can be clearly understood.
- *Naturalness and Pleasantness.* This is often related to the presence or absence of various types of distortion or artifact in the synthetic speech, such as noise, echoes, muffling, and clicking.
- *Speaker Recognizability.* Does the synthetic speech allow straightforward identification to the original speaker?

It is important to note that speech quality must be measured under various conditions, such as:

- *Dependency on Speaker.* Many low-bit-rate coders show a quality that is speaker dependent. A thorough quality evaluation should involve speech material from adult males, adult females, and children.
- *Dependency on Language.* Different languages have different speech sounds (phonemes) that may or may not be appropriately modeled by the coder. Evaluating the coder using speech material from different languages can discover a weakness toward a given idiom.
- *Dependency on Signal Levels.* Quality of a coder is not uniform with respect to variations in input power level and must be tested for various power levels so as to understand its behavior. In particular, the coder should be tested for extremely low level (or even zero) as well as extremely high level. Under these conditions, the output should be well behaved.
- *Background Noise.* Performance of the coder for speech with background noise is important due to the increasing number of portable applications. Typical noise sources include car, street, and transient noise, music, and interfering speakers.
- *Tandem Coding.* In a typical communication link, the speech signal might be encoded and decoded repeatedly; sometimes under the same coder, and other times with different coders. Performance under these conditions must be evaluated.
- *Channel Errors.* All communication channels contain some sort of errors. Quality of the coder can be measured under different bit error rates.
- *Nonspeech Signal.* The faithfulness in the reproduction of nonspeech signals, such as music or signaling tones (i.e., dual-tone multifrequency—DTMF—in modern telephone systems), might be important in certain applications. In general, it is not expected for the speech coder to be able to synthesize music, but it should not generate annoying artifacts.

19.2 OBJECTIVE QUALITY MEASUREMENTS FOR WAVEFORM CODERS

Performance of waveform coders, such as PCM and ADPCM, can simply be measured using some form of signal-to-noise ratio (SNR). Discussion of SNR and its refinement is included in this section.

Signal-to-Noise Ratio

Given the original speech $x[n]$ and the synthetic version $y[n]$, the SNR is defined by

$$\mathrm{SNR} = 10\log_{10}\left(\frac{\sum_n x[n]^2}{\sum_n (x[n]-y[n])^2}\right), \tag{19.1}$$

with the range of the time index n covering the measurement interval.

Segmental Signal-to-Noise Ratio (SSNR)

This is a refinement with respect to conventional SNR measure and is created to handle the dynamic nature of nonstationary signals such as speech. The definition of SSNR is

$$\mathrm{SSNR} = \frac{1}{N}\sum_{m=1}^{N} \mathrm{SNR}_m; \tag{19.2}$$

that is, it is an average of SNR values obtained for isolated frames, with the frame being a block of samples. This measure compensates for the underemphasis of weak-signal performance in conventional SNR measure. As we can see from (19.2), very high SNR values corresponding to well-coded high-power frames cannot camouflage performance of low-power frames, as in conventional SNR.

Poorly designed coders do not handle low-power frames appropriately, leading to a drop in SNR for these frames. This normally leads to deterioration in subjective performance, since low-power frames play an important role in perception: silence frames should be silent instead of annoying noise. Conventional SNR does not penalize the bad rendition of weak signals since strong signals dominate the measurement outcome. The SSNR measure uncovers the performance in low-power frames, leading to a more subjectively meaningful parameter.

The SNR in (19.2) is computed locally for frames of 10 to 20 ms in duration. Further refinements to the computation of SSNR include the exclusion of extremely high and extremely low SNR values from the equation, and the elimination of silence frames.

A Summary

The SNR and SSNR measures are meaningful only for waveform coders. As we can see from (19.1) and (19.2), both the SNR and SSNR are extremely sensitive toward waveform misalignments and phase distortions, which are not always perceptually relevant. On the other hand, most low bit-rate coders do not preserve the original waveform, and hence the SNR and SSNR are meaningless for the evaluation of these coders. Subjective quality measurement techniques are designed to overcome the limitations of the simple SNR approach. This is covered in the next section.

19.3 SUBJECTIVE QUALITY MEASURES

In subjective testing, speech materials are played to a group of listeners, who are asked to rate the passage just heard. The ratings are then gathered and averaged to yield the final score. The test is normally done for a wide variety of conditions (Section 19.1) so as to obtain a general performance appreciation for a particular coder. Some popular procedures to perform subjective testing are described in this section.

Absolute Category Rating (ACR)

In this test the listeners are required to make a single rating for each speech passage. Five choices exist: excellent (5), good (4), fair (3), poor (2), and bad (1). The average of all votes is known as the mean opinion score, or MOS.

Degradation Category Rating (DCR)

In this test the listeners are presented with the original signal as a reference, before they listen to the synthetic signal, and are asked to compare the two and give a rating according to the amount of degradation perceived. The choices are: not perceived (5), perceived but not annoying (4), slightly annoying (3), annoying (2), and very annoying (1). The average of all votes is known as the degradation mean opinion score, or DMOS.

Comparison Category Rating (CCR)

Due to the order by which the speech materials are presented in the DCR test, the final score might be biased. A better approach is to present two samples and ask the listeners to compare and rate. The order by which the original speech and synthetic speech is presented can be made arbitrary or random. The suggested choices are: much better (3), better (2), slightly better (1), about the same (0), slightly worse (−1), worse (−2), and much worse (−3).

A Summary

Reliability of subjective tests' outcomes depends on the number of listeners. In most tests the minimum number is 16, with the listeners recruited from the general population so as to reflect better the conditions under which the system eventually will be deployed.

It is clear that subjective tests are expensive to implement and highly time consuming. Therefore, current research efforts are being directed toward perceptually based objective measures, described in the next section.

19.4 IMPROVEMENTS ON OBJECTIVE QUALITY MEASURES

Since most objective quality measures designed for waveform-approximating coders are inadequate for the evaluation of low-bit-rate coders, and subjective tests are too costly to implement, it is natural for the speech coding community to consider the development of alternative objective quality measurement techniques. Ideally, the outcomes should be highly correlated with subjective test scores. Figure 19.1 contrasts the case of a good objective measure, where the outcomes are nearly proportional to subjective scores, to that of a bad objective measure, with almost no correlation with respect to subjective scores.

Since the 1980s, the ITU-T has been investigating many proposals for objective quality measurements. After careful comparisons, it was concluded that the *perceptual speech quality measure* (PSQM) algorithm best correlated with the subjective quality of coded speech. The effort led to the standardization of PSQM as ITU-T Recommendation P.861 in 1996 [ITU, 1996c].

Figure 19.2 shows a high-level block diagram of the PSQM algorithm. The delay of the coder under test is first estimated and compensated for. The perceptual transformation contains a simple hearing model, while the distance measure block models judgment. A mapping function is used to convert the perceptual distance into the scale of mean opinion score. Note that the algorithm resembles the DCR

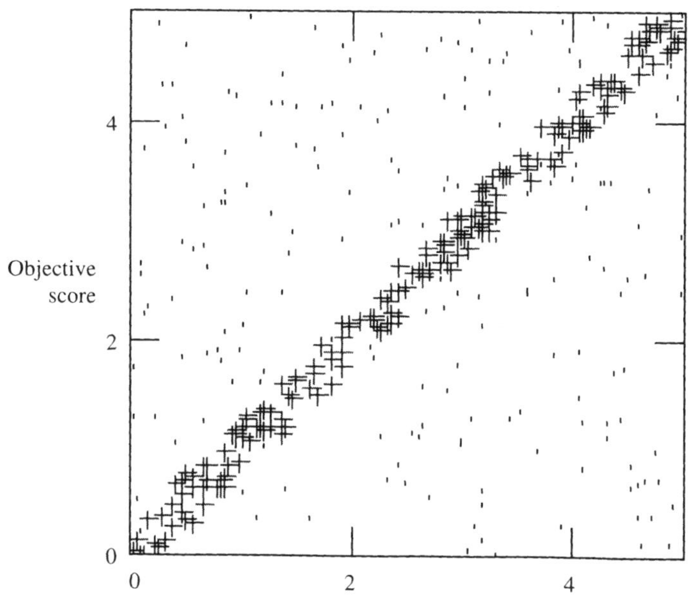

Figure 19.1 Illustration of the outcomes for a good objective quality measure (+) and a bad objective quality measure (dots).

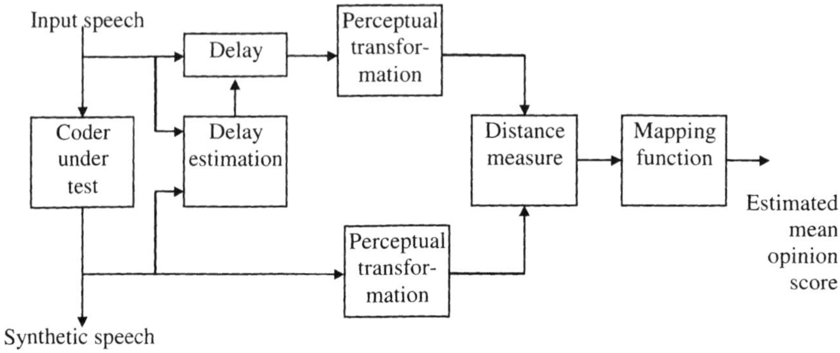

Figure 19.2 High-level block diagram of the PSQM algorithm.

test, where the reference and test signals are compared and a score is generated at the end.

Voran [1999a] addressed some limitations of the P.861 standard and came up with an improvement that eventually led to a new ITU standard [ITU, 1998a]. The technique is known as measuring normalizing block (MNB) and has found to provide significant improvements in many cases, particularly at lower bit-rates, and when bit errors or frame erasures are present.

Neither PSQM nor MNB was suitable for end-to-end measurement of telecommunication networks, where conditions such as linear filtering, variable delay, and background noise are common. To address these problems, a competition to select a new algorithm was conducted in late 1998 by ITU-T. The winner was the perceptual evaluation of speech quality (PESQ) algorithm, which was subsequently standardized by the ITU-T as recommendation P.862 [ITU, 2001]. PESQ performs much better than earlier assessment methods, such as PSQM and MNB, and has been evaluated on a very wide range of coders and network conditions. It represents a milestone in the evolution of quality assessment technology.

Additional References

See Jayant and Noll [1984] for signal-to-noise ratio types of objective measurements; subjective tests are described in Panzer et al. [1993] and Dimolitsas [1993]. Kroon [1995] contains a survey of existing objective and subjective measures up to 1995. Comprehensive discussion of problems and challenges in quality assessment can be found in Denisowski [2001]. The PSQM algorithm is described in ITU [1996c], and ITU [1998a] contains the improved version based on MNB; detailed descriptions of the MNB technique are given in Voran [1999a, 1999b]. See Rix et al. [2000, 2001] for description of the PESQ algorithm. Quality measurements of speech are often overlapped with audio; in ITU [1998c], an algorithm known as perceptual audio quality measure (PAQM) is standardized as ITU-R Recommendation BS.1387. Description for the design of a combined measurement tool for speech and audio is found in Keyhl et al. [1999].

APPENDIX A

MINIMUM-PHASE PROPERTY OF THE FORWARD PREDICTION-ERROR FILTER

The predictor as presented in Chapter 4 is known as a forward predictor, since it predicts the future based on samples from the past. In this appendix, the minimum-phase property of the forward prediction-error filter is proved. In order to accomplish that, the concept of backward linear prediction is introduced. Relations between backward and forward predictors are developed, followed by a discussion of the dependency of mean-squared prediction error on the reflection coefficients. System functions of backward and forward prediction-error filters are given, and the minimum-phase property of the forward prediction-error filter is proved using Rouché's theorem.

Backward Linear Prediction: Predictor and Prediction-Error Filter

One-step backward prediction consists of using the subset of M samples $s[n]$, $s[n-1], \ldots, s[n-M+1]$ to predict, or estimate $s[n-M]$, with the prediction given by

$$\hat{s}[n-M] = -\sum_{i=1}^{M} g_i s[n-i+1], \tag{A.1}$$

where the g_i are referred to as the backward LPCs. The backward prediction error is equal to

$$\varepsilon[n] = s[n-M] - \hat{s}[n-M] = s[n-M] + \sum_{i=1}^{M} g_i s[n-i+1]. \tag{A.2}$$

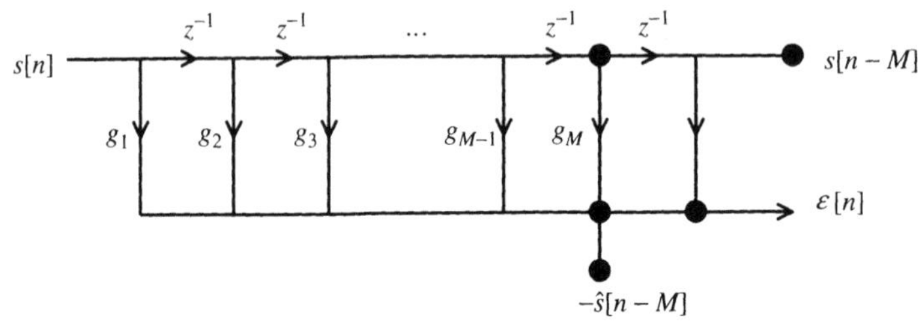

Figure A.1 The backward prediction-error filter.

Figure A.1 shows the signal flow graph implementation of (A.2) and is known as the backward prediction-error filter.

Backward Linear Prediction: Normal Equation

The backward LPCs are selected so that

$$J = E\{\varepsilon^2[n]\} = E\left\{\left(s[n-M] + \sum_{i=1}^{M} g_i s[n-i+1]\right)^2\right\} \tag{A.3}$$

is minimized. Differentiating J with respect to g_i and setting the result to zero leads to

$$\frac{\partial J}{\partial g_k} = 2E\left\{\left(s[n-M] + \sum_{i=1}^{M} g_i s[n-i+1]\right) s[n-k+1]\right\} = 0 \tag{A.4}$$

for $k = 1, 2, \ldots, M$. Rearranging the above equation gives

$$E\{s[n-M]s[n-k+1]\} + \sum_{i=1}^{M} g_i E\{s[n-i+1]s[n-k+1]\} = 0 \tag{A.5}$$

or

$$\sum_{i=1}^{M} g_i R_s[i-k] = -R_s[M-k+1], \tag{A.6}$$

which is the normal equation for backward linear prediction. In matrix form,

$$\mathbf{R}_s \mathbf{g} = -\mathbf{r}_s^B, \tag{A.7}$$

where $\mathbf{R}_s$ is the correlation matrix,

$$g = [g_1 \quad g_2 \quad \cdots \quad g_M]^T \tag{A.8}$$

and

$$\mathbf{r}_s^B = [R_s[M] \quad R_s[M-1] \quad \cdots \quad R_s[1]]^T \tag{A.9}$$

is the backward-arranged correlation vector. Equation (A.3) can be expanded to give

$$J = R_s[0] + 2\sum_{i=1}^{M} g_i R_s[M-i+1] + \sum_{i=1}^{M}\sum_{j=1}^{M} g_i g_j R_s[i-j]. \tag{A.10}$$

We are interested to find the minimum value of J, which is obtained when the LPCs satisfy (A.6) or (A.7). Substituting (A.6) in the third term on the right-hand sum of (A.10) leads to the following minimum mean-squared prediction error:

$$J_{\min} = R_s[0] + \sum_{i=1}^{M} g_i R_s[M-i+1]. \tag{A.11}$$

The above equation, combined with (A.6) and written in matrix form, gives

$$\begin{bmatrix} \mathbf{R}_s & \mathbf{r}_s^B \\ \mathbf{r}_s^{BT} & R_s[0] \end{bmatrix} \begin{bmatrix} g \\ 1 \end{bmatrix} = \begin{bmatrix} \mathbf{0} \\ J_{\min} \end{bmatrix} \tag{A.12}$$

and is known as the augmented normal equation, with $\mathbf{0}$ the $M \times 1$ zero vector.

Relations Between Backward and Forward Predictors

Arranging both sides of (A.7) in a backward manner, we come to the equivalent relation

$$\mathbf{R}_s \mathbf{g}^B = -\mathbf{r}_s. \tag{A.13}$$

Comparing with the normal equation for a forward predictor (Chapter 4), we conclude that

$$\mathbf{a} = \mathbf{g}^B; \tag{A.14}$$

that is, we can modify a backward predictor into a forward predictor by rearranging the sequence in a backward manner. Alternatively, we can write

$$a_i = g_{M+1-i} \tag{A.15}$$

or

$$g_i = a_{M+1-i} \tag{A.16}$$

for $i = 1, 2, \ldots, M$.

Reflection Coefficient and Mean-Squared Prediction Error

When applying the Levinson–Durbin algorithm, the minimum mean-squared prediction error is solved at each step of the recursion and is specified by

$$J_0 = R[0], J_l = J_{l-1}(1 - k_l^2); \quad l = 1, 2, \ldots, M, \tag{A.17}$$

with k_l being the RC. Starting with $l = 0$, and increasing the filter order by 1 at a time, we find that the repeated application of (A.17) yields

$$J_l = J_0 \prod_{l=1}^{M} (1 - k_l^2). \tag{A.18}$$

If the order of the prediction-error filter increases, the corresponding value of the mean-squared prediction error normally decreases or else remains the same. Hence, we must always have

$$0 \leq J_l \leq J_{l-1}, \quad l \geq 1. \tag{A.19}$$

The above condition is equivalent to

$$|k_l| \leq 1, \quad \text{for all } l. \tag{A.20}$$

Forward Prediction-Error Filter

For a linear predictor of order M, the prediction-error filter is described by

$$e[n] = \sum_{i=0}^{M} a_i^{(M)} s[n-i], \tag{A.21}$$

where $a_0^{(M)} = 1$. The system function is found as

$$H_f^{(M)}(z) = \sum_{i=0}^{M} a_i^{(M)} z^{-i}. \tag{A.22}$$

Backward Prediction-Error Filter

For a backward predictor of order M, the prediction-error filter is described by

$$\varepsilon[n] = \sum_{i=1}^{M+1} g_i^{(M)} s[n-i+1], \tag{A.23}$$

where $g_{M+1}^{(M)} = 1$. The system function is found to be

$$H_b^{(M)}(z) = \sum_{i=1}^{M+1} g_i^{(M)} z^{-i+1} = \sum_{i=0}^{M} g_{i+1}^{(M)} z^{-i}. \tag{A.24}$$

Using the equivalent relation between backward and forward LPCs, we can write

$$H_b^{(M)}(z) = \sum_{i=0}^{M} a_{M-i}^{(M)} z^{-i}, \tag{A.25}$$

where the system function of the backward prediction-error filter is expressed in terms of the forward LPCs.

Recursive Expression for the System Function of the Forward Prediction-Error Filter

It is desired to express the system function of the forward prediction-error filter in terms of the system functions for the corresponding filters one order below. From Chapter 4, the forward LPCs of order l satisfy

$$a_l^{(l)} = -k_l, \tag{A.26}$$

$$a_i^{(l)} = a_i^{(l-1)} - k_l a_{l-i}^{(l-1)}; \quad i = 1, 2, \ldots, l-1. \tag{A.27}$$

Equation (A.22) can first be expanded to yield

$$H_f^{(l)}(z) = 1 + \sum_{i=1}^{l-1} a_i^{(l)} z^{-i} + a_l^{(l)} z^{-l}. \tag{A.28}$$

Substituting (A.26) and (A.27) gives

$$\begin{aligned} H_f^{(l)}(z) &= 1 + \sum_{i=1}^{l-1} a_i^{(l-1)} z^{-i} - k_l \sum_{i=1}^{l-1} a_{l-i}^{(l-1)} z^{-i} - k_l z^{-l} \\ &= \sum_{i=0}^{l-1} a_i^{(l-1)} z^{-i} - k_l z^{-1} \sum_{i=0}^{l-1} a_{l-i-1}^{(l-1)} z^{-i}. \end{aligned} \tag{A.29}$$

Substituting (A.22) and (A.25) into (A.29) leads to

$$H_f^{(l)}(z) = H_f^{(l-1)}(z) - k_l z^{-1} H_b^{(l-1)}(z). \tag{A.30}$$

Thus, given the RCs k_l and the system functions of the forward and backward prediction-error filters of order $l-1$, the system function of the corresponding forward prediction-error filter of order l is uniquely determined.

Minimum-Phase Property of the Forward Prediction-Error Filter

On the unit circle in the z-plane, we find that

$$|H_f^{(l-1)}(z)| = |H_b^{(l-1)}(z)|, \quad |z| = 1, \tag{A.31}$$

which can easily be verified from the expressions of the system functions. Suppose that the RCs k_l satisfy the condition $|k_l| < 1$ for all l; then

$$|k_l z^{-1} H_b^{(l-1)}(z)| < |H_b^{(l-1)}(z)| = |H_f^{(l-1)}(z)|, \quad |z| = 1. \tag{A.32}$$

At this point, we cite Rouché's theorem [Churchill and Brown, 1990]. Suppose that two functions F and G are analytic inside and on a simple closed contour $\boldsymbol{C}$. If $|F(z)| > |G(z)|$ at each point on $\boldsymbol{C}$, then the functions $F(z)$ and $F(z) + G(z)$ have the same number of zeros, counting multiplicities, inside $\boldsymbol{C}$. Using Rouché's theorem with

$$F(z) = H_f^{(l-1)}(z) \tag{A.33}$$

and

$$G(z) = -k_l z^{-1} H_b^{(l-1)}(z) \tag{A.34}$$

and considering the unit circle $\boldsymbol{C}$ of the z-plane traversed in the clockwise direction (Figure A.2), we find that both $F(z)$ and $G(z)$ are analytic inside and on $\boldsymbol{C}$. That is, their derivatives are continuous throughout the region enclosed by $\boldsymbol{C}$, which is composed of the unit circle itself and the region outside it. Also note that

$$|G(z)| < |F(z)|, \quad |z| = 1. \tag{A.35}$$

Thus, the conditions specified by Rouché's theorem are satisfied. According to the theorem, $F(z)$ and $F(z) + G(z)$ have the same number of zeros outside the unit circle.

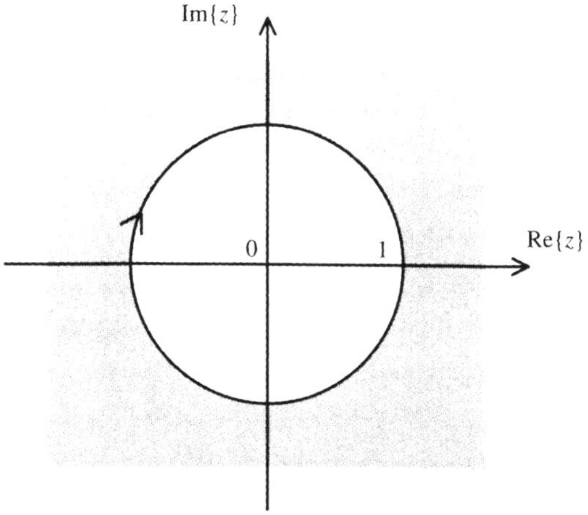

Figure A.2 The unit circle traversed in the clockwise direction.

We can examine the number of zeros outside the unit circle starting with the zero-order predictor. In this case,

$$H_f^{(0)}(z) = 1. \tag{A.36}$$

Therefore, it has no zeros at all. Then

$$H_f^{(1)}(z) = H_f^{(0)}(z) - k_1 z^{-1} H_b^{(0)}(z) = 1 - k_1 z^{-1} \tag{A.37}$$

will also have no zeros outside the unit circle. It is indeed the case from (A.37), where a zero is located at $z = k_1$, for $|k_1| < 1$, that the zero is situated inside the unit circle.

If $H_f^{(1)}(z)$ has no zeros on or outside the unit circle, then $H_f^{(2)}(z)$ will also have no zeros on or outside the unit circle provided that $|k_2| < 1$. The argument can be extended to any prediction order.

Thus, the system function $H_f^{(l)}(z)$ of a forward prediction-error filter of order l has no zeros on or outside the unit circle in the z-plane for all values of l, if and only if the reflection coefficients satisfy the condition $|k_l| < 1$ for all l. Such a system is known as minimum-phase.

APPENDIX B

SOME PROPERTIES OF LINE SPECTRAL FREQUENCY

Some properties of LSF are presented in this appendix. Consult Soong and Juang [1984] and Kim and Lee [1999] for further information.

Lemma B.1. Given the function $G(z)$ as defined in Chapter 8, if the prediction-error filter with system function $A(z)$ is minimum-phase, then

$$\begin{aligned} |G(z)| &> 1, \quad \text{if} \quad |z| < 1; \\ |G(z)| &< 1, \quad \text{if} \quad |z| > 1; \\ |G(z)| &= 1, \quad \text{if} \quad |z| = 1. \end{aligned}$$

Proof. $G(z)$ is defined by

$$G(z) = z^{-(M+1)} \frac{A(z^{-1})}{A(z)}. \tag{B.1}$$

Using the expression of $A(z)$,

$$G(z) = z^{-(M+1)} \frac{\prod_{i=1}^{M}(1 - z_i z)}{\prod_{i=1}^{M}(1 - z_i z^{-1})} = z^{-1} \prod_{i=1}^{M} \frac{1 - z_i z}{z - z_i}, \tag{B.2}$$

where the z_i are the zeros of $A(z)$. From the minimum-phase condition, we have $|z_i| < 1$. Magnitude of $G(z)$ is given by

$$|G(z)| = |z|^{-1} \prod_{i=1}^{M} \frac{|1 - z_i z|}{|z - z_i|}. \tag{B.3}$$

Each magnitude term is calculated with

$$|1 - zz_i| = \sqrt{(1 - z_i z)(1 - z_i^* z^*)} = \sqrt{1 + |zz_i|^2 - 2\,\mathrm{Re}(zz_i)}, \tag{B.4}$$

$$|z - z_i| = \sqrt{(z - z_i)(z^* - z_i^*)} = \sqrt{|z|^2 + |z_i|^2 - 2\mathrm{Re}(zz_i^*)}. \tag{B.5}$$

We need to evaluate the ratio between (B.4) and (B.5) to prove the theorem. Since the z_i form complex conjugate pairs, it is possible to find $z_i = z_j^*$ such that

$$2\,\mathrm{Re}(zz_i) = 2\,\mathrm{Re}(zz_j^*). \tag{B.6}$$

Hence, we only need to compare the magnitude of

$$\sqrt{1 + |zz_i|^2} \tag{B.7}$$

with

$$\sqrt{|z|^2 + |z_i|^2}. \tag{B.8}$$

Note that

$$(1 + |zz_i|^2) - (|z|^2 + |z_i|^2) = (1 - |z|^2)(1 - |z_i|^2). \tag{B.9}$$

Since $|z_i| < 1$ from the minimum-phase constraint on $A(z)$, it follows from (B.9) that

$$|z| > 1 \Rightarrow |z|^2 + |z_i|^2 > 1 + |zz_i|^2,$$
$$|z| < 1 \Rightarrow |z|^2 + |z_i|^2 < 1 + |zz_i|^2,$$
$$|z| = 1 \Rightarrow |z|^2 + |z_i|^2 = 1 + |zz_i|^2.$$

Substituting back in (B.3) proves the lemma.

Theorem B.1. For $A(z)$ representing a minimum-phase system, all zeros of $P(z)$ and $Q(z)$ are on the unit circle.

Proof. From Chapter 8,

$$P(z) = A(z)(1 + G(z)), \tag{B.10}$$
$$Q(z) = A(z)(1 - G(z)). \tag{B.11}$$

Thus, the polynomials can only be zero when $G(z) = \pm 1$. For $A(z)$ minimum phase, $|G(z)| = 1$ only when $|z| = 1$ (Lemma B.1). That is, the zeros of $P(z)$ and $Q(z)$ are on the unit circle.

Transfer Function, Argument, and Group Delay of *G(z)* for *A(z)* Minimum Phase

From Lemma B.1 it follows that $G(z)$ is the z-transform of an allpass filter, when $A(z)$ is minimum phase; that is,

$$|G(z)| = 1, \quad \text{if } z = e^{j\omega}.$$

Its transfer function can be written as

$$G(e^{j\omega}) = \exp(j \arg\{G(e^{j\omega})\}), \tag{B.12}$$

with $\arg\{\cdot\}$ denoting the argument or phase function. The argument function can be found by noting that

$$G(e^{j\omega}) = e^{-j\omega} \prod_{i=1}^{M} \frac{1 - r_i e^{j(\omega+\omega_i)}}{e^{j\omega} - r_i e^{j\omega_i}} = e^{-j(M+1)\omega} \prod_{i=1}^{M} \frac{1 - r_i e^{j(\omega+\omega_i)}}{1 - r_i e^{j(\omega+\omega_i)}}, \tag{B.13}$$

where

$$z_i = r_i e^{j\omega_i}; \quad i = 1, \ldots, M \tag{B.14}$$

are the zeros of $A(z)$. Therefore,

$$\begin{aligned} \arg\{G(e^{j\omega})\} = &-(M+1)\omega + \sum_{i=1}^{M} \tan^{-1}\left(\frac{-r_i \sin(\omega+\omega_i)}{1 - r_i \cos(\omega+\omega_i)}\right) \\ &- \sum_{i=1}^{M} \tan^{-1}\left(\frac{r_i \sin(\omega+\omega_i)}{1 - r_i \cos(\omega+\omega_i)}\right) \end{aligned}$$

or

$$\arg\{G(e^{j\omega})\} = -(M+1)\omega - 2\sum_{i=1}^{M} \tan^{-1}\left(\frac{r_i \sin(\omega+\omega_i)}{1 - r_i \cos(\omega+\omega_i)}\right). \tag{B.15}$$

The group delay of G is [Oppenheim and Schafer, 1989]:

$$\operatorname{grd}\{G(e^{j\omega})\} = -\frac{d}{d\omega} \arg\{G(e^{j\omega})\} = 1 + \sum_{i=1}^{M} \frac{1 - r_i^2}{1 + r_i^2 - 2r_i \cos(\omega+\omega_i)}. \tag{B.16}$$

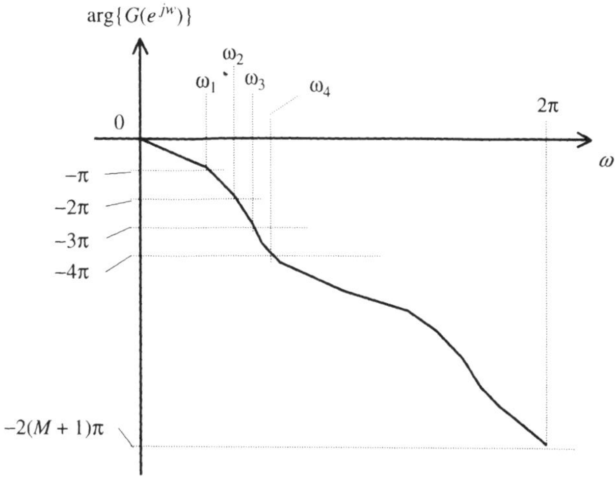

Figure B.1 A typical argument plot.

Theorem B.2. For $A(z)$ representing a minimum-phase system, the zeros of $P(z)$ and $Q(z)$ are interlaced with each other.

Proof. First, we have to show that the group delay of G is positive. Since $0 < r_i < 1$ and $|\cos(\omega + \omega_i)| \leq 1$, substitution of these extreme values in (B.16) shows that

$$1 \leq \mathrm{grd}(G(e^{j\omega})) \leq M + 1. \tag{B.17}$$

Hence, the argument function (B.15) must be a monotonically decreasing function. Furthermore, the argument values for $\omega = 0$ and $\omega = 2\pi$ are given by

$$\arg(G(e^{j\omega}))|_{\omega=0} = 0, \tag{B.18}$$

$$\arg(G(e^{j\omega}))|_{\omega=2\pi} = -2(M + 1)\pi, \tag{B.19}$$

which can be verified from the argument function directly. Therefore, the argument function crosses each $\arg(G) = n\pi$ line exactly once, resulting in $2(M + 1)$ crossing points for $0 \leq \omega < 2\pi$. These crossing points constitute the total $2(M + 1)$ zeros of $P(z)$ and $Q(z)$ alternately on the unit circle. The situation is depicted in Figure B.1. Starting from $\omega = 0$, the function crosses the $i\pi$ line at $\omega = \omega_i, i = 0, \ldots, 2(M + 1) - 1 = 2M + 1$, producing a total of $2(M + 1)$ zeros for $0 \leq \omega < 2\pi$. Since $G(\exp(j\omega_i)) = 1$ for i even and $G(\exp(j\omega_i)) = -1$ for i odd, $Q(\exp(j\omega_i)) = 0$ for i even and $P(\exp(j\omega_i)) = 0$ for i odd. The frequency values ω_i are the LSFs of $A(z)$. Note that the zeros of $P(z)$ and $Q(z)$ are interlaced with each other.

APPENDIX C

RESEARCH DIRECTIONS IN SPEECH CODING

The speech coding community has accomplished a great deal in the past thirty years. Looking ahead, there are still many challenges to be confronted. Here, we summarize some of the goals that must be conquered in order to advance the state-of-the-art in the speech coding arena.

Wide-Band Speech Coding, Handling Audio Signals

Most speech coders are targeted for telephone communications, with the bandwidth limited between 300 and 3400 Hz, requiring a sampling frequency of 8 kHz; this is referred to as *narrow-band* speech. By doubling the value of the sampling frequency to 16 kHz, it is possible to extend the bandwidth in the low- and high-frequency regions, for instance, 50 to 7 kHz. The increase in bandwidth enhances naturalness, presence, and comfort of the resultant digital speech signal, which is highly desirable for emerging applications such as teleconferencing and multimedia services. The expanded-bandwidth signal is known as *wide-band* speech.

Most standardized coders can be modified slightly to operate under a higher sampling frequency. The main problem is the increase in computational load. For instance, pitch search must be performed in a range that is doubled with respect to 8-kHz sampling. This could lead to an augmenting demand in computation by three to four times. Thus, many studies have focused on complexity reduction while dealing with wide-band speech coding. In Shoham [1993], an LD-CELP coder is developed that operates at 32 kbps; Laflamme et al. [1993] reported an ACELP coder at a bit-rate of 9.6 kbps. In Atal et al. [1993], more articles are available on wide-band speech coding. In McCree [2000], input speech is split into low-band and high-band, with the low-band signal encoded using the G.729 coder, while the high-band signal is encoded independently using a simple model, resulting in a 14-kbps coder. The work was later extended to a multimode coder [McCree et al., 2001]. A MELP coder operating around 8 kbps is covered in Lin et at. [2000].

Modern speech coding technology can encode virtually transparent wide-band speech with only a slight increase in bit-rate when compared to narrow-band speech. One of the fundamental limitations of hybrid speech coders, such as CELP, is that they do not behave well for audio signals in general, mainly due to the restriction imposed by the long-term predictor, or pitch synthesis filter. This latter technique allows great efficiency while encoding speech, but fails when facing the complex spectrum of typical audio signals, such as music.

For most wide-band applications such as teleconferencing, it is very likely that the coder has to handle occasional music segments besides speech. Thus, the challenge of low bit-rate wide-band coder design is to retain as much speech modeling as possible, while accommodating other types of signals such as music.

The ITU-T standardized a wide-band coder known as G.722 in 1986 [Maitre, 1988]. It is essentially a subband coder where the input signal is split into low- and high-frequency regions and separately encoded using ADPCM. It operates on a bit-rate of 64 kbps and does not introduce any perceptually significant distortion for speech. Due to its operational principles, its performance is rather signal independent, with good behavior under speech as well as audio in general. The G.722 standard is commonly used as a reference for comparison with other coders.

Many audio coders designed for high-fidelity quality reproduction are based on some sort of transform coding principles. The input signal is mapped to a frequency-related domain, which is then quantized according to the outcomes of a psychoacoustic model calculation, putting different weightings on various frequency regions depending on the perceptual relevance. This is the principle of the MPEG1 audio coder, with the highly popular Layer 3 scheme, or MP3 [ISO/IEC, 1993]. The problem with this approach is that it works well only at relatively high bit-rate; as it drops, performance decreases to a level that is not too bad for music, but definitely not good enough for speech communication since most parametric or hybrid speech coders work much better at similar bit-rate and below. See Noll [1993] and Painter and Spanias [2000] for a survey of major audio coding techniques.

Recently, researchers have been developing parametric modeling for audio representation with various levels of success. In a larger context, signal models offer a general framework for the solution to a wide range of conditions and provide efficient representation of general audio signals that can be explored for low bit-rate audio coding. See Levine [1998] and Verma [1999] for descriptions of the sine plus transient plus noise model; a brief tutorial on parametric audio coding appears in Purnhagen [1999]. Refinements to these models perhaps hold the key to low bit-rate audio plus speech coding.

Multispeaker

Most LP-based coders are designed to handle a single voice source; their efficiency comes from the simple model on which they rely. Since the model itself is adapted to one speaker, the performance of the coder becomes unacceptable when several speakers are involved, especially when they have comparable energy. The situation is not a problem for traditional telephone applications but becomes critical for teleconferencing, when several people might talk simultaneously. Roughly speaking, the problem can be seen as an audio coding issue since the signal spectrum when various speakers are involved mimics that of music. Thus, if the coder is capable of representing audio signals accurately, it should behave well under multispeaker conditions. Hence, the challenge is similar to that of audio coding.

Lower and Lower Bit-Rate

This is an obvious goal in speech coding and the research community has achieved a great deal of progress in this regard. Presently, it is technically feasible to produce high-quality speech at a bit-rate around 2 kbps and below. However, due to requirements related to other attributes such as delay and complexity, many standardized coders operate in the medium bit-rate range (5 to 15 kbps). Obviously, the challenge is always reducing the bit-rate with minimum deterioration of the good properties.

Many recent research activities have moved toward multimode coding, since this can generally outperform single-mode coders by delivering equivalent quality at lower rates.

A commonly asked question is in regard to the lowest bit-rate bound for coding of speech. Judging from the verbal information rate of speech, the number should lie somewhere around 100 bps. This bit-rate is likely to be achieved by a speech recognition mechanism [Rabiner and Juang, 1993] acting as the encoder, followed by a text-to-speech (TTS) system taking the role of the decoder. In Lee and Cox [2001], a very low bit-rate coder is reported that works by taking into account prosody information and speech synthesis techniques. At 800 bps, the authors claimed that the quality is equivalent to that of the MELP coder operating at 2.4 kbps. It is important to note, however, that this type of approach has inherently high coding delay that might hamper its practicality in two-way communication systems.

Objective Quality Measure

As indicated in Chapter 19, a performance comparison among various coders is frequently performed using subjective measurement techniques, which are cumbersome, expensive, and often nonrepeatable. It is clear that a reliable objective quality measurement technique that correlates well with subjective scores is necessary to improve the confidence and credibility of the final results. Several recently proposed methods have been quite successful in this aspect and are standardized by major bodies. Merging these techniques to speech encoding should elevate the quality of future generations of coders.

Scalability

The bit-stream of a scalable coder allows the decoder to reconstruct different versions of the synthetic speech at various levels of accuracy or quality. The bit-stream of a scalable coder is separated into independent units, with one unit responsible for the core, low-quality part of the synthetic signal, while additional units enhance and refine the overall accuracy. The core unit is indispensable for reconstruction, while the rest of the units are optional. Scalability has become an important issue for multimedia streaming over the internet. In that environment, transmission delay is generally unknown and varies depending on the traffic. Some advantages of scalable coders are:

- The decoder can start operating once the core unit is received. If the optional units arrive late or never reach their destination, they can be stored or discarded depending on system setting. This feature has significant impact on overall performance under variable-delay packetized networks.
- If the platform on which the decoder is deployed cannot afford the resources required to recover the high-quality signal (i.e., limitation in memory or computational capacity), it can settle with only the lower quality decoding operations.
- In case of a sudden drop in channel capacity (traffic increase) or an onset of resource demand in the platform (beginning of other operations under the same platform), the decoder can choose to switch to a lower-quality regeneration mode. Once the transients have settled and stabilized, the decoder can decide to go back to its original high-quality mode.

The point to note is that scalability permits the decoder to downgrade gracefully and predictably under unforeseen channel or system conditions. The feature enhances the performance for a wide range of real-world systems, such as packetized networks, the internet, and multitasking environments. The coders presented in this book are not scalable, that is, the decoder can operate only when all bits corresponding to the frame are recovered. In most cases, there is an all-or-nothing attitude. Many standardized coders have ways to operate under channel loss conditions, but the performance is generally hard to predict. Therefore, much more can be improved in this aspect.

APPENDIX D

LINEAR COMBINER FOR PATTERN CLASSIFICATION

This appendix discusses the application of a linear combiner for pattern classification. The patterns considered are vectors drawn from a given source, which are processed by the linear combiner so that they are classified into two classes. The mechanism can be used for voiced/unvoiced classification (LPC coder, Chapter 9), where the patterns to be classified are vectors containing parameters of the frame. Operation of the linear combiner is first explained followed by discussion of the pattern classification problem. The ultimate goal is to derive the procedure used for classifier design.

Operation of the Linear Combiner

Figure D.1 shows the structure of a linear combiner. The input vector **x** has $M + 1$ elements

$$\mathbf{x} = [x_0 \quad x_1 \quad \cdots \quad x_M]^T = [1 \quad x_1 \quad \cdots \quad x_M]^T, \tag{D.1}$$

where the first element $x_0 = 1$ is fixed and known as the bias input. The output of the linear combiner is the inner product of two vectors: input **x** and weight **w**. That is,

$$s = \mathbf{x}^T \mathbf{w} = \sum_{k=0}^{M} x_k w_k = w_0 + \sum_{k=1}^{M} x_k w_k \tag{D.2}$$

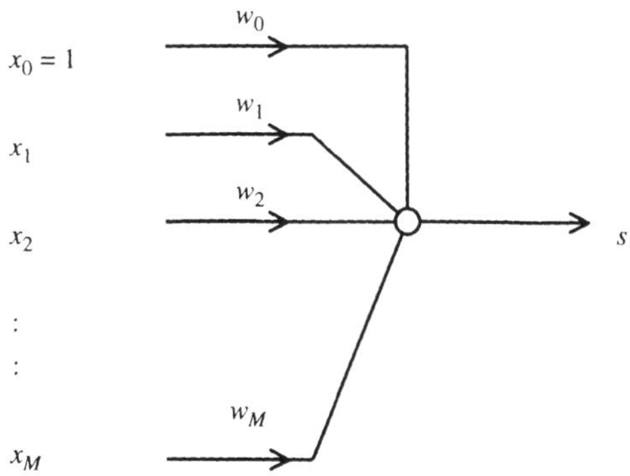

Figure D.1 Structure of a linear combiner.

with

$$\mathbf{w} = [w_0 \quad w_1 \quad \cdots \quad w_M]^T. \tag{D.3}$$

Pattern Classification

The linear combiner can be used as a pattern classifier whereby the vector space represented by the input vectors is partitioned into two subspaces, corresponding to the two classes $\boldsymbol{C}^+$ and $\boldsymbol{C}^-$. An input vector is classified as pertaining to one of two classes by

$$\mathbf{x} \in \begin{cases} \boldsymbol{C}^+, & \text{if } \mathbf{x}^T\mathbf{w} > 0, \\ \boldsymbol{C}^-, & \text{if } \mathbf{x}^T\mathbf{w} < 0. \end{cases} \tag{D.4}$$

A surface that separates the patterns into different classes is called a decision surface. In the present case, the decision surface is the hyperplane defined by

$$\mathbf{x}^T\mathbf{w} = w_0 + \sum_{k=1}^{M} x_k w_k = 0. \tag{D.5}$$

Figure D.2 illustrates an example decision surface when $M = 2$. Note how the hyperplane divides the whole space into two subspaces, corresponding to the two classes.

Linearly Separable and Nonseparable Patterns

The linear decision surface characteristic of a linear combiner limits its applicability to those cases where the patterns are naturally linearly separable. Figure D.3 shows a few cases where the linear combiner is or is not a good solution choice.

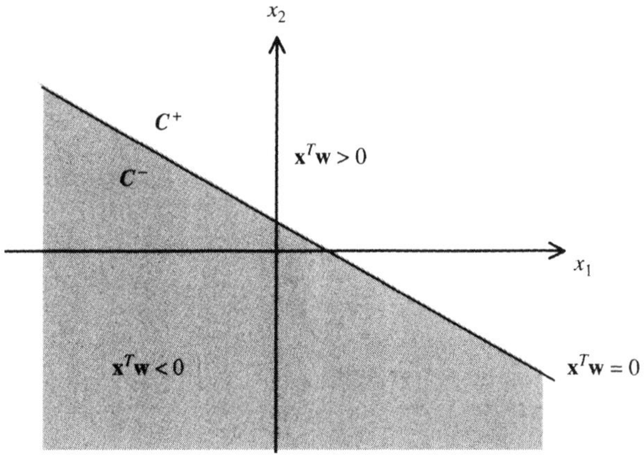

Figure D.2 Decision surface in a two-dimensional space.

For linearly separable patterns, a linear combiner provides a simple and optimal solution. Depending on the nature of the patterns, a linear combiner can be optimized to minimize the total classification error. However, there are cases where the use of a linear combiner is inadequate since excessive classification error is incurred due to boundaries of the pattern classes; differing strategies must be applied to these situations.

Classifier Design: Problem Statement

A pattern classifier* is obtained by transforming the outputs of the linear combiner (D.2) via the sign function, that is,

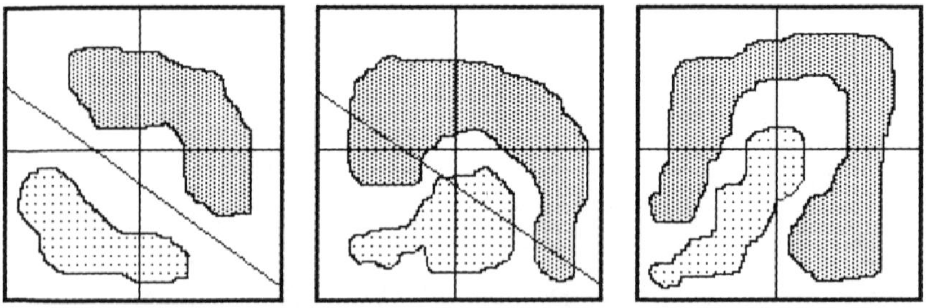

Figure D.3 Illustration of linearly separable and nonseparable patterns in a two-dimensional plane. *Left*: Linearly separable, use of linear combiner yields optimal results. *Middle*: Nonseparable but the linear combiner can be optimized to minimize classification error. *Right*: Nonseparable, use of linear combiner is inadequate due to excessive error.

*The linear combiner followed by the sign extractor is sometimes referred to as a perceptron, or single-layer perceptron, which is a simple form of neural network [Haykin, 1994].

$$y = \text{sgn}(s). \tag{D.6}$$

Then

$$\mathbf{x} \in \begin{cases} \boldsymbol{C}^+, & \text{if } y = 1, \\ \boldsymbol{C}^-, & \text{if } y = -1. \end{cases} \tag{D.7}$$

For the set of pattern vectors

$$\mathbf{x}_k; \quad k = 0, 1, \ldots, N_t - 1, \tag{D.8}$$

with N_t being the total number of vectors, each vector $\mathbf{x}_k$ has an associated desired response

$$d_k; \quad k = 0, 1, \ldots, N_t - 1. \tag{D.9}$$

The desired response d_k indicates the class to which $\mathbf{x}_k$ pertains; that is, $d_k = 1$ if $\mathbf{x}_k \in \boldsymbol{C}^+$ and $d_k = -1$ if $\mathbf{x}_k \in \boldsymbol{C}^-$. It is desired to find the classifier that can successfully classify the pattern vectors (D.8) into $\boldsymbol{C}^+$ and $\boldsymbol{C}^-$, that is,

$$y_k = d_k; \quad k = 0, 1, \ldots, N_t - 1, \tag{D.10}$$

where y_k is the classifier's output in response to the pattern vector $\mathbf{x}_k$.

Since the classifier is completely defined by the weight vector $\mathbf{w}$, solution of the classifier design problem consists of finding the weight vector $\mathbf{w}$ given the set of pattern vectors $\mathbf{x}_k$ and the associated desired response d_k. Classification error is given by

$$e_k = d_k - y_k. \tag{D.11}$$

If the set of pattern vectors is linearly separable, a weight vector can be found in such a way that $e_k = 0$ for all k. If the pattern vectors are not linearly separable, e_k will not be zero for some k. The design procedure should still be able to find a weight vector $\mathbf{w}$ such that $e_k = 0$ for the highest number of k.

To paraphrase, the problem of classifier design consists of finding the weight vector $\mathbf{w}$ so that the set of pattern vectors $\mathbf{x}_k$ is classified with minimum error. The pattern vectors associated with the desired responses are known as the training exemplars. The exemplars are used to train the classifier so as to produce the correct classification results.

Weight Vector and the Decision Surface

Consider an arbitrary weight vector $\mathbf{w}$. This vector is normal to the decision surface defined by (D.5), since by definition, if $\mathbf{x}$ and $\mathbf{w}$ are normal to each other, their inner product is

$$\mathbf{x}^T\mathbf{w} = |\mathbf{x}||\mathbf{w}|\cos(\pi/2) = 0, \tag{D.12}$$

which is the condition described by (D.5). Furthermore, if the vector $\mathbf{x}$ is positioned such that its angle θ with respect to $\mathbf{w}$ satisfies $\theta \in [-\pi/2, \pi/2]$, the inner product between $\mathbf{x}$ and $\mathbf{w}$ is greater than zero and $\mathbf{x}$ will be classified as C^+:

$$\mathbf{x}^T\mathbf{w} = |\mathbf{x}||\mathbf{w}|\cos\theta > 0 \quad \text{if} \quad \theta \in [-\pi/2, \pi/2]. \tag{D.13}$$

On the other hand, if $\theta \in [-\pi, -\pi/2]$ or $[\pi/2, \pi]$, the inner product between $\mathbf{x}$ and $\mathbf{w}$ is less than zero and $\mathbf{x}$ will be classified as C^-:

$$\mathbf{x}^T\mathbf{w} = |\mathbf{x}||\mathbf{w}|\cos\theta < 0 \quad \text{if} \quad \theta \in [-\pi, -\pi/2] \text{ or } [\pi/2, \pi]. \tag{D.14}$$

The following conclusions can be drawn:

1. The decision surface is normal to the weight vector $\mathbf{w}$.
2. $\mathbf{x} \in C^+$ if and only if the angle between $\mathbf{x}$ and $\mathbf{w}$ is such that $\theta \in [-\pi/2, \pi/2]$.
3. $\mathbf{x} \in C^-$ if and only if the angle between $\mathbf{x}$ and $\mathbf{w}$ is such that $\theta \in [-\pi, -\pi/2]$ or $[\pi/2, \pi]$.

Solution Space

For each training pattern, there is a corresponding decision surface separating the entire space into two half-spaces—positive and negative—according to whether the inner product between $\mathbf{x}$ and $\mathbf{w}$ is positive or negative. This surface is normal to the direction of the pattern vector. Any weight vector $\mathbf{w}$ lying in the positive half-space will generate correct output with zero error since $\mathbf{x}^T\mathbf{w} > 0$ if $\mathbf{x} \in C^+$

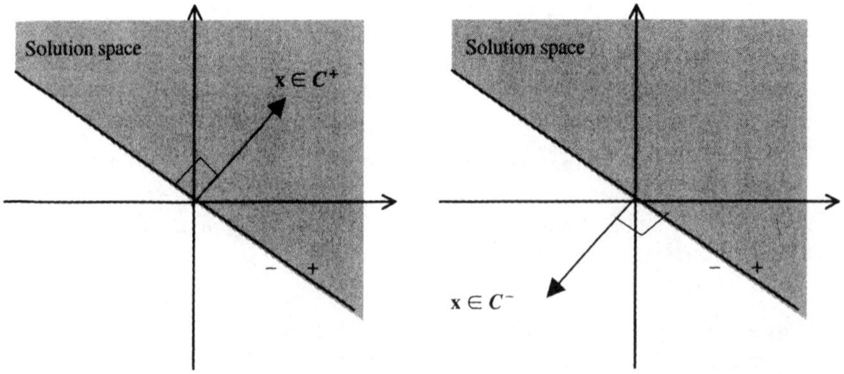

Figure D.4 *Left*: Solution space for $x \in C^+$. *Right*: Solution space for $x \in C^-$.

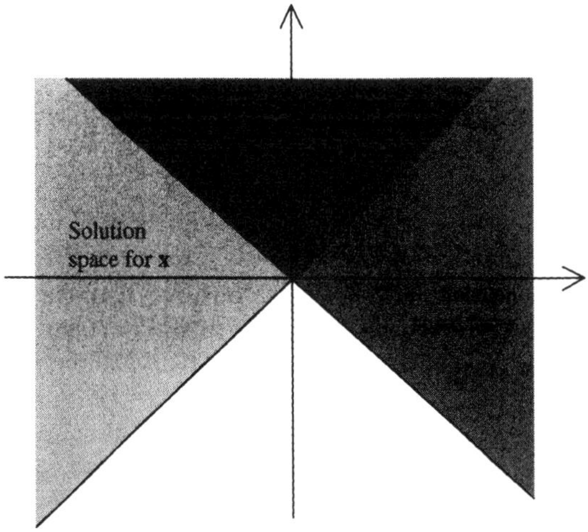

Figure D.5 Example of solution space for two patterns.

and $\mathbf{x}^T\mathbf{w} < 0$ if $\mathbf{x} \in C^-$ (Figure D.4). The positive half-space separated by the decision surface normal to the direction of $\mathbf{x}$ is called the solution space associated with $\mathbf{x}$.

For a group of patterns, there is a corresponding group of solution spaces. If the intersection of the solution spaces is nonempty, there is a solution to the classification problem. This space is called the solution space for the pattern's group, and any weight vector lying in this space will produce the desired output with zero error. An example involving two pattern vectors is shown in Figure D.5, where the intersection between the solution spaces for the individual pattern is the total solution.

Cases of Misclassification

If a pattern is misclassified, the weight vector is outside the solution space associated with the pattern; that is, it lies in the negative half-space. In order to find the correct weight vector, it is possible to move it along the direction normal to the desired decision surface toward the correct side. There are actually two different cases of misclassification considered separately as follows.

- $\mathbf{x} \in \boldsymbol{C}^+, \mathbf{x}^T\mathbf{w} < 0$

  To move $\mathbf{w}$ to the solution space or positive half-space separated by the desired decision surface, the following relation can be used to update the weight vector $\mathbf{w}$:

$$\mathbf{w}_{\text{new}} = \mathbf{w} + \mu\mathbf{x}, \tag{D.15}$$

  since $\mathbf{x}$ is normal to the desired decision surface and pointing to the positive half-space. A real positive constant μ is added to adjust the amount of weight change and is known as the step-size parameter. The classification error is

$$e = d - y = 1 - \text{sgn}(\mathbf{x}^T\mathbf{w}) = 2. \tag{D.16}$$

- $\mathbf{x} \in C^-, \mathbf{x}^T\mathbf{w} > 0$

To move $\mathbf{w}$ to the solution space or positive half-space separated by the desired decision surface, the following relation can be used to update the weight vector $\mathbf{w}$:

$$\mathbf{w}_{\text{new}} = \mathbf{w} - \mu\mathbf{x}, \tag{D.17}$$

since $\mathbf{x}$ is normal to the desired decision surface and pointing to the negative half-space. The classification error is

$$e = d - y = -1 - \text{sgn}(\mathbf{x}^T\mathbf{w}) = -2. \tag{D.18}$$

The Fixed Increment Rule

The weight update relations (D.15) and (D.17) together with the classification errors (D.16) and (D.18) can be combined together in a single equation as follows:

$$\mathbf{w}_{\text{new}} = \mathbf{w} + \tfrac{1}{2}\mu\, e\mathbf{x}, \tag{D.19}$$

where e is the classification error and μ the step-size parameter that adjusts the amount of weight change after each update. The factor of $\frac{1}{2}$ is included for convenience. Note that no change will be made to the weight vector when the error is zero. Equation (D.19) is known as the fixed increment rule for linear classifier design. It is intuitively reasonable that successive corrections to the weight vector in the direction of the solution space should eventually lead to the point where the output error is equal to zero for all the training patterns (assuming linear separability).

The algorithm for classifier design is summarized below:

Step 1. Initialization: Begin with an initial weight vector $\mathbf{w}_1$. Set $m = 1$. The initial vector can be randomly generated.

Step 2. Training: With $\mathbf{w}_{m,0} = \mathbf{w}_m$, for $k = 0$ to $N_t - 1$, perform the following:

$$y_k = \text{sgn}\left(\mathbf{x}_k^T\mathbf{w}_{m,k}\right), \tag{D.20}$$

$$e_k = d_k - y_k, \tag{D.21}$$

$$\mathbf{w}_{m,k+1} = \mathbf{w}_{m,k} + \mu e_k\mathbf{x}_k. \tag{D.22}$$

Step 3. Evaluation: For $k = 0$ to $N_t - 1$, e_k is recalculated using $\mathbf{w}_{m,Nt}$. The sum of absolute error is calculated with

$$SAE_m = \sum_k |e_k|. \tag{D.23}$$

If SAE_m is less than a certain positive threshold, stop. The final weight vector is given by $\mathbf{w}_m$. Otherwise, set $m \leftarrow m + 1$ and $\mathbf{w}_{m+1} = \mathbf{w}_m$; go to Step 2.

For the described algorithm, the weight vector is first randomly initialized. Then in Step 2, the training exemplars are processed one by one. If classification error is zero, the weight vector remains intact; otherwise, it is modified by an amount that depends on the product between the classification error and the input vector. In the evaluation step, performance of the weight vector is measured for all classification patterns. If the sum of absolute error is lower than a certain threshold, the algorithm stops. Otherwise, training resumes by executing Step 2.

So far we haven't discussed the choice of step-size parameter μ, which is a positive constant determining the speed of training. For high values of μ, the weight vector converges faster but with the possibility of oscillation or even instability. For low values of μ, convergence speed can be slow. The choice of μ can be found experimentally.

The stopping threshold is selected so that the desired performance is met upon termination of the training procedure. If the training exemplars are linearly separable, the threshold should be set to zero. Otherwise, some positive values can be used, but the algorithm might never be able to complete its mission. In this latter case an alternative stopping criterion can be used, such as counting the total number of iterations, and stop whenever it surpasses a certain value.

Many variations are obtainable from the basic scheme, including the following:

- Training time is largely dependent on the initial weight vector. A strategically well placed starting point can be estimated roughly from the training exemplars and used later in the algorithm to cut the total training time.
- A fixed step size is used throughout the algorithm. It might actually improve convergence speed and avoid unwanted oscillations by using a gradually decreasing step-size parameter.
- The sum of absolute error is used for evaluation. The sum of squared error is equally well suited for the present purpose.

Summary

Design procedures for a linear combiner in pattern classification are described. The method is relatively simple and is applicable only to those situations where the patterns are linearly separable. In general, a nonlinear system must be used to perform the classification task. Neural networks are nonlinear systems that can be applied with good classification results. In fact, a linear combiner followed by the sign extractor is a simple form of neural network known as a single-layer perceptron. Many topologies of neural networks are available for pattern classification. Most of these systems rely on the introduction of a nonlinear smooth transformation function into the network structure. The networks are then trained using a gradient-descent approach so as to minimize error. A comprehensive introduction to the vast field of neural networks can be found in Bose and Liang [1996] and Haykin [1994]; simulation experiments and practical applications are found in Freeman [1994].

The proposed design algorithm is based on a training process where the exemplars are presented consecutively with the weight adjusted after each step. It is possible to use a least-squares approach to solve for the weight vector. As we can see from the structure of the linear combiner, the weights can be jointly solved by minimizing the total sum of error squares. The procedure is similar to linear prediction analysis (Chapter 4). As a matter of fact, the linear combiner and linear predictor share the same general constraints.

APPENDIX E

CELP: OPTIMAL LONG-TERM PREDICTOR TO MINIMIZE THE WEIGHTED DIFFERENCE

This appendix contains the derivations of the relevant equations involved in the determination of the long-term predictor's parameters so as to minimize the perceptually weighted difference between the input speech and the synthetic speech. We rely on the notation from Chapters 11 and 12.

Problem Statement

Within the context for CELP (Chapter 11), it is possible to find the parameters of the long-term predictor so as to minimize the perceptually weighted error. The best way is to jointly optimize the long-term and short-term predictors resulting in the smallest error. The target parameters are excitation codevector, excitation gain, pitch period, and long-term gain. The proposition, however, is highly elaborate to implement in practice. One way of reducing the search complexity is by obtaining the long-term predictor's parameters (pitch period and gain) and the excitation codevector (including gain) in two steps. First, we assume zero excitation gain and calculate the long-term predictor's parameters such that the error is minimized. Next, the long-term predictor is held constant and the optimal excitation plus gain are searched.

The problem can be stated as follows. Find b and T so that the sum of squared error

$$J_r = \sum_{n=0}^{N-1}(u_r[n] - y2_r[n] - y3_r[n])^2 \tag{E.1}$$

is minimized. In the above equation, $u_r[n]$ (perceptually weighted speech) and $y3_r[n]$ (zero-input response of the modified formant synthesis filter) are known. Note the subscript r is added to indicate the subframe index.

Signal Relations

The signal $y2_r[n]$ is found with

$$d2_r[n] = d_{r-1}[n+N], \qquad -T \le n \le -1; \tag{E.2a}$$

$$d2_r[n] = -b.d2_r[n-T], \qquad 0 \le n \le N-1; \tag{E.2b}$$

$$y2_r[n] = 0, \qquad -M \le n \le -1; \tag{E.3a}$$

$$y2_r[n] = d2_r[n] - \sum_{i=1}^{M} a_i \gamma^i y2_r[n-i], \qquad 0 \le n \le N-1; \tag{E.3b}$$

where the short-term LPCs a_i are known. The zero-state response of the modified formant synthesis filter can also be calculated by knowing the impulse response of the filter—$h[n]$—which is an IIR causal system. Written as the convolution sum,

$$y2_r[n] = \sum_{k=0}^{n} h[k]d2_r[n-k], \quad 0 \le n \le N-1, \tag{E.4}$$

$h[n]$ can be found directly from the LPCs a_i of the current subframe.

Finding the Optimal *b* and *T*

Differentiating (E.1) with respect to b gives

$$\frac{\partial J_r}{\partial b} = (-2)\sum_{n=0}^{N-1}(u_r[n] - y2_r[n] - y3_r[n])\frac{\partial}{\partial b}y2_r[n]. \tag{E.5}$$

From (E.4),

$$\frac{\partial y2_r[n]}{\partial b} = \sum_{k=0}^{n} h[k]\frac{\partial}{\partial b}d2_r[n-k], \quad 0 \le n \le N-1. \tag{E.6}$$

Here is the tricky part. To find the derivative of J, we must have the derivative of $y2$, by which we need the derivative of $d2$. Since the signal $d2$ is unknown for the current subframe (the long-term parameters are yet to be determined), we must express $d2$ (as shown in (E.2b)) as a function of the $d2$ samples of the past (the $(r-1)$st subframe and further into the past). The computational procedure depends on the value of T.

Case 1: $T \geq N$

In this case, the values of $d2_r[n]$ for $0 \leq n \leq N-1$ depend entirely on the past, that is, for $n < 0$. Thus, from (E.2b), we have

$$\frac{\partial}{\partial b} d2_r[n] = -d2_r[n-T], \quad 0 \leq n \leq N-1. \tag{E.7}$$

Substituting in (E.6), we find

$$\frac{\partial}{\partial b} y2_r[n] = -\sum_{k=0}^{n} h[k] d2_r[n-k-T]. \tag{E.8}$$

Putting (E.4) and (E.8) into (E.5) and equating the result to zero gives

$$\sum_{n=0}^{N-1}\left(u_r[n] - y3_r[n] - \sum_{k=0}^{n} h[k] d2_r[n-k]\right)\left(\sum_{k=0}^{n} h[k] d2_r[n-k-T]\right) = 0.$$

Substituting (E.2b) into the above equation and using the definition

$$y4_r[n] = \sum_{k=0}^{n} h[k] d2_r[n-k-T] \tag{E.9}$$

yields

$$\sum_{n=0}^{N-1}(u_r[n] - y3_r[n]) y4_r[n] = -b \sum_{n=0}^{N-1}(y4_r[n])^2, \tag{E.10}$$

or

$$b = -\frac{\sum_{n=0}^{N-1}(u_r[n] - y3_r[n]) y4_r[n]}{\sum_{n=0}^{N-1}(y4_r[n])^2}, \tag{E.11}$$

which is the expression for the optimal long-term gain. Given b, we would like to find the expression for the sum of squared error as a function of T. Substituting (E.4) into (E.1), we find

$$J(T) = \sum_{n=0}^{N-1}\left(u_r[n] - y3_r[n] - \sum_{k=0}^{n} h[k] d2_r[n-k]\right)^2.$$

From (E.2b) and (E.9),

$$J(T) = \sum_{n=0}^{N-1}(u_r[n] - y3_r[n] + b y4_r[n])^2. \tag{E.12}$$

Expanding the above equation and substituting (E.11) for b gives

$$J(T) = \sum_{n=0}^{N-1} (u_r[n] - y3_r[n])^2 - \frac{\left(\sum_{n=0}^{N-1} (u_r[n] - y3_r[n]) y4_r[n]\right)^2}{\sum_{n=0}^{N-1} (y4_r[n])^2}. \tag{E.13}$$

Since $u_r[n]$ and $y3_r[n]$ are known, (E.13) is evaluated for all possible values of T. The particular value that minimizes (E.13) or maximizes

$$P(T) = \frac{\left(\sum_{n=0}^{N-1} (u_r[n] - y3_r[n]) y4_r[n]\right)^2}{\sum_{n=0}^{N-1} (y4_r[n])^2} \tag{E.14}$$

is the optimal pitch period. Note that for each T, there is a corresponding $y4_r[n]$ given by (E.9). Once T is found, b is calculated with (E.11) and all the necessary parameters are obtained.

Case 2: $N/2 \leq T < N$

From (E.2b), $d2[n]$ can be written in this case as

$$d2_r[n] = \begin{cases} -bd2_r[n-T], & 0 \leq n \leq T-1, \\ b^2 d2_r[n-2T], & T \leq n \leq N-1, \end{cases} \tag{E.15}$$

where the values for the present frame are written as a function of the last frame ($n < 0$) only. Then

$$\frac{\partial}{\partial b} d2_r[n] = \begin{cases} -d2_r[n-T_r], & 0 \leq n \leq T-1, \\ 2bd2_r[n-2T_r], & T \leq n \leq N-1. \end{cases} \tag{E.16}$$

From (E.6),

$$\frac{\partial}{\partial b} y2_r[n] = \begin{cases} -\sum_{k=0}^{n} h[k] d2_r[n-k-T], & 0 \leq n \leq T-1, \\ 2b \sum_{k=0}^{n} h[k] d2_r[n-k-2T], & T \leq n \leq N-1. \end{cases} \tag{E.17}$$

Substituting (E.17) and (E.4) into (E.5) and equating to zero, we find

$$\sum_{n=0}^{T-1} \left(u_r[n] - y3_r[n] - \sum_{k=0}^{n} h[k] d2_r[n-k] \right) \left(-\sum_{k=0}^{n} h[k] d2_r[n-k-T] \right)$$
$$+ \sum_{n=T}^{N-1} \left(u_r[n] - y3_r[n] - \sum_{k=0}^{n} h[k] d2_r[n-k] \right) \left(2b \sum_{k=0}^{n} h[k] d2_r[n-k-2T] \right) = 0.$$

Note that the sum is broken into two parts, corresponding to the two intervals of n described from (E.15) to (E.17). Substituting (E.15) into the above equations gives

$$\sum_{n=0}^{T-1}\left(u_r[n] - y3_r[n] + b\sum_{k=0}^{n} h[k]d2_r[n-k-T]\right)\left(-\sum_{k=0}^{n} h[k]d2_r[n-k-T]\right)$$
$$+\sum_{n=T}^{N-1}\left(u_r[n] - y3_r[n] - b^2\sum_{k=0}^{n} h[k]d2_r[n-k-2T]\right)\left(2b\sum_{k=0}^{n} h[k]d2_r[n-k-2T]\right)$$
$$= 0. \tag{E.18}$$

Let's define

$$y5_r[n] = \sum_{k=0}^{n} h[k]d2_r[n-k-2T]. \tag{E.19}$$

Using definitions (E.9) and (E.19) in (E.18) leads to

$$-\sum_{n=0}^{T-1}(u_r[n] - y3_r[n])y4_r[n] - b\sum_{n=0}^{T-1}(y4_r[n])^2$$
$$+ 2b\sum_{n=T}^{N-1}(u_r[n] - y3_r[n])y5_r[n] - 2b^3\sum_{n=T}^{N-1}(y5_r[n])^2 = 0 \tag{E.20}$$

Rearranging terms, we find

$$2b^3\sum_{n=T}^{N-1}(y5_r[n])^2 + b\left(\sum_{n=0}^{T-1}(y4_r[n])^2 - 2\sum_{n=T}^{N-1}(u_r[n] - y3_r[n])y5_r[n]\right)$$
$$+\sum_{n=0}^{T-1}(u_r[n] - y3_r[n])y4_r[n] = 0. \tag{E.21}$$

As we can see, for the case of $T \geq N$, b can be written in closed form; when T is less than N, however, the solution to b requires solving a cubic expression. This is obviously very costly. One solution is to adopt a trial-and-error method based on quantized values of b. In this method the sum terms are precomputed, and then each of the possible quantized values of b is substituted into the equation. The value of b that gives the smallest squared error is the desired value.

For $T < N/2$, more complicated expressions result for the solution of b. For $N/3 \leq T < N/2$, for instance, $d2[n]$ can be written as

$$d2_r[n] = \begin{cases} -bd2_r[n-T], & 0 \leq n \leq T, \\ b^2 d2_r[n-2T], & T \leq n \leq 2T-1, \\ -b^3 d2_r[n-3T], & 2T \leq n \leq N-1. \end{cases} \tag{E.22}$$

This obviously results in an even more complex expression for the solution of b and hence too complex for practical purposes.

APPENDIX F

REVIEW OF LINEAR ALGEBRA: ORTHOGONALITY, BASIS, LINEAR INDEPENDENCE, AND THE GRAM–SCHMIDT ALGORITHM

Fundamental concepts of linear algebra are reviewed here, which form the background material for the study of Chapter 13, the VSELP coder. For simplicity, many mathematical formalities are dropped. Readers pursuing a more rigorous framework are invited to consult Strang [1988], an introductory textbook; or Lancaster and Tismenetsky [1985], a more advanced reference. In Golub and Van Loan [1996], many algorithms dealing with a large array of matrix computation problems are given.

For the purpose of this appendix, the N-dimensional vector

$$\mathbf{x}[x_1 \quad x_2 \quad \cdots \quad x_N]^T$$

has real elements $x_i, i = 1$ to N.

Definition F.1: Inner Product of Two Vectors. Given the vectors $\mathbf{x}$ and $\mathbf{y}$, their inner product, denoted by ($\mathbf{x}$, $\mathbf{y}$) is defined by

$$(\mathbf{x}, \mathbf{y}) = \mathbf{y}^T\mathbf{x} = \sum_{i=1}^{N} x_i y_i. \tag{F.1}$$

Definition F.2: Orthogonal Vectors. Two vectors are said to be orthogonal if their inner product is equal to zero.

Definition F.3: Linear Independence. The set of M vectors $\mathbf{x}_1, \ldots, \mathbf{x}_M$ are said to be linearly independent if the condition

$$\sum_{i=1}^{M} \alpha_i \mathbf{x}_i = \mathbf{0} \tag{F.2}$$

implies that

$$\alpha_1 = \alpha_2 = \cdots = \alpha_M = 0,$$

where the α_i are scalars.

Definition F.4: Norm of a Vector. Given the vector $\mathbf{x}$, its norm is defined by

$$\| \mathbf{x} \| = \sqrt{(\mathbf{x}, \mathbf{x})} = \sqrt{\mathbf{x}^T \mathbf{x}}. \tag{F.3}$$

Theorem F.1: Linear Independence and Orthogonality. Given the vectors $\mathbf{x}_1, \ldots, \mathbf{x}_M$ with nonzero norm, if these vectors are mutually orthogonal, then they are linearly independent.

Proof. Suppose $\alpha_1 \mathbf{x}_1 + \cdots + \alpha_M \mathbf{x}_M = 0$. To show that α_1 must be zero, take the inner product of both sides with $\mathbf{x}_1$:

$$\mathbf{x}_1^T (\alpha_1 \mathbf{x}_1 + \cdots + \alpha_M \mathbf{x}_M) = \alpha_1 \mathbf{x}_1^T \mathbf{x}_1 = 0,$$

which is due to the orthogonality constraint of the $\mathbf{x}_i$. Because the vectors were assumed nonzero, $\mathbf{x}_1^T \mathbf{x}_1 \neq 0$ and therefore $\alpha_1 = 0$. The same is true for every α_i. Thus, the only combination of the $\mathbf{x}_i$ producing zero is the trivial one with all $\alpha_i = 0$, and the vectors are independent.

Definition F.5: Linear Space. A linear space or vector space is a set of vectors. Within these spaces, two operations are possible: we can add any two vectors, and we can multiply vectors by scalars. (See Lancaster and Tismenetsky [1985] for additional details.)

Definition F.6: Basis. A finite set of vectors $\mathbf{x}_1, \ldots, \mathbf{x}_M$ is said to be a basis of the linear space $\boldsymbol{S}$ if they are linearly independent and every element $\mathbf{x} \in \boldsymbol{S}$ is a linear combination of the basis vectors. That is,

$$\mathbf{x} = \sum_{i=1}^{M} \alpha_i \mathbf{x}_i, \tag{F.4}$$

where the α_i are scalars. We say that the basis vectors span the linear space $\boldsymbol{S}$.

Definition F.7: Orthonormal Vectors. The vectors $\mathbf{q}_1, \ldots, \mathbf{q}_M$ are orthonormal if

$$\mathbf{q}_i^T \mathbf{q}_j = \begin{cases} 0, & i \neq j, \\ 1, & i = j; \end{cases} \tag{F.5}$$

that is, they are mutually orthogonal with unit norm.

Projection of a Vector to a Line: The Projection Matrix

Given two vectors $\mathbf{a}$ and $\mathbf{b}$, where $\mathbf{a}$ indicates the direction of a straight line and $\mathbf{b}$ represents a point in space, we want to find the point $\mathbf{p}$ along the line in the direction of the vector $\mathbf{a}$ in such a way that the distance between $\mathbf{b}$ and $\mathbf{p}$ is minimum. This is known as the projection problem and the geometry is shown in Figure F.1 for an example of a 3-D space. To find $\mathbf{p}$, we use the fact that $\mathbf{p}$ must be some multiple $\mathbf{p} = \alpha\mathbf{a}$ of the given vector $\mathbf{a}$, and the problem is to compute the coefficient α. All that we need for this computation is the geometrical fact that the line from $\mathbf{b}$ to the closest point $\mathbf{p} = \alpha\mathbf{a}$ is orthogonal (perpendicular) to the vector $\mathbf{a}$:

$$\mathbf{a}^T(\mathbf{b} - \alpha\mathbf{a}) = 0.$$

Thus,

$$\alpha = \frac{\mathbf{a}^T\mathbf{b}}{\mathbf{a}^T\mathbf{a}}. \tag{F.6}$$

Therefore, the projection of $\mathbf{b}$ onto the line whose direction is given by $\mathbf{a}$ is

$$\mathbf{p} = \alpha\mathbf{a} = \frac{\mathbf{a}^T\mathbf{b}}{\mathbf{a}^T\mathbf{a}}\mathbf{a} = \frac{\mathbf{a}\mathbf{a}^T}{\mathbf{a}^T\mathbf{a}}\mathbf{b} = \mathbf{P} \cdot \mathbf{b}. \tag{F.7}$$

$\mathbf{P}$ is an $N \times N$ matrix and is the matrix that multiplies $\mathbf{b}$ to produce $\mathbf{p}$, known as the projection matrix.

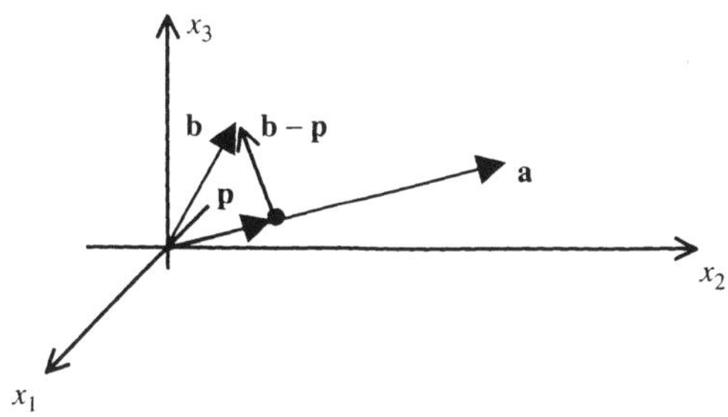

Figure F.1 A one-dimensional projection in three-dimensional space.

The Gram–Schmidt Orthogonalization Algorithm

Given a set of linearly independent vectors,

$$\mathbf{a}_1, \mathbf{a}_2, \ldots, \mathbf{a}_M,$$

it is required to find the corresponding set of orthogonal vectors,

$$\mathbf{q}_1, \mathbf{q}_2, \ldots, \mathbf{q}_M,$$

so that $\mathbf{q}_1$ is in the direction of $\mathbf{a}_1$.

The problem is solved by Gram and Schmidt and proceeds as follows. Start with $\mathbf{q}_1$; since it goes in the same direction as $\mathbf{a}_1$, we have

$$\mathbf{q}_1 = \mathbf{a}_1. \tag{F.8}$$

For $\mathbf{q}_2$, the requirement is that it must be orthogonal to $\mathbf{q}_1$. We proceed by subtracting off the component of $\mathbf{a}_2$ in the direction of $\mathbf{q}_1$:

$$\mathbf{q}_2 = \mathbf{a}_2 - \frac{\mathbf{q}_1^T \mathbf{a}_2}{\mathbf{q}_1^T \mathbf{q}_1} \mathbf{q}_1, \tag{F.9}$$

Since $(\mathbf{q}_1^T \mathbf{a}_2)\mathbf{q}_1 / (\mathbf{q}_1^T \mathbf{q}_1)$ is the projection of $\mathbf{a}_2$ in the direction of $\mathbf{q}_1$.

For $\mathbf{q}_3$, we eliminate the components of $\mathbf{a}_3$ in the direction of $\mathbf{q}_1$ and $\mathbf{q}_2$. Hence,

$$\mathbf{q}_3 = \mathbf{a}_3 - \frac{\mathbf{q}_1^T \mathbf{a}_3}{\mathbf{q}_1^T \mathbf{q}_1} \mathbf{q}_1 - \frac{\mathbf{q}_2^T \mathbf{a}_3}{\mathbf{q}_2^T \mathbf{q}_2} \mathbf{q}_2, \tag{F.10}$$

where the first and second negative term on the right-hand side are the components of $\mathbf{a}_3$ in the directions of $\mathbf{q}_1$ and $\mathbf{q}_2$, respectively. Therefore, the basic idea is to subtract from every new vector $\mathbf{a}$ its components in the directions that are already settled; and the principle is used over and over again.

To summarize, the algorithm can be written as

For $i = 1$:

$$\mathbf{q}_1 = \mathbf{a}_1. \tag{F.11}$$

For $i = 2, \ldots, M$:

$$\mathbf{q}_i = \mathbf{a}_i - \sum_{j=1}^{i-1} \frac{\mathbf{q}_j^T \mathbf{a}_i}{\mathbf{q}_j^T \mathbf{q}_j} \mathbf{q}_j. \tag{F.12}$$

In practice, it is desirable to have unit norm for the final vectors. The following algorithm includes results in a set of orthonormal vectors at the end.

```
for i ← 1 to M
        q_i ← a_i
        for j ← 1 to i - 1
                q_i ← q_i - (q_j^T a_i) q_j
        norm_i ← (q_i^T q_i)^(1/2)
        q_i ← q_i /norm_i
```

The Modified Gram-Schmidt Algorithm

The original formulation of the Gram–Schmidt algorithm has poor numerical properties in the sense that a loss of orthogonality among the output vectors is often observed. A rearrangement of the calculation, known as the modified Gram–Schmidt algorithm, yields a much sounder procedure with improved accuracy. This is specified as follows:

```
for i ← 1 to N
        norm_i ← (a_i^T a_i)^(1/2)
        q_i ← a_i /norm_i
        for j ← i + 1 to N
                a_j ← a_j - (q_i^T a_j) q_i
```

BIBLIOGRAPHY

Adoul, J-P. and C. Lamblin (1987). "A Comparison of Some Algebraic Structures for CELP Coding of Speech," *IEEE ICASSP*, pp. 1953–1956.

Adoul, J-P. and R. Lefebvre (1995). "Wideband Speech Coding," *Speech Coding and Synthesis*, W. B. Kleijn and K. K. Paliwal, eds., pp. 289–310, Elsevier Science, The Netherlands.

Adoul, J-P., P. Mabilleau, M. Delprat, and S. Morissette (1987). "Fast CELP Coding Based on Algebraic Codes," *IEEE ICASSP*, pp. 1957–1960.

Ahmed M. E. and M. I. Al-Suwaiyel (1993). "Fast Methods for Code Search in CELP," *IEEE Transactions on Speech and Audio Processing*, Vol.1, No.3, pp. 315–325, July.

Antoniou, A. (1993). *Digital Filters: Analysis, Design, and Applications*, McGraw-Hill, New York.

Atal, B. S. and J. R. Remde (1982). "A New Method of LPC Excitation for Producing Natural-Sounding Speech at Low Bit Rates," *IEEE ICASSP*, pp. 614–617.

Atal, B. S., R. V. Cox, and P. Kroon (1989). "Spectral Quantization and Interpolation for CELP Coders," *IEEE ICASSP*, pp. 69–72.

Atal, B. S., V. Cuperman, and A. Gersho, eds. (1991). *Advances in Speech Coding*, Kluwer Academic Publishers, Norwell, MA.

Atal, B. S., V. Cuperman, and A. Gersho, eds. (1993). *Speech and Audio Coding for Wireless and Network Applications*, Kluwer Academic Publishers, Norwell, MA.

Banks, J. and J. S. Carson II (1984). *Discrete-Event System Simulation*, Prentice-Hall, Englewood Cliffs, NJ.

Barnwell, T. (1981). "Recursive Windowing for Generating Autocorrelation Coefficients for LPC Analysis," *IEEE Transactions on Acoustics, Speech, and Signal Processing*, Vol. ASSP-29, No. 5, pp. 1062–1066.

Barr, M. (1999). *Programming Embedded Systems in C and C++*, O'Reilly, Sebastopol, CA.

Bose, N. K. (1993). *Digital Filters Theory and Applications*, Krieger Publishing Co., Melbourne, FL.

Bose, N. K. and P. Liang (1996). *Neural Networks Fundamentals with Graphs, Algorithms, and Applications*, McGraw-Hill, New York.

Burrus, C. S. and T. W. Parks (1985). *DFT/FFT and Convolution Algorithms*, John Wiley & Sons, Hoboken, NJ.

Buzo, A., A. H. Gray, R. M. Gray, and J. D. Markel (1980). "Speech Coding Based Upon Vector Quantization," *IEEE Transactions on Acoustics, Speech, and Signal Processing*, Vol. ASSP-28, No. 5, pp. 562–574, October.

Campbell, J. P. and T. E. Tremain (1986). "Voiced/Unvoiced Classification of Speech with Applications to the U.S. Government LPC-10E Algorithm," *IEEE ICASSP*, pp. 9.11.1–9.11.4.

Campbell, J. P., T. E. Tremain, and V. C. Welch (1991). "The DOD 4.8 KBPS Standard (Proposed Federal Standard 1016)," *Advances in Speech Coding*, B. S. Atal, V. Cuperman, and A. Gersho, eds., pp. 121–133, Kluwer Academic Publishers, Norwell, MA.

Chan, W. Y., S. Gupta, and A. Gersho (1992). "Enhanced Multistage Vector Quantization by Joint Codebook Design," *IEEE Transactions on Communications*, Vol. 40, No. 11, pp. 1693–1697, November.

Chang P-C. and R. M. Gray (1986). "Gradient Algorithms for Designing Predictive Vector Quantizers," *IEEE Transactions on Acoustics, Speech, and Signal Processing*, Vol. ASSP-34, No. 4, pp. 679–690, August.

Chen, J. H. (1990). "High-Quality 16 kb/s Speech Coding with a One-Way Delay Less Than 2 ms," *IEEE ICASSP*, pp. 453–456.

Chen, J. H. (1991). "A Robust Low-Delay CELP Speech Coder at 16 kb/s," *Advances in Speech Coding*, B. S. Atal, V. Cuperman, and A. Gersho, eds., pp. 25–35, Kluwer Academic Publishers, Norwell, MA.

Chen, J. H. (1995). "Low-Delay Coding of Speech," *Speech Coding and Synthesis*, W. B. Kleijn and K. K. Paliwal, eds., pp. 209–256, Elsevier Science, The Netherlands.

Chen, J. H., R. V. Cox, Y. C. Lin, N. Jayant, and M. J. Melchner (1992). "A Low-Delay CELP Coder for the CCITT 16 kb/s Speech Coding Standard," *IEEE Journal on Selected Areas in Communications*, Vol. 10, No. 5, pp. 830–849.

Chen, J. H. and A. Gersho (1987). "Real-Time Vector APC Speech Coding at 4800 bps with Adaptive Postfiltering," *IEEE ICASSP*, pp. 2185–2188.

Chen, J. H. and A. Gersho (1995). "Adaptive Postfiltering for Quality Enhancement of Coded Speech," *IEEE Transactions on Audio Processing*, Vol. 3, No. 1, pp. 59–70, January.

Chen, J. H., Y. C. Lin, and R. V. Cox (1991). "A Fixed-Point 16 kB/s LD-CELP Algorithm," *IEEE ICASSP*, pp. 21–24.

Chen, J. H., M. J. Melchner, R. V. Cox, and D. O. Bowker (1990). "Real-Time Implementation and Performance of a 16 kB/s Low-Delay CELP Speech Coder," *IEEE ICASSP*, pp. 181–184.

Chen, J. H. and M. S. Rauchwerk (1993). "8 kb/s Low-Delay CELP Coding of Speech," *Speech and Audio Coding for Wireless and Network Applications*, B. S. Atal, V. Cuperman, and A. Gersho, eds., pp. 25–31, Kluwer Academic Publishers, Norwell, MA.

Chitrapu, P. (1998). "Modern Speech Coding Techniques and Standards," *Multimedia Systems Design*, pp. 22–35, February.

Churchill, R. V. and J. W. Brown (1990). *Complex Variables and Applications*, McGraw-Hill, New York.

Cormen, T. H., C. E. Leiserson, and R. L. Rivest (1990). *Introduction to Algorithms*, McGraw-Hill, New York.

Cox, R. V. (1995). "Speech Coding Standards," *Speech Coding and Synthesis*, W. B. Kleijn and K. K. Paliwal, eds., pp. 49–78, Elsevier Science, The Netherlands.

Cox, R. V. (1997). "Three New Speech Coders from the ITU Cover a Range of Applications," *IEEE Communications Magazine*, pp. 40–47, September.

Das A., E. Paksoy, and A. Gersho (1995). "Multimode and Variable-Rate Coding of Speech," *Speech Coding and Synthesis*, W. B. Kleijn and K. K. Paliwal, eds., pp. 257–288, Elsevier Science, The Netherlands.

Davidson G. and A. Gersho (1986). "Complexity Reduction Methods for Vector Excitation Coding," *IEEE ICASSP*, pp. 3055–3058.

DeFatta, D. J., J. G. Lucas, and W. S. Hodgkiss (1988). *Digital Signal Processing: A System Design Approach*, John Wiley & Sons, Hoboken, NJ.

Deller, J. R., J. G. Proakis, and J. H. L. Hansen (1993). *Discrete-Time Processing of Speech Signals*, Macmillan, New York.

DeMartino, E. (1993). "Speech Quality Evaluation of the European, North-American, and Japanese Speech Coding Standards for Digital Cellular Systems," *Speech and Audio Coding for Wireless and Network Applications*, B. S. Atal, V. Cuperman, and A. Gersho, eds., pp. 55–58, Kluwer Academic Publishers, Norwell, MA.

Denisowski, P. (2001). "How Does it Sound?" *IEEE Spectrum*, pp. 60–64, February.

Dimolitsas, S. (1993). "Subjective Assessment Methods for the Measurement of Digital Speech Coder Quality," *Speech and Audio Coding for Wireless and Network Applications*, B. S. Atal, V. Cuperman, and A. Gersho, eds., pp. 43–54, Kluwer Academic Publishers, Norwell, MA.

Du, J., G. Warner, E. Vallow, and T. Hollenbach, (2000) "Using DSP16000 for GSM EFR Speech Coding—High-Performance DSPs," *IEEE Signal Processing Magazine*, pp. 16–26, March.

Dubnowski, J. J., R. W. Schafer, and L. R. Rabiner (1976). "Real-Time Digital Hardware Pitch Detector," *IEEE Transactions on Acoustics, Speech, and Signal Processing*, Vol. ASSP-24, No. 1, pp. 2–8, February.

Eckel B. (2000). *Thinking in C++*, 2nd edition, Prentice-Hall, Englewood Cliffs, NJ.

Eriksson, T., J. Linden, and J. Skoglund (1999). "Interframe LSF Quantization for Noisy Channels," *IEEE Transactions on Speech and Audio Processing*, Vol. 7, No. 5, pp. 495–509, September.

Erzin, E. and A. E. Cetin (1993). "Interframe Differential Vector Coding of Line Spectrum Frequencies," *IEEE ICASSP*, pp. II-25–II-28.

ETSI (1992a). *Recommendation GSM 6.10 Full-Rate Speech Transcoding*.

ETSI (1992b). *Recommendation GSM 6.01 European Digital Cellular Telecommunication System (Phase 1); Speech Processing Functions: General Description*.

ETSI (1992c). *Recommendation GSM 6.31 Discontinuous Transmission (DTX) for Full-Rate Speech Traffic Channels*.

ETSI (1992d). *Recommendation GSM 6.11 Substitution and Muting of Lost Frames for Full-Rate Speech Traffic Channels*.

ETSI (1992e). *Recommendation GSM 6.32 Voice Activity Detection.*

ETSI (1992f). *Recommendation GSM 6.12 Comfort Noise Aspects for Full-Rate Speech Traffic Channels.*

ETSI (1999). *Universal Mobile Telecommunications System (UMTS); Mandatory Speech Codec Speech Processing Functions AMR Speech Codec; Transcoding Fucntions*, 3G TS 26.090 Version 3.1.0, Release 1999.

Eyre, J. (2001). "The Digital Signal Processor Derby," *IEEE Spectrum*, pp. 62–68, June.

Eyre, J. and J. Bier (2000). "The Evolution of DSP Processors—From Early Architectures to the Latest Developments," *IEEE Signal Processing Magazine*, pp. 43–51, March.

Florencio, D. (1993). "Investigating the Use of Asymmetric Windows in CELP Vocoders," *IEEE ICASSP*, pp. II-427–II-430.

Freeman, J. A. (1994). *Simulating Neural Networks with Mathematica*, Addison-Wesley Publishing Co., Reading, MA.

Gardner, W. R. and B. D. Rao (1995a). "Theoretical Analysis of the High-Rate Vector Quantization of LPC Parameters," *IEEE Transactions on Speech and Audio Processing*, Vol. 3, No. 5, pp. 367–381, September.

Gardner, W. R. and B. D. Rao (1995b). "Optimal Distortion Measures for the High Rate Vector Quantization of LPC Parameters," *IEEE ICASSP*, pp. 752–755.

Gardner, W., P. Jacobs, and C. Lee (1993). "QCELP: A Variable Rate Speech Coder for CDMA Digital Cellular," *Speech and Audio Coding for Wireless and Network Applications*, B. S. Atal, V. Cuperman, and A. Gersho, eds., pp. 85–92, Kluwer Academic Publishers, Norwell, MA.

Gersho, A. and R. M. Gray (1995). *Vector Quantization and Signal Compression*, 4th printing, Kluwer Academic Publishers, Norwell, MA.

Gersho, A. and E. Paksoy (1993). "Variable Rate Speech Coding for Cellular Networks," *Speech and Audio Coding for Wireless and Network Applications*, B. S. Atal, V. Cuperman, and A. Gersho, eds., pp. 77–84, Kluwer Academic Publishers, Norwell, MA.

Gerson, I. A. and M. A. Jasiuk (1991). "Vector Sum Excited Linear Prediction (VSELP)," *Advances in Speech Coding*, B. S. Atal, V. Cuperman, and A. Gersho, eds., pp. 69–79, Kluwer Academic Publishers, Norwell, MA.

Goldberg, R. and L. Riek (2000). *A Practical Handbook of Speech Coders*, CRC Press, Boca Raton, FL.

Golub, G. H. and C. F. Van Loan (1996). *Matrix Computation*, 3rd edition, The Johns Hopkins University Press, Baltimore, MD.

Griffin, D. W. and J. S. Lim (1988). "Multiband Excitation Vocoder," *IEEE Transactions on Acoustics, Speech, and Signal Processing*, Vol. 36, No. 8, pp. 1223–1235, August.

Hagen, R. and P. Hedelin (1990). "Low Bit-Rate Spectral Coding in CELP, A New LSP-Method," *IEEE ICASSP*, pp. 189–192.

Harbison S. P. and G. L. Steele (1995). *C—A Reference Manual*, 4th edition, Prentice-Hall, Englewood Cliffs, NJ.

Hartmann, W. M. (1998). *Signals, Sound, and Sensation*, Springer-Verlag, New York.

Haykin, S. (1988). *Digital Communications*, John Wiley & Sons, Hoboken, NJ.

Haykin, S. (1991). *Adaptive Filter Theory*, Prentice-Hall, Englewood Cliffs, NJ.

Haykin, S. (1994). *Neural Networks—A Comprehensive Foundation*, Macmillan College Publishing Co., Englewood Cliffs, NJ.

Hedelin, P., P. Knagenhjelm, and M. Skoglund (1995a). "Vector Quantization for Speech Transmission," *Speech Coding and Synthesis*, W. B. Kleijn and K. K. Paliwal, eds., pp. 311–346, Elsevier Science, The Netherlands.

Hedelin, P., P. Knagenhjelm, and M. Skoglund (1995b). "Theory of Transmission of Vector Quantization Data," *Speech Coding and Synthesis*, W. B. Kleijn and K. K. Paliwal, eds., pp. 347–396, Elsevier Science, The Netherlands.

Hedelin, P. and J. Skoglund (2000). "Vector Quantization Based on Gaussian Mixture Models," *IEEE Transactions on Speech and Audio Processing*, Vol. 8, No. 4, pp. 385–401, July.

Intel Corporation (1997). *The Complete Guide to MMX Technology*, McGraw-Hill, New York.

Itakura, F. (1975). "Line Spectrum Representation of Linear Predictive Coefficients of Speech Signals," *Journal of the Acoustic Society of America*, Vol. 57, p. 535(A).

ISO/IEC (1993). *Information Technology—Coding of Moving Pictures and Associated Audio for Digital Storage Media at up to About 1.5 Mbit/s—Part 3: Audio, 11172-3*, Switzerland.

ITU (1990). *40, 32, 24, 16 kbit/s Adaptive Differential Pulse Code Modulation (ADPCM)—Recommendation G.726*, Geneva.

ITU (1992). *Coding of Speech at 16 kbit/s Using Low-Delay Code Excited Linear Prediction—Recommendation* G.728, Geneva.

ITU (1993). *Pulse Code Modulation (PCM) of Voice Frequencies—ITU-T Recommendation G.711*, Geneva.

ITU (1996a). *Coding of Speech at 8 kbit/s Using Conjugate-Structure Algebraic-Code-Excited Linear-Prediction (CS-ACELP)—ITU-T Recommendation G.729.*

ITU (1996b). *Dual Rate Speech Coder for Multimedia Communications Transmitting at 5.3 and 6.3 kbit/s—ITU-T Recommendation G.723.1.*

ITU (1996c). *Objective Quality Measurement of Telephone-Band (300–3400 Hz) Speech Codecs—ITU-T Recommendation P.861.*

ITU (1998a). *Objective Quality Measurement of Telephone-Band (300–3400 Hz) Speech Codecs Using Measuring Normalizing Blocks (MNB's)—ITU-T Recommendation P.861, App.II*, Geneva.

ITU (1998b). *Coding of Speech at 8 kbit/s Using Conjugate-Structure Algebraic-Code-Excited Linear-Prediction (CS-ACELP). Annex D: 6.4 kbit/s CS-ACELP Speech Coding Algorithm, ITU-T Recommendation G.729—Annex D.* September 1998.

ITU (1998c). *Method for Objective Measurements of Perceived Audio Quality—Recommendation ITU-R BS.1387.*

ITU (2001). *Perceptual Evaluation of Speech Quality (PESQ), An Objective Method for End-to-End Speech Quality Assessment of Narrow-Band Telephone Networks and Speech Codecs—ITU-T Recommendation P.862* (prepublication).

Jayant, N. S. and P. Noll (1984). *Digital Coding of Waveforms*, Prentice-Hall, Englewood Cliffs, NJ.

Kabal, P. and R. P. Ramachandran (1986). "The Computation of Line Spectral Frequencies Using Chebyshev Polynomials," *IEEE Transactions on Acoustics, Speech, and Signal Processing*, Vol. ASSP-34, No. 6, pp. 1419–1425, December.

Kataoka A., T. Moriya, and S. Hayashi (1993). "An 8 kbit/s Speech Coder Based on Conjugate Structure CELP," *IEEE ICASSP*, pp. II-592–II-595.

Kataoka A., T. Moriya, and S. Hayashi (1994). "Implementation and Performance of an 8 kbit/s Conjugate Structure CELP Speech Coder," *IEEE ICASSP*, pp. II-93–II-96.

Kataoka A., T. Moriya, and S. Hayashi (1996). "An 8-kb/s Conjugate Structure CELP (CS-CELP) Speech Coder," *IEEE Transactions on Speech and Audio Processing*, Vol. 4, No. 6, pp. 401–411, November.

Keyhl M., C. Schmidmer, and H. Wachter (1999). "A Combined Measurement Tool for the Objective, Perceptual Based Evaluation of Compressed Speech and Audio Signals," Preprint of the *AES 106th Convention*, Munich, Germany, May.

Kim, D. (2001). "On the Perceptually Irrelevant Phase Information in Sinusoidal Representation of Speech," *IEEE Transactions on Speech and Audio Processing*, Vol. 9, No. 8, pp. 900–905, November.

Kim, H. K. and H. S. Lee (1999). "Interlacing Properties of Line Spectrum Pair Frequencies," *IEEE Transactions on Speech and Audio Processing*, Vol. 7, No. 1, pp. 87–91, January.

Kleijn, W. B., D. J. Krasinski, and R. H. Ketchum (1988). "Improved Speech Quality and Efficient Vector Quantization in SELP," *IEEE ICASSP*, pp. 155–158.

Kleijn, W. B. and K. K. Paliwal (1995a). *Speech Coding and Synthesis*, Elsevier Science, The Netherlands.

Kleijn, W. B. and K. K. Paliwal (1995b). "An Introduction to Speech Coding," *Speech Coding and Synthesis*, W. B. Kleijn and K. K. Paliwal, eds., pp. 1–47, Elsevier Science, The Netherlands.

Kohavi, Z. (1978). *Switching and Finite Automata Theory*, 2nd edition, McGraw-Hill, New York.

Kondoz, A. M. (1994). *Digital Speech—Coding for Low Bit Rate Communication Systems*, John Wiley & Sons, Chichester, UK.

Kroon, P. (1995). "Evaluation of Speech Coders," *Speech Coding and Synthesis*, W. B. Kleijn and K. K. Paliwal, eds., pp. 467–494, Elsevier Science, The Netherlands.

Kroon, P. and B. S. Atal (1991). "On Improving the Performance of Pitch Predictors in Speech Coding Systems," *Advances in Speech Coding*, B. S. Atal, V. Cuperman, and A. Gersho, eds., pp. 321–327, Kluwer Academic Publishers, Norwell, MA.

Kroon, P., E. F. Deprettere, and R. J. Sluyter (1986). "Regular-Pulse Excitation—A Novel Approach to Effective and Efficient Multipulse Coding of Speech," *IEEE Transactions on Acoustics, Speech, and Signal Processing*, Vol. ASSP-34, No. 5, pp. 1054–1063, October.

Kroon, P. and W. B. Kleijn (1995). "Linear-Prediction Based Analysis-by-Synthesis Coding," *Speech Coding and Synthesis*, W. B. Kleijn and K. K. Paliwal, eds., pp. 79–120, Elsevier Science, The Netherlands.

Laflamme, C., R. Salami, and J-P. Adoul (1993). "9.6 kbit/s ACELP Coding of Wideband Speech," *Speech and Audio Coding for Wireless and Network Applications*, B. S. Atal, V. Cuperman, and A. Gersho, eds., pp. 147–152, Kluwer Academic Publishers, Norwell, MA.

Lancaster P. and M. Tismenetsky (1985). *The Theory of Matrices*, Academic Press, New York.

LeBlanc, W. P (1992). "Speech Coding at Low to Medium Bit Rates," Ph.D. dissertation, Carleton University, Canada.

LeBlanc, W. P., B. Bhattacharya, S. A. Mahmoud, and V. Cuperman (1993). "Efficient Search and Design Procedures for Robust Multi-Stage VQ of LPC Parameters for 4 kb/s Speech Coding," *IEEE Transactions on Speech and Audio Processing*, Vol. 1, No. 4, pp. 373–385, October.

Lee, K. and R. V. Cox (2001). "A Very Low Bit Rate Speech Coder Based on a Recognition / Synthesis Paradigm," *IEEE Transactions on Speech and Audio Processing*, Vol. 9, No. 5, pp. 482–491, July.

Leroux, J. and C. Gueguen (1979). "A Fixed Point Computation of Partial Correlation Coefficients," *IEEE Transactions on Acoustics, Speech, and Signal Processing*, Vol. ASSP-27, pp. 257–259.

Levine, S. N. (1998). "Audio Representations for Data Compression and Compressed Domain Processing," Ph.D. dissertation, Stanford University, CA.

Lim, I. and B. G. Lee (1993). "Lossless Pole-Zero Modeling of Speech Signals," *IEEE Transactions on Speech and Audio Processing*, Vol. 1, No. 3, pp. 269–276, July.

Lin, W., S. Koh, and X. Lin (2000). "Mixed Excitation Linear Prediction Coding of Wideband Speech at 8 kbps," *IEEE ICASSP*, pp. 1137–1140.

Linde, Y., A. Buzo, and R. Gray (1980). "An Algorithm for Vector Quantizer Design," *IEEE Transactions on Communications*, Vol. COM-28, No. 1, pp. 84–95, January.

Macres, J. V. (1994). "Theory and Implementation of the Digital Cellular Standard Voice Coder: VSELP on the TMS320C5x," *Texas Instruments Application Report.*

Maitre, X. (1988). "7 kHz Audio Coding Within 64 kbit/s," *IEEE Journal on Selected Areas in Communications*, Vol. 6, No. 2, pp. 283–298, February.

Maksym, J. N. (1973). "Real-Time Pitch Extraction by Adaptive Prediction of the Speech Waveform," *IEEE Transactions on Audio and Electroacoustics*, Vol. AU-21, No. 3, pp. 149–154, June.

Mano, M. (1993). *Computer System Architecture*, 3rd edition, Prentice-Hall, Englewood Cliffs, NJ.

Markel, J. D. and A. H. Gray, Jr. (1976). *Linear Prediction of Speech*, Springer-Verlag, New York.

MathSoft (2001). *Mathcad User's Guide with Reference Manual*, Cambridge, MA.

McCree, A. (2000). "A 14 kb/s Wideband Speech Coder with a Parametric Highband Model," *IEEE ICASSP*, pp. 1153–1156.

McCree, A. V. and T. P. Barnwell III (1995). "A Mixed Excitation LPC Vocoder Model for Low Bit Rate Speech Coding," *IEEE Transactions on Speech and Audio Processing*, Vol. 3, No. 4, pp. 242–250, July 1995.

McCree, A. V. and J. DeMartin (1997). "A 1.6 kb/s MELP Coder for Wireless Communications," *Proceedings of the IEEE Workshop on Speech Coding for Telecommunications*, September.

McCree, A. V. and J. DeMartin (1998). "A 1.7 kb/s MELP Coder with Improved Analysis and Quantization," *IEEE ICASSP*, pp. 593–596.

McCree, A. V., K. Truong, E. B. George, T. P. Barnwell, and V. Viswanathan (1996). "A 2.4 kbit/s MELP Coder Candidate for the New U.S. Federal Standard," *IEEE ICASSP*, pp. 200–203.

McCree, A. V., L. M. Supplee, R. P. Cohn, and J. S. Collura (1997). "MELP: The New Federal Standard at 2400 bps," *IEEE ICASSP*, pp. 1591–1594.

McCree, A., T. Unno, A. Anandakumar, A. Bernard, and E. Paksoy (2001). "An Embedded Adaptive Multi-Rate Wideband Speech Coder," *IEEE ICASSP*, pp. 761–764.

Medan, Y., E. Yair, and D. Chazan (1991). "Super Resolution Pitch Determination of Speech Signals," *IEEE Transactions on Signal Processing*, Vol. 39, No. 1, pp. 40–48, January.

Moller, U., M. Galicki, E. Baresova, and H. Witte (1998). "An Efficient Vector Quantizer Providing Globally Optimal Solutions," *IEEE Transactions on Signal Processing*, Vol. 46, No. 9, pp. 2515–2529, September.

Moore, B. C. J. (1997). *An Introduction to the Psychology of Hearing*, 4th edition, Academic Press, New York.

Moriya, T. (1992). "Two-Channel Conjugate Vector Quantizer for Noisy Channel Speech Coding," *IEEE Journal on Selected Areas in Communications*, Vol. 10, No. 5, pp. 866–874, June.

National Communications System (1992). *Details to Assist in Implementation of Federal Standard 1016 CELP*, Arlington, VA.

Noll, P. (1993). "Wideband Speech and Audio Coding," *IEEE Communications Magazine*, pp. 34–44, November.

Oppenheim, A. V. and R. W. Schafer (1989). *Discrete-Time Signal Processing*, Prentice-Hall, Englewood Cliffs, NJ.

Orfanidis, S (1988). *Optimum Signal Processing*, McGraw-Hill, New York.

Painter, T. and A. Spanias (2000). "Perceptual Coding of Digital Audio," *Proceedings of the IEEE*, Vol. 88, No. 4, pp. 451–513, April.

Paliwal, K. K. and B. S. Atal (1993). "Efficient Vector Quantization of LPC Parameters at 24 Bits/Frame," *IEEE Transactions on Speech and Audio Processing*, Vol. 1, No. 1, pp. 3–14, January.

Paliwal, K. K. and W. B. Kleijn (1995). "Quantization of LPC Parameters," *Speech Coding and Synthesis*, W. B. Kleijn and K. K. Paliwal, eds., pp. 433–466, Elsevier Science, The Netherlands.

Panzer, I. L., A. D. Sharpley, and W. D. Voiers (1993). "A Comparison of Subjective Methods for Evaluating Speech Quality," *Speech and Audio Coding for Wireless and Network Applications*, B. S. Atal, V. Cuperman, and A. Gersho, eds., pp. 59–66, Kluwer Academic Publishers, Norwell, MA.

Papamichalis, P. E. (1987). *Practical Approaches to Speech Coding*, Prentice-Hall, Englewood Cliffs, NJ.

Papoulis, A. (1991). *Probability, Random Variables, and Stochastic Processes*, McGraw-Hill, New York.

Peebles, P. (1993). *Probability, Random Variables, and Random Signal Principles*, McGraw-Hill, New York.

Perkins, M. E., K. Evans, D. Pascal, and L. A. Thorpe (1997). "Characterizing the Subjective Performance of the ITU-T 8 kb/s Speech Coding Algorithm—ITU-T G.729," *IEEE Communications Magazine*, pp. 74–81, September.

Picinbono, B. (1993). *Random Signals and Systems*, Prentice-Hall, Englewood Cliffs, NJ.

Purnhagen, H. (1999). "Advances in Parametric Audio Coding," *Proceedings IEEE Workshop on Applications of Signal Processing to Audio and Acoustics*, pp .W99-1 to W99-4, October, New York.

Rabiner, L. R. (1977). "On the Use of Autocorrelation Analysis for Pitch Detection," *IEEE Transactions on Acoustics, Speech, and Signal Processing*, Vol. ASSP-25, No. 1, pp. 24–33, February.

Rabiner, L. R., M. J. Cheng, A. E. Rosenberg, and C. A. McGonegal (1976). "A Comparative Performance Study of Several Pitch Detection Algorithms," *IEEE Transactions on Acoustics, Speech, and Signal Processing*, Vol. ASSP-24, No. 5, pp. 399–418, October.

Rabiner, L. and B. H. Juang (1993). *Fundamentals of Speech Recognition*, Prentice-Hall, Englewood Cliffs, NJ.

Rabiner, L. R. and R. W. Schafer (1978). *Digital Processing of Speech Signals*, Prentice-Hall, Englewood Cliffs, NJ.

Ramachandran, R. P. and P. Kabal (1989). "Pitch Prediction Filters in Speech Coding," *IEEE Transactions on Acoustics, Speech, and Signal Processing*, Vol. 37, No. 4, pp. 467–478, April.

Rao, S. S (1996). *Engineering Optimization*, John Wiley & Sons, Hoboken, NJ.

Rix, A. W., J. G. Beerends, M. P. Hollier, and A. P. Hekstra (2000). "PESQ—the New ITU Standard for End-to-End Speech Quality Assessment," Preprint of the *AES 109th Convention*, Los Angeles, September.

Rix, A. W., J. G. Beerends, M. P. Hollier, and A. P. Hekstra (2001), "Perceptual Evaluation of Speech Quality (PESQ)—A New Method for Speech Quality Assessment of Telephone Networks and Codecs," *IEEE ICASSP*, pp. 749–752.

Ross, M. J., H. L. Schaffer, A. Cohen, R. Freudberg, and H. J. Manley (1974). "Average Magnitude Difference Function Pitch Extractor," *IEEE Transactions on Acoustics, Speech, and Signal Processing*, Vol. ASSP-22, No. 5, pp. 353–362, October.

Salami, R., C. Laflamme, J-P. Adoul, and D. Massaloux (1994). "A Toll Quality 8 kb/s Speech Codec for the Personal Communications System (PCS)," *IEEE Transactions on Vehicular Technology*, Vol. 43, No. 3, pp. 808–816, August.

Salami, R., C. Laflamme, J-P. Adoul, K. Jarvinen, J. Vainio, P. Kapanen, T. Honkanen, and P. Haavisto (1997a). "GSM Enhanced Full Rate Speech Codec," *IEEE ICASSP*, pp. 771–774.

Salami, R., C. Laflamme, B. Bessette, and J-P. Adoul (1997b). "ITU-T G.729 Annex A: Reduced Complexity 8 kb/s CS-ACELP Codec for Digital Simultaneous Voice and Data," *IEEE Communications Magazine*, pp. 56–63, September.

Salami, R., C. Laflamme, J-P. Adoul, T. Honkanen, J. Vainio, K. Jarvinen, and P. Haavisto (1997c). "Enhanced Full Rate Speech Codec for IS-136 Digital Cellular System," *IEEE ICASSP*, pp. 731–734.

Salami, R., C. Laflamme, B. Bessette, and J-P. Adoul (1997d). "Description of ITU-T Recommendation G.729 Annex A: Reduced Complexity 8 kbit/s CS-ACELP Codec," *IEEE ICASSP*, pp. 775–778.

Salami, R., C. Laflamme, B. Bessette, and J-P. Adoul (1997e). "ITU-T G.729 Annex A: Reduced Complexity 8 kb/s CS-ACELP Codec for Digital Simultaneous Voice and Data," *IEEE Communications Magazine*, pp. 56–63, September.

Salami, R., C. Laflamme, J-P. Adoul, A. Kataoka, S. Hayashi, T. Moriya, C. Lamblin, D. Massaloux, S. Proust, P. Kroon, and Y. Shoham (1998). "Design and Description of CS-ACELP: A Toll Quality 8 kb/s Speech Coder," *IEEE Transactions on Speech and Audio Processing*, Vol. 6, No. 2, pp. 116–130, March.

Samuelsson, J. and P. Hedelin (2001). "Recursive Coding of Spectrum Parameters," *IEEE Transactions on Speech and Audio Processing*, Vol. 9, No. 5, pp. 492–503, July.

Sandige, R. S (1990). *Modern Digital Design*, McGraw-Hill, New York.

Sayood, K. (1996). *Introduction to Data Compression*, Morgan Kaufmann Publishers, San Mateo, CA.

Schroeder, M. R. and B. S. Atal (1985). "Code-Excited Linear Prediction (CELP): High-Quality Speech at Very Low Bit Rates," *IEEE ICASSP*, pp. 2511–2514.

Sedgewick, R. (1992). *Algorithms in C++*, Addison-Wesley, Reading, MA.

Shoham, Y. (1987). "Vector Predictive Quantization of the Spectral Parameters for Low Rate Speech Coding," *IEEE ICASSP*, pp. 2181–2184.

Shoham, Y. (1991). "Constrained-Stochastic Excitation Coding of Speech at 4.8 kb/s," *Advances in Speech Coding*, B. S. Atal, V. Cuperman, and A. Gersho, eds., pp. 339–348, Kluwer Academic Publishers, Norwell, MA.

Shoham, Y. (1993). "Low Delay Coding of Wideband Speech at 32 kbps Using Tree Structures," *Speech and Audio Coding for Wireless and Network Applications*, B. S. Atal, V. Cuperman, and A. Gersho, eds., pp. 133–139, Kluwer Academic Publishers.

Shoham, Y. (1999). "Coding the Line Spectral Frequencies by Jointly Optimized MA Prediction and Vector Quantization," *Proceedings of the IEEE Workshop on Speech Coding*, June 20–23, Finland.

Singhal, S. and B. S. Atal (1989). "Amplitude Optimization and Pitch Prediction in Multipulse Coders," *IEEE Transactions on Acoustics, Speech, and Signal Processing*, Vol. 37, No. 3, pp. 317–327, March.

Soliman, S. S. and M. D. Srinath (1990). *Continuous and Discrete Signals and Systems*, Prentice-Hall, Englewood Cliffs, NJ.

Sondhi, M. M (1968). "New Method of Pitch Extraction," *IEEE Transactions on Audio and Electroacoustics*, Vol. AU-16, No. 2, pp. 262–266, June.

Soong, F. K. and B. Juang (1984). "Line Spectrum Pair (LSP) and Speech Data Compression," *IEEE ICASSP*, pp. 1.10.1–1.10.4.

Soong, F. K. and B. Juang (1990). "Optimal Quantization of LSP Parameters Using Delayed Decisions," *IEEE ICASSP*, pp. 185–188.

Spanias, A. S (1994). "Speech Coding: A Tutorial Review," *Proceedings of the IEEE*, Vol. 82, No. 10, pp. 1541–1582, October 1994.

Stachurski, J., A. McCree, and V. Viswanathan (1999). "High Quality MELP Coding at Bit-Rates Around 4 KB/S," *IEEE ICASSP*.

Stearns, S. D. and D. R. Hush (1990). *Digital Signal Analysis*, Prentice-Hall, Englewood Cliffs, NJ.

Strang, G. (1988). *Linear Algebra and Its Applications*, 3rd edition, Harcourt Brace Jovanovich, Orlando, FL.

Stremler, F. G (1990). *Introduction to Communication Systems*, Addison-Wesley, Reading, MA.

Stroustrup, B. (1997). *The C++ Programming Language*, 3rd edition, Addison-Wesley, Reading, MA.

Supplee, L. M., R. P. Cohn, J. S. Collura, and A. V. McCree (1997). "MELP: The New Federal Standard at 2400 bps," *IEEE ICASSP*, pp. 1591–1594.

Texas Instruments, Inc. (1990). *Digital Signal Processing—Applications with the TMS320 Family. Theory, Algorithms, and Implementations*, Vol. 2.

Texas Instruments, Inc. (1993). *TMS320C5x User's Guide*.

Therrien, C. W. (1992). *Discrete Random Signals and Statistical Signal Processing*, Prentice-Hall, Englewood Cliffs, NJ.

Thomsen, G. and Y. Jani (2000). "Internet Telephony: Going Like Crazy," *IEEE Spectrum*, pp. 52–58, May.

TIA (1998). *Speech Service Option Standard for Wideband Spread Spectrum Systems—TIA/EIA-96C*, VA, August.

Tohkura Y., F. Itakura, and S. Hashimoto (1978). "Spectral Smoothing Technique in PARCOR Speech Analysis-Synthesis," *IEEE Transactions on Acoustics, Speech, and Signal Processing*, Vol. ASSP-26, No. 6, pp. 587–596, December.

Trancoso, I. M. and B. S. Atal (1986). "Efficient Procedures for Finding the Optimum Innovation in Stochastic Coders," *IEEE ICASSP*, pp. 2375–2378.

Tremain, T. E. (1982). "The Government Standard Linear Predictive Coding Algorithm: LPC-10," *Speech Technology*, pp. 40–49, April.

Un, C. K. and D. T. Magill (1975). "The Residual-Excited Linear Prediction Vocoder with Transmission Rate Below 9.6 kbit/s," *IEEE Transactions on Communications*, Vol. COM-23, No. 12, pp. 1466–1473, December.

Unno, T., T. P. Barnwell III, and K. Truong (1999). "An Improved Mixed Excitation Linear Prediction (MELP) Coder," *IEEE ICASSP*.

Vaidyanathan, P. P. (1993). *Multirate Systems and Filter Banks*, Prentice-Hall, Englewood Cliffs, NJ.

Vary, P., K. Hellwig, R. Hofmann, R. J. Sluyter, C. Galand, and M. Rosso (1988). "Speech Codec for the European Mobile Radio System," *IEEE ICASSP*, Vol. 1, pp. 227–230.

Veeneman, D. and B. Mazor (1993). "Efficient Multi-Tap Pitch Prediction for Stochastic Coding," *Speech and Audio Coding for Wireless and Network Applications*, B. S. Atal, V. Cuperman and A. Gersho, eds., pp. 256–229, Kluwer Academic Publishers, Norwell, MA.

Verma, T. S. (1999). "A Perceptually Based Audio Signal Model with Application to Scalable Audio Compression," Ph.D. dissertation, Stanford University, CA.

Viswanathan, R. and J. Makhoul (1975). "Quantization Properties of Transmission Parameters in Linear Predictive Systems," *IEEE Transactions on Acoustics, Speech, and Signal Processing*, Vol. ASSP-23, pp. 309–321, June.

Voran, S. (1999a). "Objective Estimation of Perceived Speech Quality—Part I: Development of the Measuring Normalizing Block Technique," *IEEE Transactions on Speech and Audio Processing*, Vol. 7, No. 4, pp. 371–382, July.

Voran, S. (1999b). "Objective Estimation of Perceived Speech Quality—Part II: Evaluation of the Measuring Normalizing Block Technique," *IEEE Transactions on Speech and Audio Processing*, Vol. 7, No. 4, pp. 383–390, July.

Walpole, R. E. and R. H. Myers (1993). *Probability and Statistics for Engineers and Scientists*, Macmillan Publishing Co., New York.

Wang, D. (1999). "QCELP Vocoders in CDMA Systems Desing," *Communications Systems Design*, pp. 40–45, April.

Wise, J. D., J. R. Caprio, and T. W. Parks (1976). "Maximum Likelihood Pitch Estimation," *IEEE Transactions on Acoustics, Speech, and Signal Processing*, Vol. ASSP-24, No. 5, pp. 418–423, October.

Xydeas, C. S. and C. Papanastasiou (1995). "Efficient Coding of LSP Parameters Using Split Matrix Quantisation," *IEEE ICASSP*, pp. 740–743.

Yong, M., G. Davidson, and A. Gersho (1988). "Encoding of LPC Spectral Parameters Using Switched-Adaptive Interframe Vector Prediction," *IEEE ICASSP*, pp. 402–405.

Zeger, K., J. Vaisey, and A. Gersho (1992). "Globally Optimal Vector Quantizer Design by Stochastic Relaxation," *IEEE Transactions on Signal Processing*, Vol. 40, No. 2, pp. 310–322, February.

Zwicker, E. and H. Fastl (1999). *Psycho-acoustics, Facts and Models*, 2nd edition, Springer-Verlag, New York.

INDEX

Printed in the USA/Agawam, MA
December 16, 2013
583137.055